Integration

To Enrique Fernández Vara,

Who, for over twenty years,
has tirelessly reread the countless versions of this work,
and has kindly suggested so many much-needed improvements.
How arid those early drafts were!

He introduced me to the fruitful alliance of work
in the *Ecuaciones Diferenciales y Analísis Numérico* group
that he directed with kindness and good humor,
of the *Feria*, temple of the *Sevillana* and *manzanilla*,
and of emotion, even for a nonbeliever, during *Semana Santa*,
at the poignant sound of the trumpets announcing the
Virgen del Baratillo.

Analysis for PDEs Set

coordinated by
Jacques Blum

Volume 4

Integration

Jacques Simon

ISTE

WILEY

First published 2026 in Great Britain and the United States by ISTE Ltd and John Wiley & Sons, Inc.

ISTE Ltd
27-37 St George's Road
London SW19 4EU
UK

www.iste.co.uk

John Wiley & Sons, Inc.
111 River Street
Hoboken, NJ 07030
USA

www.wiley.com

British Library Cataloguing-in-Publication Data
A CIP record for this book is available from the British Library
ISBN 978-1-78630-013-3

The manufacturer's authorized representative according to the EU General Product Safety Regulation is Wiley-VCH GmbH, Boschstr. 12, 69469 Weinheim, Germany, e-mail: Product_Safety@wiley.com.

Contents

Introduction

Objective. This work is the fourth volume of a series, dedicated to the resolution of partial derivative equations from physics, which could include seven volumes:

Volume 1: *Banach, Fréchet, Hilbert and Neumann spaces*
Volume 2: *Continuous functions*[1]
Volume 3: *Distributions*
Volume 4: *Integration*
Volume 5: *Lebesgue and Sobolev spaces*
Volume 6: *Traces*
Volume 7: *Partial differential equations*

The object of this fourth volume is to study the integral, both vector-valued and real-valued, and Lebesgue spaces, and to present the main properties useful for the study of partial differential equations.

Two methods make it possible to construct $L^p(\Omega; E)$ and to integrate:
— the first, classical, considers *classes of functions equal a.e.* from a *measurable* set Ω with values in a *Fréchet* space E, but no more;
— the second considers *measures* of an *open* set Ω with values in a *Neumann* space E.

These methods are equivalent when both are defined, and we emphasize the properties of the second for which E is much more general.

Contents. The first part is dedicated to integration: we define the space $L^1(\Omega; E)$ of *integrable measures*, where Ω is an open subset of $\mathbb{R}^d$ and E is a Neumann space

1 Volume 2 would do better if it had the title *Differential calculus*.

(which is to say a space endowed with semi-norms, separated, and in which every Cauchy sequence converges) and the integral $\int_\omega f$, where $f \in L^1(\Omega; E)$ and ω is a measurable subset of Ω. We provide some properties, including the image under a linear mapping, change of variable, and multivariable integration.

The second part is dedicated to the Lebesgue spaces $L^p(\Omega; E)$, of *p-integrable measures*. We study the dependence on p, on Ω, and on E, the image under linear or multilinear mappings, change of variable and separation of variables, weightings, compact subsets and duals.

The third part deals with *classes of p-integrable functions*, when the space E is Fréchet. We associate to them measures which are identical to the measures defined in Parts 1 and 2, and Lebesgue's dominated convergence theorem gives some additional results, in particular on the image under continuous functions.

Originality. We do not construct $L^p(\Omega; E)$ only for a Banach space E, like Salomon BOCHNER [5] or Nicolas BOURBAKI [8], but for a Neumann space E, which is much more general.

We construct the integral with values in a Neumann space directly, without using the real integral, which is just a special case of our construction. This is the most general property ensuring satisfactory properties for the integral, as we will see in § 5.5, p. 101.

The measures play a role usually assigned to *classes of almost everywhere equal measurable functions*. More precisely, we construct the space $L^p(\Omega; E)$ of *p-integrable measures* as a subset of the space $\mathcal{M}(\Omega; E)$ of measures. Thus, an integrable "object" is a single measure, which is better than the usual class of functions composed of an infinite number of them. Our integrable measures are "naturally" distributions like any measure or continuous function, which is suitable for PDEs.

Weightings (§ 14.4 and 14.5), which play a role analogous to the convolution for an open subset Ω (which is only defined on the entirety of $\mathbb{R}^d$), are also original.

Measure *versus* class of functions. The space of *p-integrable measures* is (Theorem 17.33, p. 366) **identical** to the space of *measures* $\overline{f}$ *associated to the classes of functions* $\dot{\mathrm{f}}$ by (Theorem 17.29, p. 363) $\langle \overline{f}, \varphi \rangle = \int_\Omega \dot{\mathrm{f}} \varphi$.

This is provided that E is a Fréchet space, otherwise the theory of measurable functions is flawed: for example, Egoroff's Theorem 17.3 is no longer satisfied.

Applications. Many possible value spaces are listed in § 5.4, p. 100.

In evolution problems, by separating the time variable from the spatial ones, we will be able to integrate with respect to time to get values in a Sobolev space $W^{m,p}(\Omega)$

or $W^{m,p}_{\rm loc}(\Omega)$ for $1 \leq p \leq \infty$, or in $W^{m,p}(\Omega)$-weak for $1 < p < \infty$ as well as in $L^\infty(\Omega)$-$*$weak. We cannot integrate with values in $L^1(\Omega)$-weak, which is not sequentially complete, but we can immerse it into the space of measures $\mathcal{M}(\Omega)$ (which we endow with its *simple* topology), which, itself, is sequentially complete.

Similarly, we could use the Lebesgue spaces $L^q(0, T; E)$, where E is any of the spaces on Ω that we have just mentioned, and even many more.

Interest of mesures. We construct the integral $\int_\Omega f$ as a limit of the Cauchy integral of uniformly continuous functions f_n. As the functions f_n converge to f in $\mathcal{M}(\Omega; E)$, we avoid the (complex, calling on filter theory) construction of their *completion* that was used by Nicolas BOURBAKI.

We thus define $L^p(\Omega; E)$ for more spaces E, since *sequentially complete* is less restrictive than *complete*. For example, if H is a finite-dimensional Hilbert space, H-weak is sequentially complete but not complete [Vol. 1, Property (4.11), p. 63].

Semi-norms. We use families of semi-norms rather than locally convex topologies, which is equivalent, since one can raise a semi-norm by a power p, but not a convex neighborhood!

The handling of semi-normed spaces is simple, although less familiar than that of topological spaces: it follows that of normed spaces, with the difference being that we work on several (semi-)norms rather than on a single norm. For example, we generate the topology of $\mathcal{K}(\Omega)$ by a family of semi-norms $\|\varphi\|_{\mathcal{K}(\Omega);q} = \sup_{x\in\Omega} q(x)|\varphi(x)|$ indexed by $q \in \mathcal{C}^+(\Omega)$, which is much simpler than its (equivalent) construction as the inductive limit of the $\mathcal{C}_K(\Omega)$.

Prerequisites. The proofs in the body of the text only call upon results established in the previous volumes, whose statement and reference are recalled.

The book is written in such a way that is can be read in any order by a non-specialist: the proofs are detailed by including what is obvious when knowing what precedes, and the numbers of the theorems used are systematically recalled.

Comments. Unlike the body of the text, comments written using small characters may refer to external or not yet established results. The appendix *Reminders* is also written using small characters since it can be assumed to be known.

History. The origin of concepts and results is indicated whenever possible as footnotes at the bottom of the page, except in *Reminders* for which the previous volumes should be referred to.

Results on integrable measures with values in a Neumann space are all new[2], to the best of the author's knowledge. They are not indicated as such, which would be tiresome; this concerns Chapters 4 to 8 and 10 to 14, as well as Theorem 17.33, p. 366.

Navigating the book:
— The **table of contents** at the front of the book lists the topics covered.
— The **index**, p. 409, provides further thematic access.
— The **list of notations and figures**, p. xv, clarifies their meaning in case of doubt.
— The hypotheses are all indicated within the theorems themselves.
— The numbering is common to all the statements so that they can be easily found by following the numbering order (for example, Propositions 1.8 and 1.9 are found between Statements 1.7 and 1.10, which are definitions).

Acknowledgements. It gives me pleasure to thank my friends Enrique FERNÁNDEZ-CARA, who suggested many improvements, and Jérôme LEMOINE, who convinced me of the need to deal with *classes of integrable functions*, who made an important contribution, including Theorem 17.12 and numerous corrections, and who carefully revised this part of the book.

Jacques SIMON
Chapdes-Beaufort, November 2025

2 Laurent SCHWARTZ has given results relating to *summable distributions* [66, § 5, p. 126–137]. They seem difficult to compare. For example, his definition of the integral (Proposition 36, p. 129) seems to have values in E even when E is not quasi-complete, which contradicts our Theorem 5.13, p. 102.

List of Notations and Figures

SPACES OF MEASURES

SPACES OF FUNCTIONS

SPACES OF MAPPINGS

SEMI-NORMED SEPARATED SPACES

PART 1

Integration

An ungrateful country constantly tells its learned men
that they are useless citizens,
and while it enjoys the fruits of their long labours,
it asks them what they have done with them.

Charles de Secondat, baron de la Brède et de MONTESQUIEU
Mes Pensées

Chapter 1

Integration of Continuous Functions

In this chapter, we define and study the Cauchy integral of a uniformly continuous function taking values in a Neumann space.

This is the first step in constructing the integral of an integrable measure, which will be defined as the limit of integrals of uniformly continuous functions in § 5.1.

We present here only the properties of the Cauchy integral that are relevant to these constructions, and in particular to the identification of continuous functions with measures through the equality $\langle f, \varphi \rangle = \int f\varphi$ for any test function φ. Each property extends to the integral of an integrable measure, as we will see in Chapter 5 onwards.

We begin with a review of Neumann spaces and continuous mappings.

1.1. Neumann spaces

Let us define separated semi-normed spaces [1] (the definitions of vector spaces and semi-norms are recalled in the appendix, in Definitions A.8 and A.9).

Definition 1.1. —*A* ***semi-normed space*** *is any vector space E endowed with a family of semi-norms, which we denote $\{\| \ \|_{E;\nu} : \nu \in \mathcal{N}_E\}$.*

Such a space is said to be ***separated*** *if $u = 0_E$ is its only element such that $\|u\|_{E;\nu} = 0$ for every $\nu \in \mathcal{N}_E$.*

A ***normed*** *space is any vector space E endowed with a norm, which we denote $\| \ \|_E$.* ▮

1 **Historical background of the notion of a semi-normed space.** John von NEUMANN introduced semi-normed spaces in 1935 [55] (with an unnecessary countability condition). He showed [55, Theorem 26, p. 19] that they coincide with the locally convex topological vector spaces that Andrey KOLMOGOROV had introduced in 1934 [41, p. 29].

Caution. Definition 1.1 of a separated semi-normed space is general, but not universal. For instance, for Laurent SCHWARTZ [67, p. 240], it is a space endowed with a *filtering* family (Definition 1.6) of semi-norms, which is equivalent since any family is equivalent to a filtering one [Volume 1, Theorem 3.15].

For Nicolas BOURBAKI [9, editions published after 1981, Chap. III, p. III.1] and Robert EDWARDS [22, p. 80], it is a space endowed with a *single* semi-norm, which represents a radical change. □

Let us define convergent sequences and Cauchy sequences[2] in a separated semi-normed space.

Definition 1.2. — *Let $(u_n)_{n\in\mathbb{N}}$ be a sequence in a separated semi-normed space E, whose family of semi-norms is denoted $\{\| \|_{E;\nu} : \nu \in \mathcal{N}_E\}$.*

(a) *We say that $(u_n)_{n\in\mathbb{N}}$* **converges to a limit** *$u \in E$, and we write $u_n \to u$, if, for every $\nu \in \mathcal{N}_E$,*

$$\|u_n - u\|_{E;\nu} \to 0 \text{ when } n \to \infty.$$

(b) *We say that $(u_n)_{n\in\mathbb{N}}$ is a* **Cauchy sequence** *if, for every $\nu \in \mathcal{N}_E$,*

$$\sup_{m\geq n} \|u_m - u_n\|_{E;\nu} \to 0 \text{ when } n \to \infty. \blacksquare$$

Let us now define the bounded subsets of a separated semi-normed space[3].

Definition 1.3. — *A subset U of a separated semi-normed space E is said to be* **bounded** *if, for every semi-norm $\| \|_{E;\nu}$ of E,*

$$\sup_{u\in U} \|u\|_{E;\nu} < \infty. \blacksquare$$

Let us define topological equalities and inclusions. A **vector subspace** of a vector space F is any subset E of F which is a vector space with respect to the addition and multiplication operations of E.

Definition 1.4. — *Let E and F be two semi-normed spaces, whose families of semi-norms are denoted $\{\| \|_{E;\nu} : \nu \in \mathcal{N}_E\}$ and $\{\| \|_{F;\mu} : \mu \in \mathcal{N}_F\}$.*

2 **Historical background of the notion of a convergent sequence.** Baron Augustin CAUCHY provided Definition 1.12 of convergence, within $\mathbb{R}$, in 1821 [14, p. 19]. Niels ABEL contributed to the emergence of this notion.

Historical background of the notion of a Cauchy sequence. Augustin CAUCHY introduced the convergence criterion of Definition 1.12 for real series in 1821 [14, p. 115–116], assuming it to be true (that is, implicitly considering $\mathbb{R}$ as the completion of $\mathbb{Q}$). Bernard BOLZANO had stated this criterion in 1817 in [6], attempting unsuccessfully to justify it due to the lack of a coherent definition of $\mathbb{R}$.

3 **Historical background of the notion of a bounded subset.** Bounded subsets in a semi-normed space were introduced in 1935 by John von NEUMANN [55]. Andrey KOLMOGOROV had introduced them in 1934 [41] for a topological vector space.

(a) *We say that* E *is* ***topologically included*** *in* F*, and we write* $E \underset{\rightarrow}{\subsetneq} F$*, if* E *is a vector subspace of* F *and if the family of semi-norms of* E *dominates the family of restrictions to* E *of the semi-norms of* F.

That is, for every $\mu \in \mathcal{N}_F$*, there exists a finite subset* N *of* $\mathcal{N}_E$ *and* $c \in \mathbb{R}$ *such that, for every* $u \in E$*, we have*

$$\|u\|_{F;\mu} \leq c \sup_{\nu \in N} \|u\|_{E;\nu}.$$

(b) *We say that* E *is* ***topologically equal*** *to* F*, and we write* $E \underset{\leftrightarrow}{=} F$*, if their families of semi-norms are* ***equivalent****, that is, if each family dominates the other. This is equivalent to saying that each space is topologically included in the other.* ▮

Caution. Suppose that $E = F$ and that, for every $\mu \in \mathcal{N}_F$, there exists $\nu \in \mathcal{N}_E$ such that, for every $u \in E$,

$$\|u\|_{F;\mu} = \|u\|_{E;\nu}.$$

This equality **does not imply the topological equality** $E \underset{\leftrightarrow}{=} F$: it only yields $E \underset{\rightarrow}{\subsetneq} F$. Indeed, it does not ensure the existence, for each $\nu \in \mathcal{N}_E$, of a μ satisfying this equality or of a finite family M of μ such that $\|u\|_{E;\nu} \leq c \sup_{\mu \in M} \|u\|_{F;\mu}$, which is necessary for the reverse inclusion $F \underset{\rightarrow}{\subsetneq} E$. □

Let us define several types of sequentially complete spaces[4].

Definition 1.5. —*A separated semi-normed space is said to be* ***sequentially complete*** *if all of its Cauchy sequences converge.*

A semi-normed space is said to be ***metrizable*** *if it is separated and if its family of semi-norms is equivalent to a countable family of semi-norms.*

A ***Neumann space*** *is any sequentially complete separated semi-normed space.*

A ***Fréchet space*** *is any metrizable sequentially complete semi-normed space.*

A ***Banach space*** *is any sequentially complete normed space.* ▮

The properties of Neumann and separated semi-normed spaces that we use in this volume were established in Volume 1 [72]. They are recalled in the text or in the appendix, with references to their proofs.

Naming of Neumann spaces. We named these spaces in Volume 1 in tribute to John VON NEUMANN, who introduced sequentially complete separated semi-normed spaces in 1935 [55]. The reader should therefore recall the definition if they choose to use them. □

---

4 **Historical background of the notion of a Neumann space.** John von NEUMANN introduced sequentially complete spaces in 1935 [55].

Why Neumann spaces? Neumann spaces are, as we shall see in § 5.5, p. 101, the most general target spaces for which the integral can be defined in a satisfactory manner. □

Examples of Neumann spaces. Examples of such spaces are given in § 5.4, p. 100. □

Completeness *versus* sequential completeness. A semi-normed space is said to be *complete* if every Cauchy filter converges [SCHWARTZ, 67, chap. XVIII, § 8, définition 1, p. 251]. We do not use this notion, because sequential completeness is much simpler and more general (completeness implies sequential completeness [SCHWARTZ, 67, p. 251]) and, above all, because certain useful spaces are sequentially complete but not complete. This is the case, for example, for any infinite-dimensional reflexive Hilbert or Banach space endowed with its weak topology (this is detailed in Volume 1 [72, Property (4.11), p. 63]). □

Completeness *versus* metrizability. For a metrizable space, *completeness* is equivalent to *sequential completeness* [SCHWARTZ, 67, théorème XVIII, 8; 1, p. 251]. This is why some authors refer to the completeness of a Banach or Fréchet space when, in fact, they only use its sequential completeness. □

Abuse of language. Definition 1.5 of metrizability is, in fact, that of a *separated countable semi-normable* space, which is equivalent to being metrizable. □

Let us define filtering families of semi-norms.

Definition 1.6. — *A family* $\{\|\ \|_\nu : \nu \in \mathcal{N}\}$ *of semi-norms on a vector space* E *is said to be* ***filtering*** *if, for every finite subset* $N \subset \mathcal{N}$*, there exists* $\mu \in \mathcal{N}$ *such that: for every* $u \in E$*,*

$$\sup_{\nu \in N} \|u\|_\nu \le \|u\|_\mu. \quad \blacksquare$$

Usefulness of filtering families. The use of filtering families simplifies certain statements by replacing the supremum over a finite number of semi-norms with a single semi-norm. This is, for example, the case in the characterization of continuous linear maps in Proposition 1.9, where both cases are considered. Definition 1.7 of continuous maps would likewise be simplified with a filtering family.

Every family of semi-norms is equivalent to a filtering family (Proposition A.23), though the latter is not necessarily convenient to use. □

Spaces endowed with filtering families. The "natural" family of certain spaces is filtering.

For example, $\mathcal{K}(\Omega)$ is endowed (Definition 3.1) with a family of semi-norms, which is filtering, indexed by $q \in \mathcal{C}^+(\Omega)$,

$$\|\varphi\|_{\mathcal{K}(\Omega);q} = \sup_{x \in \Omega} q(x)|\varphi(x)|.$$

A norm also constitutes, by itself, a filtering family of semi-norms. □

Usefulness of non-filtering families. The "natural" family of certain spaces is not filtering.

For example, $\mathcal{M}(\Omega)$ is endowed (Definition 3.2) with a family of semi-norms, which is not filtering, indexed by $\varphi \in \mathcal{K}(\Omega)$,

$$\|f\|_{\mathcal{M}(\Omega);\varphi} = |\langle f, \varphi\rangle|.$$

If one were to restrict to filtering families, one would have to endow $\mathcal{M}(\Omega)$ with semi-norms indexed by the finite subsets N of $\mathcal{K}(\Omega)$,

$$|||f|||_{\mathcal{M}(\Omega);N} = \sup_{\varphi \in N} |\langle f, \varphi\rangle|.$$

The maximum over finite subsets would disappear from Definition 1.7 of continuity but would reappear here. □

1.2. Continuous mappings

Let us define various notions of continuity[5] of a mapping from one normed space into another.

Definition 1.7. — *Let E and F be two separated semi-normed spaces, $U \subset E$ and T a mapping from U into F, and let $\{\| \, \|_{E;\nu} : \nu \in \mathcal{N}_E\}$ and $\{\| \, \|_{F;\mu} : \mu \in \mathcal{N}_F\}$ be the families of semi-norms of E and F.*

(a) *We say that T is **continuous** if, for every $u \in U$, $\mu \in \mathcal{N}_F$ and $\epsilon > 0$, there exists a finite set $N \subset \mathcal{N}_E$ et $\eta > 0$ such that:*

$$v \in U, \; \sup_{\nu \in N} \|v - u\|_{E;\nu} \leq \eta \quad \Rightarrow \quad \|T(v) - T(u)\|_{F;\mu} \leq \epsilon.$$

(b) *We say that T is **uniformly continuous** if, for every $\mu \in \mathcal{N}_F$ and $\epsilon > 0$, there exists a finite subset $N \subset \mathcal{N}_E$ and $\eta > 0$ such that:*

$$u \in U, \; v \in U, \; \sup_{\nu \in N} \|v - u\|_{E;\nu} \leq \eta \quad \Rightarrow \quad \|T(v) - T(u)\|_{F;\mu} \leq \epsilon.$$

(c) *We say that T is **sequentially continuous** if, for every $u \in U$ and every sequence $(u_n)_{n \in \mathbb{N}}$ in U:*

$$u_n \to u \text{ in } E \quad \Rightarrow \quad T(u_n) \to T(u) \text{ in } F. \; ▮$$

Recall[6] that continuity implies sequential continuity [Vol. 1, Theorem 7.2].

Proposition 1.8. — *Every continuous mapping from a subset of a separated semi-normed space into a separated semi-normed space is sequentially continuous.* ▮

---

5 **Historical background of the notion of continuity.** Augustin CAUCHY defined sequential continuity for a real-valued function on a segment in 1821, in [14]. Bernard Placidus Johann Nepomuk BOLZANO also contributed to the emergence of this notion.
Eduard HEINE defined the uniform continuity of a function defined on (a subset of) $\mathbb{R}^d$ in 1870, in [36]. It had already been used implicitly by Augustin CAUCHY in 1823 to define the integral of a real-valued function [15, p. 122–126], then explicitly by Peter DIRICHLET.

6 **Numbering of statements.** The numbering is shared by all statements — **Definition 1.1**, ..., **Definition 1.7**, **Proposition 1.8**, **Proposition 1.9**, **Definition 1.10**, etc. —, so that they can be more easily located by following the numerical order. There is no need, therefore, to look for Propositions 1.1 to 1.7; these numbers have been assigned to definitions. Nor is there any need to look for Definitions 1.8 and 1.9.
Proposition or theorem? We use the term *theorem* for any result concerning the main subject of this book, namely integrable measures, measures, and the spaces $L^p(\Omega; E)$. Other results (on related topics useful for proofs, or partial results that will be improved upon later) are called *propositions* or *lemmas*.

Let us characterize continuous linear mappings [Vol. 1, Theorem 7.14].

Proposition 1.9. — *Let L be a linear mapping from a separated semi-normed space E into a separated semi-normed space F, and $\{\| \ \|_{E;\nu} : \nu \in \mathcal{N}_E\}$ and $\{\| \ \|_{F;\mu} : \mu \in \mathcal{N}_F\}$ the families of semi-norms of E and F.*

(a) *The mapping L is continuous if, for every $\mu \in \mathcal{N}_F$, there exists a finite subset N of $\mathcal{N}_E$ and $c \geq 0$ such that: for all $u \in E$,*

$$\|Lu\|_{F;\mu} \leq c \sup_{\nu \in N} \|u\|_{E;\nu}.$$

This is also equivalent to: L is uniformly continuous.

(b) *If the family of semi-norms of E is filtering, then L is continuous if and only if, for every $\mu \in \mathcal{N}_F$, there exists $\nu \in \mathcal{N}_E$ and $c \geq 0$ such that: for all $u \in E$,*

$$\|Lu\|_{F;\mu} \leq c \, \|u\|_{E;\nu} \,. \; \blacksquare$$

Let us now turn to **function** spaces, a term we reserve for mappings defined on a subset of $\mathbb{R}^d$, generally denoted Ω. The dimension d is always an integer ≥ 1.

Let us define two spaces of continuous functions.

Definition 1.10. — *Let $\Omega \subset \mathbb{R}^d$ and let E be a separated semi-normed space whose family of semi-norms is denoted by $\{\| \ \|_{E;\nu} : \nu \in \mathcal{N}_E\}$.*

(a) *We denote by $\mathcal{C}(\Omega; E)$ the vector space of continuous functions from Ω into E endowed with the semi-norms indexed by the compact subsets $K \subset \Omega$ and by $\nu \in \mathcal{N}_E$,*

$$\|f\|_{\mathcal{C}(\Omega;E);K,\nu} \stackrel{\text{def}}{=} \sup_{x \in K} \|f(x)\|_{E;\nu}.$$

(b) *We denote by $\mathcal{C}_b(\Omega; E)$ the vector space of continuous and bounded functions from Ω into E endowed with the semi-norms indexed by $\nu \in \mathcal{N}_E$,*

$$\|f\|_{\mathcal{C}_b(\Omega;E);\nu} \stackrel{\text{def}}{=} \sup_{x \in \Omega} \|f(x)\|_{E;\nu}. \; \blacksquare$$

Let us now define two spaces of uniformly continuous functions.

Definition 1.11. — *Let $\Omega \subset \mathbb{R}^d$ and let E be a separated semi-normed space whose family of semi-norms is denoted by $\{\| \ \|_{E;\nu} : \nu \in \mathcal{N}_E\}$.*

(a) *We denote by* $\mathbf{C}_b(\Omega; E)$ *the space of uniformly continuous and bounded functions from* Ω *into* E *endowed with the semi-norms indexed by* $\nu \in \mathcal{N}_E$,

$$\|f\|_{\mathbf{C}_b(\Omega;E);\nu} = \sup_{x\in\Omega} \|f(x)\|_{E;\nu}.$$

(b) *Given a compact subset* Q *of* $\mathbb{R}^d$ *(not necessarily contained in* Ω*), we denote by* $\mathbf{C}_Q(\Omega; E)$ *the space of uniformly continuous functions from* Ω *into* E *that vanish outside of* Q *endowed with the semi-norms of* $\mathbf{C}_b(\Omega; E)$. ▌

Recall that $\mathcal{C}(\Omega; E)$, $\mathcal{C}_b(\Omega; E)$, $\mathbf{C}_b(\Omega; E)$, and $\mathbf{C}_Q(\Omega; E)$ are Neumann spaces [Vol. 2, Theorem 1.12].

Let us define two other spaces of uniformly continuous functions ($\mathcal{B}(\Omega; E)$ is not endowed with semi-norms). The definition of the support is recalled in Definition A.52. We denote

$$\mathcal{C}^+(\Omega) \stackrel{\text{def}}{=} \{q \in \mathcal{C}(\Omega) : \forall x \in \Omega,\ q(x) \geq 0\}.$$

Definition 1.12. — *Let* $\Omega \subset \mathbb{R}^d$ *and let* E *be a separated semi-normed space whose family of semi-norms is denoted by* $\{\|\ \|_{E;\nu} : \nu \in \mathcal{N}_E\}$.

(a) *We denote by* $\mathcal{B}(\Omega; E)$ *the vector space of uniformly continuous functions from* Ω *into* E *with bounded support, that is, functions vanishing outside of some bounded subset.*

(b) *We denote by* $\mathcal{K}(\Omega; E)$ *the vector space of uniformly continuous functions from* Ω *into* E *with compact support, that is, functions vanishing outside of a compact subset of* Ω*, endowed with the semi-norms indexed by* $q \in \mathcal{C}^+(\Omega)$ *and* $\nu \in \mathcal{N}_E$,

$$\|f\|_{\mathcal{K}(\Omega;E);q,\nu} \stackrel{\text{def}}{=} \sup_{x\in\Omega} q(x)\|f(x)\|_{E;\nu}.$$ ▌

We denote by $\mathcal{C}(\Omega)$, $\mathcal{C}_b(\Omega)$, $\mathbf{C}_b(\Omega)$, $\mathbf{C}_Q(\Omega)$, $\mathcal{B}(\Omega)$, and $\mathcal{K}(\Omega)$ these spaces when $E = \mathbb{R}$. We will use the same convention, without restating it, for other spaces when they take real values, for example $\mathcal{M}(\Omega)$, $L^p(\Omega)$, etc.

Let us state an elementary property.

Proposition 1.13. — *Every function in* $\mathcal{B}(\Omega; E)$ *is bounded.* ▌

Proof. Let $f \in \mathcal{B}(\Omega; E)$. Every uniformly continuous function on a bounded set is bounded [Vol. 2, Theorem 4.8 (b)]; it follows from Definition 1.12 that f is bounded on the set where is does not vanish. It is therefore bounded. □

1.3. Cauchy integral of a uniformly continuous function

Let us define the **Cauchy integral** of a uniformly continuous function with a bounded support, with values in a Neumann space.

Definition 1.14. — *Let $f \in \mathcal{B}(\Omega; E)$, that is, a uniformly continuous function from Ω into E vanishing outside a bounded subset, where Ω is an open subset of $\mathbb{R}^d$ and E is a Neumann space, and let ω be an open subset contained in Ω.*

We call the ***approximated integral***[7] *of f over ω the element of E defined, for $n \in \mathbb{N}$, by*

$$\mathbb{S}_\omega^n f \stackrel{\text{def}}{=} \sum_{s \in \mathbb{Z}^d : \Delta_{s,n} \subset \omega} 2^{-nd} f(2^{-n} s),$$

where $\Delta_{s,n}$ is the closed cube of $\mathbb{R}^d$ of side length 2^{-n} centered at $2^{-n}s$.

We call the ***integral***[8] *of f over ω the element of E defined by*

$$\int_\omega f \stackrel{\text{def}}{=} \lim_{n \to \infty} \mathbb{S}_\omega^n f. \blacksquare$$

7 **On the origin of the approximated integral.** The approximated integrals $\mathbb{S}_\omega^n f$ are sometimes called *Riemann sums*, in reference to a later generalization by Bernhard RIEMANN in his 1854 habilitation thesis [61, p. 239], published in 1867, after his death. In the case of continuous functions, according to BOURBAKI [11, p. 65], "*it would be to call* [them] *Archimedean sums, or Eudoxian sums*".

8 **History of the integral of a continuous function. Real integral.** Augustin CAUCHY defined in 1823 the integral of a real continuous function over a real interval in his *Leçons à l'école Royale Polytechnique sur le calcul infinitésimal* [15, p. 122-126], using the method employed here (our Definition 1.14). He showed [15, p. 125] that the approximated sums $\mathbb{S}_\omega^n f$ have a limit relying on the uniform continuity of the function (which would not be proven until 1872 by Eduard HEINE [38]!). He also defined the integral of a function in $\mathbb{R}^2$ in 1827 [16].

Precursors. Johann BERNOULLI had written the first treatise on differential and integral calculus in 1691 and 1692 [4], the part concerning integral calculus only being published in 1742. Leonhard EULER published in 1768 his important treatise *Institutionum calculi integralis*, in which he used the sums $\mathbb{S}^n f$ [25, p. 178], assuming their convergence. He had computed integrals in $\mathbb{R}^2$ in 1770 [26]. For more details, see *Histoire de l'intégration* by Jean-Paul PIER [58, p. 79–86].

Vector-valued integral. The integral of an integrable function with values in a Banach space E was defined in 1933 by Salomon BOCHNER [5].
Nicolas BOURBAKI defined in 1966 [8, chap. III, § 3, définition 1, p. 75] the integral of a function in $\mathcal{K}(\Omega; E\text{-weak})$, where E is a locally convex separated topological vector space, which is equivalent to a separated semi-normed space. If E-weak is sequentially complete, this integral coincides with the Cauchy integral taking values in E-weak; otherwise, it takes values in the completion $\widehat{E}$ of E [8, corollaire 2, p. 80]; this completion coincides with that of E-weak.

Historical of the integral notation. The symbol $\int$, "*a long* s, *the intitial letter of the word* summa", was introduced by Gottfried von LEIBNIZ in a 1675 letter to Henry OLDENBURG , reproduced by Florian CAJORI [13, vol. 2, p. 243]. The notation $\int_a^b f$ was introduced by Joseph FOURIER [28, p. 252] in 1822. The term *integral* was introduced in 1690 by Jacob BERNOULLI [3]; see facsimile in [58, p. 78].

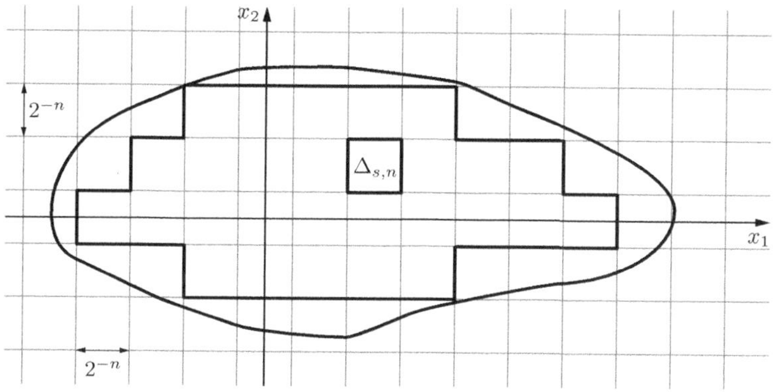

Figure 1.1. *Tiling of ω by $\Delta_{s,n}$*

A convention. The definition of the approximated integral has no meaning if there is no cube $\Delta_{s,n}$ included in ω, since a sum over no elements is undefined! To avoid this issue, we agree that in such a case, $\mathbb{S}_\omega^n f$ is zero. In particular, $\mathbb{S}_\emptyset^n f = 0_E$ for all n, and thus

$$\int_\emptyset f = 0_E. \; \square$$

Notation. We write $\int_\omega f(x)\,\mathrm{d}x$ when we wish to specify the variable with respect to which we are integrating. For example, in the expression $\int_\omega f(x,y)\,\mathrm{d}x$ for a function f of two variables. $\square$

Definition 1.14 is justified by the following result.

Lemma 1.15. — *In Definition 1.14, $\mathbb{S}_\omega^n f$ is defined and has a limit in E when $n \to \infty$.*

▮

Proof. ***1°) Reduction to a bounded open subset.*** The sum $\mathbb{S}_\omega^n f$ is well-defined because $f(2^{-n}s)$ vanishes except for at most finitely many s. More precisely, let Q be an open cube with integer side length N, centered at 0, containing $\{x \in \Omega : f(x) \neq 0_E\}$. Replacing ω by $\omega \cap Q$ does not change the value of $\mathbb{S}_\omega^n f$, which amounts to assuming, and is what we will henceforth do, that

$$\omega \subset Q.$$

2°) Decomposition of $\mathbb{S}_\omega^m f - \mathbb{S}_\omega^n f$. To show that the sequence $(\mathbb{S}_\omega^n f)_{n\in\mathbb{N}}$ has a limit, it suffices to verify that it is Cauchy, since E is sequentially complete. Let us compute $\mathbb{S}_\omega^m f - \mathbb{S}_\omega^n f$, where $m \geq n$.

Each cube $\Delta_{s,n}$ can be decomposed into $2^{(m-n)d}$ cubes $\Delta_{r,m}$, and

$$\mathbb{S}_\omega^n f = \sum_{s\,:\,\Delta_{s,n}\subset\omega} 2^{-nd} f(2^{-n}s) = \sum_{s\,:\,\Delta_{s,n}\subset\omega} 2^{-nd} \sum_{r\,:\,\Delta_{r,m}\subset\Delta_{s,n}} 2^{(n-m)d} f(2^{-n}s).$$

Additionally, there appear cubes $\Delta_{r,m} \subset \omega$ that were not contained in any of the $\Delta_{s,n} \subset \omega$, so by denoting $R_{m,n}$ the set of such indices r, we obtain

$$\mathbb{S}^m f = \sum_{r:\Delta_{r,m}\subset\omega} 2^{-md} f(2^{-m}r) =$$
$$= \sum_{s:\Delta_{s,n}\subset\omega} 2^{-nd} \sum_{r:\Delta_{r,m}\subset\Delta_{s,n}} 2^{(n-m)d} f(2^{-m}r) + \mathcal{Y}_{m,n},$$

where

$$\mathcal{Y}_{m,n} = \sum_{r\in R_{m,n}} 2^{-md} f(2^{-m}r).$$

Therefore,

$$\mathbb{S}_\omega^m f - \mathbb{S}_\omega^n f = \mathcal{X}_{n,m} + \mathcal{Y}_{m,n}, \tag{1.1}$$

where

$$\mathcal{X}_{n,m} = \sum_{s:\Delta_{s,n}\subset\omega} 2^{-nd} \sum_{r:\Delta_{r,m}\subset\Delta_{s,n}} 2^{(n-m)d}\big(f(2^{-m}r) - f(2^{-n}s)\big).$$

3o) Estimate of $\|\mathcal{X}_{n,m}\|_{E;\nu}$. Let $\|\ \|_{E;\nu}$ be a semi-norm of E. We will show that

$$\|\mathcal{X}_{n,m}\|_{E;\nu} \leq N^d \sup_{|y-x|\leq 2^{-n-1}\sqrt{d}} \|f(y) - f(x)\|_{E;\nu}. \tag{1.2}$$

On the one hand, the cube $\Delta_{s,n}$ is included in a ball centered at $2^{-n}s$ with radius $2^{-n-1}\sqrt{d}$, so when $\Delta_{r,m} \subset \Delta_{s,n}$, we have $|2^{-m}r - 2^{-n}s| \leq 2^{-n-1}\sqrt{d}$, hence

$$\|f(2^{-m}r) - f(2^{-n}s)\|_{E;\nu} \leq \sup_{|y-x|\leq 2^{-n-1}\sqrt{d}} \|f(y) - f(x)\|_{E;\nu}.$$

On the other hand, let $k_n(\omega)$ be the number of cubes $\Delta_{s,n}$ included in ω. The closed cube $\overline{Q}$ (of side length N) contains exactly $(N2^n)^d$ closed cubes $\Delta_{s,n}$ (of side length 2^{-n}), hence $k_n(\omega) \leq (N2^n)^d$, which gives

$$2^{-nd}k_n(\omega) \leq N^d. \tag{1.3}$$

From which (1.2) follows, since there are $2^{(m-n)d}$ cubes $\Delta_{r,m}$ in each $\Delta_{s,n}$.

Since f is uniformly continuous (by Definition 1.12 of $\mathcal{B}(\Omega; E)$), inequality (1.2) implies

$$\sup_{m\geq n} \|\mathcal{X}_{n,m}\|_{E;\nu} \to 0 \text{ when } n \to \infty. \tag{1.4}$$

4o) Estimate of $\|\mathcal{Y}_{n,m}\|_{E;\nu}$. There are $2^{(m-n)d}k_n(\omega)$ cubes $\Delta_{r,m}$ included in the $k_n(\omega)$ cubes $\Delta_{s,n}$ contained in ω, so there are $k_m(\omega) - 2^{(m-n)d}k_n(\omega)$ indices r in $R_{m,n}$. Thus,

$$\|\mathcal{Y}_{m,n}\|_{E;\nu} \leq (2^{-md}k_m(\omega) - 2^{-nd}k_n(\omega)) \sup_{x\in\omega} \|f(x)\|_{E;\nu}. \tag{1.5}$$

Now, on the one hand, $\sup_{x\in\omega}\|f(x)\|_{E;\nu}$ is finite, since f is bounded (Proposition 1.13).

On the other hand, $2^{-nd}k_n(\omega)$ increases with n (since $2^{-md}k_m(\omega)-2^{-nd}k_n(\omega)$ is ≥ 0 from the count above of the indices in $R_{m,n}$) and is bounded (by (1.3)), so it has a finite limit ℓ as $n\to\infty$. Hence (1.5) implies

$$\sup_{m\geq n}\|\mathcal{Y}_{n,m}\|_{E;\nu}\to 0 \ \text{ when } n\to\infty. \tag{1.6}$$

5º) Conclusion. From decomposition (1.1) and convergences (1.4) and (1.6),

$$\sup_{m\geq n}\|\mathbb{S}^m_\omega f-\mathbb{S}^n_\omega f\|_{E;\nu}\to 0 \ \text{ when } n\to\infty. \tag{1.7}$$

This shows that $(\mathbb{S}^n_\omega f)_{n\in\mathbb{N}}$ is a Cauchy sequence in E. It therefore has a limit. □

Usefulness of the Cauchy integral. The Cauchy integral is a simple tool for computing the integral of a uniformly continuous function, whether real-valued or with values in a Neumann space.

Here, it is above all a crucial step in the construction of the integral of an integrable measure. □

Usefulness of the space $\mathcal{B}(\Omega;E)$. The integral of continuous functions is often introduced on $\mathcal{K}(\Omega;E)$, see for example [BOURBAKI, 8, chap. III, § 3, définition 1, p. 75]. We prefer to define it on $\mathcal{B}(\Omega;E)$ in order to be able to apply it to functions which, like constants, are not zero near the boundary. This in particular allows us to integrate the restriction to ω of a function in $\mathcal{B}(\Omega;E)$, since it belongs to $\mathcal{B}(\omega;E)$ (whereas the restriction of a function in $\mathcal{K}(\Omega;E)$ does not necessarily belong to $\mathcal{K}(\omega;E)$). □

1.4. Some properties of the integral

Let us recall some of the properties of the Cauchy integral with values in a Neumann space that will be used to define the space $L^1(\Omega;E)$. The first of these properties is linearity, as follows.

Proposition 1.16. — *Let ω and Ω be open subsets of $\mathbb{R}^d$ such that $\omega\subset\Omega$ and let E be a Neumann space.*

Then the mapping $f\mapsto\displaystyle\int_\omega f$ is linear from $\mathcal{B}(\Omega;E)$ into E. ▮

Proof. Given f and g in $\mathcal{B}(\Omega;E)$ and $t\in\mathbb{R}$, we need to check that

$$\int_\omega f+g=\int_\omega f+\int_\omega g,\quad \int_\omega tf=t\int_\omega f.$$

These equalities are obtained by passing to the limit in the analogous equalities for the approximated integrals (Definition 1.14), which are straightforward,

$$\mathbb{S}^n_\omega(f+g)=\mathbb{S}^n_\omega f+\mathbb{S}^n_\omega g,\quad \mathbb{S}^n_\omega(tf)=t\,\mathbb{S}^n_\omega f.\ \square$$

Let us provide some upper bounds for the semi-norms of the integral, in which the semi-norms of E play the role held by the norm in the case of a Banach space.

Proposition 1.17. — *Let $f \in \mathcal{B}(\Omega; E)$, where Ω is an open subset of $\mathbb{R}^d$ and E is a Neumann space, ω an open subset included in Ω, and $\| \ \|_{E;\nu}$ a semi-norm of E.*

Then $\|f\|_{E;\nu} \in \mathcal{B}(\Omega)$, and

$$\Big\| \int_\omega f \Big\|_{E;\nu} \leq \int_\omega \|f\|_{E;\nu} \leq \int_\Omega \|f\|_{E;\nu}. \ \blacksquare$$

Proof. *1º) Property of $\|f\|_{E;\nu}$.* Every semi-norm is contracting (Proposition A.11), hence, for all x and y in Ω,

$$\big| \|f(y)\|_{E;\nu} - \|f(x)\|_{E;\nu} \big| \leq \|f(y) - f(x)\|_{E;\nu}.$$

The uniform continuity (Definition 1.7) of $\|f\|_{E;\nu}$ thus follows from that of f. Moreover, $\|f\|_{E;\nu}$ vanishes outside the bounded subset outside which f vanishes, so it belongs to $\mathcal{B}(\Omega)$, by Definition 1.12 of the latter.

2º) Inequalities. The stated inequalities are obtained by passing to the limit in the inequalities related to the approximated integrals (Definition 1.14), which are straightforward,

$$\|\mathbb{S}^n_\omega f\|_{E;\nu} \leq \mathbb{S}^n_\omega(\|f\|_{E;\nu}) \leq \mathbb{S}^n_\Omega(\|f\|_{E;\nu}). \ \square$$

Let us now give some growth properties of the real-valued integral.

Proposition 1.18. — *Let $f \in \mathcal{B}(\Omega)$ and $g \in \mathcal{B}(\Omega)$, where Ω is an open subset of $\mathbb{R}^d$, and let ω and λ be open subsets included in Ω. Then:*

(a) $$f \geq g \quad \Rightarrow \quad \int_\omega f \geq \int_\omega g;$$

(b) $$f \geq 0, \ \omega \supset \lambda \quad \Rightarrow \quad \int_\omega f \geq \int_\lambda f;$$

(c) $$f \geq 0, \ f \neq 0 \quad \Rightarrow \quad \int_\Omega f > 0. \ \blacksquare$$

Proof. ***1°) Properties*** (a) *and* (b). The stated inequalities are obtained by passing to the limit on the following inequalities related to the approximated integrals (Definition 1.14), which are evident,

$$\mathbb{S}^n_\omega f \geq \mathbb{S}^n_\omega g, \quad \mathbb{S}^n_\omega f \geq \mathbb{S}^n_\lambda f.$$

2°) Property (c). If $f \neq 0$, there exists $y \in \Omega$ such that $f(y) = a > 0$; moreover, since f is continuous, there exists an open ball ω contained in Ω on which $f \geq a/2$. This ball contains one of the cubes $\Delta_{s,n}$ for n large enough.

For $m \geq n$, this cube (of side length 2^{-n}) is dived into $2^{(m-n)d}$ cubes $\Delta_{s,m}$ (of side lengths 2^{-m}); on these cubes, $f \geq a/2$; on the others, $f \geq 0$. Definition 1.14 of the approximated integral then gives

$$\mathbb{S}^m_\Omega f = \sum_{s\,:\,\Delta_{s,m}\subset\Omega} 2^{-md} f(2^{-m}s) \geq 2^{(m-n)d} 2^{-md} \frac{a}{2} = 2^{-1-nd} a.$$

Passing to the limit as $m \to \infty$, we obtain

$$\int_\Omega f \geq 2^{-1-nd} a > 0. \ \square$$

Let us now show that the integral commutes with sequentially continuous linear maps.

Proposition 1.19. — *Let $f \in \mathcal{B}(\Omega; E)$, where Ω is an open subset of $\mathbb{R}^d$ and E is a Neumann space, and let L be a continuous, or sequentially continuous, linear mapping from E into a Neumann space F. Then $Lf \in \mathcal{B}(\Omega; F)$ and, for every open subset ω included in Ω,*

$$\int_\omega Lf = L\Big(\int_\omega f\Big),$$

where, for every $x \in \Omega$, $(Lf)(x) \stackrel{\text{def}}{=} L(f(x))$. ▮

Proof. It suffices to assume that L is sequentially continuous, since this is true of every continuous mapping (Proposition 1.8).

1°) Property of Lf. The function $Lf = L \circ f$ is uniformly continuous, as every composition of a uniformly continuous function with a sequentially continuous linear mapping is (Proposition A.43), and it vanishes outside of the bounded subset outside which f vanishes. Therefore, it belongs to $\mathcal{B}(\Omega; F)$, by Definition 1.12.

2°) Equality. The stated equality is obtained by passing to the limit in the following identity concerning the approximated integrals (Definition 1.14), which is evident,

$$\mathbb{S}^n_\omega (Lf) = L(\mathbb{S}^n_\omega f). \ \square$$

1.5. Dependence of the integral on the domain of integration

Let us define the measure[9] of an open subset of $\mathbb{R}^d$.

Definition 1.20. — *We define the* ***measure*** *of an open subset ω of $\mathbb{R}^d$ to be the real number ≥ 0 or infinity:*

$$|\omega| \stackrel{\text{def}}{=} \lim_{n\to\infty} 2^{-nd} k_n(\omega),$$

where, for $n \in \mathbb{N}$, $k_n(\omega)$ is the number of $s \in \mathbb{Z}^d$ such that the closed cube $\Delta_{s,n}$ of side length 2^{-n} centered at $2^{-n}s$ (see Figure 1.1, p. 11), is contained in ω. ▮

Justification. When passing from n to $m \geq n$, each cube $\Delta_{s,n} \subset \omega$ is divided into $(2^{-n}/2^{-m})^d$ cubes $\Delta_{r,m}$, whence $k_m(\omega) \geq 2^{(m-n)d} k_n(\omega)$, and thus

$$2^{-md} k_m(\omega) \geq 2^{-nd} k_n(\omega). \tag{1.8}$$

Being increasing, the sequence $2^{-nd} k_n(\omega)$ has a limit $|\omega|$, finite or infinite. □

Observe that (1.8) implies, for all $n \in \mathbb{N}$,

$$2^{-nd} k_n(\omega) \leq |\omega|. \tag{1.9}$$

Let us now bound the semi-norms of the integral in terms of the measure of the domain of integration.

Proposition 1.21. — *Let $f \in \mathcal{B}(\Omega; E)$, where Ω is an open subset of $\mathbb{R}^d$ and E is a Neumann space, and let ω be an open subset of $\mathbb{R}^d$ contained in Ω. Then, for any semi-norm $\| \;\|_{E;\nu}$ of E,*

$$\Big\| \int_\omega f \Big\|_{E;\nu} \leq |\omega| \sup_{x\in\omega} \|f(x)\|_{E;\nu}. \text{ ▮}$$

9 **Historical background of the notion of measure.** Definition 1.20 follows an elementary method used since ancient Greek times. Around 370 BCE, EUDOXUS proved that the volume of a cone is one third of the volume of the corresponding cylinder, and that the volume of a pyramid is one third of the volume of the prism with the same base height, results that had been stated, without proof, by DEMOCRITUS around 400 BCE. The proof appears in Book XII of EUCLID's *Elements* [24, p. 479], published around 250 BCE and attributed to EUDOXUS, as ARCHIMEDES wrote. He himself [1] calculated, among other things, the area of an ellipse and the volume of a sphere, also around 250 BCE. For more details, see Jean-Paul PIER [58, p. 4–11].

Proof. The approximated integral of f over ω is, by Definition 1.14, equal to

$$\mathbb{S}^n_\omega f = \sum_{s \in \mathbb{Z}^d \,:\, \Delta_{s,n} \subset \omega} 2^{-nd} f(2^{-n}s).$$

Therefore

$$\|\mathbb{S}^n_\omega f\|_{E;\nu} \leq 2^{-nd} k_n(\omega) \sup_{x \in \omega} \|f(x)\|_{E;\nu},$$

where $k_n(\omega)$ is the number of cubes $\Delta_{s,n}$ contained in ω. The stated inequality is obtained by passing to the limit as $n \to \infty$, since, by Definition 1.20 of the measure of an open subset, $2^{-nd} k_n(\omega) \to |\omega|$ and by Definition 1.14 of the integral, $\mathbb{S}^n_\omega f \to \int_\omega f$. □

Let us give an analogous bound for the semi-norms of a difference of integrals.

Proposition 1.22. — *Let $f \in \mathcal{B}(\Omega; E)$, where Ω is an open subset of $\mathbb{R}^d$ and E is a Neumann space, and let ω_1 and ω_2 be open subsets of $\mathbb{R}^d$ such that $\omega_1 \subset \omega_2 \subset \Omega$ and $|\omega_1| < \infty$. Then, for any semi-norm $\|\ \|_{E;\nu}$ of E,*

$$\Big\| \int_{\omega_2} f - \int_{\omega_1} f \Big\|_{E;\nu} \leq (|\omega_2| - |\omega_1|) \sup_{x \in \omega_2 \setminus \omega_1} \|f(x)\|_{E;\nu}. \ \blacksquare$$

Proof. The approximated integrals (Definition 1.14) satisfy

$$\mathbb{S}^n_{\omega_2} f - \mathbb{S}^n_{\omega_1} f = \sum_{s \in \mathcal{S}_n} 2^{-nd} f(2^{-n}s),$$

where $\mathcal{S}_n = \{s \in \mathbb{Z}^d : \Delta_{s,n} \subset \omega_2,\ \Delta_{s,n} \not\subset \omega_1\}$. Therefore,

$$\|\mathbb{S}^n_{\omega_2} f - \mathbb{S}^n_{\omega_1} f\|_{E;\nu} \leq 2^{-nd}(k_n(\omega_2) - k_n(\omega_1)) \sup_{s \in \mathcal{S}_n} \|f(2^{-n}s)\|_{E;\nu}, \qquad (1.10)$$

where $k_n(\omega)$ is the number of cubes $\Delta_{s,n}$ contained in ω. If a cube $\Delta_{s,n}$ is contained in ω_2 but not in ω_1, it contains a point $x \in \omega_2 \setminus \omega_1$; then, $|x - 2^{-n}s| \leq 2^{-n-1}\sqrt{d}$ since $\Delta_{s,n}$ is contained in the ball centered at $2^{-n}s$ with radius $2^{-n-1}\sqrt{d}$, hence

$$\sup_{s \in \mathcal{S}_n} \|f(2^{-n}s)\|_{E;\nu} \leq \sup_{x \in \omega_2 \setminus \omega_1} \|f(x)\|_{E;\nu} + \sup_{|y-x| \leq 2^{-n-1}\sqrt{d}} \|f(y) - f(x)\|_{E;\nu}.$$

As $n \to \infty$, the last upper bound tends to 0 since f is uniformly continuous, so inequality (1.10) yields, in the limit, the stated inequality because, by Definition 1.20 of the measure, $2^{-nd} k_n(\omega) \to |\omega|$. □

Let us deduce that the integral remains unchanged when restricted to the domain where the function is not zero.

Proposition 1.23. — *Let $f \in \mathcal{B}(\Omega; E)$, where Ω is an open subset of $\mathbb{R}^d$ and E is a Neumann space, and let ω and λ be open subsets such that $\lambda \subset \omega \subset \Omega$ and*

$$f = 0_E \ \textit{in}\ \omega \setminus \lambda.$$

Then,

$$\int_\omega f = \int_\lambda f. \blacksquare$$

Proof. *1°) Case where ω is bounded.* If ω is bounded, it is contained in a cube with integer side length N. This cube, and hence ω, contains at most $(2^n N)^d$ cubes $\Delta_{s,n}$ (of side lengths 2^{-n}), that is, $k_n(\omega) \le (2^n N)^d$, so Definition 1.20 of the measure gives

$$|\omega| = \lim_{n\to\infty} 2^{-nd} k_n(\omega) \le N^d < \infty.$$

The bound on the semi-norms of a difference of integrals from Proposition 1.22 then gives

$$\left\| \int_\omega f - \int_\lambda f \right\|_{E;\nu} \le (|\omega| - |\lambda|) \sup_{x\in\omega\setminus\lambda} \|f(x)\|_{E;\nu} = (|\omega| - |\lambda|) \times 0 = 0.$$

This holds for every semi-norm of E, which is separated (Definition 1.1), hence

$$\int_\omega f = \int_\lambda f. \tag{1.11}$$

2°) General case. If ω is not bounded, we introduce a bounded open subset Λ outside of which f is zero (given by Definition 1.12 of $\mathcal{B}(\Omega; E)$), and we denote Q the cube of side length 1 centered at 0. Then,

$$\int_\omega f = \int_{\omega\cap(\Lambda+Q)} f. \tag{1.12}$$

Indeed, the approximated integral (Definition 1.14) satisfies $\mathbb{S}^n_\omega f = \mathbb{S}^n_{\omega\cap(\Lambda+Q)} f$, because $f(2^{-n}s) = 0_E$ if $\Delta_{s,n} \not\subset \Lambda + Q$.

By successively using (1.12), (1.11), and (1.12) relative to λ, we obtain

$$\int_\omega f = \int_{\omega\cap(\Lambda+Q)} f = \int_{\lambda\cap(\Lambda+Q)} f = \int_\lambda f. \quad \square$$

Let us show that symmetry does not change the integral.

Proposition 1.24. — *Let $f \in \mathcal{B}(\Omega; E)$, where Ω is an open subset of $\mathbb{R}^d$ and E is a Neumann space. Then,*

$$\int_{\Omega} f(x)\,\mathrm{d}x = \int_{-\Omega} f(-y)\,\mathrm{d}y. \blacksquare$$

Proof. This equality is obtained by taking the limit of the equality concerning the approximated integrals, which is evident since to each cube $\Delta_{s,n}$ from Definition 1.14 included in Ω, there corresponds one and only one such cube included in $-\Omega$. □

Let us now express the measure of a bounded open subset in terms of an integral.

Proposition 1.25. — *For every bounded open subset ω,*

$$\int_{\omega} 1 = |\omega|. \blacksquare$$

Proof. The approximate integral of the constant function with value 1 is, by Definition 1.14,

$$\mathbb{S}^n_\omega f \overset{\text{def}}{=} \sum_{s\in\mathbb{Z}^d:\Delta_{s,n}\subset\omega} 2^{-nd} = 2^{-nd} k_n(\omega),$$

where $k_n(\omega)$ is the number of cubes $\Delta_{s,n}$ (of side length 2^{-n}, centered at $2^{-n}s$) included in ω. In the limit, using Definition 1.20 of the measure of an open subset, this yields the stated equality. □

1.6. Continuity of the integral

Let us show that the Cauchy integral is continuous on the space $\mathbf{C}_Q(\Omega; E)$ of functions that are continuous outside a given compact subset Q.

Proposition 1.26. — *Given Ω an open subset of $\mathbb{R}^d$, ω an open subset included in Ω, Q a compact subset of $\mathbb{R}^d$, and E a Neumann space, the mapping $f \mapsto \int_\omega f$ is continuous from $\mathbf{C}_Q(\Omega; E)$ into E.* ▮

Proof. Let $\{\|\ \|_{E;\nu} : \nu \in \mathcal{N}_E\}$ be the family of semi-norms of E, and let

λ be a bounded open subset of Ω containing Q.

Let $f \in \mathbf{C}_Q(\Omega; E)$. Since the domain where the function vanishes does not contribute to its integral (Proposition 1.23), we have, since $\omega \cap \lambda$ is open and f vanishes outside of it (by Definition 1.11 of $\mathbf{C}_Q(\Omega; E)$),

$$\int_\omega f = \int_{\omega \cap \lambda} f.$$

The bound on the semi-norms of the integral given in Proposition 1.21 yields, using Definition 1.11 of the semi-norms of $\mathbf{C}_Q(\Omega; E)$, for all $\nu \in \mathcal{N}_E$,

$$\Big\| \int_\omega f \Big\|_{E;\nu} = \Big\| \int_{\omega\cap\lambda} f \Big\|_{E;\nu} \leq |\omega \cap \lambda| \sup_{x \in \omega\cap\lambda} \|f(x)\|_{E;\nu} \leq |\omega \cap \lambda| \|f\|_{\mathbf{C}_Q(\Omega;E);\nu}.$$

This inequality implies, according to the characterization of continuous linear maps in Proposition 1.9 (a), that $\int_\omega$ is continuous from $\mathbf{C}_Q(\Omega; E)$ into E. □

Let us use this continuity to show that translations preserve the integral.

Proposition 1.27. — *Let $f \in \mathcal{K}(\Omega; E)$, where Ω is an open subset of $\mathbb{R}^d$ and E is a Neumann space. Then, for all $z \in \mathbb{R}^d$, denoting $\Omega - z = \{x - z : x \in \Omega\}$, we have*

$$\int_\Omega f(x)\,\mathrm{d}x = \int_{\Omega+z} f(y - z)\,\mathrm{d}y. \blacksquare$$

Proof. ***1º) Reduction to $\mathbb{R}^d$.*** The extension of f by 0_E, which we still denote by f, is also uniformly continuous with bounded support. Since the region where the function is zero does not contribute to the integral (Proposition 1.23), we may therefore replace Ω and $\Omega + z$ by $\mathbb{R}^d$.

2º) Rational translation. Suppose for the moment that $z = 2^{-k}\zeta$, where $k \in \mathbb{N}$ and $\zeta \in \mathbb{Z}^d$. Let us show that, denoting $\tau_z f(x) \stackrel{\text{def}}{=} f(x - z)$,

$$\int_{\mathbb{R}^d} f = \int_{\mathbb{R}^d} \tau_z f. \tag{1.13}$$

Indeed, if $n \geq k$, the mapping $s \mapsto s + 2^{n-k}\zeta$ is a bijection from $\mathbb{Z}^d$ onto itself and

$$(\tau_z f)(2^{-n}(s + 2^{n-k}\zeta)) = f(2^{-n}(s + 2^{n-k}\zeta) - z) = f(2^{-n}s),$$

so Definition 1.14 of the approximated integral gives

$$\mathbb{S}^n_{\mathbb{R}^d} f = \sum_{z\in\mathbb{Z}^d} 2^{-nd} f(2^{-n}s) = \sum_{s+2^{n-k}\zeta\in\mathbb{Z}^d} 2^{-nd} (\tau_z f)(2^{-n}(s+2^{n-k}\zeta)) = \mathbb{S}^n_{\mathbb{R}^d} \tau_z f.$$

Which yields (1.13) by taking the limit as $n \to \infty$.

3°) Continuity with respect to z. Let us now choose a sequence of points of the form $z_m = 2^{-k_m}\zeta_m$, where $k_m \in \mathbb{N}$ and $\zeta_m \in \mathbb{Z}^d$, which converges to z when $m \to \infty$, and let Q be a compact subset of $\mathbb{R}^d$ outside of which f and all functions $\tau_{z_m} f$ vanish.

The uniform continuity of f (Definition 1.7) implies the convergence $\tau_{z_m} f \to \tau_z f$ in $\mathbf{C}_Q(\mathbb{R}^d; E)$, so the continuity of the integral in this space (Proposition 1.26) yields, together with equality (1.13),

$$\int_{\mathbb{R}^d} f = \int_{\mathbb{R}^d} \tau_{z_m} f \to \int_{\mathbb{R}^d} \tau_z f.$$

Which, by replacing $\mathbb{R}^d$ by Ω thanks to step 1°, gives the stated equality. □

1.7. Successive integration

Let us show that the order of integration of successive variables is indifferent.

Proposition 1.28. — *Let*

$$f \in \mathbf{C}_{Q_1}(\Omega_1; \mathbf{C}_{Q_2}(\Omega_2; E)),$$

where Ω_1 is an open subset and Q_1 is a compact subset of $\mathbb{R}^{d_1}$, Ω_2 is an open subset and Q_2 is a compact subset of $\mathbb{R}^{d_2}$, and E is a Neumann space. Let $\omega_1 \subset \Omega_1$ and $\omega_2 \subset \Omega_2$. Then,

$$\int_{\omega_2} \Big(\int_{\omega_1} (f(x_1))(x_2)\, \mathrm{d}x_1 \Big)\, \mathrm{d}x_2 = \int_{\omega_1} \Big(\int_{\omega_2} (f(x_1))(x_2)\, \mathrm{d}x_2 \Big)\, \mathrm{d}x_1,$$

which makes sense because the four functions are successively integrable and belong to $\mathbf{C}_{Q_1}(\Omega_1; E)$ or $\mathbf{C}_{Q_2}(\Omega_2; E)$. ▮

Proof. We will see that this is a simple consequence of the commutation of the Cauchy integral with linear mappings.

1°) First commutation. More precisely, for any $g \in \mathbf{C}_{Q_2}(\Omega_2; E)$, define

$$Lg \stackrel{\text{def}}{=} \int_{\omega_2} g(x_2)\, \mathrm{d}x_2. \tag{1.14}$$

According to Proposition 1.26, the mapping L is a continuous linear operator from $\mathbf{C}_{Q_2}(\Omega_2; E)$, which is a Neumann space, into E.

Proposition 1.19 on the commutation of the integral, here $\int_{\omega_1}$, with continuous linear mappings, here L, thus gives $Lf \in \mathbf{C}_{Q_1}(\Omega_1; E)$ and

$$\int_{\omega_1} (Lf)(x_1)\,\mathrm{d}x_1 = L\Big(\int_{\omega_1} f(x_1)\,\mathrm{d}x_1\Big),$$

where $(Lf)(x_1) \stackrel{\text{def}}{=} L(f(x_1))$. That is, according to Definition (1.14) of L,

$$\int_{\omega_1}\Big(\int_{\omega_2} (f(x_1))(x_2)\,\mathrm{d}x_2\Big)\,\mathrm{d}x_1 = \int_{\omega_2}\Big(\int_{\omega_1} f(x_1)\,\mathrm{d}x_1\Big)(x_2)\,\mathrm{d}x_2. \tag{1.15}$$

2º) Second commutation. Fix x_2 and, for every $g \in \mathbf{C}_{Q_2}(\Omega_2; E)$, define

$$Mg \stackrel{\text{def}}{=} g(x_2). \tag{1.16}$$

The mapping M is linearly continuous from $\mathbf{C}_{Q_2}(\Omega_2; E)$ into E, so the commutation of the integral, now again $\int_{\omega_1}$, with M, gives $Mf \in \mathbf{C}_{Q_1}(\Omega_1; E)$ and

$$\int_{\omega_1} (Mf)(x_1)\,\mathrm{d}x_1 = M\Big(\int_{\omega_1} f(x_1)\,\mathrm{d}x_1\Big),$$

where $(Mf)(x_1) \stackrel{\text{def}}{=} M(f(x_1))$. That is, according to Definition (1.16) of M,

$$\int_{\omega_1} (f(x_1))(x_2)\,\mathrm{d}x_1 = \Big(\int_{\omega_1} f(x_1)\,\mathrm{d}x_1\Big)(x_2). \tag{1.17}$$

3º) Equality. By rewriting the second term of (1.15) with (1.17), we obtain

$$\int_{\omega_2}\Big(\int_{\omega_1} (f(x_1))(x_2)\,\mathrm{d}x_1\Big)\,\mathrm{d}x_2 = \int_{\omega_1}\Big(\int_{\omega_2} (f(x_1))(x_2)\,\mathrm{d}x_2\Big)\,\mathrm{d}x_1.$$

4º) Meaning of these integrals. On the left-hand side, we successively integrate:

— for fixed x_2, the function $x_1 \mapsto (f(x_1))(x_2)$, that is Mf, which belongs to the space $\mathbf{C}_{Q_1}(\Omega_1; E)$ (according to 2º);

— the function $x_2 \mapsto \int_{\omega_1} (f(x_1))(x_2)\,\mathrm{d}x_1$, that is (according to (1.17)) $\int_{\omega_1} f$, which belongs to $\mathbf{C}_{Q_2}(\Omega_2; E)$ (by Definition 1.14 of the integral).

On the right-hand side, we successively integrate:

— for fixed x_1, the function $x_2 \mapsto (f(x_1))(x_2)$, that is $f(x_1)$, which belongs to $\mathbf{C}_{Q_2}(\Omega_2; E)$ (by the definition of f);

— the function $x_1 \mapsto \int_{\omega_2} (f(x_1))(x_2)\,\mathrm{d}x_2$, that is Lf, which belongs to the space $\mathbf{C}_{Q_1}(\Omega_1; E)$ (according to 1º). □

Chapter 2

Measurable Sets

The definition and study of measurable subsets of $\mathbb{R}^d$ constitute the second step in the construction of the integral, over such a set, of an integrable measure, which will be carried out in § 5.1.

— We begin, in § 2.1, by explaining the usefulness of integration over such sets for pdes.
— In § 2.2, we present the properties of the measure of open sets that serve to define, in § 2.3, measurable sets and their measure, which is the Lebesgue measure.
— We show that measurability is preserved under finite (Theorem 2.11) or countable (Theorem 2.15) unions and intersections.
— We also show that the measure is monotonic (Theorem 2.10), that it is additive over finite (Theorem 2.13) or countable (Theorem 2.18) unions of disjoint sets, and that it is continuous with respect to increasing (Theorem 2.16) or decreasing (Theorem 2.17) sequences of sets.
— Finally, in § 2.9, we study negligible sets. These are of interest because the integral is not altered by adding such a set to the domain of integration (Theorem 6.8).

2.1. Why introduce measurable sets?

Since our goal is to solve partial differential equations, whose solutions are defined only on open sets, it might be tempting to restrict ourselves to integration over open sets which is somewhat simpler, and used in Volumes 2 and 3.

Let us show with an example that integration over an open set is not always sufficient. Moreover, the measure of a non-open set can sometimes seem counter-intuitive.

Heat diffusion in a heterogeneous medium. In an open domain Ω, the heat u is governed by the equation

$$-\nabla \cdot (\mu \nabla u) = 0,$$

where μ is the diffusion coefficient, which depends on the material. If there are two materials occupying the subdomains ω_1 and ω_2 respectively, a decomposition of the accumulated heat of the form

$$\int_\Omega u = \int_{\omega_1} u + \int_{\omega_2} u \qquad (2.1)$$

requires integrating over a non-open set if Ω is connected and is the union of ω_1 and ω_2, because then they cannot both be open (by the definition of connectedness).

Why not eliminate the interface? If ω_1 and ω_2 have "sufficiently regular shapes", one can decompose

$$\Omega = \omega_1 \cup \omega_2 \cup \sigma \tag{2.2}$$

with ω_1 and ω_2 open sets and a negligible interface σ, in which case the decomposition (2.1) is satisfied, because (Theorem 6.7)

$$\int_\sigma u = 0.$$

But the decomposition (2.2) of Ω is not always possible. For example, suppose that

$$\omega_1 = B_1 \cup B_2 \cup \ldots \cup B_n \cup \ldots,$$

where $B_n = B(a_n, r_n)$ is an open ball of volume $\leq 2^{-n}$ and where the closed balls $\overline{B_n}$ are contained in Ω and pairwise disjoint. Then ω_1 is open, its volume is ≤ 1 and, if (2.2) is satisfied and the volume of Ω is 2, then the volume of ω_2 must be ≥ 1, since σ is negligible.

Let us show that, depending on the positioning of the balls, the representations (2.2) and (2.1) are either satisfied or cannot be satisfied.

Example 1: negligible interface. If the centers of the balls B_n form a sequence converging to a point a in Ω, then, denoting by $\partial\omega$ the boundary of a set ω, the set

$$\sigma = \partial B_1 \cup \ldots \cup \partial B_n \cup \ldots \cup \{a\}, \tag{2.3}$$

is negligible and the set $\omega_2 = \Omega \setminus (\omega_1 \cup \sigma)$ is open. Thus, the decompositions (2.2) and (2.1) hold.

Example 2: non-negligible interface. We will choose the balls B_n such that *no ball can be inserted between them*, although *their total volume is arbitrarily small*. This is **counter-intuitive**, yet simple.

The set of points in Ω with rational coordinates is countable, so it can be represented by a sequence $(x_i)_{i\in\mathbb{N}^*}$. We first choose B_1 such that $\overline{B_1} \subset \Omega$, then, recursively, once $B_1, \ldots, B_{n-1}$ are fixed, we choose $B_n = B(x_{i(n)}, r_n)$, where

$$i(n) \stackrel{\text{def}}{=} \inf\{i \in \mathbb{N}^* : x_i \notin \overline{B_1} \cup \ldots \cup \overline{B_{n-1}}\} \tag{2.4}$$

and where r_n is small enough so that $\overline{B_n}$ does not intersect $\overline{B_1} \ldots \overline{B_{n-1}}$.

If (2.2) were satisfied (with ω_2 open, non-empty, and disjoint from ω_1), then ω_2 would contain a ball, which itself would contain a point with rational coordinates, say x_j. Then, (2.4) would imply $i(n) \leq j$ for all n which would contradict the fact that the $i(n)$ are all distinct. Thus, the decompositions (2.2) and (2.1) are not possible.

Position of the balls. The situations in the two examples above can be realized with the same balls, up to their positions: it suffices to choose Ω as a ball (its radius is $> 1/2$ since $|\Omega| = 2$) and to impose in the second example that $r_n \leq 2^{-n-4}$. Then, the same balls satisfy the first example if we impose $|a_n| = 2^{-n-2}$, because they are pairwise disjoint and their centers converge to $a = 0$.

The total volume of the balls is less than a given $\epsilon > 0$ if, moreover, we impose $r_n \leq \epsilon 2^{-n-1}$.

Observe that, in Example 1, $\partial\omega_1 = \sigma$ and $\partial\omega_2 = \sigma \cup \partial\Omega$. In contrast, in Example 2, we have $\partial\omega_1 \neq \sigma$ (more precisely, $\partial\omega_1 = \overline{\Omega} \setminus \omega_1$). □

Now that we have seen that it is useful to integrate over sets other than open ones, let us conversely show that it is not possible to integrate satisfactorily over just any subset of $\mathbb{R}^d$.

Example 3: the Banach–Tarski[1] decomposition. One can decompose the ball B_1 of radius 1 in $\mathbb{R}^3$ into a finite number of disjoint pieces

$$B_1 = \omega_1 \cup \cdots \cup \omega_n$$

and we can move each ω_i by a combination T_i of a translation and a rotation to obtain the ball B_2 of radius 2, that is,

$$B_2 = T_1(\omega_1) \cup \cdots \cup T_n(\omega_n).$$

This result, **very counter-intuitive** and not simple (the proof uses the axiom of choice), shows that the measure of such pieces could not be stable under disjoint unions, translations, and rotations; otherwise, B_1 and B_2 would have the same measure.

It shows that integration over such sets could not be additive with respect to disjoint domains while also being stable under translation and rotation and recovering the Cauchy integral, otherwise we would have $\int_{B_1} 1 = \int_{B_2} 1$.

2.2. Some properties of the measure of an open set

Recall that the measure of an open set is defined in § 1.5 (Definition 1.20), and let us compute the measure of a cube or an interval.

1 **Banach–Tarski decomposition.** Stefan BANACH and Alfred TARSKI established this property in 1924 [2, Théorème 24, p. 263]. They showed that it holds for $d \geq 3$ and fails for $d = 1$ or 2.

Proposition 2.1. — *The open cube* $Q \overset{\text{def}}{=} \{x \in \mathbb{R}^d : a_i < x_i < a_i + \ell\}$, *where* $a \in \mathbb{R}^d$ *and* $\ell > 0$, *has measure*

$$|Q| = \ell^d.$$

In particular, if $-\infty < a < b < \infty$,

$$\big|(b-a)\big| = |b-a|. ∎$$

Proof. Let k_n^i be the number of closed intervals of length 2^{-n} centered at $2^{-n}s_i$ (with $s_i \in \mathbb{Z}$), that are contained within the open interval $(a_i, a_i + \ell)$. Then,

$$k_n^i 2^{-n} < \ell < (k_n^i + 2)2^{-n}.$$

The cube Q contains $k_n(Q) = k_n^1 \times \cdots \times k_n^d$ closed cubes $\Delta_{s,n}$ of side lengths 2^{-n}, centered at points $2^{-n}s$ with $s \in \mathbb{Z}^d$, so that

$$(\ell - 2^{1-n})^d < 2^{-nd} k_n(Q) < \ell^d.$$

Definition 1.20 of the measure of an open set then gives

$$|Q| = \lim_{n\to\infty} 2^{-nd} k_n(Q) = \ell^d. \ \square$$

Let us show that the measure of open sets is increasing (this will be extended to measurable sets in Theorem 2.10).

Proposition 2.2. — *Let* ω_1 *and* ω_2 *be two open subsets of* $\mathbb{R}^d$. *Then:*

$$\omega_1 \subset \omega_2 \quad \Rightarrow \quad |\omega_1| \leq |\omega_2|. ∎$$

Proof. If $\omega_1 \subset \omega_2$, the number $k_n(\omega_1)$ of cubes $\Delta_{s,n}$ (from Definition 1.20 of the measure of an open set) that are included in ω_1 is less than the number $k_n(\omega_2)$ of those included in ω_2. Thus, $2^{-nd}k_n(\omega_1) \leq 2^{-nd}k_n(\omega_2)$, hence, in the limit, $|\omega_1| \leq |\omega_2|$. $\square$

Let us now show that the measure is continuous with respect to increasing sequences of open sets (this will be extended to measurable sets in Theorem 2.16).

Proposition 2.3. — *Let* $(\omega_n)_{n\in\mathbb{N}}$ *be an increasing sequence of open subsets of* $\mathbb{R}^d$ *whose union is* ω. *Then, as* $n \to \infty$,

$$|\omega_n| \to |\omega|. ∎$$

Proof. **1º)** *Limit ℓ of $|\omega_n|$.* The measure $|\omega_n|$ increases with n (Proposition 2.2), thus (Proposition A.5) it has a limit ℓ, finite if it is bounded, or infinite otherwise.

2º) *Upper bound of the limit.* The subset ω is open, as a union of open subsets, so its measure is defined. Since $\omega_n \subset \omega$, we have $|\omega_n| \leq |\omega|$ (Proposition 2.2), hence

$$\ell \leq |\omega|.$$

3º) *Lower bound of the limit: finite case.* Given $i \in \mathbb{N}$, suppose that the number $k_i(\omega)$ of cubes $\Delta_{s,i}$ from Definition 1.20 included in ω is finite. The union K_i of these cubes is then closed (since each one of them is closed) and bounded, and thus compact by the Borel–Lebesgue theorem (Proposition A.36 (a)).

Since the ω_n form an open cover of K_i, a finite number of them covers it and, because they increase with n, one of them covers it. For this n, all the $\Delta_{s,i}$ composing K_i are included in ω_n, hence $k_i(\omega) \leq k_i(\omega_n)$. Thus, by inequality (1.9),

$$2^{-id} k_i(\omega) \leq 2^{-id} k_i(\omega_n) \leq |\omega_n| \leq \ell. \tag{2.5}$$

By Definition 1.20 of the measure on an open set, this yields the limit, when $i \to \infty$,

$$|\omega| \leq \ell.$$

4º) *Lower bound of the limit: infinite case.* If $k_i(\omega) = \infty$ (for some i), by denoting K_i as the union of an arbitrary finite number k of the $\Delta_{s,i}$ included in ω, we obtain inequality (2.5) with k in place of $k_i(\omega)$. This holds for any k, hence $\ell = \infty$. Therefore, once again,

$$|\omega| \leq \ell. \ \square$$

Let us now show that the measure of open sets is subadditive, and additive with respect to disjoint open sets (this will be extended to measurable sets in Theorem 2.13).

Proposition 2.4. — *Let ω_1 and ω_2 be two open subsets of $\mathbb{R}^d$. Then,*

(a)
$$|\omega_1 \cup \omega_2| \leq |\omega_1| + |\omega_2|.$$

(b)
$$\omega_1 \cap \omega_2 = \emptyset \quad \Rightarrow \quad |\omega_1 \cup \omega_2| = |\omega_1| + |\omega_2|. \ ■$$

Proof. **1º)** *Equality* (b). When $\omega_1 \cap \omega_2 = \emptyset$, every cube $\Delta_{s,n}$ from Definition 1.20 included in $\omega_1 \cup \omega_2$ is included in either ω_1 or ω_2, and conversely. Therefore, the counts of these cubes are related by $k_n(\omega_1 \cup \omega_2) = k_n(\omega_1) + k_n(\omega_2)$. Hence,

$$2^{-nd} k_n(\omega_1 \cup \omega_2) = 2^{-nd} k_n(\omega_1) + 2^{-nd} k_n(\omega_2).$$

This gives, in the limit as $n \to \infty$,

$$|\omega_1 \cup \omega_2| = |\omega_1| + |\omega_2|.$$

2º) Inequality (a). Here, ω_1 and ω_2 are two arbitrary open subsets.

If every cube $\Delta_{s,n}$ included in $\omega_1 \cup \omega_2$ were also included in ω_1 or ω_2, we would again have $k_n(\omega_1 \cup \omega_2) \le k_n(\omega_1) + k_n(\omega_2)$ and the desired inequality would be immediate. This is not the case, however, because a cube $\Delta_{s,n}$ can be included in the union of ω_1 and ω_2 without being included in either one of them.

To avoid this difficulty, let us introduce, for $m = 1$ or 2 and $i \in \mathbb{N}^*$,

$$\omega_m^i \stackrel{\text{def}}{=} \{x \in \mathbb{R}^d : B(x, 1/i) \subset \omega_m\},$$

where $B(x, 1/i)$ denotes the closed ball. Then, ω_m^i is open and increases to ω_m, so that

$(\omega_1^i \cup \omega_2^i)_{i \in \mathbb{N}^*}$ is an increasing sequence of open subsets with union $\omega_1 \cup \omega_2$.

Fix i and n large enough so that $2^{-n}\sqrt{d} < 1/i$, and

$$\Delta_{s,n} \subset \omega_1^i \cup \omega_2^i.$$

This cube $\Delta_{s,n}$ is included in a ball of diameter $1/i$ (since its side length is 2^{-n}), so if its center belongs to ω_1^i, it is contained in ω_1. Otherwise, its center belongs to ω_2^i and it is included in ω_2. The numbers of cubes included in these sets therefore satisfy $k_n(\omega_1^i \cup \omega_2^i) \le k_n(\omega_1) + k_n(\omega_2)$. Thus,

$$2^{-nd} k_n(\omega_1^i \cup \omega_2^i) \le 2^{-nd} k_n(\omega_1) + 2^{-nd} k_n(\omega_2).$$

This gives, in the limit as $n \to \infty$,

$$|\omega_1^i \cup \omega_2^i| \le |\omega_1| + |\omega_2|.$$

Since the open subsets $\omega_1^i \cup \omega_2^i$ increase to $\omega_1 \cup \omega_2$, Proposition 2.3 yields, in the limit as $i \to \infty$,

$$|\omega_1 \cup \omega_2| \le |\omega_1| + |\omega_2|. \ \square$$

2.3. Definition of measurable sets and their measure

Let us define the measurable subsets[2] of $\mathbb{R}^d$.

---

2 **Historical background of the notion of a measurable set.** Henri LEBESGUE introduced a definition of measurable sets in $\mathbb{R}$ in 1903 [44, chap. VII, § II, p. 106], which is equivalent to Definition 2.5.
Émile BOREL had defined *measurable* sets in 1898 [7], now called *Borel* sets, which differ from those of LEBESGUE only by the "addition of a set of measure zero in the sense of Borel" [BOURBAKI, 8, chap. V, note historique, p. 139].

Definition 2.5. — *A subset ω of $\mathbb{R}^d$ is said to be* **measurable** *if, for every $\epsilon > 0$, there exists a closed set $\mathcal{F}$ and an open set $\mathcal{O}$ such that:*

$$\mathcal{F} \subset \omega \subset \mathcal{O}, \quad |\mathcal{O} \setminus \mathcal{F}| \leq \epsilon. \blacksquare$$

Justification. The measure of $\mathcal{O} \setminus \mathcal{F}$ is given by Definition 1.20, since it is open. □

Now, let us define the measure of a measurable set in $\mathbb{R}^d$.

Definition 2.6. — *The* **measure** *of a measurable subset ω of $\mathbb{R}^d$ is defined by*

$$|\omega| \stackrel{\text{def}}{=} \inf\{|\mathcal{O}| : \mathcal{O} \text{ open subset of } \mathbb{R}^d, \ \mathcal{O} \supset \omega\},$$

where $|\mathcal{O}|$ is given by Definition 1.20. Thus, $|\omega|$ is either a real number or ∞. ■

This definition is legitimate because, as we shall now verify, it recovers the measure of open sets.

Theorem 2.7. — *Every open subset of $\mathbb{R}^d$ is measurable, and its measure given by Definition 1.20 is equal to that given by Definition 2.6.*

Every closed subset of $\mathbb{R}^d$ is measurable. In particular, $\mathbb{R}^d$ and $\emptyset$ are measurable. ■

The proof for an unbounded open set will make use of the following property.

Lemma 2.8. — *A subset of $\mathbb{R}^d$ is measurable if its intersection with every bounded open set is measurable.* ■

Proof of Lemma 2.8. To avoid any ambiguity, we denote here $\|\omega\|$ the measure of an open set in the sense of Definition 1.20, and by $|\omega|$ its measure given by Definition 2.6.

Let ω be a subset of $\mathbb{R}^d$ such that, for every bounded open set B, $\omega \cap B$ is measurable.

1°) Framing of ω. Given $i \in \mathbb{N}$, consider the open annulus

$$C_i \stackrel{\text{def}}{=} \{x \in \mathbb{R}^d : i - 1 < |x| < i + 1\}.$$

Let $\epsilon > 0$. By assumption, $\omega \cap C_i$ is measurable, so, by Definition 2.5 of measurability, for all $i \in \mathbb{N}$, there exists a closed set $\mathcal{F}_i$ and an open set $\mathcal{O}_i$ such that

$$\mathcal{F}_i \subset \omega \cap C_i \subset \mathcal{O}_i, \quad \|\mathcal{O}_i \setminus \mathcal{F}_i\| \leq \epsilon 2^{-i-1}. \tag{2.6}$$

Let

$$\mathcal{F} \stackrel{\text{def}}{=} \bigcup_{i \in \mathbb{N}} \mathcal{F}_i, \quad \mathcal{O} \stackrel{\text{def}}{=} \bigcup_{i \in \mathbb{N}} \mathcal{O}_i.$$

Since the C_i cover $\mathbb{R}^d$,

$$\mathcal{F} \subset \omega \subset \mathcal{O}. \tag{2.7}$$

2º) Properties of $\mathcal{O}$ and $\mathcal{F}$. Observe that $\mathcal{O}$ is open (as any union of open sets), and let us verify that

$$\mathcal{F} \text{ is closed.}$$

Indeed, its complement $\mathcal{F}^c = \mathbb{R}^d \setminus \mathcal{F}$ is open, because for every $x \in \mathcal{F}^c$, there exists i such that $x \in C_i$; then, there exists $r > 0$ such that $B(x, r) \subset (\mathcal{F}_{i-1} \cup \mathcal{F}_i \cup \mathcal{F}_{i+1})^c$ (since $\mathcal{F}_{i-1} \cup \mathcal{F}_i \cup \mathcal{F}_{i+1}$ is closed, being the union of closed sets); and therefore, the ball $B(x, \inf\{r, 1\})$ is included in $\mathcal{F}^c$, since it does not intersect any other $\mathcal{F}_j$.

3º) Measure of $\mathcal{O} \setminus \mathcal{F}$. Observe that

$$\mathcal{O} \setminus \mathcal{F} = \Big(\bigcup_{i \in \mathbb{N}} \mathcal{O}_i\Big) \setminus \mathcal{F} = \bigcup_{i \in \mathbb{N}} \mathcal{O}_i \setminus \mathcal{F} \subset \bigcup_{i \in \mathbb{N}} \mathcal{O}_i \setminus \mathcal{F}_i. \tag{2.8}$$

The subadditivity of the measure of open sets (Proposition 2.4 (a)) gives, for all $n \in \mathbb{N}$, with (2.6) and the sum of a geometric series (Proposition A.6),

$$\|(\mathcal{O}_0 \setminus \mathcal{F}_0) \cup \cdots \cup (\mathcal{O}_n \setminus \mathcal{F}_n)\| \leq \|\mathcal{O}_0 \setminus \mathcal{F}_0\| + \cdots + \|\mathcal{O}_n \setminus \mathcal{F}_n\| \leq \epsilon.$$

As $n \to \infty$, the continuity of the measure of increasing open sets (Proposition 2.3) yields

$$\Big\|\bigcup_{i \in \mathbb{N}} \mathcal{O}_i \setminus \mathcal{F}_i\Big\| \leq \epsilon.$$

With inclusion (2.8) and the monotonicity of the measure of open sets (Proposition 2.2), it follows that

$$\|\mathcal{O} \setminus \mathcal{F}\| \leq \epsilon.$$

Thus, given inclusion (2.7), ω satisfies Definition 2.5 of a measurable set. □

Proof of Theorem 2.7. Let us again denote by $\|\omega\|$ the measure of an open set in the sense of Definition 1.20, and by $|\omega|$ its measure given by Definition 2.6.

1º) Measurability of a bounded open set. Let ω be a bounded open subset of $\mathbb{R}^d$. Let $n \in \mathbb{N}$, and let K_n be the union of the cubes $\Delta_{s,n}$ from Definition 1.20 that are included in ω. The number $k_n(\omega)$ of these $\Delta_{s,n}$ is finite, so K_n is compact by the Borel–Lebesgue theorem (Proposition A.36 (a)), since it is closed and bounded in $\mathbb{R}^d$.

For $i \geq n$, each $\Delta_{s,n}$ is divided into $2^{(i-n)d}$ sub-cubes $\Delta_{s,i}$, so K_n contains $2^{(i-n)d}k_n(\omega)$ cubes $\Delta_{s,i}$. The number $k_i(\omega \backslash K_n)$ of cubes $\Delta_{s,i}$ included in $\omega \setminus K_n$ is thus bounded above by $k_i(\omega) - 2^{(i-n)d}k_n(\omega)$. Hence,

$$2^{-id}k_i(\omega \backslash K_n) \leq 2^{-id}k_i(\omega) - 2^{-nd}k_n(\omega).$$

By letting $i \to \infty$, it follows from Definition 1.20 of the measure of an open set that,

$$\|\omega \setminus K_n\| \leq \|\omega\| - 2^{-nd}k_n(\omega).$$

For any $\epsilon > 0$, one can (again by Definition 1.20) choose n large enough so that the right-hand side is less than ϵ, that is,

$$\|\omega \setminus K_n\| \leq \epsilon. \tag{2.9}$$

Thus, ω satisfies Definition 2.5 of measurability (with $\mathcal{F} = K_n$ and $\mathcal{O} = \omega$).

2o) Measurability of an arbitrary open set. Let us now return to an arbitrary open subset ω.

Let B be a bounded open set. The intersection $\omega \cap B$ is open (as any finite intersection of open sets) and bounded, and hence measurable by point 1°.

This, according to Lemma 2.8, implies that ω is measurable.

3°) Value of the measure. Definition 2.6 gives, by taking $\mathcal{O} = \omega$,

$$|\omega| = \inf\{\|\mathcal{O}\| : \mathcal{O} \text{ ouvert}, \ \mathcal{O} \supset \omega\} \leq \|\omega\|.$$

Conversely, $\omega \subset \mathcal{O}$ implies $\|\omega\| \leq \|\mathcal{O}\|$, by the monotonicity of the measure of open sets (Proposition 2.2), so Definition 2.6 also yields $\|\omega\| \leq |\omega|$. Hence,

$$|\omega| = \|\omega\|.$$

4°) Measurability of a closed set. Now let λ be a closed subset of $\mathbb{R}^d$.

Its complement $\lambda^c \overset{\text{def}}{=} \mathbb{R}^d \setminus \lambda$ is open, so, as seen in (2.7), for all $\epsilon > 0$, there exists a closed set $\mathcal{F}$ such that

$$\mathcal{F} \subset \lambda^c, \quad \|\lambda^c \setminus \mathcal{F}\| \leq \epsilon.$$

Then, $\mathcal{O} \overset{\text{def}}{=} \mathcal{F}^c$ is open and

$$\lambda \subset \mathcal{O}, \quad \|\mathcal{O} \setminus \lambda\| = \|\lambda^c \setminus \mathcal{F}\| \leq \epsilon.$$

Therefore, λ satisfies Definition 2.5 of measurability (with $\mathcal{F} = \omega$). □

Non-measurable sets. The definition of measurable sets is relevant because there exist non-measurable sets. Indeed, according to a result by Giuseppe VITALI[3] based on the axiom of choice:

$$\textit{There exist subsets of } \mathbb{R}^d \textit{ that are not measurable.} \tag{2.10}$$

The Banach–Tarski decomposition (p. 25) provides non-measurable subsets of $\mathbb{R}^d$, for $d \geq 3$. □

2.4. First properties of the measure

Let us observe that the measure of bounded measurable sets is finite.

Theorem 2.9. — *Every measurable and bounded subset of $\mathbb{R}^d$ has finite measure.* ▮

Proof. Let ω be a measurable and bounded subset of $\mathbb{R}^d$. It is included in an open cube Q, which has finite measure (Proposition 2.1), so Definition 2.6 of the measure gives

$$|\omega| \leq |Q| < \infty. \ \square$$

Let us show that the measure is monotonic.

Theorem 2.10. — *Let ω_1 and ω_2 be two measurable subsets of $\mathbb{R}^d$. Then:*

$$\omega_1 \subset \omega_2 \quad \Rightarrow \quad |\omega_1| \leq |\omega_2|. \ ▮$$

Proof. If $\omega_1 \subset \omega_2$, then every open set $\mathcal{O}$ containing ω_2 also contains ω_1, hence

$$\inf\{|\mathcal{O}| : \mathcal{O} \text{ open}, \ \mathcal{O} \supset \omega_1\} \leq \inf\{|\mathcal{O}| : \mathcal{O} \text{ open}, \ \mathcal{O} \supset \omega_2\},$$

that is, $|\omega_1| \leq |\omega_2|$ by Definition 2.6 of the measure. □

Non-strict monotonicity. The measure is not strictly increasing, even in the case of open sets.

For example, if ω is a measurable set and x is one of its points, then $\omega \backslash \{x\}$ is measurable and $|\omega \backslash \{x\}| = |\omega|$. □

Let us show that measurability is preserved under union, intersection, and difference.

---

3 **Historical background on the existence of non-measurable sets.** Giuseppe VITALI proved in 1905 [78] that there exist subsets of $\mathbb{R}$ that are not measurable.

Theorem 2.11. — *Let ω_1 and ω_2 be two measurable subsets of $\mathbb{R}^d$. Then:*

$$\omega_1 \cup \omega_2,\ \omega_1 \cap \omega_2 \text{ and } \omega_1 \setminus \omega_2 \text{ are measurable.}$$

In particular, $\omega_1^c = \mathbb{R}^d \setminus \omega_1$ is measurable. ▌

Proof. ***1º) Measurability of the union.*** Let $\epsilon > 0$. By Definition 2.5 of measurability, there exist closed subsets $\mathcal{F}_1$ and $\mathcal{F}_2$ and open sets $\mathcal{O}_1$ and $\mathcal{O}_2$ such that

$$\mathcal{F}_1 \subset \omega_1 \subset \mathcal{O}_1,\ |\mathcal{O}_1 \setminus \mathcal{F}_1| \le \frac{\epsilon}{2},\quad \mathcal{F}_2 \subset \omega_2 \subset \mathcal{O}_2,\ |\mathcal{O}_2 \setminus \mathcal{F}_2| \le \frac{\epsilon}{2}. \tag{2.11}$$

Then,

$$\mathcal{F}_1 \cup \mathcal{F}_2 \subset \omega_1 \cup \omega_2 \subset \mathcal{O}_1 \cup \mathcal{O}_2 \tag{2.12}$$

and

$$(\mathcal{O}_1 \cup \mathcal{O}_2) \setminus (\mathcal{F}_1 \cup \mathcal{F}_2) = (\mathcal{O}_1 \setminus (\mathcal{F}_1 \cup \mathcal{F}_2)) \cup (\mathcal{O}_2 \setminus (\mathcal{F}_1 \cup \mathcal{F}_2)) \subset (\mathcal{O}_1 \setminus \mathcal{F}_1) \cup (\mathcal{O}_2 \setminus \mathcal{F}_2).$$

The monotonicity and subadditivity of the measure of open sets (Propositions 2.2 and 2.4 (a)) give

$$|(\mathcal{O}_1 \cup \mathcal{O}_2) \setminus (\mathcal{F}_1 \cup \mathcal{F}_2)| \le |(\mathcal{O}_1 \setminus \mathcal{F}_1) \cup (\mathcal{O}_2 \setminus \mathcal{F}_2)| \le |\mathcal{O}_1 \setminus \mathcal{F}_1| + |\mathcal{O}_2 \setminus \mathcal{F}_2| \le \epsilon.$$

With (2.12), this proves that $\omega \cup \omega_2$ is measurable.

2º) Measurability of the intersection. The open and closed sets satisfying (2.11) verify

$$\mathcal{F}_1 \cap \mathcal{F}_2 \subset \omega_1 \cap \omega_2 \subset \mathcal{O}_1 \cap \mathcal{O}_2 \tag{2.13}$$

and

$$(\mathcal{O}_1 \cap \mathcal{O}_2) \setminus (\mathcal{F}_1 \cap \mathcal{F}_2) = ((\mathcal{O}_1 \cap \mathcal{O}_2) \setminus \mathcal{F}_1) \cup ((\mathcal{O}_1 \cap \mathcal{O}_2) \setminus \mathcal{F}_2) \subset (\mathcal{O}_1 \setminus \mathcal{F}_1) \cup (\mathcal{O}_2 \setminus \mathcal{F}_2).$$

Whence, again by Propositions 2.2 and 2.4 (a),

$$|(\mathcal{O}_1 \cap \mathcal{O}_2) \setminus (\mathcal{F}_1 \cap \mathcal{F}_2)| \le |(\mathcal{O}_1 \setminus \mathcal{F}_1) \cup (\mathcal{O}_2 \setminus \mathcal{F}_2)| \le \epsilon.$$

This, with (2.13), proves that $\omega \cap \omega_2$ is measurable.

3º) Measurability of the complement. Again, let $\mathcal{F}_2$ and $\mathcal{O}_2$ satisfy (2.11). Then, let $\mathcal{F}_2' \stackrel{\text{def}}{=} \mathbb{R}^d \setminus \mathcal{O}_2$, which is closed, and $\mathcal{O}_2' \stackrel{\text{def}}{=} \mathbb{R}^d \setminus \mathcal{F}_2$, which is open, and they satisfy

$$\mathcal{F}_2' \subset (\mathbb{R}^d \setminus \omega_2) \subset \mathcal{O}_2',\quad |\mathcal{O}_2' \setminus \mathcal{F}_2'| = |\mathcal{O}_2 \setminus \mathcal{F}_2| \le \epsilon.$$

This proves that $\mathbb{R}^d \setminus \omega_2$ is measurable.

4º) Measurability of the difference. Observe that

$$\omega_1 \setminus \omega_2 = \omega_1 \cap (\mathbb{R}^d \setminus \omega_2).$$

Now, $\mathbb{R}^d \setminus \omega_2$ is measurable according to 3º, and its intersection with ω_1 is therefore also measurable according to 2º, which proves that $\omega_1 \setminus \omega_2$ is measurable. □

Any open set of finite measure can be *approximated in measure* by a compact set.

Theorem 2.12. — *Let ω be an open subset of $\mathbb{R}^d$ such that $|\omega| < \infty$, and let $\epsilon > 0$. Then there exists a compact set $K \subset \omega$ such that $|\omega \setminus K| \leq \epsilon$.* ▮

Proof. Let $n \in \mathbb{N}$, and let K_n be the union of the cubes $\Delta_{s,n}$ from Definition 1.20 that are contained in ω. The number k_n of these $\Delta_{s,n}$ is finite (since $2^{-nd} k_n \leq |\omega|$, see (1.9)), so K_n is compact, by the Borel–Lebesgue theorem (Proposition A.36 (a)), since it is closed and bounded in $\mathbb{R}^d$.

For $i \geq n$, each $\Delta_{s,n}$ is subdivided into $2^{(i-n)d}$ subcubes $\Delta_{s,i}$, so K_n contains $2^{(i-n)d} k_n$ cubes $\Delta_{s,i}$. The number $\ell_{i,n}$ of $\Delta_{s,i}$ included in $\omega \setminus K_n$ is therefore bounded above by $k_i - 2^{(i-n)d} k_n$. Hence, by Definition 1.20 of the measure

$$|\omega \setminus K_n| = \lim_{i \to \infty} 2^{-id} \ell_{i,n} \leq \lim_{i \to \infty} 2^{-id} (k_i - 2^{(i-n)d} k_n) = |\omega| - 2^{-nd} k_n,$$

which is less than ϵ for n large enough. □

2.5. Additivity of the measure

Let us show that the measure is additive[4].

Theorem 2.13. — *Let ω_1 and ω_2 be two measurable subsets of $\mathbb{R}^d$. Then:*

(a)
$$|\omega_1 \cup \omega_2| + |\omega_1 \cap \omega_2| = |\omega_1| + |\omega_2|;$$

(b)
$$|\omega_1 \cup \omega_2| \leq |\omega_1| + |\omega_2|;$$

(c)
$$\omega_1 \cap \omega_2 = \emptyset \quad \Rightarrow \quad |\omega_1 \cup \omega_2| = |\omega_1| + |\omega_2|;$$

(d) *if $|\omega_1|$ or $|\omega_2|$ is finite,*

$$\omega_1 \subset \omega_2 \quad \Rightarrow \quad |\omega_2 \setminus \omega_1| = |\omega_2| - |\omega_1|.$$ ▮

4 **Historical background of the additivity of measure.** Additivity of measure, that is, Theorem 2.13, is one of the three conditions required by Henri LEBESGUE for his definition of the measure of a subset of $\mathbb{R}$, given in 1903 [44, chap. VII, § II, p. 103]. The other conditions are translation invariance (Theorem 2.20) and normalization by $|(0, 1)| = 1$ (Proposition 2.1).

Proof. Method. All these properties are straightforward consequences of the additivity of the measure for disjoint sets, that is, property (c), with which we will begin. After easily bounding $|\omega_1 \cup \omega_2|$ using 1.a, we will bound it from below in three steps, by constructing disjoint open sets $\mathcal{O}_1''$ and $\mathcal{O}_2''$, respectively close to ω_1 and ω_2, and then using the additivity of the measure of such open sets.

To avoid ambiguity, we denote by $\|\mathcal{O}\|$ the measure of an open set $\mathcal{O}$ (Definition 1.20) and $|\omega|$ the measure of a measurable set ω (Definition 2.6).

1º) Equality (c). Here,

$$\omega_1 \text{ and } \omega_2 \text{ are disjoint.}$$

1.a. *Upper bound of* $|\omega_1 \cup \omega_2|$. Let $\mathcal{O}_1$ and $\mathcal{O}_2$ be two open sets containing ω_1 and ω_2, respectively. Then $\mathcal{O}_1 \cup \mathcal{O}_2$ is open and contains $\omega_1 \cup \omega_2$.

The union $\omega_1 \cup \omega_2$ is measurable (Theorem 2.11), so Definition 2.6 of the measure and the subadditivity of open sets (Proposition 2.4 (a)) give

$$|\omega_1 \cup \omega_2| \leq \|\mathcal{O}_1 \cup \mathcal{O}_2\| \leq \|\mathcal{O}_1\| + \|\mathcal{O}_2\|.$$

Taking the infimum over all such $\mathcal{O}_1$, Definition 2.6 gives $|\omega_1 \cup \omega_2| \leq |\omega_1| + \|\mathcal{O}_2\|$; then, taking the infimum over all $\mathcal{O}_2$, it follows that

$$|\omega_1 \cup \omega_2| \leq |\omega_1| + |\omega_2|. \tag{2.14}$$

1.b. *Construction of* $\mathcal{O}_1''$ *and* $\mathcal{O}_2''$ *to bound* $|\omega_1 \cup \omega_2|$ *from below.* Let $\epsilon > 0$. By Definition 2.5 of measurability, there exist closed sets $\mathcal{F}_1$ and $\mathcal{F}_2$, and open sets $\mathcal{O}_1$ and $\mathcal{O}_2$ such that

$$\mathcal{F}_1 \subset \omega_1 \subset \mathcal{O}_1, \;\; \|\mathcal{O}_1 \setminus \mathcal{F}_1\| \leq \epsilon, \quad \mathcal{F}_2 \subset \omega_2 \subset \mathcal{O}_2, \;\; \|\mathcal{O}_2 \setminus \mathcal{F}_2\| \leq \epsilon. \tag{2.15}$$

Then $\mathcal{F}_1$ and $\mathcal{F}_2$ are disjoint, so, as we will verify in Lemma 2.14, there exist two open sets $\mathcal{O}_1'$ and $\mathcal{O}_2'$ such that

$$\mathcal{F}_1 \subset \mathcal{O}_1', \quad \mathcal{F}_2 \subset \mathcal{O}_2', \quad \mathcal{O}_1' \cap \mathcal{O}_2' = \emptyset. \tag{2.16}$$

Moreover, by Definition 2.6 of the measure, there exists an open set $\mathcal{O}$ such that

$$\omega_1 \cup \omega_2 \subset \mathcal{O}, \quad \|\mathcal{O}\| \leq |\omega_1 \cup \omega_2| + \epsilon. \tag{2.17}$$

The open sets

$$\mathcal{O}_1'' \stackrel{\text{def}}{=} \mathcal{O}_1 \cap \mathcal{O}_1' \cap \mathcal{O} \;\text{ and }\; \mathcal{O}_2'' \stackrel{\text{def}}{=} \mathcal{O}_2 \cap \mathcal{O}_2' \cap \mathcal{O}$$

then satisfy

$$\mathcal{F}_1 \subset \mathcal{O}_1'', \quad \mathcal{F}_2 \subset \mathcal{O}_2'', \quad \mathcal{O}_1'' \cap \mathcal{O}_2'' = \emptyset.$$

1.c. *Approximation of* $|\omega_i|$ *by* $\|\mathcal{O}_i\|$. Since $\mathcal{F}_1 \subset \mathcal{O}_1''$ and $\omega_1 \subset \mathcal{O}_1$, we have

$$\omega_1 = \mathcal{F}_1 \cup (\omega_1 \setminus \mathcal{F}_1) \subset \mathcal{O}_1'' \cup (\mathcal{O}_1 \setminus \mathcal{F}_1).$$

Definition 2.6 of the measure of ω_1 thus gives (since $\mathcal{O}_1'' \cup (\mathcal{O}_1 \setminus \mathcal{F}_1)$ is open), using the subadditivity of the measure of open sets (Proposition 2.4 (a)) and (2.15),

$$|\omega_1| \leq \|\mathcal{O}_1'' \cup (\mathcal{O}_1 \setminus \mathcal{F}_1)\| \leq \|\mathcal{O}_1''\| + \|\mathcal{O}_1 \setminus \mathcal{F}_1\| \leq \|\mathcal{O}_1''\| + \epsilon.$$

Likewise, $|\omega_2| \leq \|\mathcal{O}_2''\| + \epsilon$, so

$$|\omega_1| + |\omega_2| \leq \|\mathcal{O}_1''\| + \|\mathcal{O}_2''\| + 2\epsilon. \tag{2.18}$$

1.d. *Lower bound for* $|\omega_1 \cup \omega_2|$. Since $\mathcal{O}_1''$ and $\mathcal{O}_2''$ are disjoint, the additivity of the measure for such open sets (Proposition 2.4 (b)) gives, along with its monotonicity (Proposition 2.2) and (2.17),

$$\|\mathcal{O}_1''\| + \|\mathcal{O}_2''\| = \|\mathcal{O}_1'' \cup \mathcal{O}_2''\| \leq \|\mathcal{O}\| \leq |\omega_1 \cup \omega_2| + \epsilon.$$

With (2.18), we obtain

$$|\omega_1| + |\omega_2| \leq |\omega_1 \cup \omega_2| + 3\epsilon.$$

Since this inequality holds for all $\epsilon > 0$, it also holds for $\epsilon = 0$. The reverse inequality having been established in (2.14), this prove that, when ω_1 and ω_2 are disjoint,

$$|\omega_1 \cup \omega_2| = |\omega_1| + |\omega_2|.$$

2º) *Equality* (a). Here,

$$\omega_1 \text{ and } \omega_2 \text{ are arbitrary measurable sets.}$$

The sets ω_1 and $\omega_2 \setminus \omega_1$ are measurable (Theorem 2.11), disjoint, and their union is $\omega_1 \cup \omega_2$, so 1º gives

$$|\omega_1 \cup \omega_2| = |\omega_1| + |\omega_2 \setminus \omega_1|.$$

Similarly, $\omega_1 \cap \omega_2$ and $\omega_2 \setminus \omega_1$ are measurable and disjoint and have ω_2 as their union, so

$$|\omega_1 \cap \omega_2| + |\omega_2 \setminus \omega_1| = |\omega_2|.$$

By adding these two equalities, we obtain

$$|\omega_1 \cup \omega_2| + |\omega_1 \cap \omega_2| = |\omega_1| + |\omega_2|.$$

3º) *Inequality* (b). It follows from equality (a), since $|\omega_1 \cap \omega_2| \geq 0$.

4º) *Equality* (d). Here,

$$\omega_1 \subset \omega_2.$$

The sets ω_1 and $\omega_2 \setminus \omega_1$ are measurable (Theorem 2.11) and disjoint, and their union is ω_2, so 1° gives

$$|\omega_2| = |\omega_1| + |\omega_2 \setminus \omega_1|.$$

If $|\omega_1|$, or, *a fortiori*, $|\omega_2|$ is finite, their difference is defined and

$$|\omega_2 \setminus \omega_1| = |\omega_2| - |\omega_1|. \ \square$$

It remains to verify the following property, which we used to establish (2.16).

Lemma 2.14. — *If $\mathcal{F}_1$ and $\mathcal{F}_2$ are two disjoint closed subsets of $\mathbb{R}^d$, then there exist two disjoint open sets $\mathcal{O}_1$ and $\mathcal{O}_2$ that contain them respectively.* ▮

Proof. For each $x \in \mathcal{F}_1$, the theorem on the separation of a closed set and a compact set (Proposition A.18) provides $r_1(x) > 0$ such that $(\mathcal{F}_2 + B(0, r_1(x))) \cap \{x\} = \emptyset$, and therefore

$$B(x, r_1(x)) \cap \mathcal{F}_2 = \emptyset.$$

The set

$$\mathcal{O}_1 \stackrel{\text{def}}{=} \bigcup_{x \in \mathcal{F}_1} B\Big(x, \frac{r_1(x)}{2}\Big)$$

is open, as any union of open sets is open, containing $\mathcal{F}_1$, which does not intersect the open set $\mathcal{O}_2 \stackrel{\text{def}}{=} \bigcup_{y \in \mathcal{F}_2} B(y, r_2(y)/2)$ obtained by switching the roles of $\mathcal{F}_1$ and $\mathcal{F}_2$.

Indeed, if they share a common point z, there would exist $x \in \mathcal{F}_1$ and $y \in \mathcal{F}_2$ such that $|x - z| \leq r_1(x)/2$ and $|y - z| \leq r_2(y)/2$, and therefore

$$|x - y| \leq |x - z| + |z - y| \leq \frac{r_1(x) + r_2(x)}{2} \leq \sup\{r_1(x), r_2(x)\},$$

which would contradict the definition of $r_1(x)$, if it is the larger of the two, or that of $r_2(y)$, otherwise. $\square$

2.6. Countable union and countable intersection of measurable sets

Let us show that any countable union and any countable intersection of measurable sets are measurable.

Theorem 2.15. — *Let $(\omega_n)_{n \in \mathbb{N}}$ be a sequence of measurable subsets of $\mathbb{R}^d$. Then,*

$$\bigcup_{n \in \mathbb{N}} \omega_n \text{ and } \bigcap_{n \in \mathbb{N}} \omega_n \text{ are measurable.} \ ▮$$

Proof. We begin with a bounded union of disjoint measurable sets before addressing a general union and then the intersection.

To avoid ambiguity, we again denote by $\|\mathcal{O}\|$ the measure of an open set $\mathcal{O}$ (Definition 1.20) and by $|\omega|$ the measure of a measurable set ω (Definition 2.6).

1º) Bounded union of disjoint measurable sets. We assume here that

$$\omega_n \cap \omega_m = \emptyset,\ \forall n,\ \forall m \quad \text{and} \quad \Omega \stackrel{\text{def}}{=} \bigcup_{n\in\mathbb{N}} \omega_n \ \text{ is bounded.}$$

We will establish the measurability of Ω by bounding in 1.a $|\omega_i| + \ldots + |\omega_j|$ by $\epsilon/4$ independently of j, for i sufficiently large, which will allow us in 1.b to construct an open set $\mathcal{U}$ of measure $\leq \epsilon/2$ containing all ω_n for $n \geq i$; we will then obtain an approximation of Ω within ϵ in 1.c, by approximating the first i sets ω_n within $\epsilon 2^{-n-2}$.

1.a. *Choice of i.* The additivity of the measure for disjoint sets (Theorem 2.13 (c)), its monotonicity (Theorem 2.10), and the fact that the measure of a bounded set is finite (Theorem 2.9) yield

$$|\omega_0| + \cdots + |\omega_i| = |\omega_0 \cup \cdots \cup \omega_i| \leq |\Omega| < \infty.$$

The first member, increasing with i, thus has a finite limit ℓ. Given $\epsilon > 0$, there exists i such that $\ell - \epsilon/4 \leq |\omega_0| + \cdots + |\omega_i| \leq \ell$. Then, for all $j \geq i$, we have

$$|\omega_i| + \cdots + |\omega_j| \leq \frac{\epsilon}{4}. \tag{2.19}$$

1.b. *Framing the residue $\bigcup_{n\geq i} \omega_n$.* By Definition 2.6 of the measure of ω_n, for every $n \in \mathbb{N}$, there exists an open set $\mathcal{U}_n$ such that

$$\omega_n \subset \mathcal{U}_n, \quad \|\mathcal{U}_n\| \leq |\omega_n| + \epsilon 2^{-n-3}. \tag{2.20}$$

The subadditivity of the measure of open sets (Proposition 2.4 (a)) gives, using (2.20), then the sum of a geometric series (Proposition A.6), and (2.19),

$$\|\mathcal{U}_i \cup \cdots \cup \mathcal{U}_j\| \leq \|\mathcal{U}_i\| + \cdots + \|\mathcal{U}_j\| \leq |\omega_i| + \cdots + |\omega_j| + \frac{\epsilon}{4} \leq \frac{\epsilon}{2}. \tag{2.21}$$

Let

$$\mathcal{U} \stackrel{\text{def}}{=} \bigcup_{n\geq i} \mathcal{U}_n.$$

It is an open set containing $\bigcup_{n\geq i} \omega_n$. Since $\mathcal{U}_i \cup \cdots \cup \mathcal{U}_j$ increases with j, the continuity of the measure of increasing open sets (Proposition 2.3) gives, together with (2.21),

$$\|\mathcal{U}\| \leq \frac{\epsilon}{2}. \tag{2.22}$$

1.c. Measurability of Ω. By Definition 2.5 of the measurability of ω_n, for every $n \in \mathbb{N}$, there exists a closed set $\mathcal{F}_n$ and an open set $\mathcal{O}_n$ such that

$$\mathcal{F}_n \subset \omega_n \subset \mathcal{O}_n, \quad \|\mathcal{O}_n \setminus \mathcal{F}_n\| \leq \epsilon 2^{-n-2}. \tag{2.23}$$

Let

$$\mathcal{F} \stackrel{\text{def}}{=} \mathcal{F}_0 \cup \cdots \cup \mathcal{F}_i, \quad \mathcal{O} \stackrel{\text{def}}{=} \mathcal{O}_0 \cup \cdots \cup \mathcal{O}_i \cup \mathcal{U}.$$

Then, $\mathcal{O} \setminus \mathcal{F} \subset (\mathcal{O}_0 \setminus \mathcal{F}_0) \cup \cdots \cup (\mathcal{O}_i \setminus \mathcal{F}_i) \cup \mathcal{U}$, so the subadditivity of the measure of open sets (Proposition 2.4 (a), again) gives, together with (2.23), (2.22), and again the sum of a geometric series,

$$\|\mathcal{O} \setminus \mathcal{F}\| \leq \|\mathcal{O}_0 \setminus \mathcal{F}_0\| + \cdots + \|\mathcal{O}_i \setminus \mathcal{F}_i\| + \|\mathcal{U}\| \leq \epsilon.$$

Since $\mathcal{F}$ and $\mathcal{O}$ are a closed set and an open set respectively, and $\mathcal{F} \subset \bigcup_{n\in\mathbb{N}} \omega_n \subset \mathcal{O}$, this proves that

$$\bigcup_{n\in\mathbb{N}} \omega_n \text{ is measurable.}$$

2º) Arbitrary union. Let us return to the union of an arbitrary sequence of measurable subsets ω_n of $\mathbb{R}^d$. Given a bounded open set B, denote

$$\Omega_n \stackrel{\text{def}}{=} (\omega_n \cap B) \setminus (\omega_1 \cup \cdots \cup \omega_{n-1}).$$

Each Ω_n is measurable, since any finite union, intersection, and difference of measurable sets is measurable (Theorem 2.11), and since B is measurable (as every open set is, see Theorem 2.7). Moreover, they are pairwise disjoint and their union is bounded.

The union of the Ω_n is therefore measurable according to 1º. That is,

$$\Big(\bigcup_{n\in\mathbb{N}} \omega_n\Big) \cap B \text{ is measurable.}$$

Since the measurability of the intersection with every bounded open set implies measurability (Lemma 2.8), it follows that

$$\bigcup_{n\in\mathbb{N}} \omega_n \text{ is measurable.}$$

3º) Intersection of sets. Now let us denote

$$\Omega_n \stackrel{\text{def}}{=} \omega_0 \cap \ldots \cap \omega_n.$$

Since each ω_n is measurable, their finite intersection Ω_n and its complement $\mathbb{R}^d \setminus \Omega_n$ are also measurable by Theorem 2.11. Moreover, $\mathbb{R}^d \setminus \Omega_n$ increases with n, so, according to 2º,

$$\bigcup_{n\in\mathbb{N}} \mathbb{R}^d \setminus \Omega_n \text{ is measurable.}$$

Its complement, which is $\bigcap_{n\in\mathbb{N}} \Omega_n$, is therefore also measurable (Theorem 2.11, again). This proves that

$$\bigcap_{n\in\mathbb{N}} \omega_n \text{ is measurable. } \square$$

The case of uncountable unions. Observe that:

An arbitrary union of measurable sets is not necessarily measurable.

Indeed, there exists (see (2.10), p. 32) a non-measurable set σ, and this set is the union of its points, which are measurable sets (since they are closed).

Likewise, an arbitrary intersection of measurable sets is not necessarily measurable. Indeed, $\mathbb{R}^d \backslash \sigma$ is not measurable (otherwise, σ would be), and it is the intersection of the $\mathbb{R}^d \backslash \{x\}$ (which are measurable, since they are open) when x runs through σ. $\square$

2.7. Continuity of the measure

Let us show that the measure is continuous with respect to increasing sequences.

Theorem 2.16. — *Let $(\omega_n)_{n\in\mathbb{N}}$ be a sequence of measurable subsets of $\mathbb{R}^d$ increasing with n. Then their union $\omega = \bigcup_{n\in\mathbb{N}} \omega_n$ is measurable and, as $n \to \infty$,*

$$|\omega_n| \to |\omega|. \blacksquare$$

Proof. Method. Since the measurability of any countable union was established in Theorem 2.15, it remains to compute its measure $|\omega|$. After first establishing a lower bound easily in 1°, we will then provide an upper bound in three steps, by including ω in a union of open sets $\mathcal{U}_n$, close to the ω_n, increasing with n, and then using the continuity of the measure for increasing sequences of open sets.

To avoid ambiguities, we denote once again by $\|\mathcal{O}\|$ the measure of an open set $\mathcal{O}$ (Definition 1.20) and by $|\omega|$ the measure of a measurable set ω (Definition 2.6).

1°) Limit ℓ of $|\omega_n|$ and lower bound of $|\omega|$. The real sequence $(|\omega_n|)_{n\in\mathbb{N}}$ is increasing (Theorem 2.10, since ω_n increases with n), so it has a limit ℓ, finite or infinite (Proposition A.5), and it is bounded above by $|\omega|$. Thus,

$$|\omega_n| \to \ell, \quad \ell \leq |\omega|. \tag{2.24}$$

2°) Construction of $\mathcal{U}_n$ to bound ω from above. Let $\epsilon > 0$. By Definition 2.6 of the measure of ω_0, there exists an open set $\mathcal{O}_0$ such that

$$\omega_0 \subset \mathcal{O}_0, \quad \|\mathcal{O}_0\| \leq |\omega_0| + \epsilon. \tag{2.25}$$

For all $n \geq 1$, the difference $\omega_n \setminus \omega_{n-1}$ is measurable (Theorem 2.11) and has measure $|\omega_n| - |\omega_{n-1}|$ (Theorem 2.13 (d)), so, again by Definition 2.6, there exists an open set $\mathcal{O}_n$ such that

$$\omega_n \setminus \omega_{n-1} \subset \mathcal{O}_n, \quad \|\mathcal{O}_n\| \leq |\omega_n| - |\omega_{n-1}| + \epsilon 2^{-n}. \tag{2.26}$$

Finally, let

$$\mathcal{U}_n \stackrel{\text{def}}{=} \mathcal{O}_0 \cup \mathcal{O}_1 \cup \cdots \cup \mathcal{O}_n.$$

3º) Measure of $\mathcal{U}_n$. The set $\mathcal{U}_n$ is open (as every union of open sets), increases with n, and contains $\omega_0 \cup (\omega_1 \setminus \omega_0) \cup \cdots \cup (\omega_n \setminus \omega_{n-1})$, which is ω_n (since ω_n increases with n). Thus,

$$\omega_n \subset \mathcal{U}_n. \tag{2.27}$$

The subadditivity of the measure for open sets (Proposition 2.4 (a)) then yields, using (2.25), (2.26), and the sum of a geometric series (Proposition A.6),

$$\|\mathcal{U}_n\| \leq \|\mathcal{O}_0\| + \|\mathcal{O}_1\| + \cdots \|\mathcal{O}_n\| \leq |\omega_n| + 2\epsilon \leq \ell + 2\epsilon. \tag{2.28}$$

4º) Upper bound of $|\omega|$. Let

$$\mathcal{U} \stackrel{\text{def}}{=} \bigcup_{n\in\mathbb{N}} \mathcal{U}_n.$$

This is an open set that contains the union ω of the ω_n (by (2.27)), so, according to Definition 2.6 of the measure of ω,

$$|\omega| \leq \|\mathcal{U}\|.$$

Now, according to the continuity of the measure with respect to increasing sequences of open sets (Theorem 2.3), the upper bound (2.28) of $\|\mathcal{U}_n\|$ yields, in the limit as $n \to \infty$,

$$\|\mathcal{U}\| \leq \ell + 2\epsilon.$$

Thus,

$$|\omega| \leq \ell + 2\epsilon.$$

Being true for all $\epsilon > 0$, this inequality also holds for $\epsilon = 0$, which proves the reverse inequality of (2.24). Thus,

$$\ell = |\omega|. \ \square$$

Let us show that the measure is continuous with respect to decreasing sequences of sets whose measure is finite from a certain index onward.

Theorem 2.17. — *Let $(\omega_n)_{n\in\mathbb{N}}$ be a decreasing sequence of measurable subsets of $\mathbb{R}^d$ such that*

$$\textit{there exists } i \in \mathbb{N} \textit{ such that } |\omega_i| < \infty. \tag{2.29}$$

Then their intersection $\omega = \bigcap_{n\in\mathbb{N}} \omega_n$ *is measurable and, as* $n \to \infty$,

$$|\omega_n| \to |\omega|. \blacksquare$$

Proof. As the measurability of any countable intersection was established in Theorem 2.15, it remains to calculate its measure $|\omega|$.

For every $n \in \mathbb{N}$, the difference $\omega_i \setminus \omega_n$ is measurable (Theorem 2.11) and increases with n towards $\omega_i \setminus \omega$, therefore, by Theorem 2.16,

$$|\omega_i \setminus \omega_n| \to |\omega_i \setminus \omega|.$$

The "subtractivity" of the measure (Theorem 2.13 (d)), which applies since the measure of ω_i is finite, thus gives

$$|\omega_n| = |\omega_i| - |\omega_i \setminus \omega_n| \to |\omega_i| - |\omega_i \setminus \omega| = |\omega|. \square$$

The case of a sequence of sets with infinite measure. Hypothesis (2.29) cannot be omitted from Theorem 2.17. For example, if ω is measure with finite measure and

$$\omega_n = \omega \cup \{x \in \mathbb{R}^d : |x| > n\},$$

then

$$|\omega_n| = \infty, \forall n, \text{ and yet } \Big|\bigcap_{n\in\mathbb{N}} \omega_n\Big| = |\omega| < \infty. \tag{2.30}$$ □

Let us bound the measure of a countable union and intersection, and show that the measure is countably additive for disjoint sets.

Theorem 2.18. — *Let* $(\omega_n)_{n\in\mathbb{N}}$ *be a sequence of measurable subsets of* $\mathbb{R}^d$*. Then:*

(a)

$$\Big|\bigcup_{n\in\mathbb{N}} \omega_n\Big| \leq \sum_{n\in\mathbb{N}} |\omega_n|;$$

(b) *if the* ω_n *are pairwise disjoint,*

$$\Big|\bigcup_{n\in\mathbb{N}} \omega_n\Big| = \sum_{n\in\mathbb{N}} |\omega_n|;$$

(c)

$$\Big|\bigcap_{n\in\mathbb{N}} \omega_n\Big| \leq \inf_{n\in\mathbb{N}} |\omega_n|. \blacksquare$$

Proof. **1°)** *Measure of a union.* The union is measurable by Theorem 2.15. For any $N \in \mathbb{N}$, the subadditivity of the measure (Theorem 2.13 (b)) gives

$$\Big|\bigcup_{0\leq n\leq N} \omega_n\Big| \leq \sum_{0\leq n\leq N} |\omega_n| \leq \sum_{n\in\mathbb{N}} |\omega_n|.$$

Since $\bigcup_{0\leq n\leq N} \omega_n$ increases with N, the limit of the left-hand side is $|\bigcup_{n\in\mathbb{N}} \omega_n|$ (Theorem 2.16), whence inequality (a).

2°) *The case of disjoint sets.* The additivity of the measure for disjoint sets (Theorem 2.13 (c)) and its continuity with respect to increasing sequences (Theorem 2.16) give successively

$$\Big|\bigcup_{0\leq n\leq N} \omega_n\Big| = \sum_{0\leq n\leq N} |\omega_n| \to \sum_{n\in\mathbb{N}} |\omega_n|.$$

Whence inequality (b), since the limit of the first term is $|\bigcup_{n\in\mathbb{N}} \omega_n|$ (Theorem 2.16).

3°) *Measure of an intersection.* The intersection is measurable (Theorem 2.15), so the monotonicity of measure (Theorem 2.10) gives

$$\Big|\bigcap_{n\in\mathbb{N}} \omega_n\Big| \leq \Big|\bigcap_{0\leq n\leq N} \omega_n\Big| \leq \inf_{0\leq n\leq N} |\omega_n|.$$

Since $\inf_{0\leq n\leq N} |\omega_n|$ decreases with N, it converges (Proposition A.5) to $\inf_{n\in\mathbb{N}} |\omega_n|$, whence inequality (c). □

Let us state an approximation property of measurable sets that will be useful, among other things, for Definition 5.1 of the integral of an integrable measure.

Lemma 2.19. — *Let ω be a measurable subset of an open set Ω of $\mathbb{R}^d$.*

(a) *There exists an increasing sequence $(\mathcal{F}_n)_{n\in\mathbb{N}}$ of closed sets and a decreasing sequence $(\mathcal{O}_n)_{n\in\mathbb{N}}$ of open sets in $\mathbb{R}^d$ such that $\mathcal{F}_n \subset \omega \subset \mathcal{O}_n \subset \Omega$ and, as $n \to \infty$,*

$$|\omega \setminus \mathcal{F}_n| \to 0, \quad |\mathcal{O}_n \setminus \omega| \to 0.$$

(b) *There exists an increasing sequence of compact sets $(K_n)_{n\in\mathbb{N}}$ such that, as $n \to \infty$,*

$$|K_n| \to |\omega|. \ \blacksquare$$

Proof. **1°)** *Property (a).* Let $n \in \mathbb{N}^*$. By Definition 2.5 of measurability, there exist a closed set $\mathcal{V}_n$ and an open set $\mathcal{U}_n$ such that

$$\mathcal{V}_n \subset \omega \subset \mathcal{U}_n, \quad |\mathcal{U}_n \setminus \mathcal{V}_n| \leq \frac{1}{n}.$$

The stated properties are then satisfied by

$$\mathcal{F}_n \stackrel{\text{def}}{=} \mathcal{V}_1 \cup \cdots \cup \mathcal{V}_n, \quad \mathcal{O}_n \stackrel{\text{def}}{=} \Omega \cap \mathcal{U}_1 \cap \cdots \cap \mathcal{U}_n.$$

Indeed:

— $\mathcal{F}_n$ is closed (as any finite union of closed sets) and increases with n;

— $\mathcal{O}_n$ is open (as any finite intersection of open sets), decreases with n, contains ω, and is included in Ω;

— $\omega \setminus \mathcal{F}_n$ is measurable, as the difference of measurable sets (Theorem 2.11, since $\mathcal{F}_n$ is measurable (Theorem 2.7));

— $\mathcal{O}_n \setminus \omega$ is measurable, as the difference of measurable sets (Theorem 2.11, since $\mathcal{O}_n$ is measurable as any open set (Theorem 2.7));

— $\omega \setminus \mathcal{F}_n$ and $\mathcal{O}_n \setminus \omega$ are included in the open set $\mathcal{U}_n \setminus \mathcal{V}_n$, hence Definition 2.6 of the measure yields

$$|\omega \setminus \mathcal{F}_n| \leq |\mathcal{U}_n \setminus \mathcal{V}_n| \leq \frac{1}{n}, \quad |\mathcal{O}_n \setminus \omega| \leq |\mathcal{U}_n \setminus \mathcal{V}_n| \leq \frac{1}{n}.$$

2o) Property (b). Let $K_n \stackrel{\text{def}}{=} \mathcal{F}_n \cap B(x, n)$, where the ball B is closed. Then K_n is compact, being closed and bounded (Borel–Lebesgue theorem, Proposition A.36 (a)). Moreover, by the measure of a difference (Theorem 2.13 (d), since $|K_n|$ is finite) and the monotonicity of the measure (Proposition 2.2),

$$|\omega| - |K_n| = |\omega \setminus K_n| \leq |\omega \setminus \mathcal{F}_n| \leq 1/n. \ \square$$

2.8. Translation invariance and the product measure

Let us show that the measure is invariant under translation.

Theorem 2.20. — *Let ω be a measurable subset of $\mathbb{R}^d$ and $x \in \mathbb{R}^d$. Then $\omega + x$ is measurable and*

$$|\omega + x| = |\omega|. \ \blacksquare$$

Proof. To avoid ambiguities, let us denote by $\|\mathcal{O}\|$ here the measure of an open set $\mathcal{O}$ (Definition 1.20), and by $|\omega|$ the measure of a measurable set ω (Definition 2.6).

1o) Measure of a translated open set. Let $\mathcal{O}$ be an open subset of $\mathbb{R}^d$ and $n \in \mathbb{N}$. The (closed) cubes $\Delta_{s,n}$ from Definition 1.20 of the measure of $\mathcal{O}$ have side length 2^{-n}, so (Proposition 2.1) $\|\mathring{\Delta}_{s,n} + x\| = \|\mathring{\Delta}_{s,n}\| = 2^{-nd}$. Since $\mathcal{O}$ contains k_n cubes $\Delta_{s,n}$, the open set $\mathcal{O} + x$ contains at least k_n (open) cubes $\mathring{\Delta}_{s,n} + x$, which are pairwise disjoint. Therefore, denoting by λ_n their union, the monotonicity and additivity of the measure of open sets (Propositions 2.2 and 2.4 (b)) give

$$\|\mathcal{O} + x\| \geq |\lambda_n| \geq k_n 2^{-nd}.$$

When $n \to \infty$, this yields in the limit, by Definition 1.20 of the measure of an open set,

$$\|\mathcal{O} + x\| \geq \|\mathcal{O}\|.$$

This inequality is in fact an equality since it also implies, conversely, that

$$\|\mathcal{O}\| = \|(\mathcal{O} + x) + (-x)\| \geq \|\mathcal{O} + x\|.$$

2º) Measurability of $\omega + x$. By Definition 2.5 of measurability, for every $\epsilon > 0$, there exists a closed set $\mathcal{F}$ and an open set $\mathcal{O}$ such that $\mathcal{F} \subset \omega \subset \mathcal{O}$ and

$$|\mathcal{O} \setminus \mathcal{F}| \leq \epsilon.$$

Then $\mathcal{F} + x$ is closed, $\mathcal{O} + x$ is open, $\mathcal{F} + x \subset \omega + x \subset \mathcal{O} + x$, and 1º gives

$$|(\mathcal{O} + x) \setminus (\mathcal{F} + x)| = |(\mathcal{O} \setminus \mathcal{F}) + x| = |\mathcal{O} \setminus \mathcal{F}| \leq \epsilon.$$

Thus, $\omega + x$ also satisfies Definition 2.5 of a measurable set.

3º) Measure of $\omega + x$. By Definition 2.6 of the measure,

$$|\omega + x| = \inf\{\|\mathcal{O}\| : \mathcal{O} \text{ open subset of } \mathbb{R}^{\mathrm{d}},\ \mathcal{O} \supset \omega + \mathrm{x}\}. \tag{2.31}$$

For such an $\mathcal{O}$, the set $\mathcal{O} - x$ is open and contains ω, hence $\|\mathcal{O} - x\| \geq |\omega|$; however $\|\mathcal{O}\| = \|\mathcal{O} - x\|$ by 1º, so (2.31) gives

$$|\omega + x| \geq |\omega|.$$

This inequality is in fact an equality since it implies, conversely, that

$$|\omega| = |(\omega + x) + (-x)| \geq |\omega + x|. \ \square$$

Now let us show that the measure of a product is the product of the measures.

Theorem 2.21. — *Let* ω_1 *be a measurable subset of* $\mathbb{R}^{d_1}$ *and* ω_2 *be a measurable subset of* $\mathbb{R}^{d_2}$. *Then,* $\omega_1 \times \omega_2$ *is measurable and*

$$|\omega_1 \times \omega_2| = |\omega_1|\,|\omega_2|. \ \blacksquare$$

Proof. To avoid any ambiguity, we will denote by $\|\mathcal{O}\|$ the measure of an open set $\mathcal{O}$ (Definition 1.20) and $|\omega|$ by the measure of a measurable set ω (Definition 2.6).

1º) Measure of a product of open sets. Let $\mathcal{O}_1$ be an open subset of $\mathbb{R}^{d_1}$ and $\mathcal{O}_2$ an open subset of $\mathbb{R}^{d_2}$. Each cube in Definition 1.20 for the measure of the open set $\mathcal{O}_1 \times \mathcal{O}_2$ decomposes into

$$\Delta_{(s_1,s_2),n} = \Delta_{s_1,n} \times \Delta_{s_2,n}.$$

The number of cubes included in $\mathcal{O}_1 \times \mathcal{O}_2$, $\mathcal{O}_1$, and $\mathcal{O}_2$, respectively, are thus related by $k_n(\mathcal{O}_1 \times \mathcal{O}_2) = k_n(\mathcal{O}_1)k_n(\mathcal{O}_2)$, whence

$$2^{-n(d_1+d_2)}k_n(\mathcal{O}_1 \times \mathcal{O}_2) = 2^{-nd_1}k_n(\mathcal{O}_1)2^{-nd_2}k_n(\mathcal{O}_2).$$

This gives in the limit, as $n \to \infty$,

$$\|\mathcal{O}_1 \times \mathcal{O}_2\| = \|\mathcal{O}_1\|\,\|\mathcal{O}_2\|. \tag{2.32}$$

2º) Measurability of $\omega_1 \times \omega_2$. **2.a.** *Case where* $\omega_1 \times \omega_2$ *is bounded.* Let ϵ be such that $0 < \epsilon \leq 1$. By Definition 2.5 of measurability, there exist closed sets $\mathcal{F}_1$ and $\mathcal{F}_2$ and open sets $\mathcal{O}_1$ and $\mathcal{O}_2$ such that (since $\|\mathcal{O}_1\| \leq |\omega_1| + 1$ and $\|\mathcal{O}_2\| \leq |\omega_2| + 1$)

$$\mathcal{F}_1 \subset \omega_1 \subset \mathcal{O}_1, \quad \|\mathcal{O}_1 \setminus \mathcal{F}_1\| \leq \frac{\epsilon}{2(|\omega_2|+1)} \leq \frac{\epsilon}{2\|\mathcal{O}_2\|},$$

$$\mathcal{F}_2 \subset \omega_2 \subset \mathcal{O}_2, \quad \|\mathcal{O}_2 \setminus \mathcal{F}_2\| \leq \frac{\epsilon}{2(|\omega_1|+1)} \leq \frac{\epsilon}{2\|\mathcal{O}_1\|}.$$

Then $\mathcal{F}_1 \times \mathcal{F}_2$ is closed, $\mathcal{O}_1 \times \mathcal{O}_2$ is open,

$$\mathcal{F}_1 \times \mathcal{F}_2 \subset \omega_1 \times \omega_2 \subset \mathcal{O}_1 \times \mathcal{O}_2 \tag{2.33}$$

and

$$(\mathcal{O}_1 \times \mathcal{O}_2) \setminus (\mathcal{F}_1 \times \mathcal{F}_2) \subset ((\mathcal{O}_1 \setminus \mathcal{F}_1) \times \mathcal{O}_2) \cup (\mathcal{O}_1 \times (\mathcal{O}_2 \setminus \mathcal{F}_2)).$$

The monotonicity and subadditivity of the measure of open sets (Propositions 2.2 and 2.4 (a)) yield, together with equality (2.32),

$$\|(\mathcal{O}_1 \times \mathcal{O}_2) \setminus (\mathcal{F}_1 \times \mathcal{F}_2)\| \leq \|\mathcal{O}_1 \setminus \mathcal{F}_1\|\,\|\mathcal{O}_2\| + \|\mathcal{O}_1\|\,\|\mathcal{O}_2 \setminus \mathcal{F}_2\| \leq \epsilon. \tag{2.34}$$

With (2.33), this proves that $\omega_1 \times \omega_2$ is measurable.

2.b. *General case.* Let $\Omega_n \stackrel{\text{def}}{=} \{x_1 \in \omega_1 : |x_1| < n\} \times \{x_2 \in \omega_2 : |x_2| < n\}$. It is a product of bounded measurable sets, because the open set $\{x_i : |x_i| < n\}$ is measurable (Theorem 2.7) and the intersection of measurable sets is measurable (Theorem 2.11). It is measurable from part 2.a. The set $\omega_1 \times \omega_2 = \bigcup_{n \in \mathbb{N}} \Omega_n$ is therefore measurable as every countable union of measurable sets (Theorem 2.15).

3°) Upper bound of $|\omega_1 \times \omega_2|$. Let $\mathcal{O}_1$ and $\mathcal{O}_2$ be two open sets containing ω_1 and ω_2 respectively. Then, the open set $\mathcal{O}_1 \times \mathcal{O}_2$ contains $\omega_1 \times \omega_2$, so Definition 2.6 of the measure and equality (2.32) yield

$$|\omega_1 \times \omega_2| \leq \|\mathcal{O}_1 \times \mathcal{O}_2\| \leq \|\mathcal{O}_1\| \, \|\mathcal{O}_2\|.$$

By taking the infimum over all the $\mathcal{O}_1$, Definition 2.6 gives $|\omega_1 \times \omega_2| \leq |\omega_1| \, \|\mathcal{O}_2\|$ then, by taking the infimum over all the $\mathcal{O}_2$,

$$|\omega_1 \times \omega_2| \leq |\omega_1| \, |\omega_2|. \tag{2.35}$$

4°) Lower bound of $|\omega_1 \times \omega_2|$. Let us return to the open and closed sets from 2°. They satisfy

$$\mathcal{O}_1 \times \mathcal{O}_2 \subset (\omega_1 \times \omega_2) \cup \big((\mathcal{O}_1 \times \mathcal{O}_2) \setminus (\mathcal{F}_1 \times \mathcal{F}_2)\big),$$

hence the monotonicity and subadditivity of the measure, now for measurable sets (Theorems 2.10 and 2.13 (b)), give

$$|\mathcal{O}_1 \times \mathcal{O}_2| \leq |\omega_1 \times \omega_2| + |(\mathcal{O}_1 \times \mathcal{O}_2) \setminus (\mathcal{F}_1 \times \mathcal{F}_2)|.$$

With (2.32) and (2.34), it follows, since (Theorem 2.7) the measure coincides with that of the open sets,

$$\|\mathcal{O}_1\| \, \|\mathcal{O}_2\| \leq |\omega_1 \times \omega_2| + \epsilon.$$

Hence, by Definition 2.6 of the measure of ω_1 and ω_2,

$$|\omega_1| \, |\omega_2| \leq |\omega_1 \times \omega_2| + \epsilon.$$

Since this inequality holds for all $\epsilon > 0$, it also holds for $\epsilon = 0$. This establishes the reverse inequality of (2.35), which is therefore an equality. □

2.9. Negligible sets

Let us define negligible sets, which are of interest because the integral is not affected by the inclusion of such a set into the domain of integration, as we will see in Theorem 6.8.

Definition 2.22. — *We say that a subset σ of $\mathbb{R}^d$ is **negligible** if it is measurable and has measure zero. That is, if:*

for every $\epsilon > 0$, there exists an open set ω containing σ such that $|\omega| \leq \epsilon$. ▮

Justification. If σ is measurable with measure zero, Definition 2.6 of the measure shows that, for every $\epsilon > 0$, there exists an open set of measure $\leq \epsilon$ containing σ.

Conversely, if, for every $\epsilon > 0$, there exists an open set ω with measure $\leq \epsilon$ containing σ, then it satisfies Definition 2.5 of measurability with $\mathcal{F} = \emptyset$, and Definition 2.6 of its measure shows that its measure is zero. □

Adding or removing negligible sets does not change the measure.

Theorem 2.23. — *Let ω be a measurable subset and σ a negligible subset of $\mathbb{R}^d$. Then $\omega \cup \sigma$ and $\omega \backslash \sigma$ are measurable and*

$$|\omega \cup \sigma| = |\omega \backslash \sigma| = |\omega|. \blacksquare$$

Proof. ***1°)*** *Union.* The monotonicity and the subadditivity of the measure (Theorems 2.10 and 2.13 (b)) give, successively,

$$|\omega| \leq |\omega \cup \sigma| \leq |\omega| + |\sigma| = |\omega|.$$

2°) *Intersection.* Since $(\omega \backslash \sigma) \cup \sigma = \omega$, 1° gives

$$|\omega \backslash \sigma| = |\omega|. \ \square$$

Let us give some conditions under which a set is negligible.

Theorem 2.24. — *In $\mathbb{R}^d$:*

(a) *every subset of a negligible set is negligible;*

(b) *every finite or countable set is negligible;*

(c) *every finite or countable union of negligible sets is negligible;*

(d) *every translate of a negligible set is negligible.* ■

Proof. ***1°)*** *Property* (a). Every subset of a negligible set is, by Definition 2.22, included in an open set of measure $\leq \epsilon$, and is therefore negligible as well.

2°) *Property* (b). Any set $\{x\}$ where $x \in \mathbb{R}^d$ is negligible, since it is included in an open cube of side length $\epsilon^{1/d}$, whose measure is ϵ according to Proposition 2.1.

Any finite or countable set, being a countable union of such sets, is therefore negligible by Property (c), we shall now verify.

3º) Property (c). Let $\{\sigma_i : i \in I\}$ be a countable family of negligible sets. By Definition A.1 of a countable set, there exists a bijection ϕ from I onto a subset of $\mathbb{N}$, so, relabeling σ_i in the place of $\sigma_{\phi(i)}$, we can assume $I \subset \mathbb{N}$.

Let $\epsilon > 0$. For each i, there exists an open set ω_i containing σ_i such that

$$|\omega_i| \leq \epsilon 2^{-i-1}.$$

To bound the measure of the union Ω of the ω_i, which is open (as any union of open sets), let us introduce, for $n \in \mathbb{N}$,

$$\Omega_n \stackrel{\text{def}}{=} \bigcup_{i \in I \,:\, i \leq n} \omega_i.$$

The subadditivity of the measure (Theorem 2.13 (b)) and the upper bound of a geometric series (Proposition A.6) give

$$|\Omega_n| \leq \sum_{i \in I \,:\, i \leq n} |\omega_i| \leq \sum_{0 \leq i \leq n} \epsilon 2^{-i-1} \leq \epsilon.$$

The Ω_n increase with n and have Ω for their union, so (Theorem 2.16) $|\Omega_n| \to |\Omega|$ as $n \to \infty$, hence

$$|\Omega| \leq \epsilon.$$

The union of the σ_i, being included in the open set Ω, thus satisfies Definition 2.22 of a negligible set.

4º) Property (d). Every translate $\sigma + x$ of a negligible set σ is negligible, as its measure is equal to that of σ (Theorem 2.20), which is zero. □

Let us observe that a set is negligible if its bounded subsets are.

Theorem 2.25. — *A subset σ of $\mathbb{R}^d$ is negligible if $\sigma \cap B$ is negligible for every bounded set B, or equivalently if there exist bounded sets $\{B_n\}_{n \in \mathbb{N}}$ whose union is $\mathbb{R}^d$, such that each $\sigma \cap B_n$ is negligible.* ▮

Proof. If each $\sigma \cap B_n$ is negligible, then their union, that is, σ when the B_n cover $\mathbb{R}^d$, is negligible by Theorem 2.24 (c).

This is in particular true if $\sigma \cap B$ is negligible for every bounded set B. □

Let us show that proper vector spaces, that is, those distinct from the entire space, are negligible.

Theorem 2.26. — *The set generated by the k vectors, v^1, v^2, …, v^k in $\mathbb{R}^d$, that is*

$$\{t_1 v^1 + \cdots + t_k v^k : t_i \in \mathbb{R},\ \forall i\},$$

is negligible if $k < d$. ▮

Proof. **1°)** *The case where v^1, …, v^k are linearly independent.* Let

$$P(v^1, \ldots, v^k) \stackrel{\text{def}}{=} \{t_1 v^1 + \cdots + t_k v^k : |t_i| < 1,\ \forall i\}.$$

It suffices to show that, for any linearly independant family v^i,

$$P(v^1, \ldots, v^k) \text{ is negligible.} \tag{2.36}$$

Indeed, $P(nv^1, \ldots, nv^k)$ is then negligible for every $n \neq 0$. Since any countable union of negligible sets is negligible (Theorem 2.24 (c)), this also holds for the set $\bigcup_{n \in \mathbb{N}^*} P(nv^1, \ldots, nv^k)$, which is $\{t_1 v^1 + \cdots + t_k v^k : t_i \in \mathbb{R}\}$.

2°) *Proof of (2.36).* Let us complete the family $v^1, \ldots, v^k$ with vectors $v^{k+1}, \ldots, v^d$ such that v^1, …, v^d are linearly independent. For $c > 0$, consider the bounded open set

$$\omega(c) \stackrel{\text{def}}{=} \{t_1 v^1 + \cdots + t_{d-1} v^{d-1} + c t_d v^d : |t_i| < 1,\ \forall i\}.$$

Let m be an integer ≥ 1. The open set $\omega(mc)$ contains the m disjoint open sets $\omega_i(c)$ obtained by translating $\omega(c)$ by $(2i - m)cv^d$, for $i = 0, 1, \ldots, m - 1$; each of them has the same measure as $\omega(c)$ (Theorem 2.20), so the monotonicity and additivity of the measure (Proposition 2.2 and Theorem 2.13 (c)) yield

$$|\omega(mc)| \geq \Big| \bigcup_{0 \leq i \leq m-1} \omega_i(c) \Big| = \sum_{0 \leq i \leq m-1} |\omega_i(c)| = m\,|\omega(c)|.$$

For $c = 1/m$, this inequality gives

$$\Big|\omega\Big(\frac{1}{m}\Big)\Big| \leq \frac{|\omega(1)|}{m}. \tag{2.37}$$

The set $P(v^1, \ldots, v^k)$, being included in $\omega(1/m)$ whose measure can be made arbitrarily small according to (2.37), is negligible (by Definition 2.22 of this notion).

3°) *The case where v^1, …, v^k are not linearly independent.* One can then eliminate at least one of the v^i without changing the set under consideration. By repeating this operation until obtaining a linearly independent family, this case reduces to 1°. □

Chapter 3

Measures

In this chapter, we define and study the space $\mathcal{M}(\Omega; E)$ of measures on an open subset Ω of $\mathbb{R}^d$, with values in a Neumann space E.

This is the third step in the construction of the space $L^1(\Omega; E)$ of integrable measures and their integral, measures that will be defined in § 4.1 as limits of sequences of continuous functions. Indeed, we let $\mathcal{M}(\Omega; E)$ play the role usually assigned to *classes of measurable functions equal almost everywhere.*

We present essentially the properties of measure that are used in these constructions, namely:
— $\mathcal{M}(\Omega; E)$ is itself a Neumann space (Theorem 3.8), that is, sequentially complete;
— we identify any continuous function f with a measure via the equality $\langle f, \varphi \rangle = \int f\varphi$, for any test function φ (Identification (3.16), p. 65, and Theorem 3.14);
— we construct continuous regularizations $R_n f$ of a measure f (Definition 3.17), which will play a crucial role in the proof of the completeness of $L^1(\Omega; E)$ in Chapter 4 (Theorem 4.12);
— we show that, when f is a uniformly continuous function with bounded support, then for every semi-norm $\| \ \|_{E;\nu}$ on E, we have $\int_\Omega \|R_n f\|_{E;\nu} \leq \int_\Omega \|f\|_{E;\nu}$ (Proposition 3.23).

3.1. Space of measures $\mathcal{M}(\Omega; E)$

Let us first define the space $\mathcal{K}(\Omega)$ [1], which is used to construct the space of measures.

Definition 3.1. — *We denote by $\mathcal{K}(\Omega)$, where Ω is an open subset of $\mathbb{R}^d$, the vector space of continuous real-valued functions on Ω with compact support, endowed with the semi-norms indexed by $q \in \mathcal{C}^+(\Omega)$,*

$$\|\varphi\|_{\mathcal{K}(\Omega);q} \stackrel{\text{def}}{=} \sup_{x\in\Omega} q(x)|\varphi(x)|. \ \blacksquare$$

1 **Historical background of the semi-norms of $\mathcal{K}(\Omega)$.** BOURBAKI endows $\mathcal{K}(\Omega)$ [8, chap. III, § 1, no. 1, p. 41] with the inductive limit topology of the $\mathcal{C}_K(\Omega)$. The semi-norms in Definition 3.1 appear to be new.

Recall that $\mathcal{K}(\Omega)$ is a Neumann space [Vol. 2, Theorem 2.24 (d)].

Notation $\mathcal{K}(\Omega)$. This is the notation used by BOURBAKI [8, chap. III, § 1, no. 1, p. 40]. □

Interest of the semi-norms of $\mathcal{K}(\Omega)$. The semi-norms of Definition 3.1 provide the properties usually obtained by means of the delicate **inductive limit topology of the** $\mathcal{C}_K(\Omega)$. These semi-norms generate the inductive limit topology [SIMON, 75, to appear], while being simpler to use. □

Now, let us define the space of measures.

Definition 3.2. — *Let Ω be an open subset of $\mathbb{R}^d$ and E a Neumann space, whose family of semi-norms is denoted $\{\| \ \|_{E;\nu} : \nu \in \mathcal{N}_E\}$.*

We call a ***measure***[2] *on Ω with values in E any continuous linear mapping from $\mathcal{K}(\Omega)$ into E.*

We denote by $\mathcal{M}(\Omega; E)$ the space of these measures endowed with the semi-norms indexed by $\varphi \in \mathcal{K}(\Omega)$ and $\nu \in \mathcal{N}_E$,

$$\|f\|_{\mathcal{M}(\Omega;E);\varphi,\nu} \stackrel{\text{def}}{=} \|\langle f, \varphi\rangle\|_{E;\nu}. \ \blacksquare$$

We denote by $\langle f, \varphi\rangle$ the value $f(\varphi)$ of a measure f applied to a function φ, and

$$\mathcal{M}(\Omega) \stackrel{\text{def}}{=} \mathcal{M}(\Omega; \mathbb{R}).$$

The case where E is not a Neumann space. If E is a separated semi-normed space that is not sequentially complete, one can still define the space $\mathcal{L}(\mathcal{K}(\Omega); E)$ of continuous linear mappings from $\mathcal{K}(\Omega)$ into E, but **it should not be** called the "space of measures", nor denoted $\mathcal{M}(\Omega; E)$, as it does not possess the expected properties. In particular, one cannot identify every continuous function with an element of $\mathcal{L}(\mathcal{K}(\Omega); E)$ when, moreover, E is metrizable (see § 5.5). □

Terminology. The measures given by Definition 3.2 are sometimes called **Radon measures**, for example in [EDWARDS, 22, § 4.3, p. 177], although Johann RADON defined them as *completely additive set functions* [59]. As for us, we follow the terminology of Nicolas BOURBAKI [8, chap. III, § 1, no. 3, p. 47]. □

Vague topology. The topology generated by the semi-norms of Definition 3.2, that is, the topology of pointwise convergence on $\mathcal{K}(\Omega)$, is called the **vague topology** of $\mathcal{M}(\Omega; E)$. □

Lebesgue measure. The mapping $\varphi \mapsto \int_{\mathbb{R}^d} \varphi$ defines a real-valued measure, called the **Lebesgue measure**, on $\mathbb{R}^d$: it is the measure associated with the function with constant value 1 by Theorem 3.10.

Henri LEBESGUE defined it, in his thesis [43], as an *additive set function* which, in the particular case of an open set ω, coincides with its measure $|\omega|$ constructed in Definition 1.20. □

Let us now characterize measures.

2 **History of measures.** The definition of measures as *linear mappings on* $\mathcal{K}(\Omega)$ appear to be due to Nicolas BOURBAKI, in 1965 [8, chap. III, § 1, no. 3, p. 47].
Previously, measures were introduced as *additive set functions*, following the works of Émile BOREL in 1894 [7], Henri LEBESGUE in 1902 [43], and Johann RADON in 1913 [59]. Their development is detailed in *Éléments d'histoire des mathématiques* by Nicolas BOURBAKI [10, p. 278–287].

Theorem 3.3. — *We have*

$$f \in \mathcal{M}(\Omega; E),$$

where Ω is an open subset of $\mathbb{R}^d$ and E is a Neumann space, if and only if: f is a linear mapping from $\mathcal{K}(\Omega)$ into E and, for every semi-norm $\| \ \|_{E;\nu}$ of E, there exist $q \in \mathcal{C}^+(\Omega)$ and $c \in \mathbb{R}$ such that, for every $\varphi \in \mathcal{K}(\Omega)$,

$$\|\langle f, \varphi \rangle\|_{E;\nu} \leq c\|\varphi\|_{\mathcal{K}(\Omega);q}. \tag{3.1}$$

That is,

$$\|\langle f, \varphi \rangle\|_{E;\nu} \leq c \sup_{x \in \Omega} q(x)|\varphi(x)|. \ \blacksquare$$

Variant of Theorem 3.3. One can omit c in the characterization (3.1), that is, replace it by

$$\|\langle f, \varphi \rangle\|_{E;\nu} \leq \|\varphi\|_{\mathcal{K}(\Omega);q}. \tag{3.2}$$

Indeed, $cq \in \mathcal{C}^+(\Omega)$ and $c\|\varphi\|_{\mathcal{K}(\Omega);q} = \|\varphi\|_{\mathcal{K}(\Omega);cq}$, so (3.1) yields (3.2) up to the choice of q. □

Proof of Theorem 3.3. Since, by Definition 3.2, a measure is a continuous linear mapping from $\mathcal{K}(\Omega)$ into E, this follows from the characterization of continuous linear mappings in Proposition 1.9 (b), because the family of semi-norms of $\mathcal{K}(\Omega)$ is filtering.

Indeed, this family, given in Definition 3.1, satisfies Definition 1.6 of a filtering family, since, given functions $q_1, \ldots, q_n$ of $\mathcal{C}^+(\Omega)$, their sum $q = q_1 + \cdots + q_n$ belongs to $\mathcal{C}^+(\Omega)$ and bounds the q_i from above, and thus

$$\sup_{1 \leq i \leq n} \|\varphi\|_{\mathcal{K}(\Omega);q_i} = \sup_{1 \leq i \leq n} \sup_{x \in \Omega} q_i(x)|\varphi(x)| \leq \sup_{x \in \Omega} q(x)|\varphi(x)| = \|\varphi\|_{\mathcal{K}(\Omega);q}. \ \square$$

Let us show that $\mathcal{M}(\Omega; E)$ is separated (we shall see in Theorem 3.8 that it is a Neumann space).

Theorem 3.4. — *The space $\mathcal{M}(\Omega; E)$ is a separated semi-normed space, for every open subset Ω of $\mathbb{R}^d$ and every Neumann space E.* $\blacksquare$

Proof. By Definition 1.1 of a separated space, we need to verify that a measure f is zero when all its semi-norms vanish. So let $f \in \mathcal{M}(\Omega; E)$ such that, for every $\varphi \in \mathcal{K}(\Omega)$ and every semi-norm $\| \ \|_{E;\nu}$ of E,

$$\|f\|_{\mathcal{M}(\Omega;E);\varphi,\nu} = \|\langle f, \varphi \rangle\|_{E;\nu} = 0.$$

Since E is separated, as is every Neumann space (Definition 1.5),

$$\langle f, \varphi \rangle = 0_E.$$

Thus,

$$f = 0. \ \square$$

Let us define the product of a measure by a continuous function.

Definition 3.5. — *Given $f \in \mathcal{M}(\Omega; E)$ and $g \in \mathcal{C}(\Omega)$, where Ω is an open subset of $\mathbb{R}^d$ and E is a Neumann space, we define $fg \in \mathcal{M}(\Omega; E)$ by: for all $\varphi \in \mathcal{K}(\Omega)$,*

$$\langle fg, \varphi \rangle \stackrel{\text{def}}{=} \langle f, g\varphi \rangle. \ ▮$$

Justification. The mapping fg is a measure because it satisfies the characterization of Theorem 3.3: indeed, the mapping $\varphi \mapsto \langle f, g\varphi \rangle$ is linear and, for every $\nu \in \mathcal{N}_E$, there exists $q \in \mathcal{C}^+(\Omega)$ and $c \in \mathbb{R}$ such that, for all $\varphi \in \mathcal{K}(\Omega)$,

$$\|\langle f, g\varphi \rangle\|_{E,\nu} \leq c \sup_{x \in \Omega} q(x)|(g\varphi)(x)| = c \sup_{x \in \Omega} (q|g|)(x)|\varphi(x)|,$$

where $q|g| \in \mathcal{C}^+(\Omega)$, since any product of continuous positive functions is continuous (Proposition A.48). □

3.2. Equicontinuity of bounded subsets of $\mathcal{M}(\Omega; E)$

Let us show that the bounded subsets of $\mathcal{M}(\Omega; E)$ are equicontinuous[3] on $\mathcal{K}(\Omega)$, which will help us to prove (in § 3.3) that it is a Neumann space.

Theorem 3.6. — *Let*

$$\mathcal{F} \text{ be a bounded subset of } \mathcal{M}(\Omega; E),$$

where Ω is an open subset of $\mathbb{R}^d$ and E is a Neumann space, and let $\| \ \|_{E;\nu}$ be a semi-norm of E. Then there exists $q \in \mathcal{C}^+(\Omega)$ such that, for every $f \in \mathcal{F}$ and $\varphi \in \mathcal{K}(\Omega)$,

$$\|\langle f, \varphi \rangle\|_{E;\nu} \leq \|\varphi\|_{\mathcal{K}(\Omega);q}.$$

That is, for every $\varphi \in \mathcal{K}(\Omega)$,

$$\sup_{f \in \mathcal{F}} \|\langle f, \varphi \rangle\|_{E;\nu} \leq \sup_{x \in \Omega} q(x)|\varphi(x)|. \ ▮$$

The proof will use the **Banach–Steinhaus theorem**, which states that any pointwise bounded set of continuous linear mappings on a Fréchet space is equicontinuous. In the case of a Banach space, it can be stated as follows [Vol. 1, Theorem 10.11].

---

3 **Historical background on the equicontinuity of bounded subsets in Theorem 3.6.** Nicolas BOURBAKI proved in 1965 [8, chap. III, § 1, Proposition 15, p. 60–61] that the subsets of $\mathcal{M}(\Omega; \mathbb{C})$ which are bounded for the vague topology, that is, the topology we use here, are equicontinuous.

Proposition 3.7. — *Let $\mathcal{T}$ be a set of continuous linear mappings from a Banach space F into a separated semi-normed space E, such that for every $u \in F$ and every semi-norm $\| \ \|_{E;\nu}$ of E,*

$$\sup_{L \in \mathcal{T}} \|L(u)\|_{E;\nu} < \infty.$$

Then, for every semi-norm $\| \ \|_{E;\nu}$ of E, there exists $c_\nu \in \mathbb{R}$ such that, for every $L \in \mathcal{T}$ and $u \in E$,

$$\|L(u)\|_{E;\nu} \leq c_\nu \|u\|_F. \ \blacksquare$$

Proof of Theorem 3.6. **1°)** *Equicontinuity in $\mathcal{C}_K(\Omega)$.* Let K be a compact subset of Ω, and let $\mathcal{C}_K(\Omega)$ be the space of continuous real-valued functions on Ω which vanish outside of K, endowed with the norm

$$|||\varphi||| = \sup_{x \in \Omega} |\varphi(x)|. \tag{3.3}$$

Let us show that $\mathcal{F}$ satisfies the three hypotheses of the Banach–Steinhaus theorem with $F = \mathcal{C}_K(\Omega)$.

— The space $\mathcal{C}_K(\Omega)$ is a Banach space (Proposition A.49).

— Every measure $f \in \mathcal{F}$ is continuous from $\mathcal{C}_K(\Omega)$ into E. Indeed, according to the characterization of measures given in Theorem 3.3, for every semi-norm $\| \ \|_{E;\mu}$ of E, there exist $q \in \mathcal{C}^+(\Omega)$ and $c \in \mathbb{R}$ such that, for every $\varphi \in \mathcal{K}(\Omega)$,

$$\|\langle f, \varphi\rangle\|_{E;\mu} \leq c \sup_{x \in \Omega} q(x)|\varphi(x)|.$$

For $\varphi \in \mathcal{C}_K(\Omega)$, it becomes, since φ vanishes outside K,

$$\|\langle f, \varphi\rangle\|_{E;\mu} \leq c \sup_{x \in K} q(x)|\varphi(x)| \leq cm|||\varphi|||,$$

where m is the supremum of q on K (which is finite since every continuous function on a compact set is bounded, see Proposition A.41), which implies the continuity of f by the characterization of continuous linear mappings in Proposition 1.9.

— The set $\mathcal{F}$ is simply bounded, that is, for every $\varphi \in \mathcal{C}_K(\Omega)$ and every semi-norm $\| \ \|_{E;\mu}$ of E,

$$\sup_{f \in \mathcal{F}} \|\langle f, \varphi\rangle\|_{E;\mu} = \sup_{f \in \mathcal{F}} \|f\|_{\mathcal{M}(\Omega;E);\varphi,\mu} < \infty$$

(this follows from Definitions 3.2 of the semi-norms of $\mathcal{M}(\Omega; E)$ and 1.3 of a bounded set in a separated semi-normed space).

The Banach–Steinhaus theorem (Proposition 3.7) then provides $c_K \in \mathbb{R}$ such that, for all functions $f \in \mathcal{F}$ and $\varphi \in \mathcal{C}_K(\Omega)$,

$$\|\langle f, \varphi\rangle\|_{E;\nu} \leq c_K|||\varphi|||. \tag{3.4}$$

2°) Construction of p. Given $n \in \mathbb{N}^*$, we define

$$Q_n \stackrel{\text{def}}{=} \{x \in \Omega : |x| < n,\ B(x, 1/n) \subset \Omega\},$$

where $B(x,r) \stackrel{\text{def}}{=} \{y \in \mathbb{R}^d : |y - x| \leq r\}$, that is, Ω truncated and with a neighborhood of width $1/n$ near its boundary removed. Also, let $Q_0 \stackrel{\text{def}}{=} \emptyset$ and α_0 be the zero function on Ω.

Still for $n \in \mathbb{N}^*$, we define a compact subset K_n contained in Ω by

$$K_n \stackrel{\text{def}}{=} \overline{Q_{n+1} \backslash Q_{n-1}}$$

and, starting from α_0, we define by recurrence on n a function $\alpha_n \in \mathcal{C}_{K_n}(\Omega)$ by

$$\alpha_n(x) \stackrel{\text{def}}{=} \begin{cases} 1 - \alpha_{n-1}(x) & \text{if } x \in Q_n \backslash Q_{n-1}, \\ 1 - n(n+1)d(x, K_n) & \text{if } x \in Q_{n+1} \backslash Q_n \text{ and } d(x, K_n) \leq \dfrac{1}{n(n+1)}, \\ 0 & \text{otherwise,} \end{cases}$$

where $d(x, K_n) \stackrel{\text{def}}{=} \inf\{|y - x| : y \in K_n\}$.

Finally, we define, for all $x \in \Omega$,

$$q(x) = \sum_{n \in \mathbb{N}^*} 2^n c_{K_n} (\alpha_{n-1} + \alpha_n + \alpha_{n+1})(x), \tag{3.5}$$

where c_{K_n} satisfies inequality (3.4) relative to K_n.

3°) Properties of q. Let us first show that

$$q \in \mathcal{C}^+(\Omega).$$

Indeed, on K_n, only α_{n-1}, α_n, and α_{n+1} are nonzero, so Definition (3.5) of q reduces there to a finite sum of continuous functions; thus, q is continuous on each open set $\mathring{K}_n$, and hence on their union Ω; and it is positive, since all the α_n are.

Observe that, if $x \in Q_i \backslash Q_{i-1}$, only $\alpha_{i-1}(x)$ and $\alpha_i(x)$ are nonzero and their sum equals 1, so

$$\sum_{n \in \mathbb{N}^*} \alpha_n(x) = 1. \tag{3.6}$$

This equality holds for all $x \in \Omega$, since the Q_i cover Ω.

Now, if $x \in K_n$, then, again since all the α_i are positive,

$$q(x) \geq 2^n c_{K_n} (\alpha_{n-1} + \alpha_n + \alpha_{n+1})(x) = 2^n c_{K_n}. \tag{3.7}$$

4°) Equicontinuity in $\mathcal{K}(\Omega)$. Let $\varphi \in \mathcal{K}(\Omega)$. This function vanishes outside a compact set included in Ω, and hence outside Q_N for N sufficiently large. Therefore, (3.6) gives

$$\varphi = \sum_{1\le n\le N} \alpha_n \varphi.$$

Each $\alpha_n\varphi$ belongs to $\mathcal{C}_{K_n}(\Omega)$, so

$$\langle f, \varphi\rangle = \sum_{1\le n\le N} \langle f, \alpha_n\varphi\rangle$$

and the equicontinuity inequality (3.4) gives

$$\|\langle f, \varphi\rangle\|_{E;\nu} \le \sum_{1\le n\le N} \|\langle f, \alpha_n\varphi\rangle\|_{E;\nu} \le \sum_{1\le n\le N} c_{K_n} |||\alpha_n\varphi|||.$$

Since α_n vanishes outside K_n and is ≤ 1, Definition (3.3) of $|||\ |||$ gives

$$\|\langle f, \varphi\rangle\|_{E;\nu} \le \sum_{1\le n\le N} c_{K_n} \sup_{x\in K_n} |\varphi(x)|. \tag{3.8}$$

Let us set $r_n = c_{K_n} \sup_{x\in K_n} |\varphi(x)|$ and bound (using the bound for a geometric series from Proposition A.6 (b))

$$\sum_{1\le n\le N} r_n = \sum_{1\le n\le N} 2^{-n}2^n r_n \le \Big(\sum_{1\le n\le N} 2^{-n}\Big) \sup_{1\le n\le N} 2^n r_n \le \sup_{1\le n\le N} 2^n r_n.$$

Thus, (3.8) gives, with the lower bound (3.7) of q,

$$\|\langle f, \varphi\rangle\|_{E;\nu} \le \sup_{1\le n\le N} 2^n c_{K_n} \sup_{x\in K_n} |\varphi(x)| \le \sup_{1\le n\le N} \sup_{x\in K_n} q(x)|\varphi(x)|.$$

Whence,

$$\|\langle f, \varphi\rangle\|_{E;\nu} \le \sup_{x\in\Omega} q(x)|\varphi(x)| = \|\varphi\|_{\mathcal{K}(\Omega);q}. \ \square$$

3.3. Sequential completeness of $\mathcal{M}(\Omega; E)$

Let us show that $\mathcal{M}(\Omega; E)$ is sequentially complete[4], that is, that its Cauchy sequences converge.

4 **History of Theorem 3.8.** If E is complete or quasi-complete (which is stronger than sequential completeness used here), then $\mathcal{M}(\mathbb{R}^d; E)$-unif (that is, endowed with the topology of uniform convergence on the bounded subsets of $\mathcal{K}(\mathbb{R}^d)$, which is stronger than the simple topology used here) is also complete or quasi-complete.
This does not imply that $\mathcal{M}(\Omega; E)$ is sequentially complete if E is, which is the new property stated by Theorem 3.8.

Theorem 3.8. — *Let Ω be an open subset of $\mathbb{R}^d$ and let E be a Neumann space. Then*

$$\mathcal{M}(\Omega; E) \text{ is a Neumann space.}$$

In particular, $\mathcal{M}(\Omega)$ is a Neumann space. ▌

Proof. Since $\mathcal{M}(\Omega; E)$ is semi-normd and separated (Theorem 3.4), it remains, by Definition 1.5 of a Neumann space, to verify that it is sequentially complete. So let

$$(f_n)_{n\in\mathbb{N}} \text{ be a Cauchy sequence in } \mathcal{M}(\Omega; E).$$

1º) Existence of a limit f. According to Definitions 1.2 (b) of a Cauchy sequence and 3.2 of the semi-norms of $\mathcal{M}(\Omega; E)$, for every $\varphi \in \mathcal{K}(\Omega)$ and $\nu \in \mathcal{N}_E$, when $n \to \infty$,

$$\sup_{m\geq n} \|f_n - f_m\|_{\mathcal{M}(\Omega;E);\varphi,\nu} = \sup_{m\geq n} \|\langle f_n - f_m, \varphi\rangle\|_{E;\nu} \to 0.$$

The sequence $(\langle f_n, \varphi\rangle)_{n\in\mathbb{N}}$ is thus a Cauchy sequence in E, and since E is a Neumann space, it has a limit, denoted $\langle f, \varphi\rangle$. That is,

$$\langle f_n, \varphi\rangle \to \langle f, \varphi\rangle \text{ in } E.$$

As this limit is unique (Proposition A.12), since E is separated, this defines a mapping f from $\mathcal{K}(\Omega)$ into E, which is linear since the f_n are.

2º) Property of the limit. Let us check that f is a measure. The sequence $(f_n)_{n\in\mathbb{N}}$ is bounded, as every Cauchy sequence is (Proposition A.12). Theorem 3.6 then shows that it is equicontinuous; more precisely, it provides, for every $\nu \in \mathcal{N}_E$, the existence of $q_\nu \in \mathcal{C}^+(\Omega)$ such that, for every $\varphi \in \mathcal{K}(\Omega)$,

$$\sup_{n\in\mathbb{N}} \|\langle f_n, \varphi\rangle\|_{E;\nu} \leq \|\varphi\|_{\mathcal{K}(\Omega);q_\nu}.$$

In the limit,

$$\|\langle f, \varphi\rangle\|_{E;\nu} \leq \|\varphi\|_{\mathcal{K}(\Omega);q_\nu}.$$

Therefore, according to the characterization of measures in Theorem 3.3,

$$f \in \mathcal{M}(\Omega; E).$$

3º) Convergence of f_n. Since f is a measure, Definition 3.2 again gives

$$\|f_n - f\|_{\mathcal{M}(\Omega;E);\varphi,\nu} = \|\langle f_n, \varphi\rangle - \langle f, \varphi\rangle\|_{E;\nu} \to 0.$$

Thus,

$$f_n \to f \text{ in } \mathcal{M}(\Omega; E).$$

This proves that $\mathcal{M}(\Omega; E)$ is indeed sequentially complete.

4º) The real case. The space $\mathcal{M}(\Omega) = \mathcal{M}(\Omega; \mathbb{R})$ is a Neumann space because $\mathbb{R}$ is a Banach space (Proposition A.35), and hence a Neumann space by Definition 1.5. □

Non-metrizability. The space $\mathcal{M}(\Omega; E)$ is not metrizable, and therefore not a Fréchet space, except if Ω is empty or if E is reduced to $\{0_E\}$ (in which case it has dimension zero) [SIMON, 75, to appear]. □

Non-completeness. The space $\mathcal{M}(\Omega; E)$ is sequentially complete, as we saw in Theorem 3.8, but it is not complete, even if E is complete, when it is endowed with the simple topology used in this volume. However, if E is quasi-complete, that is, if its bounded subsets are complete, then $\mathcal{M}(\Omega; E)$ is quasi-complete [SIMON, 75].

On the other hand, $\mathcal{M}(\Omega; E)$-unif, that is, endowed with the topology of uniform convergence on the bounded subsets of $\mathcal{K}(\Omega)$, is complete if E is complete [75]. □

3.4. Continuity of the $\langle\ ,\ \rangle$ mapping

Let us show that the mapping $(f, \varphi) \mapsto \langle f, \varphi \rangle$ is sequentially continuous.

Theorem 3.9. — *Let Ω be an open subset of $\mathbb{R}^d$ and E a Neumann space.*

(a) *For every $\varphi \in \mathcal{K}(\Omega)$, the mapping $f \mapsto \langle f, \varphi \rangle$ is linear and continuous, and hence sequentially continuous, from $\mathcal{M}(\Omega; E)$ into E.*

(b) *For every $f \in \mathcal{M}(\Omega; E)$, the mapping $\varphi \mapsto \langle f, \varphi \rangle$ is linear and continuous, and hence sequentially continuous, from $\mathcal{K}(\Omega)$ into E.*

(c) *The mapping $(f, \varphi) \mapsto \langle f, \varphi \rangle$ is bilinear and sequentially continuous from the product $\mathcal{M}(\Omega; E) \times \mathcal{K}(\Omega)$ into E.* ▌

Proof. Let $\{\| \ \|_{E;\nu} : \nu \in \mathcal{N}_E\}$ be the family of semi-norms of E.

1°) Property (a). By Definition 3.2 of the semi-norms of $\mathcal{M}(\Omega; E)$, for all $\nu \in \mathcal{N}_E$,

$$\|\langle f, \varphi \rangle\|_{E;\nu} = \|f\|_{\mathcal{M}(\Omega;E);\varphi,\nu}.$$

The mapping $f \mapsto \langle f, \varphi \rangle$ is therefore continuous, by the characterization of continuous linear mappings in Proposition 1.9 (a).

Sequentially continuity follows from continuity (Proposition 1.8).

2°) Property (b). This is Definition 3.2 of a measure.

3°) Property (c). Let $(f_n)_{n\in\mathbb{N}}$ and $(\varphi_n)_{n\in\mathbb{N}}$ be two sequences such that, as $n \to \infty$,

$$f_n \to f \ \text{ in } \mathcal{M}(\Omega; E), \quad \varphi_n \to \varphi \ \text{ in } \mathcal{K}(\Omega).$$

Decompose

$$\langle f_n, \varphi_n \rangle - \langle f, \varphi \rangle = \langle f_n - f, \varphi \rangle + \langle f_n, \varphi_n - \varphi \rangle. \tag{3.9}$$

By Definition 1.2 (a) of a convergent sequence in a separated semi-normed space, for all $\nu \in \mathcal{N}_E$,

$$\|\langle f_n - f, \varphi \rangle\|_{E;\nu} = \|f_n - f\|_{\mathcal{M}(\Omega;E);\varphi,\nu} \to 0,$$

hence

$$\langle f_n - f, \varphi \rangle \to 0_E \text{ in } E. \tag{3.10}$$

On the other hand, the sequence $(f_n)_{n\in\mathbb{N}}$ is bounded, as every convergent sequence is (Proposition A.12). Theorem 3.6 then shows that it is equicontinuous from $\mathcal{K}(\Omega)$ into E; more precisely, it provides, for every $\nu \in \mathcal{N}_E$, the existence of $q \in \mathcal{C}^+(\Omega)$ such that, for every $\phi \in \mathcal{K}(\Omega)$,

$$\sup_{n\in\mathbb{N}} \|\langle f_n, \phi \rangle\|_{E;\nu} \le \|\phi\|_{\mathcal{K}(\Omega);q}.$$

In particular,

$$\|\langle f_n, \varphi_n - \varphi \rangle\|_{E;\nu} \le \|\varphi_n - \varphi\|_{\mathcal{K}(\Omega);q},$$

which tends to 0. This holds for all $\nu \in \mathcal{N}_E$, hence

$$\langle f_n, \varphi_n - \varphi \rangle \to 0_E \text{ in } E.$$

Whence, using (3.9) and (3.10),

$$\langle f_n, \varphi_n \rangle \to \langle f, \varphi \rangle \text{ in } E. \square$$

Non-continuity of $\langle\ ,\ \rangle$. The mapping $(f, \varphi) \mapsto \langle f, \varphi \rangle$ is not continuous from $\mathcal{M}(\Omega; E) \times \mathcal{K}(\Omega)$ into E, except when Ω is empty or E is reduced to $\{0_E\}$ (in which case $\mathcal{M}(\Omega; E)$ is zero dimensional). However, it is continuous on the bounded subsets of $\mathcal{M}(\Omega; E) \times \mathcal{K}(\Omega)$. These properties are proved in [SIMON, 75, to appear].

In the case where $E = \mathbb{R}$, one can observe that the mapping $(f, \varphi) \mapsto \langle f, \varphi \rangle$ is not continuous from $\mathcal{M}(\Omega) \times \mathcal{K}(\Omega)$ into $\mathbb{R}$, because it is not even continuous when $\mathcal{M}(\Omega)$ is endowed with the dual topology of $\mathcal{K}(\Omega)$, which is stronger than the simple topology used here, since the bilinear duality form on $F' \times F$ of a semi-normed space F is continuous only if F is normable [Vol. 1, Theorem 13.22, p. 204]. $\square$

3.5. Identification of continuous functions with measures

Since integrable measures will be defined in § 4.1 as limits of sequences of continuous functions, we identify these functions with measures. Let us begin by associating a measure to each continuous function.

Theorem 3.10. — *Let $f \in \mathcal{C}(\Omega; E)$, where Ω is an open subset of $\mathbb{R}^d$ and E a Neumann space, whose family of semi-norms is denoted $\{\|\ \|_{E;\nu} : \nu \in \mathcal{N}_E\}$.*

We define $\bar{f} \in \mathcal{M}(\Omega; E)$ by: for every $\varphi \in \mathcal{K}(\Omega)$,

$$\langle \bar{f}, \varphi \rangle \stackrel{\text{def}}{=} \int_\Omega f\varphi. \blacksquare$$

Proof. ***1º) Meaning of the integral.*** The integral of $f\varphi$ is well-defined, according to Definition 1.14 of the integral with values in a Neumann space, because $f\varphi$ is uniformly continuous and vanishes outside a bounded set.

Indeed, every continuous function with compact support is uniformly continuous, according to a corollary (Proposition A.42) of Heine's theorem. This is the case for $f\varphi$, since it is continuous (as any product of continuous functions, by Proposition A.48) and has compact support (as any function that vanishes outside a compact subset of Ω, here the support of φ, by Proposition A.53).

2º) Obtaining a measure. Let us show that $\bar{f}$ satisfies the characterization of measures from Theorem 3.3. Given $\nu \in \mathcal{N}_E$, define $q \in \mathcal{C}^+(\Omega)$ by

$$q(x) = (3 + |x|)^{d+2} \|f(x)\|_{E;\nu}. \tag{3.11}$$

Let $\varphi \in \mathcal{K}(\Omega)$ and $\omega = \Omega \cap B$, where B is any open ball containing the support of φ. Only ω contributes to the integral of $f\varphi$ by Proposition 1.23, so the bound on the semi-norms of the integral from Proposition 1.17 gives

$$\Big\| \int_\Omega f\varphi \Big\|_{E;\nu} = \Big\| \int_\omega f\varphi \Big\|_{E;\nu} \leq \int_\omega \|f\|_{E;\nu} |\varphi| = \int_\omega \frac{q(x)|\varphi(x)|}{(3+|x|)^{d+2}} \,\mathrm{d}x.$$

However, as we shall verify in Lemma 3.11,

$$\int_\omega \frac{1}{(3+|x|)^{d+2}} \,\mathrm{d}x \leq 2^d, \tag{3.12}$$

so the growth of the real integral (Proposition 1.18 (a)) and its linearity give

$$\Big\| \int_\Omega f\varphi \Big\|_{E;\nu} \leq 2^d \sup_{x \in \Omega} q(x)|\varphi(x)|. \tag{3.13}$$

That is, using the definition of $\bar{f}$ and Definition 3.1 of the semi-norms of $\mathcal{K}(\Omega)$,

$$\|\langle \bar{f}, \varphi \rangle\|_{E;\nu} \leq 2^d \|\varphi\|_{\mathcal{K}(\Omega);q}.$$

Since the mapping $\bar{f}$ is linear, as well as the integral (Proposition 1.16), this inequality implies, by the characterization of measures in Theorem 3.3, that

$$\bar{f} \in \mathcal{M}(\Omega; E). \ \square$$

It remains to establish the bound (3.12) that we have just used.

Lemma 3.11. — *For every bounded open subset Ω of $\mathbb{R}^d$, $x \mapsto 1/(3+|x|)^{d+2}$ is a function of $\mathcal{B}(\Omega)$ and*

$$\int_\Omega \frac{1}{(3+|x|)^{d+2}} \,\mathrm{d}x \leq 2^d. \ \blacksquare$$

Proof. **1º)** *Belonging to* $\mathcal{B}(\Omega)$. The function $x \mapsto 1/(3+|x|)^{d+2}$ is continuous on all of $\mathbb{R}^d$, hence uniformly continuous on the compact set $\overline{\Omega}$, by Heine's theorem (Proposition A.41). Its restriction to Ω is thus uniformly continuous with bounded support, that is, it belongs to $\mathcal{B}(\Omega)$, by Definition 1.12.

2º) *Integral.* Let $C_\ell \stackrel{\text{def}}{=} \{x \in \mathbb{R}^d : \ell - 1 < |x| < \ell + 1\}$. For N large enough, Ω is included in the union $\bigcup_{0 \le \ell \le N} C_\ell$. Then,

$$\int_\Omega \frac{1}{(3+|x|)^{d+2}}\,\mathrm{d}x \le \sum_{\ell=0}^{N} \int_{C_\ell} \frac{1}{(3+|x|)^{d+2}}\,\mathrm{d}x, \tag{3.14}$$

since each of the cubes $\Delta_{s,n}$ of Definition 1.14 of the Cauchy integral included in Ω is contained in one of the C_l. Thus, the approximate integral satisfies

$$\mathbb{S}_\Omega\Big(x \mapsto \frac{1}{(3+|x|)^{d+2}}\Big) \le \sum_{\ell=0}^{N} \mathbb{S}_{C_\ell}\Big(x \mapsto \frac{1}{(3+|x|)^{d+2}}\Big).$$

The crown C_ℓ is contained in the open cube Q_ℓ of side length $2(\ell+1)$ centered at 0, hence $|C_\ell| \le |Q_\ell| = (2(\ell+1))^d$ (Propositions 2.2 and 2.1). The bound on the semi-norms (here, the absolute value) of the Cauchy integral from Proposition 1.21 then gives

$$\int_{C_\ell} \frac{1}{(3+|x|)^{d+2}}\,\mathrm{d}x \le |C_\ell| \sup_{x\in C_\ell} \frac{1}{(3+|x|)^{d+2}} \le \frac{2^d(\ell+1)^d}{(\ell+2)^{d+2}} = \frac{2^d}{(\ell+2)^2}. \tag{3.15}$$

The stated inequality follows from (3.14), (3.15), and

$$\int_{\ell=0}^{N} \frac{1}{(\ell+2)^2} \le \sum_{\ell=0}^{N} \frac{1}{(\ell+1)(\ell+2)} = \sum_{\ell=0}^{N} \frac{1}{\ell+1} - \frac{1}{\ell+2} = 1 - \frac{1}{N+2} \le 1. \ \square$$

Non-optimality. The bound in Lemma 3.11 could be improved, but is sufficient for our purpose. □

Let us show that two distinct continuous functions yield distinct measures, which will allow us to identify each continuous function with a measure.

Theorem 3.12. — *The mapping* $f \mapsto \overline{f}$ *given by Theorem 3.10 is linear, injective, and continuous, and therefore sequentially continuous, from* $\mathcal{C}(\Omega; E)$ *into* $\mathcal{M}(\Omega; E)$. ∎

Proof. ***1º)*** *Injectivity.* Let $f \in \mathcal{C}(\Omega; E)$ such that

$$\overline{f} = 0.$$

That is, by the definition of $\overline{f}$, for every $\varphi \in \mathcal{K}(\Omega)$,

$$\int_{\Omega} f\varphi = 0_E.$$

The Du Bois-Reymond lemma with values in a Neumann space (Proposition 3.13, below) then yields

$$f = 0.$$

The mapping $f \mapsto \overline{f}$, being linear, is therefore injective.

2º) *Continuity.* Let $\{\| \, \|_{E;\nu} : \nu \in \mathcal{N}_E\}$ be the family of semi-norms of E.

Let $\varphi \in \mathcal{K}(\Omega)$, $\nu \in \mathcal{N}_E$, and

$$\omega \overset{\text{def}}{=} \{x \in \Omega : \varphi(x) \neq 0\}.$$

Since only ω contributes to the integral (Proposition 1.23), Definition 3.2 of the semi-norms of $\mathcal{M}(\Omega; E)$ and the bound on the semi-norms of the integral from Proposition 1.17 yield

$$\|\overline{f}\|_{\mathcal{M}(\Omega;E);\varphi,\nu} = \|\langle \overline{f}, \varphi \rangle\|_{E;\nu} = \Big\| \int_{\Omega} f\varphi \Big\|_{E;\nu} = \Big\| \int_{\omega} f\varphi \Big\|_{E;\nu} \leq \int_{\omega} \|f\|_{E;\nu} |\varphi|.$$

The growth of the real integral (Proposition 1.18 (a)) and Definition 1.10 of the semi-norms of $\mathcal{C}(\Omega; E)$ (since $\overline{\omega}$ is compact) therefore yield, denoting $c_\varphi = \int_\Omega |\varphi|$,

$$\|\overline{f}\|_{\mathcal{M}(\Omega;E);\varphi,\nu} \leq c_\varphi \sup_{x \in \overline{\omega}} \|f(x)\|_{E;\nu} = c_\varphi \|f\|_{\mathcal{C}(\Omega;E);\overline{\omega},\nu}.$$

According to the characterization of continuous linear mappings in Proposition 1.9, this proves that the mapping $^{-}$ is continuous from $\mathcal{C}(\Omega; E)$ into $\mathcal{M}(\Omega; E)$.

3º) *Sequential continuity.* The mapping $^{-}$ is sequentially continuous, as is any continuous mapping (Proposition 1.8). □

It remains to prove the **Du Bois-Reymond lemma**[5], with values in a Neumann space.

5 **History of the Du Bois-Reymond lemma.** Friedrich Ludwig STEGMANN stated Proposition 3.13 for a real-valued function on an interval in 1854 [76], with an incorrect justification. Correct proofs were subsequently given by Eduard HEINE in 1870 [37], and then by Paul DU BOIS-REYMOND in 1879 [20]. This result is also known as the **fundamental lemma of the calculus of variations**.

Proposition 3.13. — *Let $f \in \mathcal{C}(\Omega; E)$, where Ω is an open subset of $\mathbb{R}^d$ and E is a Neumann space, such that, for every $\varphi \in \mathcal{K}(\Omega)$,*

$$\int_\Omega f\varphi = 0_E.$$

Then,

$$f = 0. \blacksquare$$

Proof. If f were not zero, there would exist $y \in \Omega$ such that

$$f(y) \neq 0_E.$$

Since E is separated (Definition 1.1), as is any Neumann space, there would exist a semi-norm $\| \ \|_{E;\nu}$ of E such that

$$\|f(y)\|_{E;\nu} = c > 0.$$

Since Ω is open and f is continuous, there would exist $r > 0$ such that $|x - y| \leq r$ implies $x \in \Omega$ and

$$\|f(x) - f(y)\|_{E;\nu} \leq \frac{c}{2}.$$

Let $\varphi \in \mathcal{K}(\Omega)$ have support in the ball $B = \{x \in \mathbb{R}^d : |y - x| \leq r\}$, such that

$$\varphi > 0 \text{ in } \mathring{B}.$$

This is for example realized by $\varphi(x) = (x - y)/r$ if $|x - y| \leq r$ and $\varphi(x) = 0$ otherwise. Then (Proposition 1.18 (c)),

$$\int_\Omega \varphi = b > 0.$$

Since only $\mathring{B}$ contributes to the integral here (Proposition 1.23), the bound on the semi-norms from Proposition 1.17, the growth of the real integral (Proposition 1.18 (a)), and its linearity (Proposition 1.16) would yield successively

$$\Big\|\int_\Omega (f - f(y))\varphi\Big\|_{E;\nu} = \Big\|\int_{\mathring{B}} (f - f(y))\varphi\Big\|_{E;\nu} \leq \int_{\mathring{B}} \|f - f(y)\|_{E;\nu}\,\varphi \leq \frac{c}{2}b$$

and

$$\Big\|\int_\Omega f(y)\varphi\Big\|_{E;\nu} = \Big\|f(y)\int_\Omega \varphi\Big\|_{E;\nu} = cb,$$

whence

$$\Big\|\int_\Omega f\varphi\Big\|_{E;\nu} \geq \frac{cb}{2}.$$

This would contradict the hypothesis $\int_\Omega f\varphi = 0_E$, since E is separated. Thus,

$$f = 0. \ \square$$

From now on:

$$\begin{cases} \textbf{\textit{We identify each continuous function }} f \in \mathcal{C}(\Omega; E) \textbf{\textit{ with the}} \\ \textbf{\textit{measure }} \overline{f} \in \mathcal{M}(\Omega; E) \textbf{\textit{ defined in Theorem 3.10.}} \end{cases} \quad (3.16)$$

This is permissible since the mapping $f \mapsto \overline{f}$ is injective by Theorem 3.12.

Let us now show that, in this way, the space of continuous functions is topologically embedded into that of measures.

Theorem 3.14. — *Given Ω an open subset of $\mathbb{R}^d$ and E a Neumann space, we have*

$$\mathcal{C}(\Omega; E) \subsetneq \mathcal{M}(\Omega; E)$$

and the identity is continuous, and a fortiori sequentially continuous, from $\mathcal{C}(\Omega; E)$ into $\mathcal{M}(\Omega; E)$. ▮

Proof. With the identification (3.16), the mapping $f \mapsto \overline{f}$ constructed in Theorem 3.10 has become the identity from $\mathcal{C}(\Omega; E)$ into $\mathcal{M}(\Omega; E)$, so the continuity properties stated are those of Theorem 3.12.

Topological inclusion is always equivalent to the continuity of the identity mapping, by Definitions 1.4 of the topological inclusion and 1.7 of continuity. □

We now have the following equalities.

Theorem 3.15. — *For every $f \in \mathcal{C}(\Omega; E)$, where Ω is an open subset of $\mathbb{R}^d$ and E is a Neumann space, and every $\varphi \in \mathcal{K}(\Omega)$,*

$$\langle f, \varphi \rangle_{\mathcal{M}(\Omega;E)\times\mathcal{K}(\Omega)} = \int_\Omega f\varphi$$

and, for every semi-norm $\| \ \|_{E;\nu}$ of E,

$$\|f\|_{\mathcal{M}(\Omega;E);\varphi,\nu} = \Big\| \int_\Omega f\varphi \Big\|_{E;\nu}. \ ▮$$

Proof. ***1°)*** *First equality.* This expression of $\langle f, \varphi \rangle$ is the rewriting of the equality from Theorem 3.10 with the identification (3.16) of f with $\overline{f}$.

2°) *Second equality.* This expression of $\|f\|_{\mathcal{M}(\Omega;E);\varphi,\nu}$ is the rewriting of Definition 3.2 of the semi-norms of $\mathcal{M}(\Omega; E)$ with the expression of $\langle f, \varphi \rangle$ above. □

Compatibility with operations on functions. To generalize an operation pertaining to functions (support, restriction, etc.) to measures, one must verify that the generalization is compatible with the identification (3.16), that is, **verify that the usual operations for functions are recovered**.

Recall that caution must be exercised with identifications, as we saw in the *Dangerous identifications* section from Volume 1 [72, § 14.6, p. 240–243]. □

Convergence of functions. Convergence in $\mathcal{C}(\Omega; E)$, that is, uniform convergence on compact subsets of Ω, implies convergence in the sense of measures, according to Theorem 3.12. Conversely, point wise convergence of continuous functions is neither stronger nor weaker than convergence in the sense of measures.

For example, the functions on $\mathbb{R}^d$ defined by

$$f_n(x) = n^d \rho(n^2 x),$$

where $\rho \in \mathcal{C}(\mathbb{R}^d)$ has support in the ball $B(0,1)$ and $\int_{\mathbb{R}^d} \rho = 1$, converge to 0 in $\mathcal{M}(\mathbb{R}^d)$ as $n \to \infty$, because, for every $\varphi \in \mathcal{D}(\mathbb{R}^d)$,

$$\Big|\int_{\mathbb{R}^d} f_n \varphi\Big| \leq c_\varphi \int_{\mathbb{R}^d} |f_n| = c_\varphi n^{-d}.$$

But they do not converge pointwise if $\rho(0) \neq 0$, since $f_n(0) = n^d \rho(0)$.

Conversely, the functions defined by

$$g_n(x) = n^{2d} \rho(nx)$$

converge pointwise to 0 if $\rho(0) = 0$, since $g_n(x) = 0$ as soon as $n \geq 1/|x|$ and $g_n(0) = 0$. But they do not converge in $\mathcal{M}(\mathbb{R}^d)$, because, if $\varphi = 1$ in $B(0,1)$,

$$\int_{\mathbb{R}^d} g_n \varphi = \int_{\mathbb{R}^d} g_n = n^d. \ \square$$

3.6. Regularization of measures

Before defining the regularizations of a measure, we define the regularizing sequences and the localizing sequences used to construct them.

Definition 3.16. — *We say that a* ***regularizing sequence*** *is a sequence of functions* $(\rho_n)_{n \in \mathbb{N}^*}$ *such that, for all* n*:*

$$\rho_n \in \mathcal{K}(\mathbb{R}^d), \quad \rho_n(x) = 0 \textit{ if } |x| \geq \frac{1}{n}, \quad \rho_n \geq 0, \quad \int_{\mathbb{R}^d} \rho_n = 1.$$

Given an open subset Ω *of* $\mathbb{R}^d$*, we say that a* ***localizing sequence*** *is a sequence of functions* $(\alpha_n)_{n \in \mathbb{N}^*}$ *such that, for all* n*:*

$$\alpha_n \in \mathcal{K}(\Omega), \quad 0 \leq \alpha_n \leq 1, \quad \alpha_n = \begin{cases} 1 & \textit{in } \Omega^n_{3/n}, \\ 0 & \textit{outside of } \Omega^n_{2/n}, \end{cases}$$

where $\Omega^n_r \overset{\text{def}}{=} \{x \in \Omega : |x| < n,\ B(x,r) \subset \Omega\}$ *and* $B(x,r) \overset{\text{def}}{=} \{y \in \mathbb{R}^d : |y| \leq r\}$. ▌

Let us construct the regularizations of a measure, which we will verify to be continuous approximations with compact support. They will serve to characterize integrable measures (Theorem 4.2) and to establish many of their properties.

Definition 3.17. — *Let $f \in \mathcal{M}(\Omega; E)$, where Ω is an open subset of $\mathbb{R}^d$ and E is a Neumann space, $(\rho_n)_{n\in\mathbb{N}^*}$ a regularizing sequence, and $(\alpha_n)_{n\in\mathbb{N}^*}$ a localizing sequence (Definition 3.16).*

*We say that the **regularizations** of f are the functions $R_n f$ defined on Ω, for $n \in \mathbb{N}^*$, by*

$$(R_n f)(x) \overset{\text{def}}{=} \langle f, \alpha_n \tau_x \rho_n \rangle_{\mathcal{M}(\Omega;E)\times\mathcal{K}(\Omega)},$$

where[6]

$$(\tau_x \rho_n)(y) \overset{\text{def}}{=} \rho_n(y - x). \blacksquare$$

Examples of localizing and regularizing functions. The properties required in Definition 3.16 are, for example, satisfied by

$$\alpha_n(x) \overset{\text{def}}{=} \begin{cases} 1 & \text{if } d(x) \geq 3/n, \\ nd(x) - 2 & \text{if } 3/n \geq d(x) \geq 2/n, \\ 0 & \text{if } 2/n \geq d(x), \end{cases} \tag{3.17}$$

where $d(x) = \sup\{r \in \mathbb{R} : B(x,r) \subset \Omega \cap \mathring{B}(0,n)\}$, and

$$\rho_n(x) \overset{\text{def}}{=} \begin{cases} c(1 - n|x|) & \text{if } |x| \leq 1/n, \\ 0 & \text{otherwise,} \end{cases}$$

where $c = 1/\int_{\mathring{B}(0,1/n)} 1 - n|x| \, dx$. □

Terminology. The term **regularized** may seem misleading for $R_n f$ since it is *a priori* only continuous. We use this term because, if we choose α_n and ρ_n to be regular, then the $R_n f$ are also regular, as we established in Volume 3 [74, Theorem 8.15] for any distribution f.

Here, we only require α_n and ρ_n to be continuous, in order to avoid unnecessarily invoking the notion of differentiation in the construction of the theory of integration. □

Let us show that regularizations are continuous functions with compact support.

Theorem 3.18. — *Any regularization of a measure $f \in \mathcal{M}(\Omega; E)$, where Ω is an open subset of $\mathbb{R}^d$ and E is a Neumann space, satisfies*

$$R_n f \in \mathbf{C}_{K_n}(\Omega; E),$$

where $K_n \overset{\text{def}}{=} \overline{\Omega^n_{1/n}}$ is compact and included within Ω, and thus

$$R_n f \in \mathcal{K}(\Omega; E). \blacksquare$$

6 **Definition of the translation.** It varies, unfortunately, according to the author. Jacques-Louis LIONS [49, 2nd edition, p. 54], Serge LANG [42, p. 277], or Kôsaku YOSIDA [81, 6th edition, p. 158], for example, use the definition above $(\tau_x f)(y) = f(y - x)$.
But Haïm BRÉZIS [12, p. 72] or George O. OKIKIOLU [56, p. 294] use $(\tau_x f)(y) = f(y + x)$.

Proof. **1º)** *Uniform continuity.* Let $f \in \mathcal{M}(\Omega; E)$ and $\| \ \|_{E;\nu}$ be a semi-norm of E. According to the characterization of measures in Theorem 3.3, there exists a function $q \in \mathcal{C}^+(\Omega)$ and a constant $c \in \mathbb{R}$ such that, for all $\varphi \in \mathcal{K}(\Omega)$,

$$\|\langle f, \varphi\rangle\|_{E;\nu} \leq c \sup_{x\in\Omega} q(x)|\varphi(x)|.$$

For all y and z in Ω, we therefore have, according to Definition 3.17 of $R_n f$,

$$\|(R_n f)(y) - (R_n f)(z)\|_{E;\nu} \leq c \sup_{x\in\Omega} q(x)\big|\alpha_n(x)\big(\rho_n(x-y) - \rho_n(x-z)\big|.$$

Since $\alpha_n = 0$ outside of $\Omega^n_{2/n}$ and $|\alpha| \leq 1$, it follows, if $|y - z| \leq \eta$, that

$$\|(R_n f)(y) - (R_n f)(z)\|_{E;\nu} \leq cm \sup_{|v-w|\leq\eta} \big|\rho_n(v) - \rho_n(w)\big|, \tag{3.18}$$

where m is the supremum of q over the compact set $\overline{\Omega^n_{2/n}}$ (which is finite since every continuous function on a compact set is bounded, see Proposition A.41).

Since ρ_n is uniformly continuous (Definition 1.7), for every $\epsilon > 0$, we can choose $\eta > 0$ such that the right-hand side of (3.18) is $\leq \epsilon$, which proves that $R_n f$ is uniformly continuous.

2º) *Vanishing.* If $x \in \Omega \setminus \Omega^n_{1/n}$, the function $\alpha_n \tau_x \rho_n$ is zero throughout Ω because:
— it is zero outside $\Omega^n_{2/n}$, since α_n vanishes there;
— it is zero within $\Omega^n_{2/n}$, since $\tau_x \rho_n$ vanishes there (because $\tau_x \rho_n$ vanishes outside $B(x, 1/n)$, which does not intersect $\Omega^n_{2/n}$).

Definition 3.17 then gives

$$R_n f(x) = \langle f, 0\rangle = 0_E. \ \square$$

Let us show that the regularizations $R_n f$ are indeed approximations of the measure f.

Theorem 3.19. — *Let $f \in \mathcal{M}(\Omega; E)$, where Ω is an open subset of $\mathbb{R}^d$ and E is a Neumann space, and let $(R_n f)_{n\in\mathbb{N}}$ be the regularizations (Definition 3.17). Then, as $n \to \infty$,*

$$R_n f \to f \ \text{ in } \mathcal{M}(\Omega; E). \ \blacksquare$$

Proof. **1º)** *Method.* According to Definitions 3.2 of the semi-norms of the space $\mathcal{M}(\Omega; E)$ and 1.2 of the convergence in a separated semi-normed space, it suffices to show that, for every $\varphi \in \mathcal{K}(\Omega)$ and every semi-norm $\| \ \|_{E;\nu}$ of E,

$$\|R_n f - f\|_{\mathcal{M}(\Omega;E);\varphi,\nu} = \|\langle R_n f - f, \varphi\rangle\|_{E;\nu} \to 0. \tag{3.19}$$

As we shall verify in 2º to 4º, for n sufficiently large,

$$\langle R_n f, \varphi\rangle = \langle f, \alpha_n \Phi_n\rangle, \quad \text{where } \Phi_n(x) \stackrel{\text{def}}{=} \int_{\mathring{B}(0,1/n)} \rho_n(z)\varphi(x-z)\,\mathrm{d}z, \tag{3.20}$$

where φ is extended by 0 outside Ω (since $x - z$ does not necessarily belong to Ω), and

$$\langle f, \varphi\rangle = \langle f, \alpha_n \phi_n\rangle, \quad \text{where } \phi_n(x) \stackrel{\text{def}}{=} \int_{\mathring{B}(0,1/n)} \rho_n(z)\varphi(x)\,\mathrm{d}z. \tag{3.21}$$

According to the characterization of measures in Theorem 3.3, there exist $q \in \mathcal{C}^+(\Omega)$ and $c \in \mathbb{R}$ such that, for every $\phi \in \mathcal{K}(\Omega)$,

$$\|\langle f, \phi\rangle\|_{E;\nu} \le c \sup_{x\in\Omega} q(x)|\phi(x)|. \tag{3.22}$$

As we shall verify in 5º, (3.20), (3.21) and (3.22) imply

$$\|\langle R_n f - f, \varphi\rangle\|_{E;\nu} \le cm \sup_{x\in\Omega,\ |z|\le 1/n} |\varphi(x-z) - \varphi(x)|, \tag{3.23}$$

where $m = \sup_{x\in K} q(x)$ and K is a compact set outside of which $\Phi_n - \phi_n$ is zero for n sufficiently large.

The second term in (3.23) tends to 0 since φ is uniformly continuous; it remains to verify (3.20), (3.21), and (3.23) to establish (3.19) and hence the theorem.

2º) An equality. The identification of the continuous function $R_n f$ with a measure (Theorem 3.15) and its Definition 3.17 yield successively

$$\langle R_n f, \varphi\rangle = \int_\Omega (R_n f)(y)\varphi(y)\,\mathrm{d}y = \int_\Omega \langle f, \alpha_n \tau_y \rho_n\rangle \varphi(y)\,\mathrm{d}y.$$

The linearity of f (to move $\varphi(y)$), then the commutation of the Cauchy integral with continuous linear maps (Proposition 1.19), here the mapping f from $\mathcal{K}(\Omega)$ into E, thus yield

$$\langle R_n f, \varphi\rangle = \int_\Omega \langle f, \alpha_n \tau_y \rho_n \varphi(y)\rangle\,\mathrm{d}y = \Big\langle f, \int_\Omega \alpha_n \tau_y \rho_n \varphi(y)\,\mathrm{d}y\Big\rangle. \tag{3.24}$$

(The integral in the second term takes values in E, whereas the integral in the third term takes values in $\mathcal{K}(\Omega)$.)

3º) Verification of (3.20). The integral in the last term of (3.24) belongs to $\mathcal{K}(\Omega)$; let us calculate its value at a point x in Ω. By using once again the commutation of the integral with continuous linear mappings, here the mapping $\phi \mapsto \phi(x)$ from $\mathcal{K}(\Omega)$ into $\mathbb{R}$, we obtain

$$\Big(\int_\Omega \alpha_n \tau_y \rho_n \varphi(y)\,\mathrm{d}y\Big)(x) = \int_\Omega \alpha_n(x)\rho_n(x-y)\varphi(y)\,\mathrm{d}y.$$

Taking $\alpha_n(x)$ out of the integral, and then making the change of variable $z = x - y$, that is, performing a symmetry (Proposition 1.24) following by a translation (Proposition 1.27, since the integrand, like φ, has compact support), we obtain

$$\Big(\int_\Omega \alpha_n \tau_y \rho_n \varphi(y)\,\mathrm{d}y\Big)(x) = \alpha_n(x)\int_{-\Omega+x} \rho_n(z)\varphi(x-z)\,\mathrm{d}z. \tag{3.25}$$

Since the domain where the function vanishes does not contribute to the integral (Proposition 1.23), we can replace $-\Omega + x$ by $\mathbb{R}^d$ (by extending φ by 0), and then replace $\mathbb{R}^d$ by the open ball $\mathring{B}(0,1/n) = \{x \in \mathbb{R}^d : |x| < 1/n\}$ since ρ_n vanishes outside of it.

The second term of (3.25) is therefore equal to $(\alpha_n \Phi_n)(x)$, which, together with (3.24) gives (3.20) for all n.

4°) Verification of (3.21). It suffices to verify that, for n large enough, we have, for all $x \in \Omega$,

$$\varphi(x) = \alpha_n(x)\varphi(x) = \alpha_n(x)\int_{\mathring{B}(0,1/n)} \rho_n(z)\varphi(x)\,\mathrm{d}z.$$

The second equality follows from $\int_{\mathring{B}(0,1/n)} \rho_n = 1$.

The first equality is true outside the support S_φ of φ, since φ vanishes there; and it holds in $\Omega^n_{3/n}$, since α_n equals 1 there; it is therefore valid *a fortiori* in S_φ, and hence throughout Ω, as soon as $n \geq n_\varphi$, where n_φ is such that

$$S_\varphi \subset \Omega^{n_\varphi}_{3/n_\varphi}. \tag{3.26}$$

Such an n_φ exists since S_φ is compact and the $\Omega^n_{3/n}$ form an open and increasing cover of Ω.

5°) Verification of (3.23). Let, for $x \in \Omega$,

$$\psi(x) \stackrel{\text{def}}{=} \int_{\mathring{B}(0,1/n)} \rho_n(z)\big(\varphi(x-z) - \varphi(x)\big)\,\mathrm{d}z. \tag{3.27}$$

From (3.20), (3.21), and (3.22), for all $n \geq n_\varphi$ satisfying (3.26),

$$\|\langle R_n f - f, \varphi\rangle\|_{E;\nu} = \|\langle f, \alpha_n \psi\rangle\|_{E;\nu} \leq c \sup_{x\in\Omega} q(x)|\alpha_n(x)\psi(x)|. \tag{3.28}$$

Let

$$K \stackrel{\text{def}}{=} S_\varphi + B(0, 1/n_\varphi).$$

The function ψ vanishes outside K because, if $x \notin K$, then on the one hand $\varphi(x) = 0$, and on the other hand $x - z \notin S_\varphi$ (since $|z| \leq 1/n \leq 1/n_\varphi$), and thus $\varphi(x - z) = 0$.

Therefore, (3.28) yields, since $0 \leq \alpha_n \leq 1$,

$$\|\langle R_n f - f, \varphi\rangle\|_{E;\nu} \leq c \sup_{x \in K} q(x)|\alpha_n(x)\psi(x)| \leq cm \sup_{x \in K} |\psi(x)|, \tag{3.29}$$

where m is the supremum of q over K, which is finite because K is a compact subset of Ω (since S_φ is included in $\Omega^{n_\varphi}_{3/n_\varphi}$) and any continuous function on a compact set is bounded (Proposition A.41).

Inequality (3.23) follows from (3.29) because, according to (3.27), the monotonicity and linearity of the real integral (Propositions 1.18 (a) and 1.16), together with the equality $\int_{B_n} |\rho_n| = 1$, yield

$$|\psi(x)| \leq \sup_{x \in \Omega,\ |z| \leq 1/n} |\varphi(x - z) - \varphi(x)|. \ \square$$

Let us state a continuity property of the regularization operators.

Theorem 3.20. — *Every regularization operator R_n defined by Definition 3.17 is sequentially continuous from $\mathcal{M}(\Omega; E)$ into $\mathbf{C}_\mathrm{b}(\Omega; E)$, and also into $\mathbf{C}_{K_n}(\Omega; E)$ where $K_n \stackrel{\text{def}}{=} \overline{\Omega^n_{1/n}}$ is compact and included in Ω.* ▮

Proof. **1°)** *Problem setting.* The continuity with values in $\mathbf{C}_{K_n}(\Omega; E)$ follows from that with values in $\mathbf{C}_\mathrm{b}(\Omega; E)$, since $R_n f$ vanishes outside K_n (Theorem 3.18) and $\mathbf{C}_{K_n}(\Omega; E)$ is endowed (Definition 1.11) with the semi-norms of $\mathbf{C}_\mathrm{b}(\Omega; E)$.

Since the operator R_n is linear, it suffices to show that, given a sequence of measures $(f_k)_{k\in\mathbb{N}}$ such that $f_k \to 0$ in $\mathcal{M}(\Omega; E)$, then $R_n f_k \to 0$ in $\mathbf{C}_\mathrm{b}(\Omega; E)$.

According to Definitions 3.2 of the semi-norms of $\mathcal{M}(\Omega; E)$ and 1.2 of convergence in a separated semi-normed space, for all $\varphi \in \mathcal{K}(\Omega)$ and all semi-norms $\| \ \|_{E;\nu}$ of E, we have

$$\|f_k\|_{\mathcal{M}(\Omega;E);\varphi,\nu} = \|\langle f_k, \varphi\rangle\|_{E;\nu} \to 0. \tag{3.30}$$

By Definition 1.11 of the semi-norms of $\mathbf{C}_\mathrm{b}(\Omega; E)$, it remains to show that

$$\|R_n f_k\|_{\mathbf{C}_\mathrm{b}(\Omega;E);\nu} = \sup_{x \in \Omega} \|(R_n f_k)(x)\|_{E;\nu} \to 0. \tag{3.31}$$

2°) Start of the proof of (3.31). For each $x \in \Omega$, Definition 3.17 of R_k and the convergence (3.30) yield

$$\|(R_n f_k)(x)\|_{E;\nu} = \|\langle f_k, \alpha_n \tau_x \rho_n \rangle\|_{E;\nu} \to 0. \tag{3.32}$$

Let us show by contradiction that this convergence is uniform in x, that is, that (3.31) holds. Otherwise, there would exist $a > 0$, a subsequence (that we still denote $(f_k)_{k\in\mathbb{N}}$) and, for each $k \in \mathbb{N}$, a point z_k in Ω such that

$$\|(R_n f_k)(z_k)\|_{E;\nu} \geq a. \tag{3.33}$$

The z_k would all belong to the compact set $\overline{\Omega^n_{1/n}}$, since the $R_n f_k$ vanish outside of it (Theorem 3.18). Therefore, by the Bolzano–Weierstrass theorem (Proposition A.37), they would have a subsequence converging to a point z in Ω. That is, by again denoting the extracted subsequences as the original sequence,

$$z_k \to z. \tag{3.34}$$

Yet,

$$(R_n f_k)(z_k) - (R_n f_k)(z) = \langle f_k, \alpha_n(\tau_{z_k}\rho_n - \tau_z \rho_n)\rangle. \tag{3.35}$$

3°) Use of equicontinuity. The sequence $(f_k)_{k\in\mathbb{N}}$ is bounded in $\mathcal{M}(\Omega; E)$ (as is any convergent sequence in a separated semi-normed space, see Proposition A.12). It is therefore equicontinuous; more precisely (Theorem 3.6), there exists $q \in \mathcal{C}^+(\Omega)$ such that, for all $k \in \mathbb{N}$ and all $\varphi \in \mathcal{K}(\Omega)$,

$$\|\langle f_k, \varphi\rangle\|_{E;\nu} \leq \sup_{x\in\Omega} q(x)|\varphi(x)|.$$

Thus, (3.35) would yield

$$\|(R_n f_k)(z_k) - (R_n f_k)(z)\|_{E;\nu} \leq \sup_{x\in\Omega} q(x)\left|\alpha_n(x)\big(\rho_n(x - z_k) - \rho_n(x - z)\big)\right|.$$

By Definition 3.16 of a localizing sequence, α_n vanishes outside the compact set $\overline{\Omega^n_{2/n}}$ and takes values between 0 and 1 throughout Ω, so we would obtain

$$\|(R_n f_k)(z_k) - (R_n f_k)(z)\|_{E;\nu} \leq \\ \leq m \sup_{y\in\mathbb{R}^d,\ y\in\mathbb{R}^d:\ |y-y'|\leq|z_k - z|} |(\rho_n(y) - \rho_n(y'))|, \tag{3.36}$$

where m is the supremum of q on $\overline{\Omega^n_{2/n}}$ (which is finite since every continuous function on a compact set is bounded, see Proposition A.41).

4°) Conclusion. For all k sufficiently large, the second term in (3.36) would be $\leq a/2$, since $z_k \to z$ (by (3.34)) and since ρ_n is uniformly continuous (by Definition 3.16, again), which, together with (3.33) would yield

$$\|(R_n f_k)(z)\|_{E;\nu} \geq \frac{a}{2}.$$

This would contradict the convergence (3.32), which proves (3.31) and thus completes the proof of Theorem 3.20. □

3.7. Regularization of functions

We present here some properties of the regularizations of a continuous function, which play a key role in the construction of $L^1(\Omega; E)$. Let us begin with integral expressions for these regularizations.

Proposition 3.21. — *If $f \in \mathcal{C}(\Omega; E)$, its regularizations given by Definition 3.17 are, for all $n \in \mathbb{N}^*$ and $x \in \Omega$,*

$$R_n f(x) = \int_{\mathbb{R}^d} \widetilde{\alpha_n f}(x+y)\rho_n(y)\,\mathrm{d}y = \int_{\mathring{B}(0,1/n)} \widetilde{\alpha_n f}(x+y)\rho_n(y)\,\mathrm{d}y,$$

where $\widetilde{\alpha_n f} \in \mathcal{K}(\mathbb{R}^d; E)$ is the extension of $\alpha_n f$ by 0_E outside Ω, and $\mathring{B}(0, 1/n)$ is the open ball centered at 0 with radius $1/n$. ▮

Proof. Definition 3.17 of $R_n f$ and the identification of f with a measure (Theorem 3.15) give

$$R_n f(x) = \langle f, \alpha_n \tau_x \rho_n \rangle = \int_\Omega f(z)\alpha_n(z)\rho_n(z-x)\,\mathrm{d}z.$$

Since the domain where the function is zero does not contribute to the integral (Proposition 1.23), one can replace Ω with $\mathbb{R}^d$ by replacing $f\alpha_n$ with its extension $\widetilde{f\alpha_n}$, because the function $z \mapsto \widetilde{f\alpha_n}(z)\rho_n(z-x)$ is uniformly continuous on $\mathbb{R}^d$ (since α_n has compact support in Ω) and vanishes outside Ω.

We deduce the first stated expression by doing the change of variables $y = z - x$, that is, by applying a translation of $-x$ (Proposition 1.27).

The second expression follows, since only $\mathring{B}(0, 1/n)$ contributes to the integral (Proposition 1.23, again), as ρ_n vanishes outside this ball. □

Let us observe that the regularizations of a continuous function with compact support all have their support included in the same compact set and converge uniformly.

Proposition 3.22. — *If $f \in \mathcal{K}(\Omega; E)$, where Ω is an open subset of $\mathbb{R}^d$ and E is a Neumann space, its regularizations $R_n f$ (Definition 3.17) are all zero outside a single compact set included in Ω, and, as $n \to \infty$,*

$$R_n f \to f \text{ in } \mathbf{C}_b(\Omega; E).$$ ▮

Proof. **1º)** *Vanishing of $R_n f$.* Let us denote by S the support of f, and show that

$$R_n f \text{ vanishes outside of } S + B(0, 1/n). \tag{3.37}$$

Since (Proposition 3.21)

$$R_n f(x) = \int_{\mathring{B}(0,1/n)} \widetilde{\alpha_n f}(x+y)\rho_n(y)\,dy, \tag{3.38}$$

where $\widetilde{\alpha_n f}$ is the extension of $\alpha_n f$ by 0_E outside of Ω, it suffices to observe that if $x \notin S + B(0, 1/n)$, then, for all $y \in B(0, 1/n)$, we have $x + y \notin S$, and thus

$$\widetilde{\alpha_n f}(x+y) = 0_E$$

(because, if $x + y \in \Omega$, then $f(x+y) = 0_E$; otherwise, $\widetilde{\alpha_n f}(x+y)$ is zero by definition).

2º) *Domain of vanishing independent of n.* The open subsets

$$\Omega^n_{2/n} = \{x \in \Omega : |x| < n,\ B(x, 2/n) \subset \Omega\}$$

increase with n and cover Ω, and therefore S, which is compact. One of the $\Omega^n_{2/n}$ thus covers S, say $\Omega^k_{2/k} \supset S$. Then, for all $n \in \mathbb{N}^*$,

$$R_n f \text{ vanishes outside of } \overline{\Omega^k_{2/k}}, \text{ which is a compact subset of } \Omega.$$

Indeed, if $n \geq k$, we have $S + B(0, 1/n) \subset \Omega^k_{2/k} + B(0, 1/k) \subset \overline{\Omega^k_{1/k}}$ and we conclude with (3.37).

And if $n \leq k$, we observe that $R_n f$ vanishes outside of $\overline{\Omega^n_{1/n}}$ (Theorem 3.18), which in this case is contained in $\overline{\Omega^k_{1/k}}$.

3º) *Uniform convergence.* For n large enough, α_n equals 1 on the support of f, and thus $\alpha_n f = f$ throughout Ω. Then, (3.38) gives, since $\int_{\mathring{B}(0,1/n)} \rho_n = 1$ (by Definition 3.16), for all $x \in \Omega$,

$$R_n f(x) - f(x) = \int_{\mathring{B}(0,1/n)} \big(\tilde{f}(x+y) - f(x)\big)\rho_n(y)\,dy.$$

For every semi-norm $\|\ \|_{E;\nu}$ of E, the bound on the semi-norms of the integral from Proposition 1.17 gives, together with the growth and linearity of the real integral (Propositions 1.18 (a) and 1.16),

$$\sup_{x\in\Omega} \|R_n f(x) - f(x)\|_{E;\nu} \leq \sup_{x\in\Omega,\ |y|\leq 1/n} \|\tilde{f}(x+y) - f(x)\|_{E;\nu}.$$

When $n \to \infty$, the second term tends to 0 because $\widetilde{f}$ is uniformly continuous, which, according to Definitions 1.11 of the semi-norms of $\mathbf{C}_b(\Omega; E)$ and 1.2 of convergence in a separated semi-normed space, proves that

$$R_n f \to f \ \text{ in } \mathbf{C}_b(\Omega; E). \ \square$$

Let us give properties of the integral of the semi-norms of the regularizations of a continuous function that play a key role in justifying the definition of the semi-norms of $L^1(\Omega; E)$ and in the proof of its completion (Theorems 4.5 and 4.12).

Proposition 3.23. — *Let $f \in \mathcal{B}(\Omega; E)$, where Ω is an open subset of $\mathbb{R}^d$ and E is a Neumann space, $(R_n f)_{n\in\mathbb{N}^*}$ the regularizations given by Definition 3.17, and $\| \ \|_{E;\nu}$ a semi-norm of E.*

(a) *For all $n \in \mathbb{N}^*$, $R_n f \in \mathcal{K}(\Omega; E)$ and*

$$\int_\Omega \|R_n f\|_{E;\nu} \le \int_\Omega \|f\|_{E;\nu}.$$

(b) *When $n \to \infty$,*

$$\int_\Omega \|R_n f - f\|_{E;\nu} \to 0,$$

$$\int_\Omega \|R_n f\|_{E;\nu} \to \int_\Omega \|f\|_{E;\nu}. \ \blacksquare$$

Proof. ***1°)*** *Inequality* (a). We have seen in Theorem 3.18 that $R_n f \in \mathcal{K}(\Omega; E)$.

The expression for $(R_n f)(x)$ from Proposition 3.21 and the bound on the semi-norms of the Cauchy integral from Proposition 1.17 give, for all $x \in \Omega$, since $\rho_n \ge 0$,

$$\|R_n f\|_{E;\nu}(x) \le \int_{|y|<1/n} \|\widetilde{\alpha_n f}\|_{E;\nu}(x+y)\,\rho_n(y)\,\mathrm{d}y.$$

By integrating and then switching the order of integration using Proposition 1.28, we obtain

$$\begin{aligned}\int_\Omega \|R_n f\|_{E;\nu}(x)\,\mathrm{d}x &\le \int_\Omega \Big(\int_{|y|<1/n} \|\widetilde{\alpha_n f}\|_{E;\nu}(x+y)\,\rho_n(y)\,\mathrm{d}y\Big)\,\mathrm{d}x = \\ &= \int_{|y|<1/n} \Big(\int_\Omega \|\widetilde{\alpha_n f}\|_{E;\nu}(x+y)\,\rho_n(y)\,\mathrm{d}x\Big)\,\mathrm{d}y = \\ &= \int_{|y|<1/n} \Big(\int_\Omega \|\widetilde{\alpha_n f}\|_{E;\nu}(x+y)\,\mathrm{d}x\Big)\rho_n(y)\,\mathrm{d}y.\end{aligned}$$

The growth properties of the real integral in Proposition 1.18 ((c) and 1.18 (a)) yield, since the integral can be restricted to the domain where the function is not zero (Proposition 1.23) and since (Definition 3.16) $\alpha_n = 0$ outside of Ω and $0 \le \alpha_n \le 1$,

$$\int_\Omega \|\widetilde{\alpha_n f}\|_{E;\nu}(x \dot{+} y)\,dx \le \int_{\mathbb{R}^d} \|\widetilde{\alpha_n f}\|_{E;\nu}(z)\,dz = \\ = \int_\Omega \|\widetilde{\alpha_n f}\|_{E;\nu}(z)\,dz \le \int_\Omega \|f\|_{E;\nu}(z)\,dz.$$

Using again Proposition 1.18 (a) and the linearity of the integral, we get, since (Definition 3.16, again) $\int_{|y|<1/n} \rho_n = 1$,

$$\int_\Omega \|R_n f\|_{E;\nu}(x)\,dx \le \int_\Omega \|f\|_{E;\nu}(z)\,dz \int_{|y|<1/n} \rho_n(y)\,dy = \int_\Omega \|f\|_{E;\nu}(z)\,dz.$$

2o) *First convergence of* (b). We will obtain it by approximating, in 2.a, f by a function $\beta_k f$ with compact support, and then by bounding, in 2.b, the integral of $R_n(\beta_k f - f)$ using the inequality (a).

2.a. *Localization of* f. Let $\epsilon > 0$ and $\mathcal{O} \stackrel{\text{def}}{=} \{x \in \Omega : f(x) \ne 0_E\}$. This is an open set (Proposition A.54, since f is continuous and Ω is open) and bounded (since f vanishes outside of a bounded set). Let β_k be a localization function in $\mathcal{O}$ such that (Definition 3.16, extended outside of $\mathcal{O}$)

$$\beta_k \in C(\Omega), \quad 0 \le \beta \le 1, \quad \beta_k = \begin{cases} 1 & \text{in } \mathcal{O}^k_{3/k}, \\ 0 & \text{outside of } \mathcal{O}^k_{2/k}. \end{cases}$$

Each set $\omega_k \stackrel{\text{def}}{=} \mathcal{O} \setminus \overline{\mathcal{O}^k_{3/k}}$ is open, bounded, decreases as k increases, and their intersection for $k \in \mathbb{N}^*$ is empty, therefore (Theorem 2.17) $|\omega_k| \to 0$ as $k \to \infty$.

Let $b \stackrel{\text{def}}{=} \sup_{x\in\Omega} \|f(x)\|_{E;\nu}$, which is finite because the functions in $\mathcal{B}(\Omega; E)$ are bounded (Proposition 1.13), and let k be such that

$$|\omega_k| \le \frac{\epsilon}{3b}.$$

Then $\beta_k f = f$ outside of ω_k (since $f = 0_E$ outside of $\mathcal{O}$ and $\beta_k = 1$ in $\overline{\mathcal{O}^k_{3/k}}$), so, since the domain where the function vanishes does not contribute to the Cauchy integral (Proposition 1.23), the supremum of the integral from Proposition 1.21 yields

$$\int_\Omega \|\beta_k f - f\|_{E;\nu} = \int_{\omega_k} \|\beta_k f - f\|_{E;\nu} \le |\omega_k| \sup_{x\in\Omega} \|(\beta_k f - f)(x)\|_{E;\nu} \le \frac{\epsilon}{3}. \quad (3.39)$$

2.b. *Estimate of* $R_n f - f$. As the number k is fixed, let us decompose, writing β in the place of β_k,

$$R_n f - f = R_n(f - \beta f) + (R_n(\beta f) - \beta f) + (\beta f - f). \quad (3.40)$$

Let us bound the integral of the semi-norm $\| \ \|_{E;\nu}$ of the three terms on the right-hand side.

For the first and third terms, observe that, according to the bound in part (a), inequality (3.39) implies

$$\int_\Omega \|R_n(f - \beta f)\|_{E;\nu} \le \int_\Omega \|f - \beta f\|_{E;\nu} \le \frac{\epsilon}{3}. \tag{3.41}$$

For the second term, observe that the support of βf is compact (since it is included in $\mathcal{O}^k_{2/k}$, which is a compact subset of Ω), which implies (Proposition 3.22) that all the $R_n(\beta f)$ and βf have their support contained in the same compact set $Q \subset \Omega$. Since the domain where the function vanishes does not contribute to the integral (Proposition 1.23, again), the bound of the integral from Proposition 1.21 gives, denoting $\omega \overset{\text{def}}{=} \mathring{Q}$,

$$\int_\Omega \|R_n(\beta f) - \beta f\|_{E;\nu} = \int_\omega \|R_n(\beta f) - \beta f\|_{E;\nu} \le |\omega| \sup_{x \in \Omega} \|(R_n(\beta f) - \beta f)(x)\|_{E;\nu}.$$

However, as $n \to \infty$, $R_n(\beta f)$ converges (Theorem 3.20) to βf in $\mathbf{C}_b(\Omega; E)$. So, for n sufficiently large,

$$\int_\Omega \|R_n(\beta f) - \beta f\|_{E;\nu} \le \frac{\epsilon}{3}. \tag{3.42}$$

The decomposition (3.40) then yields, together with (3.41), for n large enough (depending on ϵ),

$$\int_\Omega \|R_n f - f\|_{E;\nu} \le \epsilon.$$

Which proves that, as $n \to \infty$,

$$\int_\Omega \|R_n f - f\|_{E;\nu} \to 0. \tag{3.43}$$

3°) Second convergence of (b). Since every semi-norm is contracting (Proposition A.11), we have

$$\big| \|R_n f\|_{E;\nu} - \|f\|_{E;\nu} \big| \le \|R_n f - f\|_{E;\nu}.$$

Thus, convergence (3.43) implies, by the monotonicity of the real Cauchy integral (Proposition 1.18 (a)),

$$\int_\Omega \|R_n f\|_{E;\nu} \to \int_\Omega \|f\|_{E;\nu}. \ \square$$

Chapter 4

Integrable Measures

Here, we construct the space $L^1(\Omega; E)$ of integrable measures on an open subset Ω of $\mathbb{R}^d$ with values in a Neumann space.

Roughly speaking, $L^1(\Omega; E)$ is the *sequential completion of* $\mathcal{B}(\Omega; E)$ *in* $\mathcal{M}(\Omega; E)$ with respect to the semi-norms $\int_\Omega \| \ \|_{E;\nu}$. More precisely:

— we say that a measure f is *integrable* (Definition 4.1) if it is the limit in $\mathcal{M}(\Omega; E)$ of a sequence, called an *integrabilizing sequence*, of functions f_n in $\mathcal{B}(\Omega; E)$ (the space of uniformly continuous functions with bounded support) such that, for every $\nu \in \mathcal{N}_E$,

$$\sup_{m \geq n} \int_\Omega \|f_m - f_n\|_{E;\nu} \to 0 \text{ as } n \to \infty;$$

— we endow (Definition 4.4) $L^1(\Omega; E)$ with the semi-norms, indexed by $\nu \in \mathcal{N}_E$,

$$\|f\|_{L^1(\Omega;E);\nu} = \lim_{n\to\infty} \int_\Omega \|f_n\|_{E;\nu},$$

which is a limit that does not depend on the choice of integrabilizing sequence (Theorem 4.5).

We show (Theorem 4.12) that $L^1(\Omega; E)$ is a Neumann space, *i.e.* it is sequentially complete. The regularizations $R_n f$ play a key role in the proof (see the comment *Another proof*, p. 90), particularly because, as we establish beforehand:

— a measure f is integrable if and only if (Theorem 4.2), for every $\nu \in \mathcal{N}_E$,

$$\sup_{m \geq n} \int_\Omega \|R_m f - R_n f\|_{E;\nu} \to 0 \text{ as } n \to \infty;$$

— the $R_n f$ therefore converge in $L^1(\Omega; E)$ and the operators R_n on it are contracting (Theorem 4.11).

4.1. Definition of integrable measures

Let us define **integrable measures** with values in a Neumann space.

Definition 4.1. —*A measure f, on an open subset Ω of $\mathbb{R}^d$ with values in a Neumann space E, is said to be* **integrable** *if there exists a sequence $(f_n)_{n\in\mathbb{N}}$ in $\mathcal{B}(\Omega; E)$, that is, of uniformly continuous functions with bounded support, such that, as $n \to \infty$,*

$$f_n \to f \text{ in } \mathcal{M}(\Omega; E), \tag{4.1}$$

and, for every semi-norm $\| \; \|_{E;\nu}$ of E,

$$\sup_{m \geq n} \int_\Omega \|f_m - f_n\|_{E;\nu} \to 0. \tag{4.2}$$

We say that such a sequence $(f_n)_{n\in\mathbb{N}}$ is an ***integrabilizing sequence*** *for f.* ▮

Observe that (after the identification (3.16), p. 65, of continuous functions with measures) each function f_n in an integrabilizing sequence is itself an integrable measure. Indeed, the constant sequence $(f_n, f_n, \ldots)$ is an integrabilizing sequence for f_n.

Let us characterize integrable measures with the help of regularizations.

Theorem 4.2. —*A measure f, on an open subset Ω of $\mathbb{R}^d$ with values in a Neumann space E, is integrable if and only if its regularizations $R_n f$ given by Definition 3.17 satisfy, for every semi-norm $\| \; \|_{E;\nu}$ of E, as $n \to \infty$,*

$$\sup_{m \geq n} \int_\Omega \|R_m f - R_n f\|_{E;\nu} \to 0. \tag{4.3}$$

Then, $(R_n f)_{n\in\mathbb{N}^*}$ is a integrabilizing sequence and $R_n f \in \mathcal{K}(\Omega; E)$. ▮

Interest of the characterization in Theorem 4.2. This characterization shows that every integrable measure possesses an integrabilizing sequence formed of functions from $\mathcal{K}(\Omega; E)$.

This is interesting because it is simpler to show that the $R_n f$, that we know explicitly, form an integrabilizing sequence, rather than to prove the existence or non-existence of functions f_n having the same behavior, but whose eventual expression is unknown. For example, this is how we will show that $L^1(\Omega; E)$ is sequentially complete in Theorem 4.12.

The manipulation of this integrabilizing sequence will be facilitated by the inequality established for a continuous f in Proposition 3.23, namely $\int_\Omega \|R_n f\|_{E;\nu} \leq \int_\Omega \|f\|_{E;\nu}$. □

Proof of Theorem 4.2. ***1º)*** *Sufficiency.* If the $R_n f$ satisfy (4.3), then they form an integrabilizing sequence and thus f is integrable. Indeed, they are uniformly continuous with compact support (Theorem 3.18) converge to f in $\mathcal{M}(\Omega; E)$ (Theorem 3.19), and (4.3) gives (4.2).

2º) *Necessity.* Conversely, suppose f is integrable.

Given an integrabilizing sequence $(f_k)_{k\in\mathbb{N}}$ for f satisfying the conditions (4.1) and (4.2) of integrability, decompose, for integers m, n, and k,

$$R_m f - R_n f = R_m(f - f_k) + (R_m f_k - f_k) + (f_k - R_n f_k) + R_n(f_k - f).$$

Whence,

$$\int_\Omega \|R_m f - R_n f\|_{E;\nu} \leq \int_\Omega \|R_m(f - f_k)\|_{E;\nu} + \int_\Omega \|(R_m f_k - f_k)\|_{E;\nu} +$$

$$+ \int_\Omega \|(f_k - R_n f_k)\|_{E;\nu} + \int_\Omega \|R_n(f_k - f)\|_{E;\nu}. \quad (4.4)$$

Let $\epsilon > 0$ be given. We will now seek to bound the terms on the right-hand side of (4.4).

2.a. *Bounding the first and fourth terms.* Let us assume for the moment (we will verify this in Lemma 4.3) that condition (4.1), namely $f_k \to f$ in $\mathcal{M}(\Omega; E)$, implies, for all n and k,

$$\int_\Omega \|R_n(f - f_k)\|_{E;\nu} \leq \sup_{\ell \geq k} \int_\Omega \|f_\ell - f_k\|_{E;\nu}. \quad (4.5)$$

According to hypothesis (4.2), we can choose k such that the second term is bounded by $\epsilon/4$. Thus,

$$\int_\Omega \|R_m(f - f_k)\|_{E;\nu} + \int_\Omega \|R_n(f_k - f)\|_{E;\nu} \leq \frac{\epsilon}{2}. \quad (4.6)$$

2.b. *Bounding the second and third terms.* Since f_k is uniformly continuous and vanishes outside a bounded set, Proposition 3.23 (b) shows that, as $n \to \infty$,

$$\int_\Omega \|(f_k - R_n f_k)\|_{E;\nu} \to 0.$$

We can thus choose n_k such that this quantity is bounded by $\epsilon/4$ for all $n \geq n_k$. Then, for $m \geq n \geq n_k$,

$$\int_\Omega \|(R_m f_k - f_k)\|_{E;\nu} + \int_\Omega \|(f_k - R_n f_k)\|_{E;\nu} \leq \frac{\epsilon}{2}. \quad (4.7)$$

2.c. *Conclusion.* For all $m \geq n \geq n_k$, the second term in (4.4) is bounded by ϵ from (4.6) and (4.7), which proves that the regularizations $R_n f$ satisfy (4.3). □

It remains to verify inequality (4.5), *i.e.* the following property.

Lemma 4.3. — *Let $(f_k)_{k \in \mathbb{N}}$ be a sequence in $\mathcal{B}(\Omega; E)$ converging to a measure f in $\mathcal{M}(\Omega; E)$, and R_n a regularization operator given by Definition 3.17. Then, for all k and every semi-norm $\| \ \|_{E;\nu}$ of E,*

$$\int_\Omega \|R_n(f - f_k)\|_{E;\nu} \leq \sup_{\ell \geq k} \int_\Omega \|f_\ell - f_k\|_{E;\nu}. \ \blacksquare$$

Proof. By decomposing $f - f_k = (f - f_\ell) + (f_\ell - f_k)$ where $\ell \geq k$, we bound

$$\int_\Omega \|R_n(f - f_k)\|_{E;\nu} \leq \int_\Omega \|R_n(f - f_\ell)\|_{E;\nu} + \int_\Omega \|R_n(f_\ell - f_k)\|_{E;\nu}. \tag{4.8}$$

By the contracting property of R_n from Proposition 3.23 (a), the last integral is bounded by

$$\int_\Omega \|R_n(f_\ell - f_k)\|_{E;\nu} \leq \int_\Omega \|f_\ell - f_k\|_{E;\nu} \leq \sup_{\ell \geq k} \int_\Omega \|f_\ell - f_k\|_{E;\nu}. \tag{4.9}$$

It remains to choose ℓ such that the second integral in (4.8) is less than any given $\epsilon > 0$. For this, observe that:

— the functions $R_n(f - f_\ell)$ vanish outside of $\Omega^m_{1/m}$ (Theorem 3.18), so only this set contributes to their integral (Proposition 1.23);

— $R_n(f - f_\ell) \to 0$ in $\mathbf{C}_b(\Omega; E)$ as $\ell \to \infty$, since the operator R_n is sequentially continuous from $\mathcal{M}(\Omega; E)$ into $\mathbf{C}_b(\Omega; E)$ (Theorem 3.20).

The bound on the integral from Proposition 1.21 thus gives

$$\int_\Omega \|R_n(f - f_\ell)\|_{E;\nu} = \int_{\Omega^m_{1/m}} \|R_n(f - f_\ell)\|_{E;\nu} \leq |\Omega^m_{1/m}| \sup_{x \in \Omega} \|R_n(f - f_\ell)(x)\|_{E;\nu},$$

and, for ℓ large enough, the last term is less than ϵ. Together with (4.8) and (4.9), it becomes

$$\int_\Omega \|R_n(f - f_k)\|_{E;\nu} \leq \epsilon + \sup_{\ell \geq k} \int_\Omega \|f_\ell - f_k\|_{E;\nu}.$$

This holds for every $\epsilon > 0$, and hence for $\epsilon = 0$, which proves Lemma 4.3. □

4.2. Space of integrable measures $L^1(\Omega; E)$

Let us define the space $L^1(\Omega; E)$ of integrable measures[1].

Definition 4.4. — *Let Ω be an open subset of $\mathbb{R}^d$ and E a Neumann space, whose family of semi-norms is denoted $\{\| \|_{E;\nu} : \nu \in \mathcal{N}_E\}$.*

We denote by $L^1(\Omega; E)$ the vector space of integrable measures endowed with the semi-norms, indexed by $\nu \in \mathcal{N}_E$,

$$\|f\|_{L^1(\Omega;E);\nu} \stackrel{\text{def}}{=} \lim_{n \to \infty} \int_\Omega \|f_n\|_{E;\nu}, \tag{4.10}$$

where $(f_n)_{n \in \mathbb{N}}$ is any integrabilizing sequence (Definition 4.1) for f. ▮

1 **Historical background of the Lebesgue space.** It will be mentioned in Note 1, p. 184.

Definition 4.4 is well-defined by the following result.

Theorem 4.5. — *The set of integrable measures is a vector space and, in (4.10), the limit exists, does not depend on the choice of integrabilizing sequence for f, and defines a semi-norm $\| \ \|_{L^1(\Omega;E);\nu}$ on $L^1(\Omega;E)$.* ∎

Proof. ***1°)*** *Obtaining a vector space.* Measures which admit an integrabilizing sequence, that is, approximations satisfying (4.1) and (4.2), satisfy the properties that characterize a vector subspace (Definition A.8 (b)). More precisely:

— the set is non-empty, since the zero measure possesses an integrabilizing sequence (the sequence of zero functions);

— if two measures f and g have integrabilizing sequences $(f_n)_{n\in\mathbb{N}}$ and $(g_n)_{n\in\mathbb{N}}$, then $f+g$ and tf, where $t \in \mathbb{R}$, have integrabilizing sequences, namely $(f_n+g_n)_{n\in\mathbb{N}}$ and $(tf_n)_{n\in\mathbb{N}}$, respectively.

2°) *Existence of a limit of* $\int_\Omega \|f_n\|_{E;\nu}$. Let $(f_n)_{n\in\mathbb{N}}$ be a sequence in $\mathcal{B}(\Omega;E)$ such that

$$\sup_{m\geq n} \int_\Omega \|f_m - f_n\|_{E;\nu} \to 0. \tag{4.11}$$

Let us show that $\int_\Omega \|f_n\|_{E;\nu}$ is a Cauchy sequence in $\mathbb{R}$. Since every semi-norm is contracting (Proposition A.11), for all $x \in \Omega$,

$$\big|\|f_m(x)\|_{E;\nu} - \|f_n(x)\|_{E;\nu}\big| \leq \|f_m(x) - f_n(x)\|_{E;\nu},$$

so the linearity and monotonicity of the real-valued Cauchy integral (Propositions 1.17 and 1.18 (a)) yield

$$\Big|\int_\Omega \|f_m\|_{E;\nu} - \int_\Omega \|f_n\|_{E;\nu}\Big| \leq \int_\Omega \big|\|f_m\|_{E;\nu} - \|f_n\|_{E;\nu}\big| \leq \int_\Omega \|f_m - f_n\|_{E;\nu}. \tag{4.12}$$

Thus, as $n \to \infty$,

$$\sup_{m\geq n} \Big|\int_\Omega \|f_m\|_{E;\nu} - \int_\Omega \|f_n\|_{E;\nu}\Big| \to 0.$$

As every Cauchy sequence in $\mathbb{R}$ has a limit (Proposition A.35), it follows that

$$\int_\Omega \|f_n\|_{E;\nu} \text{ has a limit.}$$

3°) *Uniqueness of the limit.* Recall that the regularizations given by Definition 3.17 satisfy, according to Lemma 4.3, for all n and k,

$$\int_\Omega \|R_k(f - f_n)\|_{E;\nu} \leq \sup_{m\geq n} \int_\Omega \|f_m - f_n\|_{E;\nu}.$$

From the hypothesis (4.11), given $\epsilon > 0$, there exists n_ϵ such that the second term is bounded by $\epsilon/2$ for all $n \geq n_\epsilon$. Likewise, there exists n'_ϵ such that this also holds for the f'_n. Then, for all $n \geq \sup\{n_\epsilon, n'_\epsilon\}$,

$$\int_\Omega \|R_k(f'_n - f_n)\|_{E;\nu} \leq \int_\Omega \|R_k(f - f_n)\|_{E;\nu} + \int_\Omega \|R_k(f - f'_n)\|_{E;\nu} \leq \epsilon.$$

As $k \to \infty$, the left-hand term tends to $\int_\Omega \|f'_n - f_n\|_{E;\nu}$ by Proposition 3.23 (b), whence, in the limit

$$\int_\Omega \|f'_n - f_n\|_{E;\nu} \leq \epsilon.$$

From the bound (4.12) with f'_n in the place of f_m, it follows that

$$\left| \int_\Omega \|f'_n\|_{E;\nu} - \int_\Omega \|f_n\|_{E;\nu} \right| \leq \epsilon.$$

Thus holds for all $n \geq \sup\{n_\epsilon, n'_\epsilon\}$, thus

$$\int_\Omega \|f'_n\|_{E;\nu} \text{ has the same limit as } \int_\Omega \|f_n\|_{E;\nu}.$$

4°) Obtaining a semi-norm. The mapping $f \mapsto \lim_{n\to\infty} \int_\Omega \|f_n\|_{E;\nu}$ satisfies the three properties of Definition A.9 for a semi-norm.

More precisely, if two measures f and g have integrabilizing sequences $(f_n)_{n\in\mathbb{N}}$ and $(g_n)_{n\in\mathbb{N}}$, i.e. satisfying properties (4.1) and (4.2), and if $t \in \mathbb{R}$, then

$$\lim_{n\to\infty} \int_\Omega \|f_n\|_{E;\nu} \geq 0,$$

$$\lim_{n\to\infty} \int_\Omega \|tf_n\|_{E;\nu} = |t| \lim_{n\to\infty} \int_\Omega \|f_n\|_{E;\nu},$$

$$\lim_{n\to\infty} \int_\Omega \|f_n + g_n\|_{E;\nu} \leq \lim_{n\to\infty} \int_\Omega \|f_n\|_{E;\nu} + \lim_{n\to\infty} \int_\Omega \|g_n\|_{E;\nu}.$$

Indeed, the analogous properties for fixed n follow from the monotonicity and linearity of the real-valued Cauchy integral (Proposition 1.18). □

Other methods of constructing $L^1(\Omega; E)$. Various construction methods are described on p. 205 (in the more general framework of $L^p(\Omega; E)$ spaces). □

Non-integrable measures. An example of a non-integrable measure is the Dirac mass. Note that it is a measure that cannot be represented by a class of functions equal almost everywhere and integrable, see Theorems 17.31 and 17.33. □

4.3. Some properties of $L^1(\Omega; E)$

Let us show that $L^1(\Omega; E)$ is separated (we will see in Theorem 4.12 that it is a Neumann space).

Theorem 4.6. — *The space $L^1(\Omega; E)$, where Ω is an open subset of $\mathbb{R}^d$ and E is a Neumann space, is a separated semi-normed space.* ▮

Proof. Let us denote by $\{\| \|_{E;\nu} : \nu \in \mathcal{N}_E\}$ the family of semi-norms of E, and let $f \in L^1(\Omega; E)$ be such that, for every $\nu \in \mathcal{N}_E$,

$$\|f\|_{L^1(\Omega;E);\nu} = 0.$$

As we will verify in Lemma 4.7, for every $\varphi \in \mathcal{K}(\Omega)$,

$$\|\langle f, \varphi\rangle\|_{E;\nu} \leq \|f\|_{L^1(\Omega;E);\nu} \sup_{x\in\Omega} |\varphi(x)|.$$

Thus, $\|\langle f, \varphi\rangle\|_{E;\nu} = 0$ for every semi-norm of E. Since E is separated, as any Neumann space is (Definition 1.5), it follows that $\langle f, \varphi\rangle = 0_E$, whence

$$f = 0.$$

The space $L^1(\Omega; E)$ therefore also satisfies Definition 1.1 of a separated space. □

It remains for us to verify the following inequality.

Lemma 4.7. — *Let $f \in L^1(\Omega; E)$ and $\varphi \in \mathcal{K}(\Omega)$, where Ω is an open subset of $\mathbb{R}^d$ and E is a Neumann space, and let $\| \|_{E;\nu}$ be a semi-norm of E. Then,*

$$\|\langle f, \varphi\rangle\|_{E;\nu} \leq \|f\|_{L^1(\Omega;E);\nu} \sup_{x\in\Omega} |\varphi(x)|.$$ ▮

Proof. Let $(f_n)_{n\in\mathbb{N}}$ be an integrabilizing sequence (Definition 4.1) for f. By using the identification of the continuous function f_n with a measure (Theorem 3.15, as $f_n \in \mathcal{B}(\Omega; E)$), the bound on the semi-norms of the Cauchy integral (Proposition 1.17), and the monotonicity and linearity of the real-valued integral (Propositions 1.18 (a) and 1.16), we get

$$\|\langle f_n, \varphi\rangle\|_{E;\nu} = \left\|\int_\Omega f_n\varphi\right\|_{E;\nu} \leq \left(\int_\Omega \|f_n\|_{E;\nu}\right) \sup_{x\in\Omega} |\varphi(x)|.$$

The stated inequality then follows by passing to the limit as $n \to \infty$, since:

— $\langle f_n, \varphi\rangle \to \langle f, \varphi\rangle$ by Definition 4.1 of an integrabilizing sequence;

— $\int_\Omega \|f_n\|_{E;\nu} \to \|f\|_{L^1(\Omega;E);\nu}$ by Definition 4.4 of the semi-norms of $L^1(\Omega; E)$. □

Observe that uniformly continuous functions vanishing outside of a bounded set are integrable measures (after their identification (3.16), p. 65, with measures).

Theorem 4.8. — *Every function $f \in \mathcal{B}(\Omega; E)$, where Ω is an open subset of $\mathbb{R}^d$ and E is a Neumann space, is an integrable measure and, for every semi-norm $\| \, \|_{E;\nu}$ of E,*

$$\|f\|_{L^1(\Omega;E);\nu} = \int_\Omega \|f\|_{E;\nu}. \blacksquare$$

Proof. If $f \in \mathcal{B}(\Omega; E)$, Definitions 4.1 of an integrable measure and 4.4 of its semi-norms are satisfied with $f_n = f$. □

Let us compare $L^1(\Omega; E)$ with various other spaces.

Theorem 4.9. — *Let Ω be an open subset and Q be a compact subset of $\mathbb{R}^d$, and let E be a Neumann space. Then,*

$$\mathbf{C}_Q(\Omega; E) \subsetneq L^1(\Omega; E) \subsetneq \mathcal{M}(\Omega; E). \blacksquare$$

Proof. Denote by $\{\| \, \|_{E;\nu} : \nu \in \mathcal{N}_E\}$ the family of semi-norms of E.

1º) First inclusion. Let $f \in \mathbf{C}_Q(\Omega; E)$ and $\nu \in \mathcal{N}_E$. The expression of the semi-norms in Theorem 4.8, since the domain where the function vanishes does not contribute to the integral (Proposition 1.23), give

$$\|f\|_{L^1(\Omega;E);\nu} = \int_\Omega \|f\|_{E;\nu} = \int_{\Omega\cap\mathcal{O}} \|f\|_{E;\nu},$$

where $\mathcal{O}$ is any bounded open set containing Q.

Using the upper bound on the semi-norms of the Cauchy integral (Proposition 1.21) and Definition 1.11 (b) of the semi-norms of $\mathbf{C}_Q(\Omega; E)$, we obtain

$$\|f\|_{L^1(\Omega;E);\nu} \leq |\Omega \cap \mathcal{O}| \sup_{x\in\Omega} \|f(x)\|_{E;\nu} = |\Omega \cap \mathcal{O}| \, \|f\|_{\mathbf{C}_Q(\Omega;E);\nu}.$$

Whence, by Definition 1.4 (a) of a topological inclusion,

$$\mathbf{C}_Q(\Omega; E) \subsetneq L^1(\Omega; E).$$

2º) Second inclusion. Let $f \in L^1(\Omega; E)$, $\varphi \in \mathcal{K}(\Omega)$, and $\nu \in \mathcal{N}_E$. Definition 3.2 of the semi-norms of $\mathcal{M}(\Omega; E)$ give, together with the equality from Lemma 4.7,

$$\|f\|_{\mathcal{M}(\Omega;E);\varphi,\nu} = \|\langle f, \varphi\rangle\|_{E;\nu} \leq \|f\|_{L^1(\Omega;E);\nu} \sup_{x\in\Omega} |\varphi(x)|.$$

Whence, again by the definition of a topological inclusion,

$$L^1(\Omega; E) \subsetneq \mathcal{M}(\Omega; E). \ \square$$

Now, let us characterize integrabilizing sequences.

Theorem 4.10. — *Let $f \in L^1(\Omega; E)$, where Ω is an open subset of $\mathbb{R}^d$ and E is a Neumann space.*

A sequence $(f_n)_{n \in \mathbb{N}}$ is an integrabilizing sequence for f if and only if: each f_n is continuous from Ω into E, vanishes outside of a bounded set and, as $n \to \infty$,

$$f_n \to f \ \textit{in} \ L^1(\Omega; E). \ \blacksquare$$

Proof. **1°)** *Forward implication.* Let

$$(f_n)_{n \in \mathbb{N}} \text{ be an integrabilizing sequence for } f.$$

By Definition 4.1, $f_n \to f$ in $\mathcal{M}(\Omega; E)$ and, for every semi-norm $\| \ \|_{E;\nu}$ of E,

$$\sup_{m \geq n} \int_\Omega \|f_m - f_n\|_{E;\nu} \to 0.$$

For fixed n, $(f_m - f_n)_{m \in \mathbb{N}}$ is thus an integrabilizing sequence for $f - f_n$. It is also therefore an integrable measure, and its semi-norms are, by Definition 4.4,

$$\|f - f_n\|_{L^1(\Omega;E);\nu} = \lim_{m \to \infty} \int_\Omega \|f_m - f_n\|_{E;\nu} \leq \sup_{m \geq n} \int_\Omega \|f_m - f_n\|_{E;\nu}.$$

Whence,

$$f_n \to f \ \text{ in } L^1(\Omega; E).$$

2°) *Backward implication.* If each f_n is continuous and vanishes outside of a bounded set, and if $f_n \to f$ in $L^1(\Omega; E)$, then $\sup_{m \geq n} \int_\Omega \|f_m - f_n\|_{E;\nu} \to 0$, thus $(f_n)_{n \in \mathbb{N}}$ is an integrabilizing sequence for f. $\square$

4.4. Regularization in $L^1(\Omega; E)$

Let us show that, in $L^1(\Omega; E)$, regularizations converge and that regularization operators are contracting.

Theorem 4.11. — *Let $f \in L^1(\Omega; E)$, where Ω is an open subset of $\mathbb{R}^d$ and E is a Neumann space, and let $(R_n f)_{n \in \mathbb{N}^*}$ be the regularizations given by Definition 3.17.*

(a) *As $n \to \infty$,*

$$R_n f \to f \ \textit{in} \ L^1(\Omega; E).$$

(b) *For every $n \in \mathbb{N}$ and every semi-norm of $L^1(\Omega; E)$,*

$$\|R_n f\|_{L^1(\Omega;E);\nu} \leq \|f\|_{L^1(\Omega;E);\nu}. \blacksquare$$

Proof. ***1º)*** *Property* (a). The $R_n f$ form an integrabilizing sequence for f according to Theorem 4.2, so they converge to f in $L^1(\Omega; E)$ according to Theorem 4.10.

2º) *Property* (b). Let $(f_k)_{k\in\mathbb{N}}$ be an integrabilizing sequence for f (Definition 4.1). Decompose

$$R_n f = R_n f_k + R_n(f - f_k).$$

The function $R_n f$ is uniformly continuous and vanishes outside of the bounded open set $\Omega^n_{1/n}$ (Theorem 3.18), so its semi-norms are given (Theorem 4.8), for all $\nu \in \mathcal{N}_E$, by

$$\|R_n f\|_{L^1(\Omega;E);\nu} = \int_\Omega \|R_n f\|_{E;\nu} \leq \int_\Omega \|R_n f_k\|_{E;\nu} + \int_\Omega \|R_n(f - f_k)\|_{E;\nu}. \quad (4.13)$$

Given $\epsilon > 0$, we will bound the two integrals in the right-hand term.

2.a. *First integral.* The bound from Proposition 3.23 (a), followed by Definition 4.4 of the semi-norms of $L^1(\Omega; E)$, shows that there exists k_1 such that, for all $k \geq k_1$,

$$\int_\Omega \|R_n f_k\|_{E;\nu} \leq \int_\Omega \|f_k\|_{E;\nu} \leq \|f\|_{L^1(\Omega;E);\nu} + \frac{\epsilon}{2}. \quad (4.14)$$

2.b. *Second integral.* The bound from Theorem 4.3, followed by Definition 4.1 of an integrabilizing sequence, shows that there exists $k \geq k_1$ such that

$$\int_\Omega \|R_n(f - f_k)\|_{E;\nu} \leq \sup_{\ell \geq k} \int_\Omega \|f_\ell - f_k\|_{E;\nu} \leq \frac{\epsilon}{2}. \quad (4.15)$$

2.c. *Conclusion.* The inequalities (4.13), (4.14), and (4.15) give

$$\|R_n f\|_{L^1(\Omega;E);\nu} \leq \|f\|_{L^1(\Omega;E);\nu} + \epsilon.$$

This holds for all $\epsilon > 0$, and hence also for $\epsilon = 0$. □

4.5. Sequential completeness of $L^1(\Omega; E)$

Let us show that $L^1(\Omega; E)$ is sequentially complete.

Theorem 4.12. — *The space $L^1(\Omega; E)$, where Ω is an open subset of $\mathbb{R}^d$ and E is a Neumann space, is itself a Neumann space.* ▌

Proof. Let $(f_k)_{k\in\mathbb{N}}$ be a Cauchy sequence in $L^1(\Omega; E)$.

1°) Construction of the limit. Since $L^1(\Omega; E) \subsetneq \mathcal{M}(\Omega; E)$ (Theorem 4.9), the sequence $(f_k)_{k\in\mathbb{N}}$ is also Cauchy (Proposition A.22) in $\mathcal{M}(\Omega; E)$, which is sequentially complete (Theorem 3.8), and hence has a limit f in it. That is,

$$f_k \to f \text{ in } \mathcal{M}(\Omega; E). \tag{4.16}$$

2°) Membership of f in $L^1(\Omega; E)$. It suffices, by Definition 4.1 of $L^1(\Omega; E)$, to show that the regularizations $R_n f$ form an integrabilizing sequence for f.

Since $R_n f \in \mathcal{K}(\Omega; E)$ (Theorem 3.18) and $R_n f \to f$ in $\mathcal{M}(\Omega; E)$ (Theorem 3.19), it remains to verify criterion (4.2), that is, in this context, that for every semi-norm $\| \ \|_{E;\nu}$ of E, as $n \to \infty$,

$$\sup_{m\geq n} \int_\Omega \|R_m f - R_n f\|_{E;\nu} \to 0. \tag{4.17}$$

For this, we decompose, for m, n, and ℓ integers,

$$R_m f - R_n f = R_m(f - f_\ell) + (R_m f_\ell - R_n f_\ell) + R_n(f_\ell - f). \tag{4.18}$$

Given $\epsilon > 0$, let us estimate the terms on the right-hand side.

2.a. *First and third terms.* By Definition 1.2 of a Cauchy sequence, there exists ℓ such that

$$\sup_{k\geq \ell} \|f_k - f_\ell\|_{L^1(\Omega;E);\nu} \leq \frac{\epsilon}{3}. \tag{4.19}$$

With the expression for the semi-norms of $L^1(\Omega; E)$ for a continuous function from Theorem 4.8, and the bound on the regularizations from Theorem 4.11 (b), it follows that, for all m and all $k \geq \ell$,

$$\int_\Omega \|R_m(f_k - f_\ell)\|_{E;\nu} = \|R_m(f_k - f_\ell)\|_{L^1(\Omega;E);\nu} \leq \|f_k - f_\ell\|_{L^1(\Omega;E);\nu} \leq \frac{\epsilon}{3}. \tag{4.20}$$

Since the operator R_m is sequentially continuous from $\mathcal{M}(\Omega; E)$ into $\mathbf{C}_{K_m}(\Omega; E)$ (Theorem 3.20), the convergence (4.16) implies that, as $k \to \infty$, $R_m f_k \to R_m f$ in $\mathbf{C}_{K_m}(\Omega; E)$, and therefore

$$\|R_m(f_k - f_\ell)\|_{E;\nu} \to \|R_m(f - f_\ell)\|_{E;\nu} \text{ in } \mathbf{C}_{K_m}(\Omega).$$

Since the Cauchy integral is continuous from $\mathbf{C}_{K_m}(\Omega)$ into $\mathbb{R}$ (Proposition 1.26), the inequality (4.20) gives, in the limit, for all m,

$$\int_\Omega \|R_m(f - f_\ell)\|_{E;\nu} \leq \frac{\epsilon}{3}. \tag{4.21}$$

2.b. *Second term.* The regularizations converge in $L^1(\Omega; E)$ (Theorem 4.11, (a)), thus $R_n f_\ell - f_\ell$ converges to 0 as $n \to \infty$, so there exists n_1 such that, for all $n \geq n_1$,

$$\|R_n f_\ell - f_\ell\|_{L^1(\Omega;E);\nu} \leq \frac{\epsilon}{6}.$$

So, again using the expression of the semi-norms from Theorem 4.8, if $m \geq n \geq n_1$,

$$\int_\Omega \|R_m f_\ell - R_n f_\ell\|_{E;\nu} = \|R_m f_\ell - R_n f_\ell\|_{L^1(\Omega;E);\nu} \leq \frac{\epsilon}{3}. \tag{4.22}$$

2.c. *Integrability of f.* The decomposition (4.18) gives, with the bound (4.21) relative to m and then to n (instead of m), and (4.22) together with the linearity and monotonicity of the real-valued Cauchy integral (Propositions 1.16 and 1.18 (a)), for all $m \geq n \geq n_1$,

$$\int_\Omega \|R_m f - R_n f\|_{E;\nu} \leq \epsilon.$$

This proves (4.17), hence the $R_n f$ form an integrabilizing sequence for f, and

$$f \in L^1(\Omega; E).$$

3o) Convergence $f_k \to f$. Since f is integrable, inequality (4.21) can be written, again using Theorem 4.8, as

$$\|R_m(f - f_\ell)\|_{L^1(\Omega;E);\nu} \leq \frac{\epsilon}{3}.$$

As the regularizations converge in $L^1(\Omega; E)$ (Theorem 4.11 (a), again), it follows that, in the limit as $m \to \infty$,

$$\|f - f_\ell\|_{L^1(\Omega;E);\nu} \leq \frac{\epsilon}{3}.$$

Whence, together with (4.19), for all $k \geq \ell$,

$$\|f_k - f\|_{L^1(\Omega;E);\nu} \leq \epsilon.$$

This proves that

$$f_k \to f \text{ in } L^1(\Omega; E).$$

4o) Conclusion. This proves that $L^1(\Omega; E)$ is sequentially complete, *i.e.* it is a Neumann space, since it is semi-normed and separated (Theorem 4.6). □

Another proof. When E is normed, one can avoid the use of regularizations by applying the diagonal process to show that the limit f in $\mathcal{M}(\Omega; E)$ is integrable. This is explained in the comment *Another proof ...*, p. 202, in the more general framework of $L^p(\Omega; E)$ spaces.

This method no longer words for a general Neumann space E, as explained in the comment *Difficulty when E is not metrizable*, p. 205, hence the use of regularizations. □

Chapter 5

Integration of Integrable Measures

Here, we define the integral $\int_\omega f$ over a measurable set ω of an integrable measure f with values in a Neumann space E.

This integral is the *continuous extension*, with respect to f and ω, of the Cauchy integral of a uniformly continuous function over an open set. More precisely (Definition 5.1), $\int_\omega f$ is the limit of $\int_{\omega_n} f_n$, where:
— the f_n form an integrabilizing sequence for f, so that $\sup_{m \geq n} \int_\Omega \|f_m - f_n\|_{E;\nu} \to 0$ as $n \to \infty$;
— the ω_n are open sets decreasing to ω such that $|\omega_n \backslash \omega| \to 0$.

The key point is that the limit of $\int_{\omega_n} f_n$ does not depend on the choice of the f_n nor the ω_n (Theorem 5.2). Once this has been established, we easily obtain (Sections 5.2, 5.3, and Chapter 6), for values in a Neumann space, the properties of the integral that are classical for functions taking values in a Banach space.

In § 5.4, we present numerous examples of possible target spaces E that are useful when studying partial differential equations:
— the spaces $L^p(\Omega)$ and $W^{m,p}(\Omega)$, as well as their local analogues $L^p_{\text{loc}}(\Omega)$ and $W^{m,p}_{\text{loc}}(\Omega)$;
— their weak analogues $L^p(\Omega)$-weak and $W^{m,p}(\Omega)$-weak for $1 < p < \infty$, and $L^\infty(\Omega)$-$*$weak;
— in the place of $L^1(\Omega)$-weak, which is not a Neumann space, the space $\mathcal{M}(\Omega)$, into which it can be embedded to study of PDEs whose *right-hand side is in* L^1.

Finally, we show that it is not desirable to extend this construction to (metrizable) target spaces E that are not Neumann spaces, because then (§ 5.5) there exists a uniformly continuous function f with values in E such that $\int_\Omega f \notin E$, and one can no longer identify $\mathcal{C}(\Omega; E)$ with a subspace of $\mathcal{M}(\Omega; E)$. On the other hand, such a space E can be embedded into its sequential completion, and integration can then be performed with values in the latter.

5.1. Integral of an integrable measure

Let us define the integral[1] of an integrable measure with values in a Neumann space.

---

1 **Historical background of the integral.** In 1904, Henri LEBESGUE defined [44, ch. VII, § IV, p. 120] the integral of a real-valued *integrable function* on $\mathbb{R}$ by a method entirely different from the one presented here. He extended his definition to functions defined on $\mathbb{R}^d$ in 1910 [45].

Definition 5.1. — *Let $f \in L^1(\Omega; E)$, where Ω is an open subset of $\mathbb{R}^d$ and E is a Neumann space, and let ω be a measurable subset of Ω.*

*We say that the **integral** of f over ω is the element of E defined by*

$$\int_\omega f \stackrel{\text{def}}{=} \lim_{n\to\infty} \int_{\omega_n} f_n,$$

where $(f_n)_{n\in\mathbb{N}}$ is any integrabilizing sequence (Definition 4.1) for f, $(\omega_n)_{n\in\mathbb{N}}$ is any decreasing sequence of open sets such that $\omega \subset \omega_n \subset \Omega$ and

$$|\omega_n \setminus \omega| \to 0 \text{ as } n \to \infty, \tag{5.1}$$

and $\int_{\omega_n} f_n$ is the Cauchy integral (Definition 1.14). ∎

Recall that an integrabilizing sequence is a sequence of uniformly continuous functions with bounded support such that, as $n \to \infty$, for every semi-norm of E,

$$f_n \to f \ \text{ in } \mathcal{M}(\Omega; E), \tag{5.2}$$

$$\sup_{m\geq n} \int_\Omega \|f_m - f_n\|_{E;\nu} \to 0. \tag{5.3}$$

The term *integral* and the notation $\int_\omega f$ are legitimate, since (for $f_n = f$ and $\omega_n = \omega$)

$$\begin{cases} \textbf{we recover the Cauchy integral} \text{ when } f \\ \text{is uniformly continuous and } \omega \text{ is open.} \end{cases} \tag{5.4}$$

Definition 5.1 is justified by the following result.

Theorem 5.2. — *In Definition 5.1, there exist sequences $(f_n)_{n\in\mathbb{N}}$ and $(\omega_n)_{n\in\mathbb{N}}$, and $\int_{\omega_n} f_n$ has a limit that does not depend on the choice of f_n and ω_n.* ∎

Proof. **1°)** Existence of $(f_n)_{n\in\mathbb{N}}$ and $(\omega_n)_{n\in\mathbb{N}}$. There exists an integrabilizing sequence $(f_n)_{n\in\mathbb{N}}$ by Definition 4.1 of an integrable measure, and a sequence $(\omega_n)_{n\in\mathbb{N}}$ satisfying (5.1) according to Lemma 2.19.

Vector-valued integrals. The integral of an *integrable function* with values in a Banach space E was defined in 1933 by Salomon BOCHNER [5].

In 1966, Nicolas BOURBAKI defined [8, chap. III, § 3, définition 1, p. 75] the integral of a function in $\mathcal{K}(\Omega; E\text{-weak})$, when E is a *LCTVS*. If E-weak is sequentially complete, this integral coincides with the Cauchy integral with values in E-weak; otherwise, it takes values in the completion of $\widehat{E}$ of E, which is $\widehat{E\text{-weak}}$.

The extension of the integral to values in a Neumann space E is new.

Further reading. For other methods of constructing the integral, see *Other methods of constructing L^p...* p. 205. For more details, see *Histoire de l'intégration* by Jean-Paul PIER [58, p. 79–86].

2°) Existence of a limit. Let us show that $(\int_{\omega_n} f_n)_{n\in\mathbb{N}}$ is a Cauchy sequence in E. For $m \geq n \geq k$, using the linearity of the Cauchy integral, decompose

$$\int_{\omega_m} f_m - \int_{\omega_n} f_n = \int_{\omega_m} (f_m - f_k) + \Big(\int_{\omega_m} f_k - \int_{\omega_n} f_k\Big) + \int_{\omega_n} (f_k - f_n). \quad (5.5)$$

Given a semi-norm $\| \|_{E;\nu}$ of E and $\epsilon > 0$, we bound the integrals of the semi-norms of the three terms on the right-hand side by $\epsilon/3$, for k large enough.

2.a. *Third term.* Observe that the bound for the semi-norms of the Cauchy integral from Proposition 1.17 gives

$$\Big\|\int_{\omega_n} f_n - f_k\Big\|_{E;\nu} \leq \int_\Omega \|f_n - f_k\|_{E;\nu} \quad (5.6)$$

and that, according to hypothesis (5.3), the supremum for $n \geq k$ of the right-hand side tends to 0 as $k \to \infty$. We can therefore choose k such that, for all $n \geq k$,

$$\Big\|\int_{\omega_n} f_n - f_k\Big\|_{E;\nu} \leq \frac{\epsilon}{3}. \quad (5.7)$$

2.b. *Second term.* Introduce a bounded open set B containing the support of f_k. Since the exterior of B does not contribute to the integral of f_k according to Proposition 1.23, the bound on the semi-norms of a difference of Cauchy integrals from Proposition 1.22 gives (since $\omega_m \subset \omega_n$ by the hypothesis of the decreasing sequence ω_n),

$$\begin{aligned}\Big\|\int_{\omega_m} f_k - \int_{\omega_n} f_k\Big\|_{E;\nu} &= \Big\|\int_{\omega_m\cap B} f_k - \int_{\omega_n\cap B} f_k\Big\|_{E;\nu} \leq \\ &\leq \big||\omega_n \cap B| - |\omega_m \cap B|\big| \sup_{x\in\Omega} \|f_k(x)\|_{E;\nu}.\end{aligned} \quad (5.8)$$

By the monotonicity and additivity of the measure of open sets (Propositions 2.2 and 2.4 (a)) yield, since $\omega_n \cap B \subset (\omega_m \cap B) \cup (\omega_n \setminus \omega)$ (because $\omega \subset \omega_m \subset \omega_n$),

$$\big||\omega_n \cap B| - |\omega_m \cap B|\big| \leq |\omega_n \setminus \omega|,$$

which tends to 0 as $n \to \infty$ by hypothesis. Thus, there exists $n_\epsilon \geq k$ such that, for all $n \geq n_\epsilon$ and $m \geq n$,

$$\Big\|\int_{\omega_m} f_k - \int_{\omega_n} f_k\Big\|_{E;\nu} \leq \frac{\epsilon}{3}. \quad (5.9)$$

2.c. *First term.* We use the bound (5.7) with m in the place of n.

2.d. *Conclusion.* With inequalities (5.7), (5.9), and (5.7) with m in the place of n, the decomposition (5.5) gives, for all $m \geq n \geq n_\epsilon$,

$$\Big\|\int_{\omega_m} f_m - \int_{\omega_n} f_n\Big\|_{E;\nu} \leq \epsilon.$$

Thus, $(\int_{\omega_n} f_n)_{n\in\mathbb{N}}$ is a Cauchy sequence in E, which is sequentially complete by hypothesis. It indeed therefore has a limit.

3⁰) Uniqueness of the limit. Let $(f'_n)_{n\in\mathbb{N}}$ be another integrabilizing sequence for f and $(\omega'_n)_{n\in\mathbb{N}}$ another decreasing sequence of open sets such that $\omega \subset \omega'_n \subset \Omega$ and satisfy (5.1). For $n \geq k$, decompose

$$\int_{\omega'_n} f'_n - \int_{\omega_n} f_n = \tag{5.10}$$
$$= \int_{\omega'_n} (f'_n - f_n) + \int_{\omega'_n} (f_n - f_k) + \Big(\int_{\omega'_n} f_k - \int_{\omega_n} f_k\Big) + \int_{\omega_n} (f_k - f_n).$$

Again, given a semi-norm $\| \ \|_{E;\nu}$ of E and $\epsilon > 0$, let us bound the integral of the semi-norm of the four terms on the right-hand side by $\epsilon/4$, for k large enough.

3.a. *First term.* Recall that the regularizations defined in Definition 3.17 satisfy, according to Lemma 4.3, for all n and ℓ,

$$\int_\Omega \|R_\ell(f - f_n)\|_{E;\nu} \leq \sup_{m\geq n} \int_\Omega \|f_m - f_n\|_{E;\nu}.$$

According to hypothesis (5.3), there exists n_1 such that the right-hand side is bounded by $\epsilon/8$ for all $n \geq n_1$. Similarly, there exists n'_1 such that this is also satisfied by the f'_n for all $n \geq n'_1$. Then, for all $n \geq \sup\{n_1, n'_1\}$,

$$\int_\Omega \|R_\ell(f'_n - f_n)\|_{E;\nu} \leq \int_\Omega \|R_\ell(f - f_n)\|_{E;\nu} + \int_\Omega \|R_\ell(f - f'_n)\|_{E;\nu} \leq \frac{\epsilon}{4}.$$

When $\ell \to \infty$, the left-hand side tends to $\int_\Omega \|f'_n - f_n\|_{E;\nu}$ by Proposition 3.23 (b), hence, in the limit

$$\int_\Omega \|f'_n - f_n\|_{E;\nu} \leq \frac{\epsilon}{4}.$$

The bound on the semi-norms of the integral from Proposition 1.17 then gives

$$\Big\|\int_{\omega'_n} f'_n - f_n\Big\|_{E;\nu} \leq \int_\Omega \|f'_n - f_n\|_{E;\nu} \leq \frac{\epsilon}{4}. \tag{5.11}$$

3.b. *Second term.* Proposition 1.17 again, together with hypothesis (5.3), shows that one can choose $k \geq n_1$ such that, for all $n \geq k$,

$$\Big\|\int_{\omega'_n} f_n - f_k\Big\|_{E;\nu} \leq \int_\Omega \|f_n - f_k\|_{E;\nu} \leq \frac{\epsilon}{4}. \tag{5.12}$$

3.c. *Third term.* For this term, we cannot directly use inequality (5.8) because ω'_n is not necessarily included in ω_n. We therefore compare these two open sets by considering $\lambda_n = \omega'_n \cap \omega_n$. Decompose

$$\int_{\omega'_n} f_k - \int_{\omega_n} f_k = \Big(\int_{\omega'_n} f_k - \int_{\lambda_n} f_k\Big) + \Big(\int_{\lambda_n} f_k - \int_{\omega_n} f_k\Big). \tag{5.13}$$

Then, given again a bounded open set B containing the support of f_k, (5.8) gives, denoting $b_k = \sup_{x\in\Omega} \|f_k(x)\|_{E;\nu}$,

$$\Big\| \int_{\omega_n} f_k - \int_{\lambda_n} f_k \Big\|_{E;\nu} \leq b_k \big| |\omega_n \cap B| - |\lambda_n \cap B| \big|.$$

The second term tends to 0 as $n \to \infty$ (according to hypothesis (5.1), since then $|\omega \cap B| \leq |\lambda_n \cap B| \leq |\omega_n \cap B|$), so there exists $n_3 \geq k$ such that, for all $n \geq n_3$,

$$\Big\| \int_{\omega_n} f_k - \int_{\lambda_n} f_k \Big\|_{E;\nu} \leq \frac{\epsilon}{8}.$$

Similarly, there exists n'_3 such that, for all $n \geq n'_3$, the same inequality is satisfied with ω'_n in the place of ω_n. Then, (5.13) gives, for all $n \geq \sup\{n_3, n'_3\}$,

$$\Big\| \int_{\omega'_n} f_k - \int_{\omega_n} f_k \Big\|_{E;\nu} \leq \frac{\epsilon}{4}. \tag{5.14}$$

3.d. *Fourth term.* We apply the bound (5.11) with respect to ω_n instead of ω'_n, which holds for all $n \geq n_4$.

3.e. *Conclusion.* With the inequalities (5.11), (5.12), (5.14), and (5.11) with respect to ω_n instead of ω'_n, the decomposition (5.10) gives, for all $n \geq \sup\{n_3, n'_3, n_4\}$,

$$\Big\| \int_{\omega'_n} f'_n - \int_{\omega_n} f_n \Big\|_{E;\nu} \leq \epsilon.$$

This proves that $\int_{\omega'_n} f_n$ has the same limit as $\int_{\omega_n} f_n$. □

5.2. Linearity and continuity of the integral

Let us show that the integral is linear.

Theorem 5.3. — *Let $f \in L^1(\Omega; E)$ and $g \in L^1(\Omega; E)$, where Ω is an open subset of $\mathbb{R}^d$, E is a Neumann space, ω is a measurable subset of Ω, and $t \in \mathbb{R}$. Then,*

$$\int_\omega f + g = \int_\omega f + \int_\omega g, \quad \int_\omega tf = t \int_\omega f. \blacksquare$$

Proof. As the Cauchy integral is linear on $\mathcal{B}(\Omega; E)$ (Proposition 1.16), the approximations of f, g, and ω given by Definition 5.1 of the integral satisfy

$$\int_{\omega_n} f_n + g_n = \int_{\omega_n} f_n + \int_{\omega_n} g_n, \quad \int_{\omega_n} tf_n = t_n \int_{\omega_n} f.$$

The stated equalities then follow by passing to the limit as $n \to \infty$. □

Let us provide a bound for the semi-norms of the integral (here, the semi-norms play the role typically held by the norm when E is a Banach space).

Theorem 5.4. — *Let $f \in L^1(\Omega; E)$, where Ω is an open subset of $\mathbb{R}^d$, E is a Neumann space, ω is a measurable subset of Ω, and $\{\| \ \|_{E;\nu} : \nu \in \mathcal{N}_E\}$ is a family of semi-norms of E. Then, for all $\nu \in \mathcal{N}_E$,*

$$\Big\| \int_\omega f \Big\|_{E;\nu} \leq \|f\|_{L^1(\Omega;E);\nu}. \ \blacksquare$$

Proof. The approximations given by Definition 5.1 of the integral satisfy (Proposition 1.17)

$$\Big\| \int_{\omega_n} f_n \Big\|_{E;\nu} \leq \int_\Omega \|f_n\|_{E;\nu}.$$

The stated inequality then follows by passing to the limit as $n \to \infty$ since, by Definition 4.4 of the semi-norms of $L^1(\Omega; E)$, $\int_\Omega \|f_n\|_{E;\nu} \to \|f\|_{L^1(\Omega;E);\nu}$. □

Supplement to the Theorem 5.4. We shall see in Theorems 13.5 and 13.4 that

$$\Big\| \int_\omega f \Big\|_{E;\nu} \leq \int_\omega \|f\|_{E;\nu}, \quad \int_\Omega \|f\|_{E;\nu} = \|f\|_{L^1(\Omega;E);\nu}.$$

This requires us to define $\|f\|_{E;\nu}$, which will be done in Theorem 13.1. □

Let us now show that the integral is continuous[2].

Theorem 5.5. — *Given Ω an open subset of $\mathbb{R}^d$, E a Neumann space, and ω a measurable subset of Ω, the mapping $f \mapsto \int_\omega f$ is continuous, uniformly continuous, and sequentially continuous from $L^1(\Omega; E)$ into E.* ▮

Proof. Continuity follows from the bound on the semi-norms of the integral from Theorem 5.4, and by the characterization of continuous linear mappings in Proposition 1.9.

Continuity and linearity together imply uniform continuity (Proposition 1.9).

Continuity implies sequential continuity (Proposition 1.8). □

Let us express the measure of a bounded measurable set in terms of an integral (this will be established for any set of finite measure later in Theorem 6.20).

2 **Historical background of the linearity and continuity of the integral.** Linearity and continuity with respect to an increasing sequence of functions were two of the six conditions required by Henri LEBESGUE in his definition of the integral in one dimension given in 1904 [44, chap. VII, § 1, p. 98–99]. The other conditions were additivity with respect to the domain of integration (*i.e.* Chasles' relation), translation invariance, positivity, and normalization by $\int_0^1 1 = 1$. These properties are proven in [44, p. 100–102].

Proposition 5.6. — *For any bounded measurable set* ω*,*

$$\int_\omega 1 = |\omega|. \blacksquare$$

Proof. By Definition 2.6 of a measurable set, there exists a sequence of open sets $(\omega_n)_{n\in\mathbb{N}}$ containing ω such that $|\omega_n| \to |\omega|$. By Definition 5.1 of the integral, together with the calculation of the measure of an open set (Proposition 1.25),

$$\int_\omega 1 = \lim_{n\to\infty} \int_{\omega_n} 1 = \lim_{n\to\infty} |\omega_n| = |\omega|. \ \square$$

Generalization of Proposition 5.6. It extends to the case $|\omega| < \infty$, see Theorem 6.20. □

5.3. Positive measures, real-valued integrals

In order to state the monotonicity properties of the real-valued integral, let us first define positive real measures.

Definition 5.7. — *A measure* $f \in \mathcal{M}(\Omega)$*, where* Ω *is an open subset of* $\mathbb{R}^d$*, is said to be* ***positive****, and we write* $f \geq 0$*, if, for every* $\varphi \in \mathcal{K}(\Omega)$ *such that* $\varphi \geq 0$*,*

$$\langle f, \varphi \rangle \geq 0.$$

(That is, if this holds for every $\varphi \in \mathcal{K}(\Omega)$ *such that* $\varphi(x) \geq 0$ *for all* $x \in \Omega$*.)*

We then write $f \geq g$*, or* $g \leq f$ *if* $f - g \geq 0$*.* ▮

Let us verify that this definition is compatible with that of a positive function.

Theorem 5.8. — *If* $f \in \mathcal{C}(\Omega)$*, then positivity in the sense of measures given by Definition 5.7 coincides with that in the sense of functions.*

That is, with the identification (3.16), p. 65, of f *with a measure, Definition 5.7 is equivalent to:*

$$f(x) \geq 0, \ \ \forall x \in \Omega. \blacksquare$$

Proof. Here, let $\overline{f}$ denote the measure associated to the function f by Theorem 3.10, and $\succeq$ the positivity in the sense of distributions. It suffices to verify that $\overline{f} \succeq 0$ is equivalent to $f \geq 0$.

1º) *Forward implication.* Suppose

$$\overline{f} \succeq 0.$$

If f were not positive, there would exist $x \in \Omega$ such that $f(x) = a < 0$, and hence there would exist $r > 0$ such that the open ball $\mathring{B} = \{y \in \mathbb{R}^d : |y - x| < r\}$ is included in Ω by continuity, and

$$f \leq \frac{a}{2} \ \text{ in } \mathring{B}.$$

Let $\varphi \in \mathcal{K}(\Omega)$ have its support in $\mathring{B}$ such that

$$\varphi \geq 0, \quad \varphi \neq 0.$$

Then the definition of $\overline{f}$ from Theorem 3.10 would give, together with the properties of the real-valued Cauchy integral from Theorems 1.16 and 1.18 (a) and (c),

$$\langle \overline{f}, \varphi \rangle = \int_\Omega f\varphi = \int_{\mathring{B}} f\varphi \leq \frac{a}{2} \int_{\mathring{B}} \varphi < 0,$$

which would contradict $\overline{f} \succeq 0$. Thus,

$$f \geq 0.$$

2º) *Backward implication.* Now, suppose that

$$f \geq 0.$$

Then, for every $\varphi \in \mathcal{K}(\Omega)$ such that $\varphi \geq 0$, we have $f\varphi \geq 0$, whence, again according to Theorem 3.10 and Proposition 1.18 (a),

$$\langle \overline{f}, \varphi \rangle = \int_\Omega f\varphi \geq 0.$$

Thus,

$$\overline{f} \succeq 0. \ \square$$

Let us show that the L^1 norm of a positive integrable measure is its integral.

Lemma 5.9. — *Let $f \in L^1(\Omega)$ such that $f \geq 0$, where Ω is an open subset of $\mathbb{R}^d$. Then,*

$$\|f\|_{L^1(\Omega)} = \int_\Omega f. \ \blacksquare$$

Proof. The regularizations $R_n f$ form an integrabilizing sequence for f (Theorem 4.2). From Definition 3.17, since α_n and ρ_n are positive (Definition 3.16), we get

$$R_n f(x) = \langle f, \alpha_n \tau_x \rho_n \rangle \geq 0. \tag{5.15}$$

Definitions 4.4 of the semi-norms of $L^1(\Omega; \mathbb{R})$ and 5.1 of the integral therefore yield

$$\|f\|_{L^1(\Omega)} = \lim_{n\to\infty} \int_\Omega |R_n f| = \lim_{n\to\infty} \int_\Omega R_n f = \int_\Omega f. \ \square$$

Let us show that the integral of an integrable real measure increases with the measure[3].

Theorem 5.10. — *Let $f \in L^1(\Omega)$ and $g \in L^1(\Omega)$, where Ω is an open subset of $\mathbb{R}^d$, and let ω be a measurable subset of Ω. Then:*

(a) $$f \geq g \quad \Rightarrow \quad \int_\omega f \geq \int_\omega g;$$

(b) $$f \neq g,\ f \geq g \quad \Rightarrow \quad \int_\Omega f > \int_\Omega g. \ ∎$$

Proof. ***1°) Property (a).*** Let $R_n f$ be a regularization for f (Definition 3.17) and ω_n an approximation of ω given by Definition 5.1 of the integral.

If $f \geq g$, then $f - g$ is positive, so $R_n f$ is also positive (as established in (5.15)). Thus, the monotonicity of the real-valued Cauchy integral (Proposition 1.18 (a)) gives

$$\int_{\omega_n} R_n(f - g) \geq 0.$$

By passing to the limit as $n \to \infty$, Definition 5.1 gives, since the $R_n f$ form an integrabilizing sequence for f (Theorem 4.2),

$$\int_\omega f - g \geq 0.$$

2°) Property (b). If $f \geq g$, then $f - g \geq 0$, so (Lemma 5.9)

$$\int_\Omega f - g = \|f - g\|_{L^1(\Omega)}.$$

This is strictly positive if $f \neq g$, since $L^1(\Omega)$ is normed (because it has a unique semi-norm and is separated, according to Theorem 4.6). $\square$

---

3 **Historical background of the monotonicity of the integral.** Positivity, which forms the basis of Theorem 5.10 (a), was one of the six conditions required by Henri LEBESGUE for his definition of the integral given in 1904 [44, p. 126 and 122], see Note 2, p. 96.

Let us show that the integral of a real-valued positive integrable measure increases with the domain of integration. (Another growth property is given in Theorem 6.3.)

Theorem 5.11. — *Let $f \in L^1(\Omega)$ such that $f \geq 0$, where Ω is an open subset of $\mathbb{R}^d$, and let ω_1 and ω_2 be two measurable subsets of Ω. Then,*

$$\omega_1 \subset \omega_2 \quad \Rightarrow \quad \int_{\omega_1} f \leq \int_{\omega_2} f. \blacksquare$$

Proof. Let $R_n f$ be a regularization of f (Definition 3.17), and let ω_1^n and ω_2^n be approximations of ω_1 and ω_2 given by Definition 5.1 of the integral.

Then, the $\lambda_1^n = \omega_1^n \cap \omega_2^n$ are also approximations for ω_1 in the sense of Definition 5.1, because (like ω_1^n and ω_2^n) they are open, decrease with n, are contained between ω and Ω and, finally, by the monotonicity of the measure (Theorem 2.10),

$$|\lambda_1^n \setminus \omega_1| \leq |\omega_n^1 \setminus \omega_1| \to 0 \text{ when } n \to \infty.$$

Furthermore, $\lambda_1^n \subset \omega_2^n$, so the monotonicty of the Cauchy integral of a positive function relative to the open set (Proposition 1.18 (b)) gives, since $R_n f \geq 0$ (as established in (5.15)),

$$\int_{\lambda_1^n} R_n f \leq \int_{\omega_2^n} R_n f.$$

Taking the limit as $n \to \infty$, Definition 5.1 gives, since the $R_n f$ form an integrabilizing sequence for f (Theorem 4.2),

$$\int_{\omega_1} f \leq \int_{\omega_2} f. \ \Box$$

5.4. Examples of value spaces

The following spaces are Neumann spaces, which makes it possible, for instance in evolution problems, to use spaces of the form $L^1((0,T); E)$ and integrals $\int_\omega f$ of their elements f, where E is one of them.

— The Lebesgue spaces $L^p(\Omega)$ (Theorem 10.16), for $1 \leq p \leq \infty$.

— The spaces $L^p(\Omega)$-weak, for $1 < p < \infty$ (Theorem 16.2).

— The space $L^\infty(\Omega)$-$*$weak (Theorem 16.17).

— The local spaces $L^p_{\text{loc}}(\Omega)$ (Theorem 11.6), for $1 \leq p \leq \infty$.

— The Sobolev spaces $W^{m,p}(\Omega)$, for $m \in \mathbb{Z}$ and $1 \leq p \leq \infty$ [Vol. 5].

— The spaces $W^{m,p}(\Omega)$-weak, for $1 < p < \infty$ and $m \in \mathbb{Z}$ [Vol. 5].

— The local spaces $W^{m,p}_{\text{loc}}(\Omega)$, for $m \in \mathbb{Z}$ and $1 \leq p \leq \infty$ [Vol. 5].

— The Beppo Levi spaces $\dot{W}_0^{m,p}(\Omega)$, if Ω^c contains a ball for $m \geq 0$, $m \in \mathbb{Z}$, and $1 \leq p \leq \infty$ [Vol. 5].

— The space of measures $\mathcal{M}(\Omega)$ (Theorem 3.8).

— The spaces of continuously differentiable functions $\mathcal{C}^m(\Omega)$, $\mathcal{C}_b(\Omega)$, $\mathbf{C}_b(\Omega)$, $\mathcal{C}^m_K(\Omega)$, and $\mathcal{K}^m(\Omega)$, for $m \in \mathbb{N}$ or $m = \infty$ [Vol. 2, Theorem 2.24].

— The space of distributions $\mathcal{D}'(\Omega)$ [Vol. 3, Theorem 4.5].

— The space of test functions $\mathcal{D}(\Omega)$ [Vol. 3, Theorem 2.8].

The following spaces are also Neumann spaces.

— Banach, Fréchet, and Hilbert spaces (Definitions 1.5 and A.30).

— The space E-weak, where E is either a Hilbert, or reflexive, or semi-reflexive space [Vol. 1, Theorems 17.7 and 17.12].

— The duals E', E'', E''', ... of a normed or semi-normed metrizable space E [Vol. 1, Theorem 13.8 (a) and Property (17.3), p. 248].

— The weak duals E'-weak, E''-weak, E'''-weak, ... of either a Hilbert or reflexive space E [Vol. 1, Property (17.6), p. 248].

— The weak$*$ duals E'-$*$weak, E''-$*$weak, E'''-$*$weak, ... of either a Hilbert, or Banach, or Fréchet, or reflexive space [Vol. 1, Theorem 13.8 (b) and Properties (17.4) and (17.6)].

— All of their closed or sequentially closed subspaces [Vol. 1, Theorem 4.12].

When it is not possible to integrate within a separated semi-normed space that is not a Neumann space, such a space can be embedded into a Neumann space, for example its sequential completion (Definition 5.12), in the following manner.

— One can immerse $L^1(\Omega)$-weak into $\mathcal{M}(\Omega)$ (Theorem 16.13) which is a Neumann space. (If $1 < p < \infty$, this immersion is not useful for $L^p(\Omega)$-weak since it is a Neumann space.)

— One can immerse $L^\infty(\Omega)$-weak into $L^\infty(\Omega)$-$*$weak, which is a Neumann space.

— One can immerse E'-weak into E'-$*$weak when E is a non-reflexive either Banach or Fréchet space, since E'-weak $\underset{\neq}{\subset}$ E'-$*$weak [Vol. 1, Theorem 13.5]. (If E is reflexive, this immersion is not useful since E'-weak is a Neumann space).

Complete spaces. Any (separated semi-normed) complete space is a Neumann space. We do not use this notion of *completeness* for a space, as it is less straightforward (apart from its name!) and less general than *sequential completeness*, see the comment *Completeness versus sequential completeness*, p. 6. □

5.5. The case where E is not a Neumann space

We aim to show that it is not desirable to define the integral of functions or measures f with values in a separated semi-normed space E which is not a Neumann space, since then $\int_\Omega f$ would not always belong to E.

To be able to state this precisely, recall that every separated semi-normed space E possesses [Vol. 1, Theorem 4.24] a sequential completion.

Definition 5.12. — *The* ***sequential completion*** $\widehat{E}$ *of a semi-normed space* E *is:*

— *a Neumann space in which* E *is included and dense;*

— *whose semi-norms extend those of* E *(or equivalent semi-norms);*

— *which has no proper sequentially closed vector subspace containing* E. ∎

Every space has infinitely many sequential completions, but one can pass from one to another via a bijection that is sequentially continuous, in addition to its inverse; they are all included in the *completions*. Recall that E is sequentially dense in $\widehat{E}$ if E is metrizable [Vol. 1, Property (4.10), p. 63], but this is not always the case.

Now let us extend the definition of the Cauchy integral to a separated semi-normed space E, not necessarily sequentially complete.

$$\begin{cases} \text{Given } f \in \mathcal{B}(\Omega; E)\text{, its } \textbf{completed integral } \widehat{\int_\omega} f \text{ is the limit} \\ \text{in } \widehat{E} \text{ of the approximated integrals } \mathbb{S}^n_\omega f \text{ from Definition 1.14.} \end{cases} \tag{5.16}$$

This definition makes sense, since $(\mathbb{S}^n_\omega f)_{n \in \mathbb{N}}$ is a Cauchy sequence in E (the proof of (1.7), p. 13, does not rely on the sequential completion of E), and hence also in $\widehat{E}$, so it also converges there. **It is not ideal** to denote it by $\int_\omega f$, nor to call it "the integral of f", since it does not necessarily belong to E and furthermore $\widehat{E}$ is not unique.

We are now in the position to state a precise result, as follows.

Theorem 5.13. — *If* E *is a metrizable separated semi-normed space that is not sequentially complete, and if* Ω *is a non-empty open subset of* $\mathbb{R}^d$, *there exists* f *in* $\mathcal{K}(\Omega; E)$ *such that*

$$\widehat{\int_\Omega} f \in \widehat{E} \setminus E. \ ∎$$

Proof. **1°)** *Construction of* f. Since E is not sequentially complete, it possesses a Cauchy sequence $(u_n)_{n \in \mathbb{N}}$ that does not converge. This sequence is also Cauchy in $\widehat{E}$, which is sequentially complete, and therefore has a limit u in it. Thus,

$$u_n \to u \text{ in } \widehat{E}, \quad u \in \widehat{E} \setminus E. \tag{5.17}$$

Let $B(z, r)$ be a closed ball included in Ω, and let $(B_n)_{n \in \mathbb{N}}$ a sequence of pairwise disjoint balls included in $B(z, r)$ such that $B_n = B(z_n, r_n)$, with

$$r_n > 0, \quad r_n \to 0, \quad z_n \to z.$$

Since E is metrizable, it can (Proposition A.26) be endowed with an increasing sequence of semi-norms, say $(\| \|_m)_{m\in\mathbb{N}}$. Let $(\||\ \||_m)_{m\in\mathbb{N}}$ denote the semi-norms of $\widehat{E}$ that extend those of E.

After selecting a subsequence of u_n if required, we can assume that, for $n \geq 1$,

$$\|u_n - u_{n-1}\|_n \leq \frac{1}{n}|r_n|^d; \tag{5.18}$$

Indeed, it suffices to consider a subsequence $(u_{\sigma(n)})_{n\in\mathbb{N}}$ such that, for each n,

$$\||u_{\sigma(n)} - u\||_n \leq \frac{1}{2}\inf\Big\{\frac{1}{n}|r_n|^d, \frac{1}{n+1}|r_{n+1}|^d\Big\}.$$

Finally, let $\phi \in \mathcal{C}(\mathbb{R}^d)$ be such that

$$\operatorname{supp}\phi \subset \{x \in \mathbb{R}^d : |x| < 1\} \text{ and } \int_{\mathbb{R}^d}\phi = 1. \tag{5.19}$$

Define f on Ω by

$$f(x) = \begin{cases} |r_0|^{-d}\phi((x-z_0)/r_0)\, u_0 & \text{if } x \in B_0, \\ |r_n|^{-d}\phi((x-z_n)/r_n)(u_n - u_{n-1}) & \text{if } x \in B_n,\ n \geq 1, \\ 0_E & \text{if } x \notin \bigcup_{n\geq 0} B_n. \end{cases} \tag{5.20}$$

This function is continuous away from the point z. It also is continuous at z, since $f(z) = 0_E$ and (5.18) and (5.20) give, when $n \geq m$,

$$\sup_{x\in B_n}\|f(x)\|_m \leq \sup_{x\in B_n}\|f(x)\|_n \leq \frac{1}{n}\sup_{y\in\mathbb{R}^d}|\phi(y)|,$$

which tends to 0 as $n \to \infty$.

Therefore, f is continuous on all of Ω, and its support is included in $B(z, r)$.

2°) Integral of f. Let us denote by $\widehat{f}$ the same function whose values $\widehat{f}(x) = f(x)$ are considered to be elements of $\widehat{E}$. The latter is a Neumann space according to Definition 5.12, which makes it possible to use the properties of the integral with values in such a space.

First, using the translation and homothety properties of the integral (Proposition 1.27 and Theorem 7.23), (5.20) and (5.19) yield

$$\begin{aligned} \int_{\mathring{B}_n} \widehat{f} = |r_n|^{-d}\Big(\int_{\mathring{B}(z_n,r_n)} \phi\Big(\frac{x-z_n}{r_n}\Big)\,\mathrm{d}x\Big)(u_n - u_{n-1}) = \\ = \Big(\int_{\mathring{B}(0,1)}\phi\Big)(u_n - u_{n-1}) = u_n - u_{n-1} \end{aligned} \tag{5.21}$$

and similarly,

$$\int_{\mathring{B}_0} \widehat{f} = u_0.$$

Then, after setting $\omega_n = \bigcup_{0 \le i \le n} B_i$, the additivity of the integral with respect to disjoint open subsets (Theorem 6.1 (b)) yields

$$\int_{\omega_n} \widehat{f} = \sum_{0 \le i \le n} \int_{\mathring{B}_i} \widehat{f} = u_n. \tag{5.22}$$

Furthermore, after setting $\omega = \bigcup_{i \in \mathbb{N}} \mathring{B}_i$, the continuity of the integral with respect to increasing open sets (Theorem 6.5) yields, as $n \to \infty$,

$$\int_{\omega_n} \widehat{f} \to \int_{\omega} \widehat{f}. \tag{5.23}$$

With (5.17) and (5.22), we get

$$\int_{\omega} \widehat{f} = u.$$

Finally, since the domain where a function vanishes does not affect the integral (Proposition 1.23), we can replace ω with Ω. Definition (5.16) of the completed integral then gives

$$\widehat{\int_{\Omega}} f = \int_{\Omega} \widehat{f} = u.$$

Since u does not belong to E by (5.17), this proves that

$$\widehat{\int_{\Omega}} f \notin E. \ \square$$

To integrate a function taking values in a separated semi-normed space E which is not a Neumann space, **it suffices** to embed E into its sequential completion $\widehat{E}$ (Definition 5.12).

Thus $f \in \mathcal{K}(\Omega; E)$ can be identified with a function $\widehat{f} \in \mathcal{K}(\Omega; \widehat{E})$, for which one can define the integral $\int_{\Omega} \widehat{f}$, which plays the role expected of the integral of f without the ambiguity regarding the space where it takes its values, namely $\widehat{E}$.

Anticipation. We have used Theorems 7.23, 6.1, and 6.5 (to establish (5.21), (5.22), and (5.23)) concerning the homothety effect, the additivity and the continuity of the integral, even though they have not yet been proven. We did so to lighten the proof of Theorem 5.13; otherwise, we would have had to *contrive* or relocate these results. This has no fundamental consequence, since the proofs of these theorems do not rely on Theorem 5.13 (which we do not use anywhere). □

The case where E is not metrizable. Theorem 5.13 can be extended to certain non-metrizable spaces E [75]. □

Remarks on the methods for constructing $L^1(\Omega; E)$. The reader is referred to the comments at the end of § 10.2, where the more general case involving $L^p(\Omega; E)$ spaces will be considered. □

Chapter 6

Properties of the Integral

In this chapter, we demonstrate that the integral is additive with respect to disjoint domains (Theorem 6.1), that it depends continuously on the domain of integration (§ 6.2), that the contribution from negligible sets is zero (§ 6.3), and that it commutes with continuous linear mappings (Theorem 6.13) and differentiation (Theorem 6.21).

We also show that the domain where the function vanishes does not contribute to its integral (§ 6.6).

6.1. Additivity with respect to the domain of integration

Let us show that the integral is additive with respect to disjoint measurable sets [1].

Theorem 6.1. — *Let $f \in L^1(\Omega; E)$, where Ω is an open subset of $\mathbb{R}^d$ and E is a Neumann space, and let ω_1 and ω_2 be two measurable subsets of Ω. Then:*

(a)
$$\int_{\omega_1 \cup \omega_2} f + \int_{\omega_1 \cap \omega_2} f = \int_{\omega_1} f + \int_{\omega_2} f;$$

(b)
$$\omega_1 \cap \omega_2 = \emptyset \quad \Rightarrow \quad \int_{\omega_1 \cup \omega_2} f = \int_{\omega_1} f + \int_{\omega_2} f;$$

(c)
$$\omega_2 \subset \omega_1 \quad \Rightarrow \quad \int_{\omega_1 \setminus \omega_2} f = \int_{\omega_1} f - \int_{\omega_2} f. \ \blacksquare$$

1 **Historical background of the additivity with respect to the domain of integration.** Additivity with respect to disjoint intervals was one of the six conditions required by Henri LEBESGUE for his definition of the integral in one dimension, given in 1904 [44, p. 105 and 122].

Proof. **1°)** *Property* (a). Here, $f \in L^1(\Omega; E)$, and ω_1 and ω_2 are measurable.

1.a. *Reduction of the problem.* Let $(\omega_1^n)_{n\in\mathbb{N}}$ be an approximating sequence for ω_1 given by Definition 5.1 of the integral, that is: ω_1^n is open and decreasing with n, $\omega_1 \subset \omega_1^n \subset \Omega$ and, as $n \to \infty$,

$$|\omega_1^n \setminus \omega_1| \to 0.$$

Additionally, let $(\omega_2^n)_{n\in\mathbb{N}}$ be an approximating sequence for ω_2.

Then, $(\omega_1^n \cup \omega_2^n)_{n\in\mathbb{N}}$ is an approximating sequence for $\omega_1 \cup \omega_2$. Indeed:

— $\omega_1^n \cup \omega_2^n$ is open, decreasing with n, and is contained between $\omega_1 \cup \omega_2$ and Ω;

— $(\omega_1^n \cup \omega_2^n) \setminus (\omega_1 \cup \omega_2)$ is included in $(\omega_1^n \setminus \omega_1) \cup (\omega_2^n \setminus \omega_2)$, so its measure is bounded (Theorem 2.13 (b)) by $|\omega_1^n \setminus \omega_1| + |\omega_2^n \setminus \omega_2|$, and thus

$$|(\omega_1^n \cup \omega_2^n) \setminus (\omega_1 \cup \omega_2)| \to 0 \text{ when } n \to \infty.$$

Similarly, $(\omega_1^n \cap \omega_2^n)_{n\in\mathbb{N}}$ is an approximating sequence for $\omega_1 \cap \omega_2$.

Finally, let $(f_n)_{n\in\mathbb{N}}$ be an integrabilizing sequence for f. Let us assume for the moment that

$$\int_{\omega_1^n \cup \omega_2^n} f_n + \int_{\omega_1^n \cap \omega_2^n} f_n = \int_{\omega_1^n} f_n + \int_{\omega_2^n} f_n. \tag{6.1}$$

Then, by Definition 5.1 of the integral, as $n \to \infty$, we obtain the equality (a).

Since $f_n \in \mathcal{B}(\Omega; E)$, it suffices therefore to prove the equality (a) when

$$f \in \mathcal{B}(\Omega; E), \text{ and } \omega_1 \text{ and } \omega_2 \text{ are open.} \tag{6.2}$$

Let us do this by *localizing* the equality (a), and then passing to the limit.

1.b. *Localization.* So let us assume (6.2), and let $(\alpha_1^n)_{n\in\mathbb{N}}$ be a sequence of functions such that, for all $n \in \mathbb{N}^*$,

$$\alpha_1^n \in \mathcal{C}(\mathbb{R}^d), \quad 0 \leq \alpha_1^n \leq 1, \quad \alpha_1^n = \begin{cases} 1 & \text{on } (\omega_1)_{1/n}, \\ 0 & \text{outside } \omega_1, \end{cases} \tag{6.3}$$

where $(\omega_1)_{1/n} = \{x \in \mathbb{R}^d : B(x, 1/n) \subset \omega_1\}$; also let $(\alpha_2^n)_{n\in\mathbb{N}}$ be a sequence of functions satisfying the analogous property with respect to ω_2.

Then,

$$\alpha_1^n + \alpha_2^n = \alpha_1^n \alpha_2^n + \beta^n, \quad \text{where } \beta^n = 1 - (1 - \alpha_1^n)(1 - \alpha_2^n),$$

whence

$$\int_{\omega_1 \cup \omega_2} \alpha_1^n f + \int_{\omega_1 \cup \omega_2} \alpha_2^n f = \int_{\omega_1 \cup \omega_2} \alpha_1^n \alpha_2^n f + \int_{\omega_1 \cup \omega_2} \beta^n f. \tag{6.4}$$

Since $\alpha_1^n f$ vanishes outside ω_1, we have (Proposition 1.23) $\int_{\omega_1 \cup \omega_2} \alpha_1^n f = \int_{\omega_1} \alpha_1^n f$. Similarly, $\alpha_2^n f$, $\alpha_1^n \alpha_2^n f$, and β^n vanish outside ω_2 and $\omega_1 \cap \omega_2$ respectively, so (6.4) yields

$$\int_{\omega_1} \alpha_1^n f + \int_{\omega_2} \alpha_2^n f = \int_{\omega_1 \cap \omega_2} \alpha_1^n \alpha_2^n f + \int_{\omega_1 \cup \omega_2} \beta^n f. \tag{6.5}$$

1.c. *Convergence.* We take the limit as $n \to \infty$ for all four terms in equality (6.5) using Lemma 6.2 below, noting that the localization functions involved all satisfy hypothesis (6.6) of said lemma. Indeed:

— for the first and second terms, this is hypothesis (6.3) and the analogous property for α_2^n;

— for the third term, observe that $\alpha_1^n \alpha_2^n$ equals 1 in $(\omega_1)_{1/n} \cap (\omega_2)_{1/n}$, which increases to $\omega_1 \cap \omega_2$, and equals 0 outside the latter;

— for the fourth term, observe that β_n equals 1 in $\omega_1^n \cup \omega_2^n$, which increases to $\omega_1 \cup \omega_2$, and equals 0 outside the latter.

Thus, according to Lemma 6.2, equality (6.5) yields, in the limit, equality (a). This, together with hypothesis (6.2), establishes (6.1) and therefore equality (a) in the general case.

2º) *Property* (b). Here, $\omega_1 \cap \omega_2 = \emptyset$. Equality (a) then reduces to equality (b) because Definition 5.1 of the integral with $\omega = \omega_n = \emptyset$ gives

$$\int_{\emptyset} f = 0_E.$$

3º) *Property* (c). Here, $\omega_2 \subset \omega_1$. The sets $\omega_1 \setminus \omega_2$ and ω_2 have ω_1 as their union, and they are disjoint and measurable (Theorem 2.11), so Property (b) gives

$$\int_{\omega_1} f = \int_{\omega_1 \setminus \omega_2} f + \int_{\omega_2} f,$$

and thus equality (c). □

It remains to prove the following result.

Lemma 6.2. — *Let $f \in \mathcal{B}(\Omega; E)$, where Ω is an open subset of $\mathbb{R}^d$, E is a Neumann space, ω is an open subset included in Ω, and $(\omega^n)_{n\in\mathbb{N}}$ is a sequence of open sets increasing with n, such that*

$$\omega = \bigcup_{n\in\mathbb{N}} \omega_n.$$

Furthermore, let $(\alpha^n)_{n\in\mathbb{N}}$ *be a sequence of functions such that, for all* $n \in \mathbb{N}$*:*

$$\alpha^n \in \mathcal{C}(\mathbb{R}^d), \quad 0 \le \alpha^n \le 1, \quad \alpha^n = \begin{cases} 1 & \text{on } \omega^n, \\ 0 & \text{outside } \omega. \end{cases} \tag{6.6}$$

Then, $\alpha^n f \in \mathcal{B}(\Omega; E)$ *and, as* $n \to \infty$,

$$\int_\omega \alpha^n f \to \int_\omega f. \blacksquare$$

Proof. First of all, observe that, since the domain where a uniformly continuous function vanishes does not contribute to its Cauchy integral (Proposition 1.23),

$$\int_\omega (\alpha^n f - f) = \int_{(\omega\setminus\overline{\omega_n})\cap\mathcal{O}} (\alpha_n f - f),$$

where $\mathcal{O}$ is a bounded open subset of $\mathbb{R}^d$ outside which f vanishes. Indeed, such an $\mathcal{O}$ exists by Definition 1.12 of $\mathcal{B}(\Omega; E)$ and, by hypothesis, α^n equals 1 in ω_n and hence in $\overline{\omega_n}$.

The bound on the semi-norms of the Cauchy integral from Proposition 1.21 therefore give, for every semi-norm $\| \ \|_{E;\nu}$ of E,

$$\Big\| \int_\omega (\alpha^n f - f) \Big\|_{E;\nu} \le |(\omega \setminus \overline{\omega_n}) \cap \mathcal{O}| \sup_{x\in\omega} \|(\alpha_n f - f)(x)\|_{E;\nu}.$$

As $n \to \infty$, the right-hand side tends to 0, whence we obtain the stated convergence, because:

— on the one hand, $|(\omega \setminus \overline{\omega_n}) \cap \mathcal{O}| \to 0$ according to Theorem 2.17, since the $(\omega \setminus \overline{\omega_n}) \cap \mathcal{O}$ decrease with n and their intersections are empty;

— on the other hand, $\sup_{x\in\omega} \|(\alpha_n f - f)(x)\|_{E;\nu} \le \sup_{x\in\omega} \|f(x)\|_{E;\nu}$ which is finite, since f is bounded (Proposition 1.13). □

Observe that the integral of a positive measure is subadditive with respect to the domain.

Theorem 6.3. — *Let* $f \in L^1(\Omega)$ *such that* $f \ge 0$, *where* Ω *is an open subset of* $\mathbb{R}^d$, *and* ω_1 *and* ω_2 *are two measurable subsets of* Ω. *Then,*

$$\int_{\omega_1\cup\omega_2} f \le \int_{\omega_1} f + \int_{\omega_2} f. \blacksquare$$

Proof. This follows from the equality in Theorem 6.1 (a) because, here (Theorem 5.10 (a)),

$$\int_{\omega_1\cap\omega_2} f \ge 0. \ \square$$

6.2. Continuity with respect to the domain of integration

Let us show that, if the measures of a sequence of sets tend to 0, then the same is true for the integrals over these sets.

Theorem 6.4. — *Let $f \in L^1(\Omega; E)$, where Ω is an open subset of $\mathbb{R}^d$, E is a Neumann space, and let $(\omega_n)_{n\in\mathbb{N}}$ be a sequence of measurable subsets of Ω such that, as $n \to \infty$,*

$$|\omega_n| \to 0.$$

Then,

$$\int_{\omega_n} f \to 0_E. \blacksquare$$

Prior to proving this property, let us state some of its consequences, starting with the continuity of the integral with respect to increasing or decreasing sequences of the domain of integration.

Theorem 6.5. — *Let $f \in L^1(\Omega; E)$, where Ω is an open subset of $\mathbb{R}^d$, E is a Neumann space, ω is a measurable set, and $(\omega_n)_{n\in\mathbb{N}}$ is a sequence of measurable subsets of Ω such that, as $n \to \infty$, either*

$$\omega_n \text{ increases with } n \text{ and } \omega = \bigcup_{n\in\mathbb{N}} \omega_n$$

or

$$\omega_n \text{ decreases with } n \text{ and } \omega = \bigcap_{n\in\mathbb{N}} \omega_n.$$

Then,

$$\int_{\omega_n} f \to \int_\omega f. \blacksquare$$

Approximating the integral using bounded sets. In particular, Theorem 6.5 shows that, for every $f \in L^1(\Omega; E)$ and measurable subset ω of Ω,

$$\int_\omega f = \lim_{n\to\infty} \int_{\omega\cap B(0,n)} f. \ \square$$

Let us state another consequence of Theorem 6.4: *uniform* convergence with respect to measure discrepancies of the domains of integration.

Theorem 6.6. — *Let $f \in L^1(\Omega; E)$, where Ω is an open subset of $\mathbb{R}^d$, E is a Neumann space, and let $(\omega_n)_{n\in\mathbb{N}}$ and $(\lambda_n)_{n\in\mathbb{N}}$ be two sequences of measurable subsets of Ω such that, as $n \to \infty$,*

$$|\omega_n \setminus \lambda_n| + |\lambda_n \setminus \omega_n| \to 0.$$

Then,

$$\int_{\omega_n} f - \int_{\lambda_n} f \to 0_E. \blacksquare$$

Let us now turn to the proofs.

Proof of Theorem 6.4. Here,

$$|\omega_n| \to 0.$$

1°) Convergence of the integral. Let $(f_k)_{n\in\mathbb{N}}$ be an integrabilizing sequence for f given by Definition 4.1 of an integrable measure. Decompose

$$\int_{\omega_n} f = \int_{\omega_n} (f - f_k) + \int_{\omega_n} f_k. \tag{6.7}$$

Let $\| \ \|_{E;\nu}$ be a semi-norm of E and $\epsilon > 0$. The sequence $(f_k)_{k\in\mathbb{N}}$ converges to f in $L^1(\Omega; E)$ (Theorem 4.10), so, according to the bound for the semi-norms of the integral from Theorem 5.4, we can choose k such that

$$\left\| \int_{\omega_n} (f - f_k) \right\|_{E;\nu} \leq \|f - f_k\|_{L^1(\Omega;E);\nu} \leq \frac{\epsilon}{2}. \tag{6.8}$$

Since $f_k \in \mathcal{B}(\Omega; E)$ (by definition of an integrabilizing sequence), we have, as we will verify in 2°,

$$\left\| \int_{\omega_n} f_k \right\|_{E;\nu} \leq |\omega_n| \sup_{x\in\Omega} \|f_k(x)\|_{E;\nu}. \tag{6.9}$$

As every function in $\mathcal{B}(\Omega; E)$ is bounded (Proposition 1.13),

$$m \stackrel{\text{def}}{=} \sup_{x\in\Omega} \|f_k(x)\|_{E;\nu} \text{ is finite.}$$

Since $|\omega_n|$ tends to 0 by hypothesis, there therefore exists n_1 such that, for all $n \geq n_1$,

$$|\omega_n| \leq \frac{\epsilon}{2m}.$$

Hence, the decomposition (6.7) and inequalities (6.8) and (6.9) yield

$$\left\| \int_{\omega_n} f \right\|_{E;\nu} \leq \epsilon.$$

Which proves that, as $n \to \infty$,

$$\int_{\omega_n} f \to 0_E.$$

2º) Verification of (6.9). Fix an index n, and let $(\lambda_\ell)_{\ell \in \mathbb{N}}$ be a decreasing sequence of open subsets (given by Lemma 2.19) such that $\omega_n \subset \lambda_\ell \subset \Omega$ and, as $\ell \to \infty$,

$$|\lambda_\ell \setminus \omega_n| \to 0.$$

According to the bound for the semi-norms of the Cauchy integral from Proposition 1.21,

$$\Big\| \int_{\lambda_\ell} f_k \Big\|_{E;\nu} \leq |\lambda_\ell| \sup_{x \in \Omega} \|f_k(x)\|_{E;\nu}. \tag{6.10}$$

From Definition 5.1 of the integral (with $(f_k)_\ell = f_k$), as $\ell \to \infty$,

$$\int_{\lambda_\ell} f_k \to \int_{\omega_n} f_k.$$

Furthermore, $|\lambda_\ell| \to |\omega_n|$ since $|\lambda_\ell| - |\omega_n| = |\lambda_\ell \setminus \omega_n|$ by the additivity of the measure (Theorem 2.13 (d)). Hence, in the limit, inequality (6.10) yields (6.9). □

Proof of Theorem 6.6. Here,

$$|\omega_n \setminus \lambda_n| + |\lambda_n \setminus \omega_n| \to 0.$$

The set ω_n is the union of the sets $\omega_n \setminus \lambda_n$ and $\omega_n \cap \lambda_n$, which are measurable (Theorem 2.11) and disjoint, hence the additivity of the integral with respect to disjoint sets (Theorem 6.1 (b)) yields

$$\int_{\omega_n} f = \int_{\omega_n \setminus \lambda_n} f + \int_{\omega_n \cap \lambda_n} f.$$

Similarly,

$$\int_{\lambda_n} f = \int_{\lambda_n \setminus \omega_n} f + \int_{\omega_n \cap \lambda_n} f.$$

By subtraction,

$$\int_{\omega_n} f - \int_{\lambda_n} f = \int_{\omega_n \setminus \lambda_n} f - \int_{\lambda_n \setminus \omega_n} f,$$

which tends to 0 as $n \to \infty$, as the two integrals on the right-hand side tend to 0_E according to Theorem 6.4, since $|\omega_n \setminus \lambda_n| \to 0$ and $|\lambda_n \setminus \omega_n| \to 0$. □

Proof of Theorem 6.5. **1º)** *First condition.* Here,

$$\omega_n \text{ increases with } n \text{ and } \omega \text{ is the union of the } \omega_n.$$

Theorem 6.6 with $\lambda_n = \omega$ therefore gives

$$\int_{\omega_n} f - \int_{\omega} f \to 0_E \tag{6.11}$$

since, on the one hand, $|\omega_n \setminus \omega| = |\emptyset| = 0$ and, on the other hand, the additivity of the measure (Theorem 2.13 (d)) and its continuity with respect to increasing sequences (Theorem 2.16) yield

$$|\omega \setminus \omega_n| = |\omega| - |\omega_n| \to 0.$$

2º) *Second condition.* Here,

$$\omega_n \text{ decreases with } n \text{ and } \omega \text{ is the intersection of the } \omega_n.$$

2.a. *A simple case.* If one of the ω_n (and thus all of the subsequent ones) has a finite measure, then (Theorem 2.17) $|\omega_n| \to |\omega|$, thus $|\omega_n \setminus \omega| \to 0$, and Theorem 6.6 again gives (6.11).

However, if none of the ω_n has finite measure, then $|\omega_n|$ does not necessarily tend to $|\omega|$ (see (2.30), p. 42) and one must proceed in an alternative way.

2.b. *The general case.* Here $\omega \subset \omega_n$, so (Theorem 6.1 (c))

$$\int_{\omega_n} f - \int_{\omega} f = \int_{\omega_n \setminus \omega} f. \tag{6.12}$$

Let $(f_k)_{k \in \mathbb{N}}$ be an integrabilizing sequence (Definition 4.1) for f, and decompose

$$\int_{\omega_n \setminus \omega} f = \int_{\omega_n \setminus \omega} (f - f_k) + \int_{\omega_n \setminus \omega} f_k. \tag{6.13}$$

Let $\| \ \|_{E;\nu}$ be a semi-norm of E and $\epsilon > 0$. The sequence $(f_k)_{k \in \mathbb{N}}$ converges to f in $L^1(\Omega; E)$ (Theorem 4.10) so, according to the bound for the semi-norms of the integral from Theorem 5.4, we can choose k such that

$$\Big\| \int_{\omega_n \setminus \omega} (f - f_k) \Big\|_{E;\nu} \leq \|f - f_k\|_{E;\nu} \leq \frac{\epsilon}{2}. \tag{6.14}$$

By Definition 4.1 of an integrabilizing sequence, f_k is a function that vanishes outside a bounded set B (that one can choose to be open and thus measurable), so only this set contributes to the Cauchy integral (Proposition 1.23). The bound (6.9), p. 110, for the semi-norms of the integral (relative to $(\omega_n \setminus \omega) \cap B$ in the place of ω_n) can then be written as

$$\Big\| \int_{\omega_n \setminus \omega} f_k \Big\|_{E;\nu} = \Big\| \int_{(\omega_n \setminus \omega) \cap B} f_k \Big\|_{E;\nu} \leq |(\omega_n \setminus \omega) \cap B| \sup_{x \in \Omega} \|f_k(x)\|_{E;\nu}. \quad (6.15)$$

As every function in $\mathcal{B}(\Omega; E)$ is bounded (Proposition 1.13),

$$m \stackrel{\text{def}}{=} \sup_{x \in \Omega} \|f_k(x)\|_{E;\nu} \text{ is finite.}$$

The $(\omega_n \setminus \omega) \cap B$ form a decreasing sequence of measurable sets with empty intersection. Since the measure of each one is finite (as for any bounded set, see Theorem 2.9), we can therefore use Theorem 2.17 on the monotonicity of the measure with respect to such sequences; it gives

$$\big|(\omega_n \setminus \omega) \cap B\big| \to |\emptyset| = 0.$$

Therefore, there exists n_1 such that, for every $n \geq n_1$,

$$|(\omega_n \setminus \omega) \cap B| \leq \frac{\epsilon}{2m}.$$

Thus, the decompositions (6.12) and (6.13), together with inequalities (6.14) and (6.15), give

$$\Big\| \int_{\omega_n} f - \int_{\omega} f \Big\|_{E;\nu} \leq \epsilon.$$

Which again yields the convergence (6.11). □

6.3. Contribution of negligible sets

Let us show that the integral over a negligible set is zero.

Theorem 6.7. — *Let $f \in L^1(\Omega; E)$, where Ω is an open subset of $\mathbb{R}^d$ and E is a Neumann space, and let σ be a negligible subset of Ω. Then,*

$$\int_{\sigma} f = 0_E. \blacksquare$$

Proof. By Definition 2.22 of a negligible set, for all $n \in \mathbb{N}$, there exists an open set ω_n containing σ such that $|\omega_n| \leq 1/n$. Thus, according to Theorem 6.4, as $n \to \infty$,

$$\int_{\omega_n} f \to 0_E.$$

Since the measure is increasing (Proposition 2.2), $|\omega_n \backslash \sigma| \leq |\omega_n| \leq 1/n$, and, by Theorem 6.6,

$$\int_{\omega_n} f \to \int_\sigma f.$$

The uniqueness of the limit (Proposition A.12) thus gives the stated equality. □

Let us observe that the integral does not change when a negligible set is added to or removed from the domain of integration.

Theorem 6.8. — *Let $f \in L^1(\Omega; E)$, where Ω is an open subset of $\mathbb{R}^d$, E is a Neumann space, ω is a measurable set, and σ is a negligible subset of Ω. Then,*

$$\int_{\omega \cup \sigma} f = \int_\omega f = \int_{\omega \backslash \sigma} f. \blacksquare$$

Proof. The set $\sigma \backslash \omega$ is negligible, as any subset of a negligible set is negligible (Theorem 2.24 (a)). Therefore, by Theorem 6.7,

$$\int_{\sigma \backslash \omega} f = 0_E.$$

The additivity of the integral with respect to disjoint sets (Theorem 6.1 (b)) thus gives, since $\omega \cup \sigma = \omega \cup (\sigma \backslash \omega)$,

$$\int_{\omega \cup \sigma} f = \int_\omega + f \int_{\sigma \backslash \omega} f = \int_\omega f.$$

The second stated equality then follows from $\omega = (\omega \backslash \sigma) \cup \sigma$. □

6.4. Image of a measure under a linear mapping

Let us define the image of a measure under a linear mapping.

Definition 6.9. — *Let $f \in \mathcal{M}(\Omega; E)$, where Ω is an open subset of $\mathbb{R}^d$, E is a Neumann space, and let L be a continuous linear mapping from E into a Neumann space F. We define the **image** $Lf \in \mathcal{M}(\Omega; F)$ by: for all $\varphi \in \mathcal{K}(\Omega)$,*

$$\langle Lf, \varphi \rangle \stackrel{\text{def}}{=} L(\langle f, \varphi \rangle). \blacksquare$$

Justification. By Definition 3.2 of a measure, f is continuous linear from $\mathcal{K}(\Omega)$ into E. The mapping $Lf = L \circ f$ is therefore continuous, as is any composition of continuous mappings (Proposition A.39); it therefore is indeed a measure. □

Let us show that the mapping $f \mapsto Lf$ is continuous on the space of measures.

Theorem 6.10. — *Let L be a continuous linear mapping from a Neumann space E into a Neumann space F, and let Ω be an open subset of $\mathbb{R}^d$.*

Then, the mapping $f \mapsto Lf$ is continuous linear, and therefore sequentially continuous, from $\mathcal{M}(\Omega; E)$ into $\mathcal{M}(\Omega; F)$.

If L is injective, then the mapping $f \mapsto Lf$ is also injective. ▮

Proof. ***1°) Linearity.*** Given two measures f_1 and f_2, Definitions 3.2 of a measure and 6.9 of Lf, together with the linearity of L give, for every $\varphi \in \mathcal{K}(\Omega)$,

$$\langle L(f_1 + f_2), \varphi\rangle = L(\langle f_1 + f_2, \varphi\rangle) = L(\langle f_1, \varphi\rangle) + L(\langle f_2, \varphi\rangle) = \langle Lf_1, \varphi\rangle + \langle Lf_2, \varphi\rangle$$

and, for all $t \in \mathbb{R}$,

$$\langle L(tf), \varphi\rangle = L(\langle tf, \varphi\rangle) = tL(\langle f, \varphi\rangle) = \langle tLf, \varphi\rangle.$$

This proves that the mapping $f \mapsto Lf$ is linear.

2°) Continuity. According to the characterization of continuous linear mappings given in Proposition 1.9, for every semi-norm $\| \ \|_{F;\mu}$ of F, there exist $c \in \mathbb{R}$ and a finite set $\{\| \ \|_{E;\nu} : \nu \in N\}$ of semi-norms of E such that, for all $u \in E$,

$$\|Lu\|_{F;\mu} \leq c \sup_{\nu \in N} \|u\|_{E;\nu}.$$

Definition 3.2 of the semi-norms of $\mathcal{M}(\Omega; F)$ therefore gives, for all $\varphi \in \mathcal{K}(\Omega)$,

$$\|Lf\|_{\mathcal{M}(\Omega;F);\varphi,\mu} = \|\langle Lf, \varphi\rangle\|_{F;\mu} \leq c \sup_{\nu \in N} \|\langle f, \varphi\rangle\|_{E;\nu} = c \sup_{\nu \in N} \|f\|_{\mathcal{M}(\Omega;E);\varphi,\nu}.$$

This implies the continuity of the mapping $f \mapsto Lf$ from $\mathcal{M}(\Omega; E)$ into $\mathcal{M}(\Omega; F)$, again by the characterization from Proposition 1.9.

3°) Sequential continuity. This is always follows from continuity (Proposition 1.8).

4°) Injectivity. If L is injective and $Lf = Lg$, then, for all $\varphi \in \mathcal{K}(\Omega)$,

$$\langle Lf - Lg, \varphi\rangle = L(\langle f - g, \varphi\rangle) = 0_E.$$

Thus $\langle f - g, \varphi\rangle$, hence

$$f = g. \ \square$$

6.5. Image under a linear mapping

Let us show that the image of an integrable measure under a continuous linear mapping is integrable.

Theorem 6.11. — *Let $f \in L^1(\Omega; E)$ and $L \in \mathcal{L}(E; F)$, where Ω is an open subset of $\mathbb{R}^d$ and E and F are Neumann spaces. Then,*

$$Lf \in L^1(\Omega; F).$$

If $(f_n)_{n\in\mathbb{N}}$ is an integrabilizing sequence for f, then $(Lf_n)_{n\in\mathbb{N}}$ is an integrabilizing sequence for Lf. ∎

Let us complete this result with a bound for the semi-norms.

Theorem 6.12. — *Let $L \in \mathcal{L}(E; F)$, where E and F are Neumann spaces, $\| \ \|_{F;\mu}$ is a semi-norm of F, $\{\| \ \|_{E;\nu} : \nu \in N\}$ is a finite set of semi-norms of E, and $c \in \mathbb{R}$ such that, for every $u \in E$,*

$$\|Lu\|_{F;\mu} \leq c \sup_{\nu\in N} \|u\|_{E;\nu}. \tag{6.16}$$

Then, for every $f \in L^1(\Omega; E)$,

$$\|Lf\|_{L^1(\Omega;F);\mu} \leq c \sup_{\nu\in N} \|f\|_{L^1(\Omega;E);\nu}. \ ∎$$

Proofs of Theorems 6.11 and 6.12. **1°)** *Preliminary.* According to the characterization of continuous linear mappings from Proposition 1.9, for every semi-norm of F, there exist c and N such that (6.16) is satisfied.

2°) *integrabilizing sequence for Lf.* Let $f \in L^1(\Omega; E)$, and let $(f_n)_{n\in\mathbb{N}}$ be an integrabilizing sequence for f, that is, by Definition 4.1, such that f_n is uniformly continuous and vanishes outside a bounded set and, as $n \to \infty$, $f_n \to f$ in $\mathcal{M}(\Omega; E)$ and, for every semi-norm of E,

$$\sup_{m\geq n} \int_\Omega \|f_m - f_n\|_{E;\nu} \to 0. \tag{6.17}$$

Let us verify that $(Lf_n)_{n\in\mathbb{N}}$ satisfies the three criteria of an integrabilizing sequence for Lf.

2.a. Each Lf_n is uniformly continuous, as is any composition of uniformly continuous mappings (Proposition A.39), and vanishes outside the same bounded set as f_n.

2.b. The mapping $f \mapsto Lf$ is continuous from $\mathcal{M}(\Omega; E)$ into $\mathcal{M}(\Omega; F)$ (Theorem 6.10), and thus sequentially continuous (Proposition 1.8). The convergence $f_n \to f$ in $\mathcal{M}(\Omega; E)$ therefore gives

$$Lf_n \to Lf \ \text{ in } \mathcal{M}(\Omega; F).$$

2.c. Inequality (6.16), the monotonicity of the real-valued Cauchy integral (Proposition 1.18 (a)), and convergence (6.17) give

$$\sup_{m \geq n} \int_\Omega \|Lf_m - Lf_n\|_{F;\mu} \leq c \sup_{m \geq n} \sup_{\nu \in N} \int_\Omega \|f_m - f_n\|_{E;\nu} \to 0.$$

3º) Integrability of Lf. By Definition 4.1 of an integrable measures, 2º proves that

$$Lf \in L^1(\Omega; F).$$

4º) Semi-norms of Lf. Again, according to inequality (6.16) and the monotonicity of the real-valued Cauchy integral (Proposition 1.18 (a)),

$$\int_\Omega \|Lf_n\|_{F;\mu} \leq c \sup_{\nu \in N} \int_\Omega \|f_n\|_{E;\nu}.$$

Since $(Lf_n)_{n\in\mathbb{N}}$ and $(f_n)_{n\in\mathbb{N}}$ are integrabilizing sequences for Lf and f, Definition 4.4 of the semi-norms of $L^1(\Omega; F)$ and $L^1(\Omega; E)$ thus gives

$$\|Lf\|_{L^1(\Omega;F);\mu} = \lim_{n\to\infty} \int_\Omega \|Lf_n\|_{F;\mu} \leq$$

$$\leq c \sup_{\nu \in N} \lim_{n\to\infty} \int_\Omega \|f_n\|_{E;\nu} = c \sup_{\nu \in N} \|f\|_{L^1(\Omega;E);\nu}. \ \square$$

Let us show that the integral commutes with continuous linear mappings.

Theorem 6.13. — *Let $f \in L^1(\Omega; E)$ and $L \in \mathcal{L}(E; F)$, where Ω is an open subset of $\mathbb{R}^d$, E and F are Neumann spaces, and ω is a measurable subset of Ω. Then,*

$$\int_\omega Lf = L\Big(\int_\omega f\Big). \ \blacksquare$$

Proof. Let $(f_n)_{n\in\mathbb{N}}$ be an integrabilizing sequence for f, and $(\omega_n)_{n\in\mathbb{N}}$ be a sequence of open sets approximating ω given by Definition 5.1 of the integral. When $n \to \infty$,

$$\int_{\omega_n} f_n \to \int_\omega f.$$

Since the Cauchy integral commutes with continuous linear mappings (Proposition 1.19),

$$\int_{\omega_n} Lf_n = L\Big(\int_{\omega_n} f_n\Big).$$

Passing to the limit as $n \to \infty$, we obtain the stated equality since, on the one hand $(Lf_n)_{n\in\mathbb{N}}$ is an integrabilizing sequence for Lf (Theorem 6.11) and, on the other hand, L is sequentially continuous, as is any continuous mapping (Proposition 1.8). □

Let us show that the mapping $f \mapsto Lf$ is continuous linear.

Theorem 6.14. — *Let L be a continuous linear mapping from a Neumann space E into a Neumann space F, and let Ω be an open subset of $\mathbb{R}^d$.*

Then, the mapping $f \mapsto Lf$ is continuous, and thus sequentially continuous, linear from $L^1(\Omega; E)$ into $L^1(\Omega; F)$. ▮

Proof. ***1°) Linearity.*** Being linear on $\mathcal{M}(\Omega; E)$ by Theorem 6.10, it follows that the mapping $f \mapsto Lf$ is also linear on the vector subspace $L^1(\Omega; E)$.

2°) Continuity. According to the characterization of continuous linear mappings from Proposition 1.9, for every semi-norm $\|\ \|_{F;\mu}$ of F, there exist $c \in \mathbb{R}$ and a finite set $\{\|\ \|_{E;\nu} : \nu \in N\}$ of semi-norms of E such that, for every $u \in E$,

$$\|Lu\|_{F;\mu} \le c \sup_{\nu\in N} \|u\|_{E;\nu}.$$

Theorem 6.12 then gives, for every $f \in L^1(\Omega; E)$,

$$\|Lf\|_{L^1(\Omega;F);\mu} \le c \sup_{\nu\in N} \|f\|_{L^1(\Omega;E);\nu}.$$

So, again by the characterization of continuous linear mappings from Proposition 1.9, the mapping $f \mapsto Lf$ is continuous from $L^1(\Omega; E)$ into $L^1(\Omega; F)$.

3°) Sequential continuity. Continuity always implies sequential continuity (Proposition 1.8). □

When F is a topological inclusion, we get the following properties.

Theorem 6.15. — *Let Ω be an open subset of $\mathbb{R}^d$, and E and F be two Neumann spaces such that*

$$E \subsetneq F.$$

Then,

$$\mathcal{M}(\Omega; E) \subsetneq \mathcal{M}(\Omega; F),$$

$$L^1(\Omega; E) \subsetneq L^1(\Omega; F). \blacksquare$$

Proof. Topological inclusion is equivalent to the identity being continuous by Definitions 1.4 of a topological inclusion and 1.7 of continuity.

The hypothesis $E \subsetneq F$ is therefore equivalent to the continuity of the identity from E into F. This implies the continuity of the identity from $\mathcal{M}(\Omega; E)$ into $\mathcal{M}(\Omega; F)$ and from $L^1(\Omega; E)$ into $L^1(\Omega; F)$, according to Theorems 6.10 and 6.14. The continuity of these mappings is equivalent to the stated topological inclusions. □

6.6. Restriction and support

Let us define the restriction of a measure.

Definition 6.16. — *Let $f \in \mathcal{M}(\Omega; E)$, where Ω is an open subset of $\mathbb{R}^d$, E is a Neumann space, and let ω be an open set included in Ω.*

We define the ***restriction*** $f|_\omega \in \mathcal{M}(\omega; E)$ *by: for every $\varphi \in \mathcal{K}(\omega)$,*

$$\langle f|_\omega, \varphi \rangle_\omega \stackrel{\text{def}}{=} \langle f, \widetilde{\varphi} \rangle_\Omega,$$

where $\widetilde{\varphi} \in \mathcal{K}(\Omega)$ is the extension of φ by 0. $\blacksquare$

Justification. The linear mapping $f|_\omega$ is indeed a measure on ω because, according to the characterization of measures from Theorem 3.3, for every semi-norm $\| \ \|_{E;\nu}$ of E, there exist $q \in \mathcal{C}^+(\Omega)$ and $c \in \mathbb{R}$ such that

$$\|\langle f|_\omega, \varphi \rangle_\omega\|_{E;\nu} = \|\langle f, \widetilde{\varphi} \rangle_\Omega\|_{E;\nu} \leq c\|\widetilde{\varphi}\|_{\mathcal{K}(\Omega);q} = c\|\varphi\|_{\mathcal{K}(\omega);q|_\omega}.$$

The last equality follows from the fact that, by Definition 3.1 of the semi-norms of $\mathcal{K}(\Omega)$,

$$\|\widetilde{\varphi}\|_{\mathcal{K}(\Omega);q} = \sup_{x \in \Omega} q(x)|\widetilde{\varphi}(x)| = \sup_{x \in \omega} q(x)|\varphi(x)| = \|\varphi\|_{\mathcal{K}(\omega);q|_\omega}. \ \square$$

The restriction of a measure is continuous.

Theorem 6.17. — *Let Ω and ω be open subsets of $\mathbb{R}^d$ such that $\omega \subset \Omega$, and let E be a Neumann space.*

Then, the mapping $f \mapsto f|_\omega$ is continuous linear from $\mathcal{M}(\Omega; E)$ into $\mathcal{M}(\omega; E)$. ▮

Proof. According to the characterization of continuous linear mappings from Proposition 1.9, this follows from the fact that, for every $\varphi \in \mathcal{M}(\omega; E)$,

$$\|f|_\omega\|_{\mathcal{M}(\omega;E);\varphi,\nu} = \|\langle f|_\omega, \varphi\rangle_\omega\|_{E;\nu} = \|\langle f, \widetilde{\varphi}\rangle_\Omega\|_{E;\nu} = \|f\|_{\mathcal{M}(\Omega;E);\widetilde{\varphi},\nu}. \quad \square$$

Let us define the support of a measure.

Definition 6.18. — *We say that the **support** of $f \in \mathcal{M}(\Omega; E)$, where Ω is an open subset of $\mathbb{R}^d$, and E is a Neumann space, is the set*

$$\operatorname{supp} f \stackrel{\text{def}}{=} \Omega \setminus \mathcal{O},$$

where $\mathcal{O}$ is the largest open subset such that $f = 0_E$ on $\mathcal{O}$.

We denote by $\mathcal{M}_D(\Omega; E)$ the set of measures whose support is included in a subset D of Ω. ▮

Let us show that the integral remains unchanged when restricting to a domain outside which the integrated measure is zero.

Theorem 6.19. — *Let*

$$f \in L^1(\Omega; E), \quad \operatorname{supp} f \subset \omega \subset \Omega,$$

where Ω and ω are two open subsets of $\mathbb{R}^d$ and E is a Neumann space. Then,

$$\int_\Omega f = \int_\omega f. \text{ ▮}$$

Proof. Let $\mathcal{O} = \Omega \setminus \operatorname{supp} f$. If $\operatorname{supp} f \subset \omega$, we have $\omega \cup \mathcal{O} = \Omega$, whence (Theorem 6.1 (a))

$$\int_\Omega f + \int_{\omega \cap \mathcal{O}} f = \int_\omega f + \int_\mathcal{O} f. \tag{6.18}$$

By Definition 6.18 of the support, $f = 0$ on $\mathcal{O}$, hence

$$\int_{\mathcal{O}} f = 0.$$

As the restriction is continuous linear (Theorem 6.17), $f = 0_E$ on $\omega \cap \mathcal{O}$, thus

$$\int_{\omega \cap \mathcal{O}} f = 0.$$

Therefore, (6.18) reduces to the stated equality. □

Complement. If the support of f is bounded and E is a Fréchet space, it is enough that $\mathrm{supp}\, f \subset \mathring{\omega}$ for $\int_\Omega f = \int_\omega f$ to hold. A proof can be modeled after the one given for $E = \mathbb{R}$ by Laurent SCHWARTZ [65, chap. III, § 7, théorème XXVIII]. □

Let us express the measure of a set of finite measure in terms of an integral.

Theorem 6.20. — *For every measurable set ω of finite measure,*

$$\int_\omega 1 = |\omega|. \blacksquare$$

Proof. Let

$$\omega_n \stackrel{\text{def}}{=} \{x \in \omega,\ |x| < n\}.$$

The open ball $\{x : |x| < n\}$ is measurable (Theorem 2.7), so its intersection ω_n with ω is measurable (Theorem 2.11). Since ω_n is bounded (Proposition 5.6),

$$\int_{\omega_n} 1 = |\omega_n|.$$

The set ω_n increases with n and $\omega = \bigcup_{n\in\mathbb{N}} \omega_n$, thus (Theorem 2.16) $|\omega_n| \to |\omega|$ and (Theorem 6.5) $\int_{\omega_n} 1 \to \int_\omega 1$. □

6.7. Differentiation under the integral sign

Let us observe that the derivative commutes with the integral[2] (the definition of differentiation from one separated semi-normed space into another is recalled in Definition A.78).

2 **Historical background of the theorem of differentiation under the integral sign.** Gottfried Wilhelm von LEIBNIZ stated formula (6.21) for the derivative of a real-valued integral with respect to a real parameter u. Augustin CAUCHY established it in his 33rd lecture [15] for a continuously differentiable f.
The analogous result for an integral with values in $L^1(\Omega; E)$, i.e. Theorem 6.21, is new, even for $E = \mathbb{R}$, although it is easier than (6.21).

Theorem 6.21. — *Let f be a mapping from an open subset U of a separated semi-normed space into $L^1(\Omega; E)$, where Ω is an open subset of $\mathbb{R}^d$ and E is a Neumann space, such that*

$$f \text{ is differentiable at a point } u \text{ in } U \text{ with values in } L^1(\Omega; E),$$

and let ω be an open set included in Ω.

Then, the mapping $v \mapsto \int_\omega f(v)$ is differentiable at the point u with values in E, and

$$\mathrm{d}\Big(\int_\omega f\Big)(u) = \int_\omega \mathrm{d}f(u).$$ ∎

Proof. This is a particular case of Proposition A.81 on the commutation of differentiation with a continuous linear mapping, namely $\int_\omega$, which is continuous from $L^1(\Omega; E)$ into E (Theorem 5.5). □

If U is an open subset of $\mathbb{R}$, differentiability can be replaced by derivability [Vol. 1, Definition 22.1], and the formula in Theorem 6.21 would then state

$$\Big(\int_\omega f\Big)'(u) = \int_\omega f'(u). \tag{6.19}$$

If U is an open subset of $\mathbb{R}^\ell$, whose variable we denote by v, the formula in Theorem 6.21 would then state [Vol. 2, Theorem 3.16, Formula (3.19)]

$$\frac{\partial}{\partial v_j}\Big(\int_\omega f\Big)(u) = \int_\omega \frac{\partial f}{\partial v_j}(u). \tag{6.20}$$

Differentiation under the integral sign without using vector values. Theorem 6.21 utilizes vector-valued differentiation even for a real-valued function. Indeed, in that case, f is differentiable with values in $L^1(\Omega)$ when $E = \mathbb{R}$. However, one can nonetheless assign a meaning to (and prove) the formula

$$\frac{\mathrm{d}}{\mathrm{d}u}\Big(\int_\omega f(u,x)\,\mathrm{d}x\Big) = \int_\omega \frac{\mathrm{d}}{\mathrm{d}u} f(x,u)\,\mathrm{d}x \tag{6.21}$$

without using vector-valued integration. In order to define the right-hand side, this requires:

1°) assigning a meaning, for fixed x, to the mapping $v \mapsto f(x,v)$, which we shall denote by f_x;

2°) showing that f_x is differentiable at the point u;

3°) showing that $x \mapsto \dfrac{\mathrm{d}f_x}{\mathrm{d}u}(u)$ belongs to $L^1(\Omega)$. □

Interest of vector values. Vector-valued differentiation simplifies the proof of (6.21) and its statement by reducing it to a particular case of the derivative commuting with a linear mapping. □

Chapter 7

Change of Variables

Here, we show that, if T is a C^1-diffeomorphism with bounded derivatives and if f is integrable, then

$$\int_{T(\lambda)} f = \int_{\lambda} (f \circ T)\, |\det[\nabla T]|.$$

We shall accomplish this in steps.

— We first establish this equality for the Cauchy integral of a continuous function with compact support (Theorem 7.9), by carrying out a laborious limiting process on the approximated integrals.

— We then define the image $f \circ T$ of a measure f (not necessarily integrable) under a C^1-diffeomorphism, and show that the mapping $f \mapsto f \circ T$ is an isomorphism from $\mathcal{M}(\Omega; E)$ onto $\mathcal{M}(\Lambda; E)$ (Theorem 7.13), and from $L^1(\Omega; E)$ onto $L^1(\Lambda; E)$ (Theorem 7.16), if $\Omega = T(\Lambda)$.

— To give meaning to $(f \circ T)\, |\det[\nabla T]|$, we define the product $f\alpha$ of a measure f with a continuous function α, and show that it is integrable provided that f is integrable and α is bounded (§ 7.7).

— Finally, we establish the general result (Theorem 7.20) by reducing it to the Cauchy integral of an integrabilizing sequence over a decreasing sequence of open sets converging to λ in measure.

We complete this result by applying it to various affine transformations (§ 7.9).

We begin with the following preliminary results:

— the measure of the image of a measurable set under a Lipschitz mapping (Theorem 7.1);

— a review of determinants (§ 7.2), which we use to compute the volume of a parallelepiped (Theorem 7.6), which later serves for the change of variables in the Cauchy integral.

7.1. Image of a measurable set

Let us give a measurability property of the image of a measurable set under a Lipschitz mapping.

Theorem 7.1. — *Let S be a Lipchitz bijection from an open subset Ω of $\mathbb{R}^d$ onto another, i.e. there exists $c \in \mathbb{R}$ such that, for every x and y in Ω,*

$$|S(x) - S(y)| \leq c|x - y|.$$

Further suppose that S^{-1} is continuous, and let $\omega \subset \Omega$.

(a) *If ω is measurable, then $S(\omega)$ is measurable and*

$$|S(\omega)| \leq (c\sqrt{d})^n |\omega|.$$

(b) *If ω is negligible, then $S(\omega)$ is negligible.* ▮

Proof. **1°)** *Image of an open subset $\mathcal{O}$ of Ω.* Thus, $S(\mathcal{O})$ is open, as is any image of an open set under an inverse mapping, here $(S^{-1})^{-1}$, of a continuous mapping (Proposition A.46). Therefore, it remains to show that its measure is bounded.

Each closed cube $\Delta_{s,n}$ in Definition 1.20 of the measure of an open set has side length 2^{-n}, hence it is included in a ball of radius $\sqrt{d}2^{-n}$. Its image $S(\Delta_{s,n})$ is included in a ball of radius $c\sqrt{d}2^{-n}$ (since S is Lipchitz), and thus in a cube with the same side length, whence

$$|S(\Delta_{s,n})| \leq (c\sqrt{d}2^{-n})^d. \tag{7.1}$$

(This follows from the value of an open cube given in Proposition 2.1, which extends to closed cubes, as is the case here, according to Definition 2.6 of the general measure.)

Let K_n be the union of the cubes $\Delta_{s,n}$ included in $\mathcal{O}$, Q_n be the union of their images $S(\Delta_{s,n})$, and $k_n(\mathcal{O})$ be their number. The subadditivity of the measure (Theorem 2.13 (b)) and inequality (7.1) give

$$|Q_n| \leq \sum_{s\in\mathbb{Z}^d:\Delta_{s,n}\subset\mathcal{O}} |S(\Delta_{s,n})| \leq (c\sqrt{d})^d 2^{-nd} k_n(\mathcal{O}). \tag{7.2}$$

Continuity of the measure with respect to increasing sequences (Theorem 2.16) gives, since the Q_n increase with n (as do the K_n) and have $S(\mathcal{O})$ for their union (since the K_n have $\mathcal{O}$ for their union), as $n \to \infty$,

$$|Q_n| \to |S(\mathcal{O})|.$$

On the other hand, $2^{-nd}k_n(\mathcal{O}) \to |\mathcal{O}|$ by Definition 1.20, so (7.2) gives, in the limit, for any open set $\mathcal{O}$,

$$|S(\mathcal{O})| \leq (c\sqrt{d})^n |\mathcal{O}|. \tag{7.3}$$

2°) *Image of a measurable set ω.* **2.a.** *Measurability of $S(\omega)$.* Let $\epsilon > 0$. By Definition 2.5 of measurability, there exists a closed subset $\mathcal{F}$ and an open set $\mathcal{O}$ such that $\mathcal{F} \subset \omega \subset \mathcal{O}$, and

$$|\mathcal{O} \setminus \mathcal{F}| \leq \frac{\epsilon}{(c\sqrt{d})^n}.$$

Thus, $S(\mathcal{F})$ is closed and $S(\mathcal{O})$ is open, as is any inverse image of a closed or open set, respectively, under a continuous mapping (Proposition A.46, again).

Moreover, $S(\mathcal{F}) \subset S(\omega) \subset S(\mathcal{O})$ and, by 1°,

$$|S(\mathcal{F}) \setminus S(\mathcal{O})| = |S(\mathcal{F} \setminus \mathcal{O})| \leq (c\sqrt{d})^n |\mathcal{F} \setminus \mathcal{O}| \leq \epsilon.$$

So, $S(\omega)$ also satisfies Definition 2.5 of measurability.

2.b. *Calculation of the measure.* For any open set $\mathcal{O}$ containing ω, $S(\mathcal{O})$ is an open set containing $S(\omega)$, so Definition 2.6 of the measure of $S(\omega)$ gives, together with (7.3),

$$|S(\omega)| \leq |S(\mathcal{O})| \leq (c\sqrt{d})^n |\mathcal{O}|.$$

Which, again by Definition 2.6, now with respect to ω, gives

$$|S(\omega)| \leq (c\sqrt{d})^n |\omega|.$$

3°) *Image of a negligible set* ω. By Definition 2.22 of a negligible set, ω is measurable and its measure is zero. According to 2°, the same holds for $S(\omega)$, which is therefore itself negligible. □

7.2. Determinant of *d* vectors

First, let us define the permutation of d vectors and its signature.

Definition 7.2. — (a) *We say the **permutation** of* $(1, \ldots, d)$ *is any family, say* $p = (p_1, \ldots, p_d)$*, composed of the same elements arranged in a different (or identical) order.*

(b) *We say that two integers* i *and* j*, such that* $1 \leq i < j \leq d$*, are **inverted** by such a permutation if* $p_i > p_j$*.*

(c) *We say that the **signature** of such a permutation, denoted* $\varepsilon(p)$*, is the number* 1 *if it inverts an even number of integers, and* -1 *otherwise. So,*

$$\varepsilon(p) = \prod_{1 \leq i < j \leq d} \text{sign}\,(p_j - p_i). \blacksquare$$

Now we come to the determinant of d vectors in $\mathbb{R}^d$.

Definition 7.3. — *We say that the **determinant** of* d *vectors* $v^1, v^2, \ldots, v^d$ *in* $\mathbb{R}^d$ *is the real number*

$$\det[v^1, \ldots, v^d] \stackrel{\text{def}}{=} \sum_{p \in \mathcal{P}_d} \varepsilon(p)\, v^1_{p_1} \cdots v^d_{p_d},$$

where $\mathcal{P}_d$ *is the set of permutations of* $(1, \ldots, d)$*,* $\varepsilon(p)$ *is the signature of* p*, and* v^j_i *is the* i*-th component of* v^j*, i.e.* $v^j = (v^j_1, \ldots, v^j_d)$*.* ■

The data of d vectors in $\mathbb{R}^d$ is equivalent to the data of a $d \times d$ **matrix**, i.e. of a family $[v] = [v_i^j]_{1\leq i\leq d,\, 1\leq j\leq d}$ of real numbers. The v^j make up the **columns** of the matrix. Denote

$$\det[v] \stackrel{\text{def}}{=} \det[v^1, \ldots, v^d].$$

Matrix notation. We use the brackets [ ] for $d \times d$ matrices and also often for families of d vectors in $\mathbb{R}^d$, to emphasize that these are families that can be represented by a square array. This convention is somewhat unusual. □

Let us state some properties of the determinant, which were proven in Volume 2 [73, Theorem 5.14].

Proposition 7.4. — *Let v^1, v^2, ..., v^d be vectors in $\mathbb{R}^d$.*

(a) *The mapping $(v^1, \ldots, v^d) \mapsto \det[v^1, \ldots, v^d]$ is multilinearly continuous from $(\mathbb{R}^d)^d$ into $\mathbb{R}$.*

(b) *If $\mathbf{e}_j$ is the j-th basis vector of $\mathbb{R}^d$, namely $(\mathbf{e}_j)_j = 1$ and $(\mathbf{e}_j)_i = 0$ if $i \neq j$, then*

$$\det[\mathbf{e}_1, \ldots, \mathbf{e}_d] = 1.$$

(c) *If one exchanges v^k and v^ℓ, where $k \neq \ell$, then $\det[v^1, \ldots, v^d]$ changes sign.*

(d) *If there exist $k \neq \ell$ such that $v^k = v^\ell$, then $\det[v^1, \ldots, v^d] = 0$.* ▮

The **product** $[v] = [a]\,[b]$ of two $d \times d$ matrices is defined by

$$v_i^j = \sum_{k=1}^{d} a_i^k b_k^j. \tag{7.4}$$

Recall that the determinant of a product equals the product of the determinants [73, Theorem 5.15].

Proposition 7.5. — *Let $[a]$ and $[b]$ be two $d \times d$ matrices. Then,*

$$\det([a][b]) = (\det[a])(\det[b]).$$ ▮

7.3. Measure of a parallelepiped

Let us calculate the measure of an open **parallelepiped**.

Theorem 7.6. — *Let v^1, v^2, …, v^d be linearly independent vectors in $\mathbb{R}^d$, and*

$$P(v^1, \dots, v^d) \stackrel{\text{def}}{=} \{t_1 v^1 + \cdots + t_d v^d : 0 < t_i < 1,\ 1 \le i \le d\}.$$

Then,

$$|P(v^1, \dots, v^d)| = |\det[v^1, \dots, v^d]|. \;\blacksquare$$

The proof will use the following two lemmas.

Lemma 7.7. — *Let v^1, …, v^d be linearly independent vectors in $\mathbb{R}^d$, and $c \in \mathbb{R}$. Then,*

$$|P(cv^1, v^2, \dots, v^d)| = |c|\,|P(v^1, v^2, \dots, v^d)|. \;\blacksquare$$

Proof. Denote $P(v) = P(v, v^2, \dots, v^d)$.

1°) The case when $c = n \ge 1$ is an integer. The open set $P(nv)$ is "almost" equal to the open set

$$P' = \bigcup_{i=0}^{n-1} P(v) + iv.$$

More precisely, $P(nv) \setminus P'$ is the union of $n - 1$ portions of hyperplanes separating the $P(v) + iv$, a union which is negligible (Theorems 2.26 and 2.24 (a), (c), and (d)). Since the addition of a negligible set does not affect the measure (Theorem 2.23),

$$|P(nv)| = |P'|.$$

As the measure is translation invariant (Theorem 2.20), $|P(v) + iv| = |P(v)|$. The additivity of the measure with respect to disjoint open sets (Theorem 2.13 (c)) then gives $|P'| = n|P(v)|$. Thus,

$$|P(nv)| = n|P(v)|.$$

2°) The case when $c = m/n$ is rational. For m and n integers ≥ 1, we have

$$m|P(v)| = |P(mv)| = \left|P\left(n\left(\frac{m}{n}v\right)\right)\right| = n\left|P\left(\frac{m}{n}v\right)\right|.$$

Thus,

$$\left|P\left(\frac{m}{n}v\right)\right| = \frac{m}{n}|P(v)|.$$

3o) The case when $c > 0$ is real. For some integer $n \geq 1$, let m be the unique integer such that $m/n \leq c < (m+1)/n$. Since $P(c)$ increases with c, $|P(cv)|$ is increasing (Theorem 2.10), so

$$\frac{m}{n}|P(v)| = \left|P\Big(\frac{m}{n}v\Big)\right| \leq |P(cv)| \leq \left|P\Big(\frac{m+1}{n}v\Big)\right| = \frac{m+1}{n}|P(v)|.$$

As $n \to \infty$, this gives

$$c|P(v)| \leq |P(cv)| \leq c|P(v)|.$$

And therefore,

$$|P(cv)| = c|P(v)| = |c|\,|P(v)|. \tag{7.5}$$

4o) The case when c is negative or zero. For $c < 0$, the open set $P(cv)$ is a translate of $P(-cv) + 2cv$, so Theorem 2.20 (again) and equality (7.5) give

$$|P(cv)| = |P(-cv)| = |-c|\,|P(v)| = |c|\,|P(v)|.$$

For $c = 0$, $P(0v) = \emptyset$, whence $|P(0v)| = 0 = |0|\,|P(v)|$. □

Lemma 7.8. — *Let $v^1, \ldots, v^d$ be linearly independent vectors in $\mathbb{R}^d$. Then,*

$$|P(v^1 + v^2, v^2, \ldots, v^d)| = |P(v^1, v^2, \ldots, v^d)|. \blacksquare$$

Proof. Again, denote $P(v) = P(v, v^2, \ldots, v^d)$. Consider the semi-parallelepiped

$$D = \{t_1 v^1 + t_2 v^2 + \cdots + t_d v^d : 0 < t_i < 1,\ t_2 < t_1\}.$$

Then, $P(v^1 + v^2) \cup D$ is "almost" equal to $P(v^1) \cup (D + v^2)$. More precisely (see Figure 7.1), there exist portions of hyperplanes R and R' such that

$$P(v^1 + v^2) \cup D \cup R = P(v^1) \cup (D + v^2) \cup R'.$$

Indeed, both terms are equal to $P(v^1 + v^2) \cup P(v^1)$.

Since R and R' are negligible (Theorems 2.26 and 2.24 (a) and (d)), and since the addition of a negligible set does not affect the measure (Theorem 2.23), it follows that

$$|P(v^1 + v^2) \cup D| = |P(v^1) \cup (D + v^2)|.$$

The additivity of the measure with respect to disjoint open sets (Theorem 2.13 (c)) gives

$$|P(v^1 + v^2)| + |D| = |P(v^1)| + |D + v^2|.$$

Since the measure is translation invariant (Theorem 2.20), $|D + v^2| = |D|$, and thus

$$|P(v^1 + v^2)| = |P(v^1)|. \ \square$$

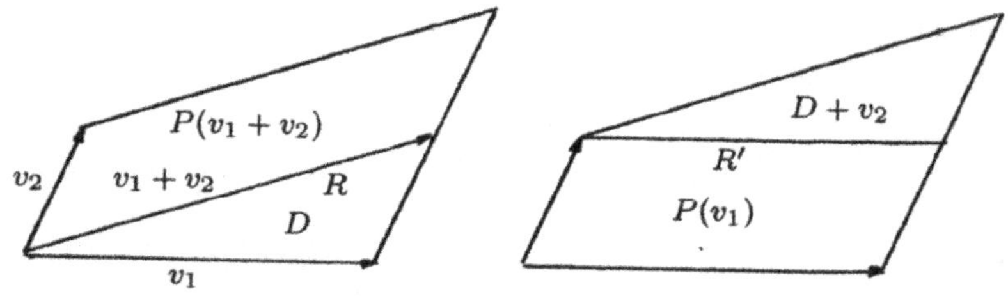

Figure 7.1. *Equivalent decompositions, to "measure" a parallelepiped*

We arrive at the computation of the measure of a parallelepiped.

Proof of Theorem 7.6. Denote

$$\phi(v^1,\ldots,v^d) \stackrel{\text{def}}{=} |P(v^1,\ldots,v^d)| - |\det[v^1,\ldots,v^d]|.$$

1º) Effect of a linear combination of the v^i. According to Lemma 7.8 and the properties of the determinant (Proposition 7.4 (a) and (d)),

$$\phi(v^1+v^2,v^2,\ldots,v^d) = \phi(v^1,v^2,\ldots,v^d).$$

From Lemma 7.7 and Proposition 7.4 (a),

$$\phi(cv^1,v^2,\ldots,v^d) = |c|\,\phi(v^1,v^2,\ldots,v^d).$$

Thus, for $c \neq 0$,

$$\begin{aligned}\phi(v^1+cv^2,v^2,\ldots,v^d) &= |c|\,\phi(c^{-1}v^1+v^2,v^2,\ldots,v^d) = \\ &= |c|\,\phi(c^{-1}v^1,v^2,\ldots,v^d) = \phi(v^1,v^2,\ldots,v^d).\end{aligned}$$

Similarly, $\phi(v^1+cv^i,v^2,\ldots,v^d) = \phi(v^1,v^2,\ldots,v^d)$, for all $i \neq 1$, as $\phi(v^1,\ldots,v^d)$ does not depend on the order of the v^i (by the definition of P and Proposition 7.4 (c)). Whence,

$$\begin{aligned}\phi\Big(\sum_{i=1}^{d} c_i v^i, v^2,\ldots,v^d\Big) &= \phi\Big(\sum_{i=1}^{d-1} c_i v^i, v^2,\ldots,v^d\Big) = \\ &= \phi\Big(\sum_{i=1}^{d-2} c_i v^i, v^2,\ldots,v^d\Big) = \cdots = \\ &= \phi(c_1v^1,v^2,\ldots,v^d) = |c_1|\,\phi(v^1,v^2,\ldots,v^d). \quad (7.6)\end{aligned}$$

2º) Effect of multiple linear combinations of the v^i. Let us show by induction on k that

$$\phi\Big(\sum_{i=1}^{d} c_i^1 v^i,\ldots,\sum_{i=1}^{d} c_i^k v^i, v^{k+1},\ldots,v^d\Big) = |c_1^1\cdots c_k^k|\,\phi(v^1,v^2,\ldots,v^d). \quad (7.7)$$

Suppose that this is true for fixed k. With equality (7.6), it implies (since $\phi(v^1, \ldots, v^d)$ does not depend on the order of the v^j),

$$\phi\Big(\sum_{i=1}^{d} c_i^1 v^i, \ldots, \sum_{i=1}^{d} c_i^{k+1} v^i, v^{k+2}, \ldots, v^d\Big) =$$

$$= |c_1^1 \cdots c_k^k|\, \phi\Big(v^1, \ldots, v^k, \sum_{i=1}^{d} c_i^{k+1} v^i, v^{k+2}, \ldots, v^d\Big) =$$

$$= |c_1^1 \cdots c_k^k||c_{k+1}^{k+1}|\, \phi(v^1, \ldots, v^d).$$

This is Property (7.7) with respect to $k+1$. Being true for $k = 1$ from (7.6), it is therefore also true for all $k \in [\![1, d]\!]$.

*3*o*) Conclusion.* Since $v^j = \sum_i v_i^j \mathbf{e}_i$ for every j, equality (7.7) with respect to $k = d$ gives

$$\phi(v^1, \ldots, v^d) = |v_1^1 \cdots v_d^d|\, \phi(\mathbf{e}_1, \ldots, \mathbf{e}_d).$$

However, the definition of ϕ gives

$$\phi(\mathbf{e}_1, \ldots, \mathbf{e}_d) = 0,$$

because $|P(\mathbf{e}_1, \ldots, \mathbf{e}_d)| = 1$ (Proposition 2.1, as $P(\mathbf{e}_1, \ldots, \mathbf{e}_d)$ is a cube of side length 1) and $\det[\mathbf{e}_1, \ldots, \mathbf{e}_d] = 1$ (Proposition 7.4 (b)). Thus,

$$\phi(v^1, \ldots, v^d) = 0.$$

That is,

$$|P(v^1, \ldots, v^d)| = |\det[v^1, \ldots, v^d]|. \quad \square$$

7.4. Change of variable in the Cauchy integral

Let us perform the change of variable in the integral of a uniformly continuous function, which is a step in the proof of the change of variable for an integrable measure, which will be carried out in Theorem 7.20.

Theorem 7.9. — *Let $f \in \mathcal{K}(\Omega; E)$, where Ω is an open subset of $\mathbb{R}^d$ and E is a Neumann space, and let T be a bijection from an open subset Λ of $\mathbb{R}^d$ onto Ω such that*

$$T \in C^1(\Lambda; \Omega) \text{ and } T^{-1} \in C^1(\Omega; \Lambda).$$

Then, denoting by $\det[\nabla T]$ *the determinant of the matrix* $[\partial_i T_j]$*:*

(a) $(f \circ T)\,|\det[\nabla T]| \in \mathcal{K}(\Lambda; E)$ *and, for any open subset* λ *of* Λ, $T(\lambda)$ *is open and*

$$\int_{T(\lambda)} f = \int_{\lambda} (f \circ T)\,|\det[\nabla T]|;$$

(b) $f\,|\det[\nabla T^{-1}]| \in \mathcal{K}(\Omega; E)$ *and, for any open subset* λ *of* Λ,

$$\int_{\lambda} f \circ T = \int_{T(\lambda)} f\,|\det[\nabla T^{-1}]|. \ \blacksquare$$

The partial derivatives $\partial_i T$ and gradient ∇T are defined in Definition A.83, and

$$\mathcal{C}^1(\Lambda; E) \stackrel{\text{def}}{=} \left\{ \begin{array}{l} \text{continuous functions from } \Lambda \text{ into } E \\ \text{with continuous partial derivatives} \end{array} \right\}.$$

By definition, $\mathcal{C}^1(\Lambda; \Omega)$ is the set of functions from Λ into Ω which are continuously differentiable from Λ into $\mathbb{R}^d$. (This is not a vector space because tT takes values in $t\Omega$, which is not always included in Ω.)

Terminology. The function $\det[\nabla T]$ is called the **Jacobian determinant** or **Jacobian** of T. □

Remark. If one knew how to approximate a continuous bijection with a $\mathcal{C}^1$-inverse using $\mathcal{C}^1$-bijections with $\mathcal{C}^1$-inverses, then one could assume T^{-1} to be merely $\mathcal{C}(\Lambda; \Omega)$ in (a), and T to be merely $\mathcal{C}(\Omega; \Lambda)$ in (b), which would allow all the results of this chapter to be extended. □

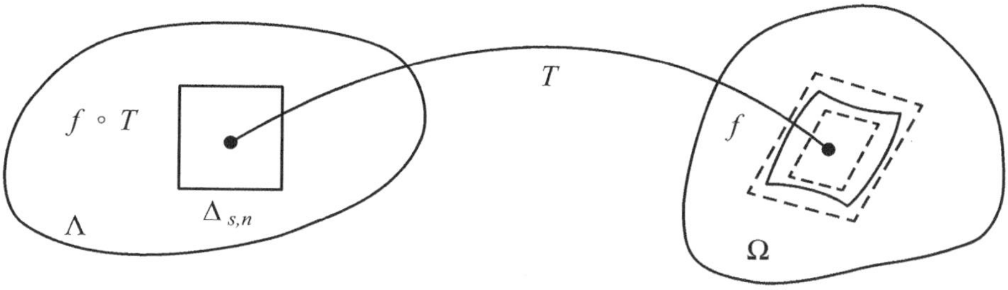

Figure 7.2. *Change of variable. The boundary of $T(\Delta_{s,n})$ is in bold, while the boundaries of the parallelepipeds $P_{s,n}^{1-\eta}$ and $P_{s,n}^{1+\eta}$ that enclose it are hatched*

Proof of Theorem 7.9. **Property** (a). **Belonging to** $\mathcal{K}(\Lambda; E)$. The function $f \circ T$ is continuous, as is any composition of continuous functions (Proposition A.39), since f and T are by hypothesis. The function $|\det[\nabla T]|$ is continuous as well, since it is the composition of the following functions:

— ∇T, which is continuous by hypothesis;

— the determinant, which is multilinearly continuous (Proposition 7.4 (a));
— the absolute value, which is continuous (since $\big||t| - |s|\big| \leq |t-s|$).

The product $(f \circ T)|\det[\nabla T]|$ is therefore itself continuous by Proposition A.39, now applied to the composition of the product from $E \times \mathbb{R}$ into E, which is continuous (Proposition A.58).

Moreover, f vanishes outside a compact set K contained in Ω (Definition 1.12). So $(f \circ T)\,|\det[\nabla T]|$ vanishes outside $T^{-1}(K)$, which is a compact set (as is any image of a compact set under a continuous mapping, see Proposition A.45 (b)) contained in Λ.

This proves that $(f \circ T)\,|\det[\nabla T]| \in \mathcal{K}(\Lambda; E)$, by Definition 1.12 of this space.

Meaning of the integrals. The integrals in the statement are well-defined because the domains of integration are open: λ is open by hypothesis, and $T(\lambda)$ is open as any inverse image $S^{-1}\lambda$ of an open set under a continuous mapping is (Proposition A.46), here $S = T^{-1}$.

Equality of the integrals. We shall establish this in six steps.

1º) Removal of a neighborhood of the boundary. Denote by $\mathcal{A}_n$ the set of the $s \in \mathbb{Z}^d$ such that the closed cube $\Delta_{s,n}$ of side length 2^{-n} centered at $2^{-n}s$ is included in λ, and

$$K_n \stackrel{\text{def}}{=} \bigcup_{s \in \mathcal{A}_n} \Delta_{s,n}, \qquad \lambda_n \stackrel{\text{def}}{=} \mathring{K}_n.$$

Then, $\bigcup_{n \in \mathbb{N}} \lambda_n = \lambda$, therefore $\bigcup_{n \in \mathbb{N}} T(\lambda_n) = T(\lambda)$. Moreover, $T(\lambda_n)$ is open and increases with n, so (Theorem 6.5)

$$\int_{T(\lambda)} f = \lim_{n \to \infty} \int_{T(\lambda_n)} f. \tag{7.8}$$

2º) Removal of the frame. Let

$$\lambda'_n \stackrel{\text{def}}{=} \bigcup_{s \in \mathcal{A}_n} \mathring{\Delta}_{s,n}.$$

Then, $\lambda_n = \lambda'_n \cup \sigma_n$, where σ_n is included in $\bigcup_{s \in \mathcal{A}_n} \partial\Delta_{s,n}$. Each boundary $\partial\Delta_{s,n}$ is negligible, since it is the union of the 2^d faces of $\Delta_{s,n}$, which are negligible by Theorem 2.26. Their union over $s \in \mathcal{A}_n$, and thus the set σ_n of this union, are negligible (Theorem 2.24 (c) and (a)). Its image $T(\sigma_n)$ is negligible by Theorem 7.1 (b) and $T(\lambda_n) = T(\lambda'_n) \cup T(\sigma_n)$, thus the additivity of the integral with respect to disjoint domains (Theorem 6.1 (b)) gives

$$\int_{T(\lambda_n)} f = \int_{T(\lambda'_n)} f. \tag{7.9}$$

On the other hand, $T(\lambda_n')$ is the union over $s \in \mathcal{A}_n$ of the $T(\mathring{\Delta}_{s,n})$. As these are pairwise disjoint, the additivity of the integral with respect to the domain of integration (Theorem 6.1 (b)) gives

$$\int_{T(\lambda_n')} f = \sum_{s\in\mathcal{A}_n} \int_{T(\mathring{\Delta}_{s,n})} f. \tag{7.10}$$

*3*o*) Approximation of* f. As the cube $\Delta_{s,n}$ is included in the ball of radius $2^{-n}\sqrt{d}$ centered at $2^{-n}s$, the bound for the semi-norms of the integral from Proposition 1.21 gives, for every semi-norm $\| \ \|_{E;\nu}$ of E,

$$\Big\| \int_{T(\mathring{\Delta}_{s,n})} (f - f(T(2^{-n}s))) \Big\|_{E;\nu} \leq \gamma_n \, |T(\mathring{\Delta}_{s,n})|,$$

where

$$\gamma_n \stackrel{\text{def}}{=} \sup_{|x'-x|\leq 2^{-n}\sqrt{d}} \|f(x') - f(x)\|_{E;\nu}. \tag{7.11}$$

Moreover, $\int_{T(\mathring{\Delta}_{s,n})} f(T(2^{-n}s)) = |T(\mathring{\Delta}_{s,n})| \, f(T(2^{-n}s))$ (according to the linearity of the integral and Proposition 1.25), so

$$\Big\| \sum_{s\in\mathcal{A}_n} \int_{T(\mathring{\Delta}_{s,n})} f - \sum_{s\in\mathcal{A}_n} |T(\mathring{\Delta}_{s,n})| \, f(T(2^{-n}s)) \Big\|_{E;\nu} \leq \gamma_n \sum_{s\in\mathcal{A}_n} |T(\mathring{\Delta}_{s,n})|. \tag{7.12}$$

Finally, as the $T(\mathring{\Delta}_{s,n})$ are pairwise disjoint and included in $T(\lambda)$, the additivity and monotonicity of the measure (Propositions 2.4 (b) and 2.2) give

$$\sum_{s\in\mathcal{A}_n} |T(\mathring{\Delta}_{s,n})| \leq |T(\lambda)|. \tag{7.13}$$

*4*o*) Approximation of* $|T(\mathring{\Delta}_{s,n})|$. The image $T(\mathring{\Delta}_{s,n})$ of the cube $\mathring{\Delta}_{s,n}$ is almost the parallelepiped $P^1_{s,n}$, where, for $r > 0$,

$$P^r_{s,n} \stackrel{\text{def}}{=} \Big\{ T(2^{-n}s) + \sum_{i=1}^{d} t_i 2^{-n-1} \partial_i T(2^{-n}s) : |t_i| < r \Big\}. \tag{7.14}$$

More precisely, as we will verify in Lemma 7.10 below, for every $\eta > 0$, there exists $n_\eta \in \mathbb{N}$ such that, for every $n \geq n_\eta$, we have, for all $s \in \mathcal{A}_n$, see Figure 7.2, p. 131,

$$P^{1-\eta}_{s,n} \subset T(\mathring{\Delta}_{s,n}) \subset P^{1+\eta}_{s,n}, \tag{7.15}$$

Using the notation of Theorem 7.6 for calculating the measure of a parallelepiped,

$$P^r_{s,n} = a_{s,n} + P(r2^{-n}\partial_1 T(2^{-n}s), \ldots, r2^{-n}\partial_d T(2^{-n}s)),$$

where $a_{s,n}$ is a translation vector (equal to $T(2^{-n}s) - r2^{-n-1}\nabla T(2^{-n}s)$, since P is not centered at 0 while $P^r_{s,n}$ is centered at $T(2^{-n}s)$, but this is not important). So,

$$|P^r_{s,n}| = |\det[r2^{-n}\partial_1 T(2^{-n}s), \dots, r2^{-n}\partial_d T(2^{-n}s)]| = r^d 2^{-nd}|\det[\nabla T(2^{-n}s)|.$$

(The second equality follows from the multilinearity of the determinant by Proposition 7.4 (a).) Denoting $\tau = |T(\mathring{\Delta}_{s,n})|$ and $\delta = 2^{-nd}|\det[\nabla T(2^{-n}s)|$, the inclusion (7.15) therefore implies $(1-\eta)^d\delta \leq \tau \leq (1+\eta)^d\delta$, whence $|\tau - \delta| \leq ((1+\eta)^d - 1)\delta \leq c_\eta \tau$, where

$$c_\eta \overset{\text{def}}{=} \frac{((1+\eta)^d - 1)}{(1-\eta)^d}. \tag{7.16}$$

That is,

$$\big|\,|T(\mathring{\Delta}_{s,n})| - 2^{-nd}|\det[\nabla T(2^{-n}s)|\,\big| \leq c_\eta\,|T(\mathring{\Delta}_{s,n})|.$$

From which, by summing over s,

$$\Big\| \sum_{s\in\mathcal{A}_n} \big(|T(\mathring{\Delta}_{s,n})| - 2^{-nd}|\det[\nabla T(2^{-n}s)|\big)\, f(T(2^{-n}s))\Big\|_{E;\nu} \leq$$

$$\leq c_\eta \sum_{s\in\mathcal{A}_n} |T(\mathring{\Delta}_{s,n})|. \tag{7.17}$$

***5º) Integral of** $(f\circ T)|\det[\nabla T]|$.* By Definition 1.14 of the approximated integral,

$$\sum_{s\in\mathcal{A}_n} 2^{-nd}|\det[\nabla T(2^{-n}s)|\, f(T(2^{-n}s)) = \mathbb{S}^n_\lambda((f\circ T)|\det[\nabla T]|) \tag{7.18}$$

and

$$\int_\lambda (f\circ T)|\det[\nabla T]| = \lim_{n\to\infty} \mathbb{S}^n_\lambda((f\circ T)|\det[\nabla T]|). \tag{7.19}$$

6º) Conclusion. The bounds (7.12), (7.17), and (7.13), yield

$$\Big\| \sum_{s\in\mathcal{A}_n} \int_{T(\mathring{\Delta}_{s,n})} f - \sum_{s\in\mathcal{A}_n} 2^{-nd}|\det[\nabla T(2^{-n}s)|\, f(T(2^{-n}s))\Big\|_{E;\nu} \leq$$

$$\leq (\gamma_n + c_\eta)|T(\lambda)|,$$

where γ_n and c_η are defined by (7.11) and (7.16). Using equalities (7.9), (7.10), and (7.18), this can be written as

$$\Big\| \Big(\int_{T(\lambda_n)} f\Big) - \mathbb{S}^n_\lambda((f\circ T)|\det[\nabla T]|)\Big\|_{E;\nu} \leq (\gamma_n + c_\eta)|T(\lambda)|.$$

For every $\epsilon > 0$, there exists $n_\epsilon > 0$ such that, for all $n \geq n_\epsilon$, the right-hand side is less than ϵ, since:

— one the one hand, f is uniformly continuous, so there exists n_γ such that $\gamma_n \leq \epsilon/2$

for all $n \geq n_\gamma$;

— on the other hand, we can choose $\eta > 0$ such that $c_\eta \leq \epsilon/(2|T(\lambda)|)$;

— and finally, $n_\epsilon = \sup\{n_\gamma, n_\eta\}$ is suitable.

This holds for every semi-norm of E, thus

$$\lim_{n\to\infty} \Big(\Big(\int_{T(\lambda_n)} f \Big) - \mathbb{S}^n_\lambda \big((f \circ T) | \det[\nabla T] | \big) \Big) = 0_E.$$

With the convergences (7.8) and (7.19), we obtain the stated equality,

$$\int_{T(\lambda)} f = \int_\lambda (f \circ T) | \det[\nabla T] |.$$

Property (b). Property (b) is the same as Property (a) with respect to the function $f \circ T$ and the change of variable T^{-1}, since $(f \circ T) \circ T^{-1} = f$. Here, (a) applies because $f \circ T$ is uniformly continuous with bounded support, as we have seen at the beginning of this proof, *i.e.* $f \circ T \in \mathcal{B}(\Lambda; E)$. □

It remains to establish the inclusions in (7.15), *i.e.* to enclose the image $T(\mathring{\Delta}_{s,n})$, which is the subject of the following lemma.

Lemma 7.10. — *With the hypotheses of Theorem 7.9, for every $\eta > 0$, there exists $n_\eta \in \mathbb{N}$ such that, for all $n \geq n_\eta$ and for all $s \in \mathbb{Z}^d$ such that $\Delta_{s,n} \subset \Lambda$,*

$$P^{1-\eta}_{s,n} \subset T(\mathring{\Delta}_{s,n}) \subset P^{1+\eta}_{s,n}, \tag{7.20}$$

where $\Delta_{s,n}$ is the closed cube of side length 2^{-n} centered at $2^{-n}s$, and $P^r_{s,n}$ is the open parallelepiped defined by (7.14). ▌

Proof. ***1º)*** *Notation.* Let $a \overset{\text{def}}{=} 2^{-n}s$, and let L be the mapping T *linearized* in a neighborhood of a, that is, for every $x \in \mathbb{R}^d$,

$$L(x) \overset{\text{def}}{=} T(a) + (x - a) \cdot \nabla T(a), \tag{7.21}$$

and let

$\Delta^r_{s,n}$ be the closed cube of side length $r2^{-n}$ centered at a.

Then, $P^r_{s,n} = L(\mathring{\Delta}^r_{s,n})$.

2º) *First inclusion.* This is equivalent to $L(\mathring{\Delta}^{1-\eta}_{s,n}) \subset T(\mathring{\Delta}^1_{s,n})$, and hence to

$$T^{-1}(L(\mathring{\Delta}^{1-\eta}_{s,n})) \subset \mathring{\Delta}^1_{s,n}.$$

As $\mathring{\Delta}^1_{s,n}$ contains every ball of radius $\eta 2^{-n-1}$ centered at a point x of $\mathring{\Delta}^{1-\eta}_{s,n}$, it suffices that, for such an x,

$$|T^{-1}(L(x)) - x| \leq \eta 2^{-n-1}. \tag{7.22}$$

However, if $x \in \mathring{\Delta}^1_{s,n}$, then this means in particular that $x \in \mathring{\Delta}^{1-\eta}_{s,n}$, and therefore $|x - a| \leq 2^{-n-1}\sqrt{d}$. Taking equality (7.21) into account, the mean value theorem (inequality (b) of Proposition A.85 with respect to T) therefore gives

$$|L(x) - T(x)| = |T(a) - T(x) - (a - x) \cdot \nabla T(a)| \leq \delta_n |a - x|,$$

where

$$\delta_n = \sup_{|x-a|\leq 2^{-n-1}\sqrt{d}} |\nabla T(x) - \nabla T(a)|.$$

Whence, again using the mean value theorem (now inequality (a) of Proposition A.85 with respect to T^{-1}),

$$|T^{-1}(L(x)) - x| = |T^{-1}(L(x)) - T^{-1}(T(x))| \leq \theta |L(x) - T(x)| \leq \theta \delta_n 2^{-n-1}\sqrt{d},$$

where

$$\theta \stackrel{\text{def}}{=} \sup_{y\in\Omega} |\nabla T^{-1}(y)|.$$

So, (7.22) and, *a fortiori*, the first inclusion of (7.20), are realized as soon as n is sufficiently large so that $\delta_n \leq \eta/(\theta\sqrt{d})$.

3º) Second inclusion. This is equivalent to $T(\mathring{\Delta}^1_{s,n}) \subset L(\mathring{\Delta}^{1+\eta}_{s,n})$. However, as we will verify, L is invertible and

$$\nabla L^{-1}(y) = \nabla T^{-1}(T(a)). \tag{7.23}$$

The inclusion above is therefore equivalent to $L^{-1}(T(\mathring{\Delta}^1_{s,n})) \subset \mathring{\Delta}^{1+\eta}_{s,n}$, which is realized if $|L^{-1}(T(x)) - x| \leq \eta 2^{-n-1}$ for every $x \in \mathring{\Delta}^1_{s,n}$. As above, we bound

$$|L^{-1}(T(x)) - x| = |L^{-1}(T(x)) - L^{-1}(L(x))| \leq \theta' \delta_n 2^{-n-1}\sqrt{d},$$

where $\theta' = \sup_{y\in\Omega} |\nabla L^{-1}(y)| = |\nabla T^{-1}(T(a))|$. Hence, $\theta' \leq \theta$ so that the second inclusion of (7.20) is itself realized once $\delta_n \leq \eta/(\theta\sqrt{d})$.

4º) Verification of (7.23). It suffices to show that $L^{-1} = M$, where

$$M(y) \stackrel{\text{def}}{=} a + (y - b) \cdot \nabla T^{-1}(b)$$

and $b \stackrel{\text{def}}{=} T(a)$. However, by Definition (7.21), $L_j(x) - b_j = \sum_{i=1}^d (x_i - a_i)\, \partial_i T_j(a)$, so

$$M_k(L(x)) = a_k + \sum_{j=1}^d (L_j(x) - b_j)\, \partial_j T_k^{-1}(b) =$$

$$= a_k + \sum_{i=1}^d \sum_{j=1}^d (x_i - a_i)\, \partial_i T_j(a)\, \partial_j T_k^{-1}(b) = x_k,$$

since, by the differential formula for a composed function from Proposition A.84, here with respect to $(T_k^{-1} \circ T)(y) = y_k$,

$$\sum_{j=1}^{d} \partial_j T_k^{-1}(b)\, \partial_i T_j(a) = \begin{cases} 1 & \text{if } i = k, \\ 0 & \text{otherwise.} \end{cases}$$

This proves that $M = L^{-1}$, hereby establishing (7.23). $\square$

Non-bijective change of variable in an integral. Although the formula in Theorem 7.9 (a) does not explicitly involve T^{-1}, it does not extend to cases where T is not invertible.

For example, if $\Lambda = (-2, 2)$, $T(x) = x^3 - x$, $\Omega = (-6, 6)$, and $f \equiv 1$, then $\int_\Omega f = \int_{-6}^{6} 1 = 12$ while $\int_\Lambda (f \circ T)\, |\det[\nabla T]| = \int_{-2}^{2} |3x^2 - 1|\, dx = 12 + 8\sqrt{3}/9$. $\square$

7.5. Change of variable in a measure

Let us define the image of a measure under a continuous bijection with a continuously differentiable inverse function.

Definition 7.11. — *Let $f \in \mathcal{M}(\Omega; E)$, where Ω is an open subset of $\mathbb{R}^d$ and E is a Neumann space, and let T be a bijection from an open subset Λ of $\mathbb{R}^d$ onto Ω such that*

$$T \in \mathcal{C}(\Lambda; \Omega) \ \text{ and } \ T^{-1} \in \mathcal{C}^1(\Omega; \Lambda).$$

We define the ***image*** *$f \circ T \in \mathcal{M}(\Lambda; E)$ by: for every $\phi \in \mathcal{K}(\Lambda)$,*

$$\langle f \circ T, \phi \rangle_\Lambda \overset{\text{def}}{=} \langle f, |\det[\nabla T^{-1}]| \phi \circ T^{-1} \rangle_\Omega. \tag{7.24}$$ ∎

Justification. ***1º)*** *Meaning of the right-hand side.* Since T^{-1} is continuously differentiable, we have $[\nabla T^{-1}] \in \mathcal{C}(\Omega; \mathbb{R}^{d \times d})$. Furthermore, the determinant is continuous (Proposition 7.4 (a)), as well as the absolute value, so the composed mapping $|\det[\nabla T^{-1}]|$ is also continuous (Proposition A.39); the same applies to $\phi \circ T^{-1}$ and to their product $|\det[\nabla T^{-1}]| \phi \circ T^{-1}$.

Moreover, ϕ vanishes outside a compact set Q included in Λ, so $\phi \circ T^{-1}$ vanishes outside $T(Q)$, which is compact (as is any image of a compact set under a continuous mapping, see Proposition A.45 (b)) and included in Ω. Thus,

$$|\det[\nabla T^{-1}]| \phi \circ T^{-1} \in \mathcal{K}(\Omega). \tag{7.25}$$

The right-hand side of (7.24) is therefore well-defined.

2°) Obtaining a measure. According to the characterization of measures from Theorem 3.3, for every semi-norm $\| \|_{E;\nu}$ of E, there exist $q \in \mathcal{C}^+(\Omega)$ and $c \in \mathbb{R}$ such that, for every $\varphi \in \mathcal{K}(\Omega)$,

$$\|\langle f, \varphi \rangle\|_{E;\nu} \leq c \sup_{x\in\Omega} q(x)|\varphi(x)|.$$

Thus,

$$\|\langle f \circ T, \phi \rangle_\Lambda\|_{E;\nu} = \|\langle f, |\det[\nabla T^{-1}]| \phi \circ T^{-1} \rangle_\Omega\|_{E;\nu} \leq$$
$$\leq c \sup_{x\in\Omega} q(x) |\det[\nabla T^{-1}(x)]| \, |(\phi \circ T^{-1})(x)| = c \sup_{y\in\Lambda} q_T(y) |\phi(y)|,$$

where $q_T \stackrel{\text{def}}{=} (q|\det[\nabla T^{-1}]|) \circ T$. Since the function q_T is continuous and positive, it follows, again by the characterization from Theorem 3.3, that

$$f \circ T \in \mathcal{M}(\Lambda; E). \ \square$$

Now that we have verified that $f \circ T$ is a measure, let us reassure ourselves that, when f is a function, we recover the classical definition.

Theorem 7.12. — *If $f \in \mathcal{C}(\Omega; E)$, $T \in \mathcal{C}^1(\Lambda; \Omega)$, and $T^{-1} \in \mathcal{C}^1(\Omega; \Lambda)$, then its image $f \circ T$ in the sense of measures given by Definition 7.11 coincides with that in the sense of functions.*

That is to say that, with the identification (3.16), p. 65, of f with a measure, Definition 7.11 also gives $f \circ T \in \mathcal{C}(\Lambda; E)$, and, for every $y \in \Lambda$,

$$(f \circ T)(y) = f(T(y)). \ ▮$$

Proof. It is a matter of verifying that the measure $\overline{f}$ associated with the function f using Theorem 3.10 satisfies

$$\overline{f} \circ T = \overline{f \circ T}. \tag{7.26}$$

Definition 7.11 of $\overline{f} \circ T$ and the expression for $\langle \overline{f}, \varphi \rangle$ from Theorem 3.10 yield, for every $\phi \in \mathcal{K}(\Lambda)$,

$$\langle \overline{f} \circ T, \phi \rangle_\Lambda = \langle \overline{f}, |\det[\nabla T^{-1}]| \, \phi \circ T^{-1} \rangle_\Omega = \int_\Omega f \, |\det[\nabla T^{-1}]| \, \phi \circ T^{-1}.$$

According to Theorem 7.9 (b) of the change of variable in the Cauchy integral,

$$\int_\Omega f \, |\det[\nabla T^{-1}]| \, \phi \circ T^{-1} = \int_\Lambda (f \circ T) \, \phi. \tag{7.27}$$

These equalities give, again using the expression from Theorem 3.10, here for $\overline{f \circ T}$,

$$\langle \bar{f} \circ T, \phi \rangle_\Lambda = \int_\Lambda (f \circ T)\, \phi = \langle \overline{f \circ T}, \phi \rangle_\Lambda,$$

which proves (7.26). □

Let us prove that a continuously differentiable change of variables, along with its inverse function, is an isomorphism of spaces of measures.

Theorem 7.13. — *Let T be a bijection from an open subset Λ of $\mathbb{R}^d$ onto another, Ω, such that $T \in \mathcal{C}^1(\Lambda; \Omega)$ and $T^{-1} \in \mathcal{C}^1(\Omega; \Lambda)$, and let E be a Neumann space.*

Then, the mapping $f \mapsto f \circ T$ is an isomorphism from $\mathcal{M}(\Omega; E)$ onto $\mathcal{M}(\Lambda; E)$, whose inverse mapping is $g \mapsto g \circ T^{-1}$. ▮

Proof. ***1°) Bijectivity.*** Definition 7.11 of the change of variable with respect to T^{-1} and T successively gives, for every $f \in \mathcal{M}(\Omega; E)$ and $\varphi \in \mathcal{K}(\Omega)$,

$$\begin{aligned} \langle (f \circ T) \circ T^{-1}, \varphi \rangle_\Omega &= \langle f \circ T, |\det[\nabla T]|\, \varphi \circ T \rangle_\Lambda = \\ &= \langle f, |\det[\nabla T^{-1}]|\, (|\det[\nabla T]| \circ T^{-1})\, \varphi \rangle_\Omega = \langle f, \varphi \rangle_\Omega, \end{aligned} \tag{7.28}$$

as the product of the Jacobian determinants equals 1, as we shall verify in Lemma 7.14. Hence,

$$(f \circ T) \circ T^{-1} = f.$$

Similarly, for every measure $g \in \mathcal{M}(\Lambda; E)$,

$$(g \circ T^{-1}) \circ T = g,$$

which proves that the mapping

$$f \mapsto f \circ T \text{ is bijective and has } g \mapsto g \circ T^{-1} \text{ as its inverse.}$$

2°) Continuity. Denote by $\{\| \ \|_{E;\nu} : \nu \in \mathcal{N}_E\}$ the family of semi-norms of E. According to Definitions 3.2 of the semi-norms of $\mathcal{M}(\Lambda; E)$ and 7.11 of $f \circ T$, we have, for every $\phi \in \mathcal{K}(\Lambda)$ and every $\nu \in \mathcal{N}_E$,

$$\begin{aligned} \|f \circ T\|_{\mathcal{M}(\Lambda;E);\phi,\nu} &= \|\langle f \circ T, \phi \rangle_\Lambda\|_{E;\nu} = \\ &= \|\langle f, |\det[\nabla T^{-1}]|\, \phi \circ T^{-1} \rangle_\Omega\|_{E;\nu} = \|f\|_{\mathcal{M}(\Omega;E);\varphi,\nu}, \end{aligned}$$

where $\varphi = |\det[\nabla T^{-1}]|\, \phi \circ T^{-1}$, which belongs to $\mathcal{K}(\Omega)$ as we have established in (7.25). Therefore, according to the characterization of continuous linear mappings from Proposition 1.9,

the mapping $f \mapsto f \circ T$ is continuous from $\mathcal{M}(\Omega; E)$ into $\mathcal{M}(\Lambda; E)$.

*3*o*) Continuity of the inverse mapping.* Part 2o relative to T^{-1} shows that

the mapping $g \mapsto g \circ T^{-1}$ is continuous from $\mathcal{M}(\Lambda; E)$ into $\mathcal{M}(\Omega; E)$. □

Let us verify the sign property of the Jacobian determinant that we used to establish (7.28).

Lemma 7.14. — *Let T be a bijection from an open subset Λ of $\mathbb{R}^d$ onto a connected open subset Ω of $\mathbb{R}^d$, such that $T \in \mathcal{C}^1(\Lambda; \Omega)$ and $T^{-1} \in \mathcal{C}^1(\Omega; \Lambda)$. Then, for every $y \in \Lambda$,*

$$\det[\nabla T^{-1}(T(y))] \det[\nabla T(y)] = 1,$$

and $\det[\nabla T^{-1}]$ *does not change sign on* Ω. ▮

Proof. ***1*o*) Product of determinants.*** Let us differentiate equality $(T_k^{-1} \circ T)(y) = y_k$. With the expression for the change of variable in a derivative from Proposition A.84, we obtain

$$\partial_i (T_k^{-1} \circ T)(y) = \sum_{j=1}^{d} \partial_j T_k^{-1}(T(y))\, \partial_i T_j(y) = \mathrm{I}_k^i,$$

where $\mathrm{I}_k^i = 0$ if $i \neq k$ and $\mathrm{I}_i^i = 1$. That is, by Definition (7.4), p. 126, of the matrix product,

$$[\nabla T^{-1}(T(y))]\, [\nabla T(y)] = [\mathrm{I}] = [\mathbf{e}_1, \ldots, \mathbf{e}_d].$$

Since the determinant of a product equals the product of the determinants (Proposition 7.5) and since the determinant of the identity matrix equals 1 (Proposition 7.4 (b)),

$$\det[\nabla T^{-1}(T(y))] \det[\nabla T(y)] = 1.$$

2*o*) Sign of $\det[\nabla T^{-1}]$. So,

$$\det[\nabla T^{-1}(T(y))] \neq 0. \tag{7.29}$$

Consider the set

$$U \stackrel{\text{def}}{=} \det[\nabla T^{-1}](\Omega).$$

It is connected in $\mathbb{R}$, as is any image of a connected set under a continuous mapping (Proposition A.45 (c)). It cannot contain both a point $a < 0$ and another $b > 0$, otherwise the disjoint open sets $(-\infty, 0)$ and $(0, +\infty)$ would cover it, since it does not contain 0 by (7.29), which would contradict its connectedness (Proposition A.17). Therefore,

$$\det[\nabla T^{-1}] \text{ is either } > 0 \text{ throughout } \Omega, \text{ or } < 0 \text{ throughout } \Omega. \quad \square$$

7.6. Change of variable in an integrable measure

Let us show that the image of an integral measure under a change of variable T is integrable, provided that $|\det[\nabla T^{-1}]|$ is bounded.

Theorem 7.15. — *Let $f \in L^1(\Omega; E)$, where Ω is an open subset of $\mathbb{R}^d$, E is a Neumann space with the semi-norms $\{\| \ \|_{E;\nu} : \nu \in \mathcal{N}_E\}$, and let T be a bijection from an open subset Λ of $\mathbb{R}^d$ onto Ω such that*

$$T \in \mathcal{C}^1(\Lambda; \Omega), \ T^{-1} \in \mathcal{C}^1(\Omega; \Lambda) \ \textit{and} \ |\det[\nabla T^{-1}]| \ \textit{is bounded.}$$

Then,

$$f \circ T \in L^1(\Lambda; E)$$

and, for every $\nu \in \mathcal{N}_E$ and $c \stackrel{\text{def}}{=} \sup_{x \in \Omega} |\det[\nabla T^{-1}(x)]|$,

$$\|f \circ T\|_{L^1(\Lambda;E);\nu} \leq c\|f\|_{L^1(\Omega;E);\nu}.$$

If $(f_n)_{n\in\mathbb{N}}$ is an integrabilizing sequence for f, then $(f_n \circ T)_{n\in\mathbb{N}}$ is an integrabilizing sequence for $f \circ T$. ▮

Proof. **1o)** *Integrabilizing sequence for $f \circ T$.* Let $(f_n)_{n\in\mathbb{N}}$ be an integrabilizing sequence for f, *i.e.* by Definition 4.1, that f_n is uniformly continuous and vanishes outside a bounded set, that $f_n \to f$ in $\mathcal{M}(\Omega; E)$ when $n \to \infty$ and, for every semi-norm of E,

$$\sup_{m \geq n} \int_\Omega \|f_m - f_n\|_{E;\nu} \to 0. \tag{7.30}$$

Let us verify that $(f_n \circ T)_{n\in\mathbb{N}}$ is an integrabilizing sequence for $f \circ T$.

— On the one hand, $f_n \circ T$ is continuous, as is any composition of continuous mappings (Proposition A.39), and vanishes outside the compact set $T^{-1}(S_n)$, where S_n is the support of f, which is included in Λ.

— On the other hand, since the mapping $f \mapsto f \circ T$ is continuous from $\mathcal{M}(\Omega; E)$ into $\mathcal{M}(\Lambda; E)$ (Theorem 7.13), and hence sequentially continuous (Proposition 1.8), the convergence $f_n \to f$ in $\mathcal{M}(\Omega; E)$ means that

$$f_n \circ T \to f \circ T \ \text{ in } \mathcal{M}(\Lambda; E).$$

— Finally, Theorem 7.9 (b) on the change of variable in the Cauchy integral gives, denoting $c \stackrel{\text{def}}{=} \sup_{x\in\Omega} |\det[\nabla T^{-1}(x)]|$,

$$\int_\Lambda \|f_m \circ T - f_n \circ T\|_{E;\nu} = \int_\Omega \|f_m - f_n\|_{E;\nu} |\det[\nabla T^{-1}]| \leq c \int_\Omega \|f_m - f_n\|_{E;\nu},$$

which, together with the convergence (7.30), implies

$$\sup_{m\geq n} \int_\Lambda \|f_m \circ T - f_n \circ T\|_{E;\nu} \to 0.$$

Therefore, $(f_n \circ T)_{n\in\mathbb{N}}$ is indeed an integrabilizing sequence for $f \circ T$.

2o) *Integrability of $f \circ T$.* By Definition 4.1 of an integrable measure, this proves that

$$f \circ T \in L^1(\Lambda; E).$$

3o) *Semi-norms of $f \circ T$.* By Definition 4.4 of the semi-norms of $L^1(\Omega; E)$,

$$\|f\|_{L^1(\Omega;E);\nu} = \lim_{n\to\infty} \int_\Omega \|f_n\|_{E;\nu}.$$

Again, according to Theorem 7.9 (b) on the change of variable,

$$\int_\Lambda \|f_n \circ T\|_{F;\mu} = \int_\Omega \|f_n\|_{E;\nu} |\det[\nabla T^{-1}]| \leq c \int_\Omega \|f_n\|_{E;\nu}.$$

Since $(f_n \circ T)_{n\in\mathbb{N}}$ is an integrabilizing sequence for $f \circ T$, in the limit, we obtain

$$\|f \circ T\|_{L^1(\Lambda;E);\nu} \leq c\|f\|_{L^1(\Omega;E);\nu}. \ \square$$

Let us show that the change of variable is an isomorphism of the spaces of integrable measures if it is continuously differentiable with bounded derivatives and if the same holds for its inverse.

Theorem 7.16. — *Let E be a Neumann space and T a bijection from an open subset Λ of $\mathbb{R}^d$ onto another, Ω, such that $T \in \mathcal{C}^1(\Lambda; \Omega)$ and $T^{-1} \in \mathcal{C}^1(\Omega; \Lambda)$.*

If $|\det[\nabla T^{-1}]|$ is bounded, then the mapping $f \mapsto f \circ T$ is continuous linear, and thus sequentially continuous, from $L^1(\Omega; E)$ into $L^1(\Lambda; E)$.

Moreover, if $|\det[\nabla T]|$ is bounded, then the mapping $f \mapsto f \circ T$ is an isomorphism from $L^1(\Omega; E)$ onto $L^1(\Lambda; E)$. In particular, this is the case if $[\nabla T]$ and $[\nabla T^{-1}]$ are bounded. ▮

Proof. **1°)** *Continuity of $f \mapsto f \circ T$.* This follows from the bound for the semi-norms from Theorem 7.15, according to the characterization (Proposition 1.9) of continuous linear mappings.

2°) *Sequential continuity.* As always, this follows from continuity (Proposition 1.8).

3°) *Isomorphism.* If $|\det[\nabla T]|$ is bounded, then the mapping $g \mapsto g \circ T^{-1}$, which is the inverse of the mapping $f \mapsto f \circ T$, is continuous from $L^1(\Lambda; E)$ onto $L^1(\Omega; E)$ by 1° (with respect to T^{-1} in the place of T). □

7.7. Product of a measure with a continuous function

Let us define the product so that we can give a meaning to $(f \circ T)\,|\det[\nabla T]|$.

Definition 7.17. — *Let $f \in \mathcal{M}(\Omega; E)$, where Ω is an open subset of $\mathbb{R}^d$, E is a Neumann space, and $\alpha \in \mathcal{C}(\Omega)$.*

*We define the **product** $f\alpha \in \mathcal{M}(\Omega; E)$ by: for every $\varphi \in \mathcal{K}(\Omega)$,*

$$\langle f\alpha, \varphi \rangle \overset{\text{def}}{=} \langle f, \alpha\varphi \rangle. \tag{7.31}$$ ▮

Justification. **1°)** *Meaning of the right-hand side.* The function $\alpha\varphi$ is continuous (as is any product of continuous functions, see Proposition A.48), and it vanishes outside the support of φ, which is a compact subset included in Ω. Whence,

$$\alpha\varphi \in \mathcal{K}(\Omega) \tag{7.32}$$

and the right-hand side of (7.31) is well-defined.

2°) *Obtaining a measure.* According to the characterization of measures from Theorem 3.3, for every semi-norm $\|\;\|_{E;\nu}$ of E, there exist $q \in \mathcal{C}^+(\Omega)$ and $c \in \mathbb{R}$ such that, for every $\varphi \in \mathcal{K}(\Omega)$,

$$\|\langle f, \varphi \rangle\|_{E;\nu} \leq c \sup_{x \in \Omega} q(x)|\varphi(x)|.$$

Thus,

$$\|\langle f\alpha, \varphi \rangle\|_{E;\nu} = \|\langle f, \alpha\varphi \rangle\|_{E;\nu} \leq c \sup_{x \in \Omega} q(x)|(\alpha\varphi)(x)| = c \sup_{x \in \Omega} (q|\alpha|)(x)\,|\varphi(x)|.$$

Since the function $q|\alpha|$ is continuous and positive, it follows, again by the characterization of Theorem 3.3, that

$$f\alpha \in \mathcal{M}(\Omega; E). \; \square$$

Now that we have verified that $f\alpha$ is a measure, let us assure ourselves that, when f is a function, we recover the classical definition.

Theorem 7.18. — *If $f \in \mathcal{C}(\Omega; E)$, then the product $f\alpha$ in the sense of measures (Definition 7.17) coincides with that in the sense of functions.* ▮

Proof. We need to verify that the measure $\bar{f}$ associated with the function f by Theorem 3.10 satisfies

$$\bar{f}\alpha = \overline{f\alpha}. \tag{7.33}$$

Definition 7.17 of $\bar{f}\alpha$ and the expression of $\langle \bar{f}, \varphi \rangle$ from Theorem 3.10 give, for every $\varphi \in \mathcal{K}(\Omega)$,

$$\langle \bar{f}\alpha, \varphi \rangle = \langle \bar{f}, \alpha\varphi \rangle = \int_\Omega f \times \alpha\varphi = \int_\Omega f\alpha \times \varphi = \langle \overline{f\alpha}, \varphi \rangle,$$

which proves (7.33). □

Let us show that the product with a continuous function is continuous on the space of measures.

Theorem 7.19. — *Let Ω be an open subset of $\mathbb{R}^d$ and E a Neumann space.*

(a) *If $\alpha \in \mathcal{C}(\Omega)$, then the mapping $f \mapsto f\alpha$ is continuous, and thus sequentially continuous, from $\mathcal{M}(\Omega; E)$ into itself.*

(b) *If $\alpha \in \mathcal{C}_b(\Omega)$, then the mapping $f \mapsto f\alpha$ is continuous, and thus sequentially continuous, from $L^1(\Omega; E)$ into itself.* ▮

Proof. Denote by $\{\| \ \|_{E;\nu} : \nu \in \mathcal{N}_E\}$ the family of semi-norms of E.

1°) Continuity in $\mathcal{M}(\Omega; E)$. According to Definitions 3.2 of the semi-norms of the space $\mathcal{M}(\Omega; E)$ and 7.17 of the product $f\alpha$, we have, for every $\varphi \in \mathcal{K}(\Omega)$ and every $\nu \in \mathcal{N}_E$,

$$\|f\alpha\|_{\mathcal{M}(\Omega;E);\varphi,\nu} = \|\langle f\alpha, \varphi \rangle\|_{E;\nu} = \|\langle f, \alpha\varphi \rangle\|_{E;\nu} = \|f\|_{\mathcal{M}(\Omega;E);\alpha\varphi,\nu},$$

since $\alpha\varphi$ belongs to $\mathcal{K}(\Omega)$, as we have established in (7.32).

This, according to the characterization of continuous linear mappings from Proposition 1.9, implies the continuity of the mapping $f \mapsto f\alpha$ from $\mathcal{M}(\Omega; E)$ into itself.

2°) Sequential continuity. As always, this follows from continuity (Proposition 1.8).

3°) Continuity in $L^1(\Omega; E)$. Moreover, α is bounded.

3.a. *Integrabilizing sequence for $f\alpha$.* The regularizations $(R_n f)_{n\in\mathbb{N}}$ of f (Definition 3.17) belong to $\mathcal{K}(\Omega; E)$ and form an integrabilizing sequence for f (Theorem 4.2). Therefore (Definition 4.1), when $n \to \infty$, $R_n f \to f$ in $\mathcal{M}(\Omega; E)$ and, for every $\nu \in \mathcal{N}_E$,

$$\sup_{m\geq n} \int_\Omega \|R_m f - R_n f\|_{E;\nu} \to 0. \tag{7.34}$$

Let us verify that $((R_n f)\alpha)_{n\in\mathbb{N}}$ is an integrabilizing sequence for $f\alpha$.

— On the one hand, $(R_n f)\alpha$ is continuous, as is any product of continuous functions (Proposition A.48), and vanishes outside a compact set where $R_n f$ is non-zero. Therefore, $(R_n f)\alpha$ is uniformly continuous according to a corollary (Proposition A.42) of Heine's theorem.

— On the other hand, since the mapping $f \mapsto f\alpha$ is sequentially continuous from $\mathcal{M}(\Omega; E)$ into $\mathcal{M}(\Omega; E)$ by 1° and 2°, the convergence $R_n f \to f$ in $\mathcal{M}(\Omega; E)$ gives

$$(R_n f)\alpha \to f\alpha \text{ in } \mathcal{M}(\Omega; E).$$

— Finally, the monotonicity and linearity of the real-valued Cauchy integral (Propositions 1.16 and 1.18 (a)), together with the convergence (7.34), give, by denoting $c \overset{\text{def}}{=} \sup_\Omega |\alpha|$,

$$\sup_{m\geq n} \int_\Omega \|(R_m f)\alpha - (R_n f)\alpha\|_{E;\nu} \leq c \sup_{m\geq n} \int_\Omega \|R_m f - R_n f\|_{E;\nu} \to 0.$$

Thus, $((R_n f)\alpha)_{n\in\mathbb{N}}$ is indeed an integrabilizing sequence for $f\alpha$.

3.b. *Integrability of $f\alpha$.* By Definition 4.1 of an integrable measure, this proves that

$$f\alpha \in L^1(\Omega; E).$$

3.c. *Continuity of $f \mapsto f\alpha$.* Since $((R_n f)\alpha)_{n\in\mathbb{N}}$ is an integrabilizing sequence for $f\alpha$, Definition 4.4 of the semi-norms of $L^1(\Omega; E)$ give, again denoting $c \overset{\text{def}}{=} \sup_\Omega |\alpha|$,

$$\|f\alpha\|_{L^1(\Omega;E);\nu} = \lim_{n\to\infty} \int_\Omega \|(R_n f)\alpha\|_{E;\nu} \leq c \lim_{n\to\infty} \int_\Omega \|R_n f\|_{E;\nu} = c\|f\|_{L^1(\Omega;E);\nu}.$$

This, once again according to the characterization of Proposition 1.9, implies the continuity of the mapping $f \mapsto f\alpha$ from $L^1(\Omega; E)$ into itself. □

7.8. Change of variable in an integral

Let us now demonstrate the **change of variable** in an integral of an integrable measure. Recall that $\det[\nabla T]$ denotes the determinant of the matrix $[\partial_i T_j]$.

Theorem 7.20. — *Let $f \in L^1(\Omega; E)$, where Ω is an open subset of $\mathbb{R}^d$, E is a Neumann space, T is a bijection from an open subset Λ of $\mathbb{R}^d$ onto Ω, and $\lambda \subset \Lambda$.*

(a) *If*

$$T \in \mathcal{C}(\Lambda; \Omega),\ T^{-1} \in \mathcal{C}^1(\Omega; \Lambda) \text{ and } |\det[\nabla T^{-1}]| \text{ is bounded,}$$

then $f \circ T \in L^1(\Lambda; E)$ and $f\,|\det[\nabla T^{-1}]| \in L^1(\Omega; E)$.

Furthermore, if $T(\lambda)$ is measurable, then λ is measurable and

$$\int_\lambda f \circ T = \int_{T(\lambda)} f\,|\det[\nabla T^{-1}]|.$$

(b) *Furthermore, if*

$$|\det[\nabla T]| \text{ is bounded,}$$

then $(f \circ T)\,|\det[\nabla T]| \in L^1(\Lambda; E)$.

Furthermore, if either λ or $T(\lambda)$ is measurable, then they both are, and

$$\int_{T(\lambda)} f = \int_\lambda (f \circ T)\,|\det[\nabla T]|. \ ∎$$

Domain properties. Here, it is necessary to assume that Ω and Λ are open, since differentiability (here, that of T and T^{-1}) is *a priori* only defined on an open set (this can be extended, for example, to the closure of a bounded open set, as was done in Volume 2 [73, Definition 2.26]). □

Generalization to a Lipschitz change of variable. In Theorem 7.20, all the assumptions on T and T^{-1} can be replaced by the condition that T and T^{-1} are Lipschitz, which is equivalent to the existence of constants $c_1 > 0$ and c_2 such that, for all x and y in Λ,

$$c_1|x - y| \le |T(x) - T(y)| \le c_2|x - y|.$$

Then, we can define $\nabla T \in L^\infty(\Lambda; \mathbb{R}^{d\times d})$ and $\nabla T^{-1} \in L^\infty(\Omega; \mathbb{R}^{d\times d})$, and each stated property is satisfied (we will see this in Volume 5). □

Proof of Theorem 7.20. **1°)** Property (a). **1.a.** *Integrability of $f \circ T$.* This has been established in Theorem 7.15.

1.b. *Integrability of $f\,|\det[\nabla T^{-1}]|$.* This is given by Theorem 7.19 (b), since f is integrable and $|\det[\nabla T^{-1}]|$ is continuous and bounded.

1.c. *Measurability of* λ. The mean value theorem (Proposition A.85 (a)) gives, for all x and y in Ω, and by denoting $\gamma \stackrel{\text{def}}{=} \sup_\Omega |\det[T^{-1}]|$,

$$|T^{-1}(x) - T^{-1}(y)| \leq \gamma\,|x - y|.$$

The inverse image of $T(\lambda)$ under T^{-1}, *i.e.* λ, is therefore measurable by Theorem 7.1 (a).

1.d. *Approximated formula.* Let $(f_n)_{n\in\mathbb{N}}$ be an integrabilizing sequence for f, and let $(\omega_n)_{n\in\mathbb{N}}$ be a sequence of open sets such that $T(\lambda) \subset \omega_n \subset \Omega$ for all n and, as $n \to \infty$,

$$|\omega_n \setminus T(\lambda)| \to 0. \tag{7.35}$$

By Theorem 4.2, we can assume, for all n, that

f_n vanishes outside a compact subset K_n included in Ω.

The set $\lambda_n \stackrel{\text{def}}{=} T^{-1}(\omega_n)$ is open and the change of variable formula in the Cauchy integral from Theorem 7.9 (b) therefore gives

$$\int_{\lambda_n} f_n \circ T = \int_{T(\lambda_n)} f_n\,|\det[\nabla T^{-1}]|. \tag{7.36}$$

1.e. *Limit of the left-hand side.* According to Theorem 7.15, $(f_n \circ T)_{n\in\mathbb{N}}$ is an integrabilizing sequence for $f \circ T$.

On the other hand, by the bound from Theorem 7.1 and the convergence (7.35),

$$|\lambda_n \setminus \lambda| = |T^{-1}(\omega_n \setminus \omega)| \leq c|\omega_n \setminus \omega| \to 0.$$

Definition 5.1 of the integral therefore gives

$$\int_{\lambda_n} f_n \circ T \to \int_\lambda f \circ T. \tag{7.37}$$

1.f. *Limit of the right-hand side of (7.36).* According to the characterization of integrabilizing sequences from Theorem 4.10, $f_n \to f$ in $L^1(\Omega; E)$. The sequential continuity of the product (Theorem 7.19 (b)) therefore gives, since $|\det[\nabla T^{-1}]|$ is continuous and bounded,

$$f_n\,|\det[\nabla T^{-1}]| \to f\,|\det[\nabla T^{-1}]| \ \text{ in } L^1(\Omega; E).$$

Moreover, $f_n\,|\det[\nabla T^{-1}]|$ is continuous and vanishes outside K_n so, again according to the characterization from Theorem 4.10,

$(f_n\,|\det[\nabla T^{-1}]|)_{n\in\mathbb{N}}$ is an integrable sequence for $f\,|\det[\nabla T^{-1}]|$.

Definition 5.1 of the integral therefore gives, since $|\omega_n \setminus T(\lambda)| \to 0$ by (7.35),

$$\int_{T(\lambda_n)} f_n\,|\det[\nabla T^{-1}]| \to \int_{T(\lambda)} f\,|\det[\nabla T^{-1}]|. \tag{7.38}$$

1.g. *Formula.* With the convergences (7.37) and (7.38), in the limit, equality (7.36) yields

$$\int_\lambda f \circ T = \int_{T(\lambda)} f\,|\det[\nabla T^{-1}]|.$$

2º) Property (b). By Theorem 7.15,

$$f \circ T \in L^1(\Lambda; E).$$

Property (a) with $f \circ T$, T^{-1} and λ in the place of f, T and $T(\lambda)$, respectively, then gives Property (b) if λ is measurable; which is the case when $T(\lambda)$ is measurable, as we have seen in 1.c. □

7.9. Affine change of variables

Let us cover some particular changes of variable, denoting the measure $f \circ T$ by $x \mapsto f(T(x))$.

First of all, observe that the integral is preserved under translations [1].

Theorem 7.21. — *Let $f \in L^1(\Omega; E)$, where Ω is an open subset of $\mathbb{R}^d$, E is a Neumann space, $z \in \mathbb{R}^d$, and ω is a measurable subset of Ω.*

Then, the translated measure $y \mapsto f(y-z)$ belongs to the space $L^1(\Omega + z)$, the set $\omega + z = \{x + z : x \in \omega\}$ is measurable, and

$$\int_\omega f(x)\,\mathrm{d}x = \int_{\omega+z} f(y-z)\,\mathrm{d}y. \;\blacksquare$$

Proof. This follows from Theorem 7.20 (b) with respect to the change of variable $T(x) = x - z$, hence $\det[\nabla T] = \det[\mathbf{e}_1, \ldots, \mathbf{e}_d] = 1$ (Proposition 7.4 (b)), and the domain $\lambda = \omega + z$. □

Let us perform a linear change of variable in an integral.

1 **Historical background of the translation invariance of the integral.** The translation invariance of the integral is similar to the 8th axiom of EUCLID's *Elements* [24], the equality of congruent parts.

Theorem 7.22. — *Let $f \in L^1(\Omega; E)$, where Ω is an open subset of $\mathbb{R}^d$, E is a Neumann space, ω is a measurable subset of Ω, and*

$$T \text{ is a linear bijection from } \mathbb{R}^d \text{ onto itself.}$$

Then, $f \circ T \in L^1(T^{-1}(\Omega); E)$, $T^{-1}(\omega)$ is measurable, and

$$\int_\omega f = \kappa \int_{T^{-1}(\omega)} f \circ T,$$

where $\kappa = |\det[T(\mathbf{e}_1), \dots, T(\mathbf{e}_d)]|$. ▮

Proof. This follows from Theorem 7.20 (b) regarding the change of variable with $\lambda = T^{-1}(\Omega)$ because T, being linear, is differentiable and

$$\nabla T(x) = (T(\mathbf{e}_1), \dots, T(\mathbf{e}_d)).$$

Indeed, Definition A.78 of differentiability is satisfied here because, for all $z \in \mathbb{R}^d$,

$$T(x+z) - T(x) - \sum_{i=1}^{d} z_i T(\mathbf{e}_i) = 0,$$

since $T(z) = T(\sum_{i=1}^d z_i \mathbf{e}_i) = \sum_{i=1}^d z_i T(\mathbf{e}_i)$. □

In particular, a homothety has the following effect on an integral.

Theorem 7.23. — *Let $f \in L^1(\Omega; E)$, where Ω is an open subset of $\mathbb{R}^d$, E is a Neumann space, ω is a measurable subset of Ω, and $t > 0$.*

Then, the measure $y \mapsto f(ty)$ is integrable, the set $\omega/t = \{x/t : x \in \omega\}$ is measurable, and

$$\int_\omega f(x)\,\mathrm{d}x = t^d \int_{\omega/t} f(ty)\,\mathrm{d}y. \text{ ▮}$$

Proof. This is a particular case of Theorem 7.22 with $T(x) = tx$, and therefore $\kappa = |\det[t\mathbf{e}_1, \dots, t\mathbf{e}_d]| = t^d |\det[\mathbf{e}_1, \dots, \mathbf{e}_d]| = t^d$ (Proposition 7.4 (a) and (b)). □

A **permutation of variables** does not effect the integral.

Theorem 7.24. — *Let $f \in L^1(\Omega_1 \times \Omega_2; E)$, where Ω_1 is an open subset of $\mathbb{R}^{d_1}$, Ω_2 is an open subset of $\mathbb{R}^{d_2}$, E is a Neumann space, and ω is a measurable subset of $\Omega_1 \times \Omega_2$.*

Then, the measure $(y_1, y_2) \mapsto f(y_2, y_1)$, which we denote by $\breve{f}$, belongs to the space $L^1(\Omega_2 \times \Omega_1; E)$, the set $\breve{\omega} = \{(x_2, x_1) : (x_1, x_2) \in \omega\}$ is measurable, and

$$\int_\omega f(x_1, x_2)\, \mathrm{d}(x_1, x_2) = \int_{\breve{\omega}} f(y_2, y_1)\, \mathrm{d}(y_1, y_2). \blacksquare$$

In other words, for every $\varphi \in \mathcal{K}(\Omega_2 \times \Omega_1)$,

$$\langle \breve{f}, \varphi \rangle_{\Omega_2 \times \Omega_1} \stackrel{\text{def}}{=} \langle f, \breve{\varphi} \rangle_{\Omega_1 \times \Omega_2},$$

where $\breve{\varphi}(x_1, x_2) = \varphi(x_2, x_1)$, and

$$\int_\omega f = \int_{\breve{\omega}} \breve{f}. \blacksquare$$

Proof. This is a particular case of Theorem 7.22 regarding the image under a linear change of variable.

More precisely, $\breve{f} = f \circ T$, where $T(x_2, x_1) = (x_1, x_2)$ and $\Omega = \Omega_1 \times \Omega_2$. Indeed, $(T(\mathbf{e}_1), \ldots, T(\mathbf{e}_d))$, where $d = d_1 + d_2$, is the family $(\mathbf{e}_1, \ldots, \mathbf{e}_d)$ rearranged in a different order, so (Proposition 7.4 (a) and (b))

$$\kappa = |\det[T(\mathbf{e}_1), \ldots, T(\mathbf{e}_d)]| = |\det[\mathbf{e}_1, \ldots, \mathbf{e}_d]| = 1. \square$$

Chapter 8

Multivariable Integration

The aim of this chapter is to show that integrating with respect to several variables simultaneously or successively (in any order) amounts to the same thing, which is the context of Fubini's theorem.

We show this via the following steps.

— After characterizing measures of measures (§ 8.1), we interchange their integrals and permute their variables when they are integrable (§ 8.2).

— We establish Fubini's theorem for continuous functions (Proposition 8.6).

— We separate the variables of a measure f on a product $\Omega_1 \times \Omega_2$ (§ 8.4), and then show that the resulting measure of measures $\underline{f}$ is integrable provided f is (Theorem 8.12).

— We then establish Fubini's general theorem, that is, for integrable measures (Theorem 8.13), by reducing the situation via approximations to the case involving continuous functions.

8.1. Permutation of variables in a measure of measures

The space $\mathcal{M}(\Omega_1; \mathcal{M}(\Omega_2; E))$ of **measures of measures** is well-defined (by Definition 3.2 of a measure), since $\mathcal{M}(\Omega_2; E)$ is a Neumann space according to Theorem 3.8.

Also according to Definition 3.2, it is endowed with the semi-norms indexed by $\varphi_1 \in \mathcal{K}(\Omega_1)$, $\varphi_2 \in \mathcal{K}(\Omega_2)$, and $\nu \in \mathcal{N}_E$,

$$\|f\|_{\mathcal{M}(\Omega_1;\mathcal{M}(\Omega_2;E));\varphi_1,\varphi_2,\nu} \stackrel{\text{def}}{=} \|\langle\langle f, \varphi_1\rangle_{\Omega_1}, \varphi_2\rangle_{\Omega_2}\|_{E;\nu}. \tag{8.1}$$

Let us characterize these measures of measures, denoting by $\operatorname{Lin}(E; F)$ the vector space of linear mappings from E into F.

Theorem 8.1. — *We have*

$$f \in \mathcal{M}(\Omega_1; \mathcal{M}(\Omega_2; E)),$$

where Ω_1 is an open subset of $\mathbb{R}^{d_1}$, Ω_2 is an open subset of $\mathbb{R}^{d_2}$, and E is a Neumann space whose family of semi-norms is $\{\| \ \|_{E;\nu} : \nu \in \mathcal{N}_E\}$, if and only if:
— $f \in \mathrm{Lin}(\mathcal{K}(\Omega_1); \mathrm{Lin}(\mathcal{K}(\Omega_2); E))$;
— *for any $\varphi_1 \in \mathcal{K}(\Omega_1)$, there exist $q_2 \in \mathcal{C}^+(\Omega_2)$ and $c_2 \in \mathbb{R}$ such that: $\forall \varphi_2 \in \mathcal{K}(\Omega_2)$,*

$$\|\langle\langle f, \varphi_1\rangle_{\Omega_1}, \varphi_2\rangle_{\Omega_2}\|_{E;\nu} \leq c_2 \|\varphi_2\|_{\mathcal{K}(\Omega_2);q_2};$$

—*for any $\varphi_2 \in \mathcal{K}(\Omega_2)$, there exist $q_1 \in \mathcal{C}^+(\Omega_1)$ and $c_1 \in \mathbb{R}$ such that: $\forall \varphi_1 \in \mathcal{K}(\Omega_1)$,*

$$\|\langle\langle f, \varphi_1\rangle_{\Omega_1}, \varphi_2\rangle_{\Omega_2}\|_{E;\nu} \leq c_1 \|\varphi_1\|_{\mathcal{K}(\Omega_1);q_1}. \ \blacksquare$$

Proof. According to the characterization of measures (Theorem 3.3), it is necessary and sufficent to show:
— that f is a linear mapping from $\mathcal{K}(\Omega_1)$ into $\mathcal{M}(\Omega_2; E)$, which is equivalent to the first condition and to $\langle f, \varphi_1\rangle_{\Omega_1} \in \mathcal{M}(\Omega_2; E)$, *i.e.* to the second condition;
— that, for every $\varphi_2 \in \mathcal{K}(\Omega_2)$ and $\nu \in \mathcal{N}_E$,

$$\|\langle f, \varphi_1\rangle_{\Omega_1}\|_{\mathcal{M}(\Omega_2;E);\varphi_2,\nu} \leq c_1 \|\varphi_1\|_{\mathcal{K}(\Omega_1);q_1},$$

which is the last condition. □

Let us study the permutation of variables.

Theorem 8.2. — *The permutation of variables, defined for every $\varphi_1 \in \mathcal{K}(\Omega_1)$ and $\varphi_2 \in \mathcal{K}(\Omega_2)$, by*

$$\langle\langle \check{f}, \varphi_2\rangle_{\Omega_2}, \varphi_1\rangle_{\Omega_1} = \langle\langle f, \varphi_1\rangle_{\Omega_1}, \varphi_2\rangle_{\Omega_2}, \tag{8.2}$$

is an isomorphism from $\mathcal{M}(\Omega_1; \mathcal{M}(\Omega_2; E))$ onto $\mathcal{M}(\Omega_2; \mathcal{M}(\Omega_1; E))$, where Ω_1 is an open subset of $\mathbb{R}^{d_1}$, Ω_2 is an open subset of $\mathbb{R}^{d_2}$, and E is a Neumann space. ■

Proof. This follows immediately from the characterization of measures of measures from Theorem 8.1. □

8.2. Integration of an integrable measure of measures

Let us begin by showing that changing the **order of successive integrals** of an integrable measure of measures does not affect the result[1].

---

1 **Historical background of Theorems 8.3 and 8.5 on changing the order of successive integrals.** The fact that the result remains unchanged when reversing the order of successive real-valued integrals, *i.e.* (8.25), is an immediate corollary of (8.24), *i.e.* of the theorem which Guido FUBINI proved in 1907 [31] and bears his name, and which Salomon BOCHNER extended to values in a Banach space in 1933 [5].
The analogous results for the integral of an integral with values in $L^1(\Omega_2; E)$, namely **Theorems 8.3 and 8.5, are uncommon**, even for $E = \mathbb{R}$, although they are straightforward consequences of the properties of vector-valued integrals.

Theorem 8.3. — *Let*

$$f \in L^1(\Omega_1; L^1(\Omega_2; E)),$$

where Ω_1 is an open subset of $\mathbb{R}^{d_1}$, Ω_2 is an open subset of $\mathbb{R}^{d_2}$, E and is a Neumann space, and let ω_1 be a measurable subset of Ω_1 and ω_2 be a measurable subset of Ω_2. Then,

$$\int_{\omega_2} \int_{\omega_1} f = \int_{\omega_1} \int_{\omega_2} f. \blacksquare$$

Meaning of the integrals in Theorem 8.3. *On the left-hand side:* $\int_{\omega_1} f$ is the integral over ω_1 of f, which takes values in $L^1(\Omega_2; E)$. Thus, $\int_{\omega_1} f \in L^1(\Omega_2; E)$, and $\int_{\omega_2} \int_{\omega_1} f \in E$ is its integral over ω_2.

On the right-hand side: $\int_{\omega_2} f$ is the image of f under the mapping $\int_{\omega_2}$, which is continuous linear from $L^1(\Omega_2; E)$ into E. Thus, $\int_{\omega_2} f \in L^1(\Omega_1; E)$, and $\int_{\omega_1} \int_{\omega_2} f \in E$ is its integral over ω_1. □

Proof of Theorem 8.3. This is a particular case of Theorem 6.13 regarding the commutation of the integral, here $\int_{\omega_1}$, with a continuous linear mapping, here $\int_{\omega_2}$, since $\int_{\omega_2}$ is continuous from $L^1(\Omega_2; E)$ into E (Theorem 5.5). □

Let us now turn to the **permutation of variables**.

Theorem 8.4. — *The permutation of variables $f \mapsto \check{f}$ is an isomorphism from the space $L^1(\Omega_1; L^1(\Omega_2; E))$ onto $L^1(\Omega_2; L^1(\Omega_1; E))$, where Ω_1 is an open subset of $\mathbb{R}^{d_1}$, Ω_2 is an open subset of $\mathbb{R}^{d_2}$, and E is a Neumann space.*

Denoting by $\{\| \|_{E;\nu} : \nu \in \mathcal{N}_E\}$ the family of semi-norms of E, we have, for every $f \in L^1(\Omega_1; L^1(\Omega_2; E))$ and $\nu \in \mathcal{N}_E$,

$$\|\check{f}\|_{L^1(\Omega_2; L^1(\Omega_1; E));\nu} = \|f\|_{L^1(\Omega_1; L^1(\Omega_2; E));\nu}. \blacksquare \tag{8.3}$$

Proof. Since the proof is rather laborious, we refer the reader to the proof for the more general result when $1 \leq p < \infty$ (Theorem 14.5). Note that Theorem 8.4 is used solely to state Theorem 8.5, which is not used in what follows. □

Let us show that the **permutation of variables** does not affect the value of the integral, whether it involves two or more variables.

Theorem 8.5. — *Let*

$$f \in L^1(\Omega_1; L^1(\Omega_2; E)),$$

where Ω_1 is an open subset of $\mathbb{R}^{d_1}$, Ω_2 is an open subset of $\mathbb{R}^{d_2}$, E is a Neumann space, and $\check{f} \in L^1(\Omega_2; L^1(\Omega_1; E))$ is the measure obtained by permuting the variables.

Then, for any measurable subset ω_1 of Ω_1 and ω_2 of Ω_2:

(a) $$\int_{\omega_1} \check{f} = \int_{\omega_1} f;$$

(b) $$\int_{\omega_1}\int_{\omega_2} \check{f} = \int_{\omega_2}\int_{\omega_1} \check{f} = \int_{\omega_2}\int_{\omega_1} f = \int_{\omega_1}\int_{\omega_2} f. \blacksquare$$

Meaning of the integrals in Theorem 8.5 (a). *On the right-hand side:* $\int_{\omega_1} f$ is the integral over ω_1 of f. *On the left-hand side:* $\int_{\omega_1} \check{f}$ is the image of $\check{f}$ under the mapping $\int_{\omega_1}$ from $L^1(\Omega_1; E)$ into E. □

Proof of Theorem 8.5. **1o)** *An equality.* Let $\varphi_2 \in \mathcal{K}(\Omega_2)$. Observe that the mapping $g \mapsto \langle g, \varphi_2\rangle_{\Omega_2}$ is continuous linear from $\mathcal{M}(\Omega_2; E)$ into E (Theorem 3.9 (a)). The image $\langle f, \varphi_2\rangle_{\Omega_2}$ of the measure f under this mapping satisfies, by Definition 6.9, for every $\varphi_1 \in \mathcal{K}(\Omega_1)$,

$$\langle\langle f, \varphi_2\rangle_{\Omega_2}, \varphi_1\rangle_{\Omega_1} = \langle\langle f, \varphi_1\rangle_{\Omega_1}, \varphi_2\rangle_{\Omega_2}.$$

The right-hand side here is equal to $\langle\langle \check{f}, \varphi_2\rangle_{\Omega_2}, \varphi_1\rangle_{\Omega_1}$ according to Definition (8.2) of $\check{f}$, thus

$$\langle \check{f}, \varphi_2\rangle_{\Omega_2} = \langle f, \varphi_2\rangle_{\Omega_2}. \tag{8.4}$$

2o) *Equality* (a). As the integral commutes with continuous linear mappings (Theorem 6.13),

$$\left\langle \int_{\omega_1} \check{f}, \varphi_2 \right\rangle_{\Omega_2} = \int_{\omega_1} \langle \check{f}, \varphi_2\rangle_{\Omega_2}.$$

Using (8.4), as well as the commutation with the mapping $g \mapsto \langle g, \varphi_2\rangle_{\Omega_2}$ again, it follows that

$$\left\langle \int_{\omega_1} \check{f}, \varphi_2 \right\rangle_{\Omega_2} = \int_{\omega_1} \langle f, \varphi_2\rangle_{\Omega_2} = \left\langle \int_{\omega_1} f, \varphi_2 \right\rangle_{\Omega_2}.$$

This holds for all $\varphi_2 \in \mathcal{K}(\Omega_2)$, and therefore

$$\int_{\omega_1} \check{f} = \int_{\omega_1} f.$$

3o) *Equalities* (b). The intermediate equality follows from (a), whereas the other two follow from Theorem 8.3, relative to $\check{f}$ and f, respectively. □

8.3. Separation of variables in an integral of a continuous function

We shall now prove Fubini's theorem by reducing it to the case involving a continuous function and, more precisely, by using the following property of the Cauchy integral.

Proposition 8.6. — *Let $f \in \mathbf{C}_{Q_1\times Q_2}(\Omega_1 \times \Omega_2; E)$, where Ω_1 is an open subset and Q_1 is a compact subset of $\mathbb{R}^{d_1}$, Ω_2 is an open subset and Q_2 is a compact subset of $\mathbb{R}^{d_2}$, and E is a Neumann space. Let $\underline{f}$ be the function obtained via separation of variables, namely,*

$$(\underline{f}(x_1))(x_2) = f(x_1, x_2).$$

Then, $\underline{f} \in \mathbf{C}_{Q_1}(\Omega_1; \mathbf{C}_{Q_2}(\Omega_2; E))$ and, for every open set $\omega_1 \subset \Omega_1$ and $\omega_2 \subset \Omega_2$,

$$\int_{\omega_1\times\omega_2} f = \int_{\omega_2}\int_{\omega_1} \underline{f} = \int_{\omega_1}\int_{\omega_2} \underline{f}. \blacksquare$$

Meaning of the integrals in Proposition 8.6. *In the middle term:* $\int_{\omega_1} \underline{f}$ is the integral of the function $\underline{f}$, which takes values in $\mathbf{C}_{Q_2}(\Omega_2; E)$. Thus, $\int_{\omega_1} \underline{f} \in \mathbf{C}_{Q_2}(\Omega_2; E)$, and $\int_{\omega_2}\int_{\omega_1} \underline{f} \in E$ is its integral.

On the right-hand side: $\int_{\omega_2} \underline{f}$ is the image of $\underline{f}$ under the mapping $\int_{\omega_2}$, which is continuous linear from $\mathbf{C}_{Q_2}(\Omega_2; E)$ into E. Thus, $\int_{\omega_2} \underline{f} \in \mathbf{C}_{Q_1}(\Omega_1; E)$, and $\int_{\omega_1}\int_{\omega_2} \underline{f} \in E$ is its integral. □

Let us recall the property of separation of variables for a continuous function that we used in the statement [Vol. 2, Theorem 1.27 (c)].

Proposition 8.7. — *With the hypotheses of Proposition 8.6, the mapping $f \mapsto \underline{f}$ is an isomorphism from $\mathbf{C}_{Q_1\times Q_2}(\Omega_1 \times \Omega_2; E)$ onto $\mathbf{C}_{Q_1}(\Omega_1; \mathbf{C}_{Q_2}(\Omega_2; E))$.* $\blacksquare$

The proof of Proposition 8.6 will also make use of the following bounds, which we shall verify immediately afterward.

Lemma 8.8. — *Let $f \in \mathcal{B}(\Omega; E)$, where Ω is an open subset of $\mathbb{R}^d$, E is a Neumann space, ω is an open subset of Ω, and $\mathbb{S}^n_\omega f$ is the approximated integral (Definition 1.14). Then, for every semi-norm $\| \ \|_{E;\nu}$ of E:*

(a) $$\|\mathbb{S}^n_\omega f\|_{E;\nu} \leq |\omega| \sup_{x\in\omega} \|f(x)\|_{E;\nu};$$

(b) $$\left\| \int_\omega f - \mathbb{S}^n_\omega f \right\|_{E;\nu} \leq |\omega| \sup_{|y-x|\leq 2^{-n-1}\sqrt{d}} \|f(y) - f(x)\|_{E;\nu} + \\ + (|\omega| - 2^{-nd}k_n(\omega)) \sup\nolimits_{x\in\omega} \|f(x)\|_{E;\nu},$$

where $k_n(\omega)$ is the number of closed cubes $\Delta_{s,n}$ of side length 2^{-n} centered at $2^{-n}s$, where $s \in \mathbb{Z}^d$, which are included in ω. ∎

Proof of Proposition 8.6. Since the membership of $\underline{f}$ to $\mathbf{C}_{Q_1}(\Omega_1; \mathbf{C}_{Q_2}(\Omega_2; E))$ and the equality $\int_{\omega_2}\int_{\omega_1}\underline{f} = \int_{\omega_1}\int_{\omega_2}\underline{f}$ were established by Propositions 8.7 and 1.28, it remains to verify, which we shall do in two steps, the equality

$$\int_{\omega_1\times\omega_2} f = \int_{\omega_2}\int_{\omega_1}\underline{f}.$$

1°) Calculation of the approximated integrals. By Definition 1.14 of the appoximated integral,

$$\mathbb{S}^n_{\omega_1}\underline{f} = \sum_{s_1\in\mathbb{Z}^{d_1}:\Delta_{s_1,n}\subset\omega_1} 2^{-nd_1}\underline{f}(2^{-n}s_1)$$

in $\mathbf{C}_{Q_2}(\Omega_2; E)$, where $\Delta_{s_1,n}$ is the closed cube in $\mathbb{R}^{d_1}$ with side length 2^{-n} centered at $2^{-n}s_1$. Therefore, in E,

$$\mathbb{S}^n_{\omega_2}(\mathbb{S}^n_{\omega_1}\underline{f}) = \\ = \sum_{s_2\in\mathbb{Z}^{d_2}:\Delta_{s_2,n}\subset\omega_2} 2^{-nd_2} \sum_{s_1\in\mathbb{Z}^{d_1}:\Delta_{s_1,n}\subset\omega_1} 2^{-nd_1}(\underline{f}(2^{-n}s_1))(2^{-n}s_2) = \\ = \sum_{(s_1,s_2)\in\mathbb{Z}^{d_1+d_2}:\Delta_{(s_1,s_2),n}\subset\omega_1\times\omega_2} 2^{-n(d_1+d_2)} f(2^{-n}(s_1,s_2)) = \mathbb{S}^n_{\omega_1\times\omega_2} f.$$

Indeed, a cube $\Delta_{(s_1,s_2),n}$ is included in $\omega_1\times\omega_2$ if and only if we have both inclusions $\Delta_{s_1,n}\subset\omega_1$ and $\Delta_{s_2,n}\subset\omega_2$.

2°) Convergence. By Definition 1.14 of the integral, $\mathbb{S}^n_{\omega_1\times\omega_2} f \to \int_{\omega_1\times\omega_2} f$, so it suffices to show that, as $n\to\infty$,

$$\mathbb{S}^n_{\omega_2}(\mathbb{S}^n_{\omega_1}\underline{f}) \to \int_{\omega_2}\int_{\omega_1}\underline{f}.$$

Decompose

$$\mathbb{S}^n_{\omega_2}(\mathbb{S}^n_{\omega_1}\underline{f}) - \int_{\omega_2}\Big(\int_{\omega_1}\underline{f}\Big) = \mathbb{S}^n_{\omega_2}\Big(\mathbb{S}^n_{\omega_1}\underline{f} - \int_{\omega_1}\underline{f}\Big) + \Big(\mathbb{S}^n_{\omega_2} - \int_{\omega_2}\Big)\Big(\int_{\omega_1}\underline{f}\Big).$$

Since $\int_{\omega_1} \underline{f} \in \mathbf{C}_{Q_2}(\Omega_2; E)$, we have $(\mathbb{S}^n_{\omega_2} - \int_{\omega_2})(\int_{\omega_1} \underline{f}) \to 0$, so it remains to establish that

$$\mathbb{S}^n_{\omega_2}\left(\mathbb{S}^n_{\omega_1}\underline{f} - \int_{\omega_1} \underline{f}\right) \to 0. \tag{8.5}$$

Given that $\mathbb{S}^n_{\omega_1}\underline{f}$ and $\int_{\omega_1} \underline{f}$ both belong to $\mathbf{C}_{Q_2}(\Omega_2; E)$, which is endowed (Definition 1.11 (b)) with the semi-norms of $\mathcal{C}_b(\Omega; E)$, the bound from Lemma 8.8 (a) with values in this space, states that, for every semi-norm $\| \, \|_{E;\nu}$ of E,

$$\left\|\mathbb{S}^n_{\omega_2}\left(\mathbb{S}^n_{\omega_1}\underline{f} - \int_{\omega_1} \underline{f}\right)\right\|_{E;\nu} \le |\omega_2| \left\|\mathbb{S}^n_{\omega_1}\underline{f} - \int_{\omega_1} \underline{f}\right\|_{\mathcal{C}_b(\Omega_2;E);\nu}.$$

Furthermore, the bound from Lemma 8.8 (b) gives

$$\begin{aligned}\left\|\mathbb{S}^n_{\omega_1}\underline{f} - \int_{\omega_1} \underline{f}\right\|_{\mathcal{C}_b(\Omega_2;E);\nu} \le |\omega_1| \sup_{|y_1 - x_1| \le 2^{-n-1}\sqrt{d_1}} \|\underline{f}(y_1) - \underline{f}(x_1)\|_{\mathcal{C}_b(\Omega_2;E);\nu} + \\ + \left(|\omega_1| - 2^{-nd_1} k_n(\omega_1)\right) \sup_{x_1 \in \omega_1} \|\underline{f}(x_1)\|_{\mathcal{C}_b(\Omega_2;E);\nu}.\end{aligned}$$

The right-hand side tends to 0, since $\underline{f}$ is uniformly continuous and bounded with values in $\mathbf{C}_{Q_2}(\Omega_2; E)$, and $2^{-nd_1} k_n(\omega_1) \to |\omega_1|$ by Definition 1.20 of the measure. This implies (8.5), and therefore

$$\int_{\omega_2} \int_{\omega_1} \underline{f} = \int_{\omega_1 \times \omega_2} f. \;\square$$

Proof of Lemma 8.8 ***1º)*** *Inequality* (a). By Definition 1.14 of the approximated integral,

$$\|\mathbb{S}^n_\omega f\|_{E;\nu} = \left\| \sum_{s \in \mathbb{Z}^d : \Delta_{s,n} \subset \omega} 2^{-nd} f(2^{-n}s) \right\|_{E;\nu} \le 2^{-nd} k_n(\omega) \sup_{x \in \omega} \|f(x)\|_{E;\nu},$$

where $k_n(\omega)$ is the number of cubes $\Delta_{s,n}$ included in ω. This provides the stated inequality, since $2^{-nd} k_n(\omega) \le |\omega|$ (inequality (1.9), p. 16).

2º) *Inequality* (b). The decomposition $\mathbb{S}^m_\omega f - \mathbb{S}^n_\omega f = \mathcal{X}_{m,n} + \mathcal{Y}_{m,n}$ of a difference of approximated integrals, carried out in (1.1), p. 12, along with the bounds (1.2) and (1.5) for their semi-norms $\|\mathcal{X}_{m,n}\|_{E;\nu}$ and $\|\mathcal{Y}_{m,n}\|_{E;\nu}$, give, for $m \ge n$,

$$\begin{aligned}\|\mathbb{S}^m_\omega f - \mathbb{S}^n_\omega f\|_{E;\nu} &\le \|\mathcal{X}_{m,n}\|_{E;\nu} + \|\mathcal{Y}_{m,n}\|_{E;\nu} \le \\ &\le 2^{-nd} k_n(\omega) \sup_{|y - x| \le 2^{-n-1}\sqrt{d}} \|f(y) - f(x)\|_{E;\nu} + \\ &\quad + \left(2^{-md} k_m(\omega) - 2^{-nd} k_n(\omega)\right) \sup_{x \in \omega} \|f(x)\|_{E;\nu}.\end{aligned}$$

This gives the stated inequality by taking the limit as $m \to \infty$, as $2^{-nd} k_n(\omega) \le |\omega|$ (inequality (1.9), again), and $2^{-md} k_m(\omega) \to |\omega|$ (Definition 1.20). $\square$

8.4. Separation of variables of a measure

Let us define the separation of variables of a measure.

Definition 8.9. — *Let*

$$f \in \mathcal{M}(\Omega_1 \times \Omega_2; E),$$

where Ω_1 *is an open subset of* $\mathbb{R}^{d_1}$, Ω_2 *is an open subset of* $\mathbb{R}^{d_2}$, *and* E *is a Neumann space. We define*

$$\underline{f} \in \mathcal{M}(\Omega_1; \mathcal{M}(\Omega_2; E))$$

by: for every $\varphi_1 \in \mathcal{K}(\Omega_1)$ *and* $\varphi_2 \in \mathcal{K}(\Omega_2)$,

$$\langle\langle \underline{f}, \varphi_1\rangle_{\Omega_1}, \varphi_2\rangle_{\Omega_2} \stackrel{\text{def}}{=} \langle f, \varphi_1 \otimes \varphi_2\rangle_{\Omega_1\times\Omega_2},$$

where $(\varphi_1 \otimes \varphi_2)(x_1, x_2) \stackrel{\text{def}}{=} \varphi_1(x_1)\varphi_2(x_2)$. ∎

Justification. ***1°) Setting of the problem.*** We must verify our claim that

$$\underline{f} \in \mathcal{M}(\Omega_1; \mathcal{M}(\Omega_2; E)).$$

According to the characterization of measures provided by Theorem 8.1, this amounts to verifying that:

— the mapping $\underline{f}$ is linear from $\mathcal{K}(\Omega_1)$ into $\mathrm{Lin}(\mathcal{K}(\Omega_2); E)$, which is clear since f is linear;

— for every $\varphi_1 \in \mathcal{K}(\Omega_1)$, there exist $q_2 \in \mathcal{C}^+(\Omega_2)$ and $c_2 \in \mathbb{R}$ such that: for every $\varphi_2 \in \mathcal{K}(\Omega_2)$,

$$\|\langle\langle \underline{f}, \varphi_1\rangle_{\Omega_1}, \varphi_2\rangle_{\Omega_2}\|_{E;\nu} \leq c_2 \|\varphi_2\|_{\mathcal{K}(\Omega_2);q_2}; \tag{8.6}$$

— for every $\varphi_2 \in \mathcal{K}(\Omega_2)$, there exist $q_1 \in \mathcal{C}^+(\Omega_1)$ and $c_1 \in \mathbb{R}$ such that: for every $\varphi_1 \in \mathcal{K}(\Omega_1)$,

$$\|\langle\langle \underline{f}, \varphi_1\rangle_{\Omega_1}, \varphi_2\rangle_{\Omega_2}\|_{E;\nu} \leq c_1 \|\varphi_1\|_{\mathcal{K}(\Omega_1);q_1}. \tag{8.7}$$

2°) Verification of (8.7). By the definition of $\underline{f}$, this can be written as

$$\|\langle f, \varphi_1 \otimes \varphi_2\rangle_{\Omega_1\times\Omega_2}\|_{E;\nu} \leq c_1 \|\varphi_1\|_{\mathcal{K}(\Omega_1);q_1}.$$

According to the characterization of measures from Theorem 3.3, there exist $q \in \mathcal{C}^+(\Omega_1 \times \Omega_2)$ and a constant $c \in \mathbb{R}$ such that the right-hand side is bounded by $c\|\varphi_1 \otimes \varphi_2\|_{\mathcal{K}(\Omega_1\times\Omega_2);q}$.

Therefore, it suffices to show that, for every $q \in \mathcal{C}^+(\Omega_1 \times \Omega_2)$, $\varphi_2 \in \mathcal{K}(\Omega_2)$, and $\nu \in \mathcal{N}_E$, there exists $q_1 \in \mathcal{C}^+(\Omega_1)$ such that, for every $\varphi_1 \in \mathcal{K}(\Omega_1)$,

$$\|\varphi_1 \otimes \varphi_2\|_{\mathcal{K}(\Omega_1\times\Omega_2);q} \leq \|\varphi_1\|_{\mathcal{K}(\Omega_1);q_1}. \tag{8.8}$$

3º) Construction of q_1. By Definition 3.1 of the semi-norms of $\mathcal{K}(\Omega)$, we can rewrite (8.8) as

$$\sup_{(x_1,x_2)\in\Omega_1\times\Omega_2} q(x_1,x_2)|\varphi_1(x_1)\varphi_2(x_2)| \le \sup_{x_1\in\Omega_1} q_1(x_1)|\varphi_1(x_1)|.$$

This inequality is satisfied by the function q_1, defined by

$$q_1(x_1) \stackrel{\text{def}}{=} \sup_{x_2\in\Omega_2} q(x_1,x_2)|\varphi_2(x_2)|.$$

It therefore suffices to verify that it belongs to $\mathcal{C}^+(\Omega_1)$, and more precisely that it is continuous since it is positive (as q is positive).

4º) Continuity of q_1. Let $y_1 \in \Omega_1$. The function $(x_1,x_2) \mapsto q(x_1,x_2)|\varphi_2(x_2)|$ is continuous (as is any product of continuous functions, see Proposition A.48), thus, by Heine's theorem (Proposition A.41), it is uniformly continuous on the compact set $B(y_1,r_1) \times K_2$, where r_1 is small enough so that the closed ball $B(y_1,r_1)$ is included in Ω_1, and where K_2 is the support of φ_2.

Given $\epsilon > 0$, there exists $\eta > 0$ such that

$$|(x_1,x_2)-(y_1,y_2)| \le \eta \quad \Rightarrow \quad \big||q(x_1,x_2)|\varphi_2(x_2)| - q(y_1,y_2)|\varphi_2(y_2)|\big| \le \epsilon.$$

For $y_2 = x_2$, it follows then, if $|x_1 - y_1)| \le \eta$, that

$$|q_1(x_1) - q_1(y_1)| \le \epsilon.$$

This proves that q_1 is continuous at the point y_1.

5º) Verification of (8.6). It is sufficient for us to exchange the roles of φ_1 and φ_2 in the proof of (8.7). This finishes the justification of Definition 8.9. □

Let us reassure ourselves that we recover the separation of variables for functions.

Theorem 8.10. — *If $f \in \mathcal{C}(\Omega_1 \times \Omega_2; E)$, the separation of variables in the sense of measures given by Definition 8.9 coincides with that in the sense of functions.*

That is, with the identification (3.16), p. 65, of a continuous function with a measure, Definition 8.9 also states that $\underline{f} \in \mathcal{C}(\Omega_1;\mathcal{C}(\Omega_2;E))$ and, for all $x_1 \in \Omega_1$ and $x_2 \in \Omega_2$,

$$(\underline{f}(x_1))(x_2) = f(x_1,x_2). \ ▮$$

Proof. Let $\bar{f}$ be the measure associated with the function f by Theorem 3.10, and let $\underline{\bar{f}}$ be the measure obtained by separating the variables from Definition 8.9. It remains to verify that

$$\underline{\bar{f}} = \overline{\overline{\underset{\sim}{f}}^{1}}^{2}, \tag{8.9}$$

where $\underset{\sim}{f}$ is the function obtained by separating the variables of f, whereas $\overline{\ }^{1}$ and $\overline{\ }^{2}$ are the mappings that associate a measure with a continuous function (by Theorem 3.10) on Ω_1 and Ω_2, respectively.

Let $\varphi_1 \in \mathcal{K}(\Omega_1)$ and $\varphi_2 \in \mathcal{K}(\Omega_2)$. Definition 8.9 of the separation of variables (of $\bar{f}$), then the definition of $\bar{f}$ from Theorem 3.10, and then that of the tensor product, successively give

$$\langle\langle \underline{\bar{f}}, \varphi_1\rangle_{\Omega_1}, \varphi_2\rangle_{\Omega_2} = \langle\langle \bar{f}, \varphi_1 \otimes \varphi_2\rangle_{\Omega_1\times\Omega_2} =$$
$$= \int_{\Omega_1\times\Omega_2} f\,\varphi_1 \otimes \varphi_2 = \int_{\Omega_1\times\Omega_2} f(x_1,x_2)\varphi_1(x_1)\varphi_2(x_2)\,\mathrm{d}(x_1,x_2).$$

By separating the variables in the Cauchy integral (Proposition 8.6), and then in the function f, it becomes

$$\langle\langle \underline{\bar{f}}, \varphi_1\rangle_{\Omega_1}, \varphi_2\rangle_{\Omega_2} = \int_{\Omega_2}\Big(\int_{\Omega_1} f(x_1,x_2)\varphi_1(x_1)\varphi_2(x_2)\,\mathrm{d}x_1\Big)\,\mathrm{d}x_2 =$$
$$= \int_{\Omega_2}\Big(\int_{\Omega_1} (\underset{\sim}{f}(x_1))(x_2)\varphi_1(x_1)\,\mathrm{d}x_1\Big)\varphi_2(x_2)\,\mathrm{d}x_2.$$

Using the definitions of $\overline{\ }^{1}$ and $\overline{\ }^{2}$ from Theorem 3.10, we obtain

$$\langle\langle \underline{\bar{f}}, \varphi_1\rangle_{\Omega_1}, \varphi_2\rangle_{\Omega_2} = \int_{\Omega_2} \langle \overline{\underset{\sim}{f}}^{1}, \varphi_1\rangle_{\Omega_1}(x_2)\,\varphi_2(x_2)\,\mathrm{d}x_2 = \Big\langle \overline{\langle \overline{\underset{\sim}{f}}^{1}, \varphi_1\rangle_{\Omega_1}}^{2}, \varphi_2\Big\rangle_{\Omega_2}.$$

This proves (8.9), *i.e.* we obtain the same measure by separating the variables in the sense of measures, as we do by separating them in the sense of functions.

Using Proposition 8.6 above is permitted here, because the function $f\varphi_1 \otimes \varphi_2$ belongs to $\mathbf{C}_{K_1\times K_2}(\Omega_1 \times \Omega_2; E)$, where K_i is the support of φ_i. □

Let us show that the separation of variables is continuous.

Theorem 8.11. — *The separation of variables is continuous linear, and thus sequentially continuous, from $\mathcal{M}(\Omega_1 \times \Omega_2; E)$ into $\mathcal{M}(\Omega_1; \mathcal{M}(\Omega_2; E))$, where Ω_1 is an open subset of $\mathbb{R}^{d_1}$, Ω_2 is an open subset of $\mathbb{R}^{d_2}$, and E is a Neumann space.* ▌

Proof. **1o)** *Linearity.* The linearity (Definition A.55) of the mapping $f \mapsto \underline{f}$ is an immediate consequence of Definition 8.9 of $\underline{f}$.

2o) *Continuity.* The semi-norms of $\underline{f}$ are given by (8.1), that is, again by Definition 8.9 of $\underline{f}$, for every $\varphi_1 \in \mathcal{K}(\Omega_1)$, $\varphi_2 \in \mathcal{K}(\Omega_2)$, and $\nu \in \mathcal{N}_E$,

$$\|\underline{f}\|_{\mathcal{M}(\Omega_1;\mathcal{M}(\Omega_2;E));\varphi_1,\varphi_2,\nu} = \|\langle\langle \underline{f}, \varphi_1\rangle_{\Omega_1}, \varphi_2\rangle_{\Omega_2}\|_{E;\nu} =$$

$$= \|\langle f, \varphi_1 \otimes \varphi_2\rangle_{\Omega_1 \times \Omega_2}\|_{E;\nu} = \|f\|_{\mathcal{M}(\Omega_1 \times \Omega_2;E));\varphi_1 \otimes \varphi_2,\nu}.$$

From the characterization of continuous linear mappings from Proposition 1.9, the mapping $f \mapsto \underline{f}$ is thus continuous from $\mathcal{M}(\Omega_1 \times \Omega_2; E)$ into $\mathcal{M}(\Omega_1; \mathcal{M}(\Omega_2; E))$.

3o) *Sequential continuity.* Continuity always implies sequential continuity (Proposition 1.8). □

Bijectivity? Is the separation of variables a bijection from $\mathcal{M}(\Omega_1 \times \Omega_2; E)$ onto $\mathcal{M}(\Omega_1; \mathcal{M}(\Omega_2; E))$? We have not delved into this delicate question, as we do not have an application for it.

Separation of variables **is** a bijection from $L^1(\Omega_1 \times \Omega_2; E)$ onto $L^1(\Omega_1; L^1(\Omega_2; E))$ (by Theorem 14.2 (b), which further shows that it is an isomorphism of these spaces), but it **is not** bijective from $L^\infty(\Omega_1 \times \Omega_2; E)$ onto $L^\infty(\Omega_1; L^\infty(\Omega_2; E))$ (Theorem 14.2 (c)).

The question may seem similar to that which concerns distributions, since the simple topology of $\mathcal{M}(\Omega; E)$ (Theorem 3.2) bears a resemblance to that of $\mathcal{D}'(\Omega; E)$ used in Volume 3 [74, Definition 3.1]. Recall that the separation of variables **is** a bijection from $\mathcal{D}'(\Omega_1 \times \Omega_2; E)$ onto $\mathcal{D}'(\Omega_1; \mathcal{D}'(\Omega_2; E))$; this is the diabolical kernel theorem [74, Theorem 15.10]. However, for measures, the analogue is lacking an ingredient that played an important role in the study of distributions, namely that *every bounded sequence in $\mathcal{D}(\Omega)$ has a convergent subsequence* [74, Theorem 2.14], a property which does not extend to sequences in $\mathcal{K}(\Omega)$. □

8.5. Separation of variables

Observe that the space $L^1(\Omega_1; L^1(\Omega_2; E))$ of **integrable measures of measures** is well-defined (by Definition 4.4 of the space of integrable measures), since $L^1(\Omega_2; E)$ is a Neumann space by Theorem 4.12.

Let us now turn to the separation of variables of an integrable measure.

Theorem 8.12. — *Let*

$$f \in L^1(\Omega_1 \times \Omega_2; E),$$

where Ω_1 is an open subset of $\mathbb{R}^{d_1}$, Ω_2 is an open subset of $\mathbb{R}^{d_2}$, and E is a Neumann space. Then, the measure of measures obtained in Theorem 8.9 via separation of variables satisfy

$$\underline{f} \in L^1(\Omega_1; L^1(\Omega_2; E)).$$

If $(f_n)_{n\in\mathbb{N}}$ is an integrabilizing sequence for f, then $(\underline{f_n})_{n\in\mathbb{N}}$ is an integrabilizing sequence for $\underline{f}$.

Denoting by $\{\| \ \|_{E;\nu} : \nu \in \mathcal{N}_E\}$ the family of semi-norms of E, for every $\nu \in \mathcal{N}_E$,

$$\|\underline{f}\|_{L^1(\Omega_1;L^1(\Omega_2;E));\nu} = \|f\|_{L^1(\Omega_1\times\Omega_2;E);\nu}. \blacksquare$$

Isomorphism. In Theorem 14.2, we shall show in the more general framework of $L^p(\Omega; E)$ spaces, that **the separation of variables is an isomorphism from** $L^1(\Omega_1 \times \Omega_2; E)$ **onto** $L^1(\Omega_1; L^1(\Omega_2; E))$. □

Proof of Theorem 8.12. **1°)** *integrabilizing sequence.* Let $(f_n)_{n\in\mathbb{N}}$ be an integrabilizing sequence for f, i.e. by Definition 4.1, formed of uniformly continuous functions with bounded support such that, as $n \to \infty$,

$$f_n \to f \ \text{ in } \mathcal{M}(\Omega_1 \times \Omega_2; E), \tag{8.10}$$

and, for every $\nu \in \mathcal{N}_E$,

$$\sup_{m\geq n} \int_{\Omega_1\times\Omega_2} \|f_m - f_n\|_{E;\nu} \to 0. \tag{8.11}$$

Let Q_1^n and Q_2^n be compact subsets of $\mathbb{R}^{d_1}$ and $\mathbb{R}^{d_2}$ respectively, such that f_n vanishes outside $Q_1^n \times Q_2^n$. By separating the variables in f_n, we obtain (Proposition 8.7)

$$\underline{f_n} \in \mathbf{C}_{Q_1^n}(\Omega_1; \mathbf{C}_{Q_2^n}(\Omega_2; E)).$$

Let us verify that these functions satisfy the three criteria of an integrabilizing sequence for $\underline{f}$, i.e. in the space of integrable measures on Ω_1 with values in $L^1(\Omega_2; E)$.

1.a. The function $\underline{f_n}$ is uniformly continuous with values in $L^1(\Omega_2; E)$ (Proposition A.40 (a), since $\mathbf{C}_{Q_2^n}(\Omega_2; E)$ is topologically included in the latter, see Theorem 4.9); also, its support is bounded (since it is included in Q_1^n).

1.b. *Criterion (4.2).* The expression of the semi-norms in $L^1(\Omega_2; E)$ of a continuous function (Theorem 4.8) states, for all $x_1 \in \Omega_1$,

$$\|(\underline{f_m} - \underline{f_n})(x_1)\|_{L^1(\Omega_2;E);\nu} = \int_{\Omega_2} \|(\underline{f_m} - \underline{f_n})(x_1)\|_{E;\nu}.$$

The function that maps x_1 to the left-hand term is uniformly continuous (as is any composition of such functions, see Proposition A.39, which here is $f_m - f_n$ with the semi-norm $\| \ \|_{L^1(\Omega_2;E);\nu}$); after integrating it, and using Fubini's theorem for such functions (Proposition 8.6), we obtain

$$\int_{\Omega_1} \|\underline{f_m} - \underline{f_n}\|_{L^1(\Omega_2;E);\nu} = \int_{\Omega_1}\int_{\Omega_2} \|\underline{f_m} - \underline{f_n}\|_{E;\nu} = \int_{\Omega_1\times\Omega_2} \|f_m - f_n\|_{E;\nu}. \tag{8.12}$$

The convergence (8.11) thus implies

$$\sup_{m\geq n}\int_{\Omega_1}\|\underline{f_m}-\underline{f_n}\|_{L^1(\Omega_2;E);\nu}\to 0. \tag{8.13}$$

1.c. Criterion (4.1). According to the expression of the semi-norms in $L^1(\Omega_2;E)$ of a continuous function from Theorem 4.8 again, the convergence (8.13) can be written as

$$\sup_{m\geq n}\|\underline{f_m}-\underline{f_n}\|_{L^1(\Omega_1;L^1(\Omega_2;E));\nu}\to 0.$$

The $\underline{f_n}$ thus form a Cauchy sequence in $L^1(\Omega_1;L^1(\Omega_2;E))$, which is sequentially complete (by Theorem 4.12, since $L^1(\Omega_2;E)$ is sequentially complete according to the aforementioned theorem). It therefore has a limit in it, which we denote by g.

Since the space $L^1(\Omega_1;L^1(\Omega_2;E))$ is, by Theorem 4.9, topologically included in $\mathcal{M}(\Omega_1;L^1(\Omega_2;E))$,

$$\underline{f_n}\to g \text{ in } \mathcal{M}(\Omega_1;L^1(\Omega_2;E)). \tag{8.14}$$

It remains to see that $g=\underline{f}$. As the space $L^1(\Omega_2;E)$ is topologically included in $\mathcal{M}(\Omega_2;E)$ (Theorem 4.9), the space $\mathcal{M}(\Omega_1;L^1(\Omega_2;E))$ is topologically included in $\mathcal{M}(\Omega_1;\mathcal{M}(\Omega_2;E))$ by Theorem 6.15, and so

$$\underline{f_n}\to g \text{ in } \mathcal{M}(\Omega_1;\mathcal{M}(\Omega_2;E)). \tag{8.15}$$

Since the mapping $f\mapsto\underline{f}$ is sequentially continuous from $\mathcal{M}(\Omega_1\times\Omega_2;E)$ into $\mathcal{M}(\Omega_1;\mathcal{M}(\Omega_2;E))$ by Theorem 8.11, the convergence (8.10) gives, for its part,

$$\underline{f_n}\to\underline{f} \text{ in } \mathcal{M}(\Omega_1;\mathcal{M}(\Omega_2;E)).$$

This, together with (8.15), proves that

$$g=\underline{f}. \tag{8.16}$$

1.d. Conclusion. Properties 1.a, (8.13), and (8.14) (together with (8.16)) prove that

$$(\underline{f_n})_{n\in\mathbb{N}} \text{ is an integrabilizing sequence for } \underline{f} \text{ in } L^1(\Omega_1;L^1(\Omega_2;E)).$$

2°*) Integrability.* By Definition 4.1 of an integrable measure, 1.d shows that

$$\underline{f}\in L^1(\Omega_1;L^1(\Omega_2;E)).$$

3°*) Semi-norms.* Definition 4.4 of the semi-norms of $L^1(\Omega;E)$ gives, since the f_n and the $\underline{f_n}$ form integrabilizing sequences for f and $\underline{f}$, respectively,

$$\|f\|_{L^1(\Omega_1\times\Omega_2;E);\nu}=\lim_{n\to\infty}\int_{\Omega_1\times\Omega_2}\|f_n\|_{E;\nu}$$

and

$$\|\underline{f}\|_{L^1(\Omega_1;L^1(\Omega_2;E));\nu} = \lim_{n\to\infty} \int_{\Omega_1} \|\underline{f_n}\|_{L^1(\Omega_2;E);\nu}.$$

However, in (the proof of) equality (8.12), we can replace f_m by 0, giving us

$$\int_{\Omega_1} \|\underline{f_n}\|_{L^1(\Omega_2;E);\nu} = \int_{\Omega_1\times\Omega_2} \|f_n\|_{E;\nu}.$$

From which, in the limit, together with Definition 4.4, we obtain

$$\|\underline{f}\|_{L^1(\Omega_1;L^1(\Omega_2;E));\nu} = \|f\|_{L^1(\Omega_1\times\Omega_2;E);\nu}. \quad \square$$

8.6. Fubini's theorem

Here, let us show that integrating several variables either simultaneously or successively amounts to the same result, which is the content of **Fubini's theorem**[2].

Theorem 8.13. — *Let*

$$f \in L^1(\Omega_1 \times \Omega_2; E),$$

where Ω_1 is an open subset of $\mathbb{R}^{d_1}$, Ω_2 is an open subset of $\mathbb{R}^{d_2}$, E is a Neumann space, and $\underline{f} \in L^1(\Omega_1; L^1(\Omega_2; E))$ is the measure of measures obtained via separation of variables.

Then, for every measurable subset ω_1 of Ω_1 and ω_2 of Ω_2,

$$\int_{\omega_1\times\omega_2} f = \int_{\omega_2}\int_{\omega_1} \underline{f} = \int_{\omega_1}\int_{\omega_2} \underline{f}. \quad \blacksquare$$

Meaning of the integrals in Theorem 8.13. *In the middle term:* $\int_{\omega_1} \underline{f}$ is the integral of the function $\underline{f}$, with values in $L^1(\Omega_2; E)$. Thus, $\int_{\omega_1} \underline{f} \in L^1(\Omega_2; E)$, and $\int_{\omega_2}\int_{\omega_1} \underline{f} \in E$ is its integral.

On the right-hand side: $\int_{\omega_2} \underline{f}$ is the image of $\underline{f}$ under the mapping $\int_{\omega_2}$, which is continuous from $L^1(\Omega_2; E)$ into E. Thus, $\int_{\omega_2} \underline{f} \in L^1(\Omega_1; E)$, and $\int_{\omega_1}\int_{\omega_2} \underline{f} \in E$ is its integral. $\square$

---

2 **Historical background of Fubini's theorem.** Guido FUBINI proved in 1907 [31] the equivalence between a double integral and a reiterated integral of real-valued integrable functions, *i.e.* in the form (8.24). More precisely, he proved it in the more general form (8.26). Salomon BOCHNER extended (8.26) to functions taking values in a Banach space in 1933 [5].

Precursor. Leonhard EULER computed double integrals, *i.e.* in $\mathbb{R}^2$, in 1770 [26] by integrating each variable successively. An example of his calculations is presented by Jean-Paul PIER [58, p. 50]. The integral was not defined rigorously at that time (it would only be done so in 1823, see Note 8, p. 10).

Originality of Theorem 8.13. The result on the equivalence between a double integral with values in E with an integral of an integral with values in $L^1(\Omega_2; E)$ is uncommon, even for $E = \mathbb{R}$.

Traditional formulation. The first equality of Theorem 8.13 is traditionally stated in the form

$$\int_{\omega_1 \times \omega_2} f(x_1, x_2)\,\mathrm{d}(x_1, x_2) = \int_{\omega_1} \Big(\int_{\omega_2} f(x_1, x_2)\,\mathrm{d}x_2 \Big)\,\mathrm{d}x_1,$$

which involves only (reiterated) integrals with values in E: first, the integral of $x_1 \mapsto f(x_1, x_2)$ for fixed x_2, and then that of $x_2 \mapsto \int_{\omega_1} f(x_1, x_2)\,\mathrm{d}x_1$. In particular, for $E = \mathbb{R}$, only real-valued integrals are involved. □

Originality of Theorem 8.13. This statement is original due, on the one hand, to the use of *measures* in the place of *classes of functions equal almost everywhere*, and, on the other hand, to the use of a *measure of measures* $\underline{f}$ and its integral with values in $L^1(\Omega_2; E)$, rather than using reiterated integrals with values in E. □

Proof of Theorem 8.13. We will take the limit in the equalities related to an integrabilizing sequence of f and to the open sets approximating ω_1 and ω_2, in two steps: first, by approximating f and ω_1 when ω_2 is open; then, by approximating ω_2.

1°) First equality: *case where*

$$\omega_2 \text{ is open.}$$

1.a. *Approximation of* ω_1. Let $(\omega_1^n)_{n\in\mathbb{N}}$ be an approximating sequence for ω_1 given by Definition 5.1 of the integral, that is: ω_1^n is open, decreases with n, $\omega_1 \subset \omega_1^n \subset \Omega$ and, as $n \to \infty$,

$$|\omega_1^n \setminus \omega_1| \to 0.$$

Then, the $\omega_1^n \times \omega_2$ form an approximating sequence for $\omega_1 \times \omega_2$ because they are open, decrease with n, contained between $\omega_1 \times \omega_2$ and $\Omega_1 \times \Omega_2$, and, as $n \to \infty$,

$$|(\omega_1^n \times \omega_2) \setminus (\omega_1 \times \omega_2)| \to 0. \qquad (8.17)$$

Indeed, the difference $(\omega_1^n \times \omega_2) \setminus (\omega_1 \times \omega_2)$ is equal to the product $(\omega_1^n \setminus \omega_1) \times \omega_2$, so its measure is equal to the product $|\omega_1^n \setminus \omega_1|\,|\omega_2|$ of measures (Theorem 2.21).

1.b. *Convergence of the double integral.* Moreover, let $(f_n)_{n\in\mathbb{N}}$ be an integrabilizing sequence for f. Then, by Definition 5.1 of the integral, as $n \to \infty$,

$$\int_{\omega_1^n \times \omega_2} f_n \to \int_{\omega_1 \times \omega_2} f \text{ in } E. \qquad (8.18)$$

1.c. *Convergence of reiterated integrals.* The $\underline{f_n}$ form an integrabilizing sequence for $\underline{f}$ according to Theorem 8.12, so, again by Definition 5.1 of the integral,

$$\int_{\omega_1^n} \underline{f_n} \to \int_{\omega_1} \underline{f} \text{ in } L^1(\Omega_2; E).$$

Given that the mapping $f \mapsto \int_{\omega_2} f$ is (Theorem 5.5) sequentially continuous from $L^1(\Omega_2; E)$ into E, we have

$$\int_{\omega_2} \underline{\int_{\omega_1^n} f_n} \to \int_{\omega_2} \underline{\int_{\omega_1} f} \text{ in } E. \tag{8.19}$$

1.d. *Equality.* By Fubini's theorem for continuous functions (Proposition 8.6),

$$\int_{\omega_1^n \times \omega_2} f_n = \int_{\omega_2} \underline{\int_{\omega_1^n} f_n}.$$

If we pass to the limit when $n \to \infty$ in the convergences (8.18) and (8.19), we get: when (ω_1 is measurable and) ω_2 is open,

$$\int_{\omega_1 \times \omega_2} f = \int_{\omega_2} \underline{\int_{\omega_1} f}. \tag{8.20}$$

2°) *First equality: general case.* Now, ω_2 is measurable (like ω_1).

2.a. *Convergence of reiterated integrals.* Let $(\omega_2^n)_{n \in \mathbb{N}}$ be an approximating sequence for ω_2 given by Definition 5.1 of the integral. Then,

$$|\omega_2^n \setminus \omega_2| \to 0. \tag{8.21}$$

The integral converges when the difference between the measures of the domains of integration tends to 0 (Theorem 6.6), thus

$$\int_{\omega_2^n} \underline{\int_{\omega_1} f} \to \int_{\omega_2} \underline{\int_{\omega_1} f}. \tag{8.22}$$

2.b. *Convergence of the double integral.* The convergence (8.21) implies (this follows in the same way as (8.17))

$$|(\omega_1 \times \omega_2^n) \setminus (\omega_1 \times \omega_2)| \to 0.$$

Whence, again by Theorem 6.6,

$$\int_{\omega_1 \times \omega_2^n} f \to \int_{\omega_1 \times \omega_2} f \text{ in } E. \tag{8.23}$$

2.c. *Equality.* Since ω_2^n is open, equality (8.20) gives

$$\int_{\omega_1 \times \omega_2^n} f = \int_{\omega_2^n} \underline{\int_{\omega_1} f}.$$

By passing to the limit as $n \to \infty$, together with the convergences (8.22) and (8.23), we obtain

$$\int_{\omega_1 \times \omega_2} f = \int_{\omega_2} \underline{\int_{\omega_1} f}.$$

3°) Second equality. Theorem 6.13 on the commutation of the integral, here $\int_{\omega_1}$, with a continuous linear mapping, here $\int_{\omega_2}$ which is continuous linear from $L^1(\Omega_2; E)$ into E (Theorem 5.5), gives us, since (Theorem 8.12) $\underline{f} \in L^1(\Omega_1; L^1(\Omega_2; E))$,

$$\int_{\omega_1}\int_{\omega_2} \underline{f} = \int_{\omega_2}\int_{\omega_1} \underline{f}. \ \Box$$

Reiterated integration without using vector values. Theorem 8.13 uses vector values even when a measure f is real-valued, *i.e.* when $E = \mathbb{R}$. Indeed, the measure $\underline{f}$ with separated variables that is integrated belongs to $L^1(\Omega_1; L^1(\Omega_2))$.

It is not straightforward to make sense of (and prove) the formula

$$\int_{\omega_1\times\omega_2} f \, dx = \int_{\omega_1} dx_1 \int_{\omega_2} f(x_1, x_2) \, dx_2 \tag{8.24}$$

without using vector-valued integration. This requires:

1°) giving a meaning, for fixed x_1, to the measure $x_2 \mapsto f(x_1, x_2)$, which we denote by f_{x_1};

2°) showing that f_{x_1} belongs to $L^1(\Omega_2)$;

3°) showing that $x_1 \mapsto f_{x_1}$ belongs to $L^1(\Omega_1)$.

The difficulty lies in the fact that an integrable measure does not, *a priori*, have a value at a single point. By using a Lebesgue *integrable function*, one can indeed define its value *almost everywhere*, and prove 1° and 2° for almost all x_1. The proof is delicate, since f_{x_1} is the restriction of f to the set $\{x_1\} \times \Omega_2$, which has measure zero on $\Omega_1 \times \Omega_2$.

The same difficulties arise when attempting to make sense of (and establish) the following equality without using vector-valued integrals:

$$\int_{\omega_2} dx_2 \int_{\omega_1} f(x_1, x_2) \, dx_1 = \int_{\omega_1} dx_1 \int_{\omega_2} f(x_1, x_2) \, dx_2. \tag{8.25}$$ $\Box$

Interest of vector-valued integration. Vector-valued integration simplifies the proof of certain properties of multiple integrals, even for real-valued measures. For instance, it allows one to reduce the commutation of integrals to a special case of the commutation of the integral with a linear mapping (see the proof of Theorem 8.3), which is elementary. Using only real-valued integrals requires more complicated proofs (see the justification of (8.24)). $\Box$

Separation of variables on a general open set. Let Ω be an open subset of $\mathbb{R}^2$ and $f \in L^1(\Omega; E)$. Its extension by 0 satisfies (Theorem 13.6) $\widetilde{f} \in L^1(\mathbb{R}^2; E)$, hence, by separating its variables using Theorem 8.13, it follows that

$$\int_{\Omega} f \, dx = \int_{\mathbb{R}^2} \widetilde{f} \, dx = \int_{\mathbb{R}} dx_1 \int_{\mathbb{R}} \widetilde{f}_1 \, dx_2.$$

If, for a given x_1, one can give meaning to $f_1(x_1)$, that is, to the measure $x_2 \mapsto f(x_1, x_2)$, then we have $\int_{\mathbb{R}} \widetilde{f}_1(x_1) \, dx_2 = \int_{\Omega_{x_1}} f_1(x_1) \, dx_2$ where $\Omega_{x_1} = \{x_2 \in \mathbb{R} : (x_1, x_2) \in \Omega\}$, and it follows that

$$\int_{\Omega} f \, dx = \int_{\mathbb{R}} dx_1 \int_{\Omega_{x_1}} f(x_1, x_2) \, dx_2. \tag{8.26}$$

Such a meaning is, for example, immediate if $f \in \mathcal{C}(\Omega; E) \cap L^1(\Omega; E)$.

For real-valued functions that are integrable in the Lebesgue sense, one may observe that $\widetilde{f}_{x_1} \in L^1(\mathbb{R})$ for almost every x_1 (these are 1° and 2° in the justification of (8.24)), and that its restriction satisfies $f_{x_1} \in L^1(\Omega_{x_1})$ and $\int_{\Omega_{x_1}} f_{x_1} = \int_{\mathbb{R}} \widetilde{f}_{x_1}$. Hence, we have (8.26) for an integrable function, which is **Fubini's theorem.** $\Box$

PART 2

Lebesgue Spaces

Beware of engineers,
it starts with the sewing machine,
and ends with the atomic bomb.

Marcel PAGNOL
Critique des critiques

Chapter 9

Inequalities

Let us present some inequalities that will be used in the next chapter, in order to construct the space $L^p(\Omega; E)$.

We begin with some real-valued inequalities (§ 9.1), then we present Hölder's and Minkowski's inequalities for functions (§ 9.2), followed by Young's inequality (§ 9.3). We conclude with some regularization properties of continuous functions, that will serve to justify the definition of the semi-norms of $L^p(\Omega; E)$ (§ 9.4).

9.1. Elementary inequalities

Here, we present the Young's, Hölder's, and Minkowski's inequalities, which are elementary. They will be used in Section 9.2 to establish the same inequalities for continuous functions, which will then be used in Chapter 10 when constructing Lebesgue spaces.

Let us begin with **Young's inequality** [1].

Proposition 9.1. — *Let $1 < p < \infty$ and $1 < p' < \infty$ be the* ***conjugate exponent*** *defined by*

$$\frac{1}{p} + \frac{1}{p'} = 1.$$

Then, for every $a \geq 0$ and $b \geq 0$,

$$ab \leq \frac{a^p}{p} + \frac{b^{p'}}{p'}. \blacksquare$$

1 **Historical background of Young's inequality.** This inequality was established by William Henry YOUNG in 1912 [82].

Proof. The function defined by $f(a) \stackrel{\text{def}}{=} a^p/p + b^{p'}/p' - ab$ is differentiable, and furthermore (Proposition A.82), since $p - 1 = p/p'$, its derivative equals

$$f'(a) = a^{p-1} - b = a^{p/p'} - b.$$

Therefore, f is positive everywhere because:

— it vanishes at $b^{p'/p}$ (since $f(b^{p'/p}) = b^{p'} - b^{1+p'/p} = 0$, as $p' = 1 + p'/p$);

— for $a \geq b^{p'/p}$, we have $f'(a) \geq 0$, so (Proposition A.80 (a)) f is increasing;

— for $a \leq b^{p'/p}$, we have $f'(a) \leq 0$, so (Proposition A.80 (b)) f is decreasing. □

From this, we may deduce **Hölder's inequality**[2] for finite sums.

Proposition 9.2. — *Let* $1 \leq p < \infty$, $1 < p_1 < \infty$, *and* $1 < p_2 < \infty$, *such that*

$$\frac{1}{p} = \frac{1}{p_1} + \frac{1}{p_2}.$$

Then, for all $x \in \mathbb{R}^d$ *and* $y \in \mathbb{R}^d$,

$$\left(\sum_{i=1}^{d} |x_i y_i|^p\right)^{1/p} \leq \left(\sum_{i=1}^{d} |x_i|^{p_1}\right)^{1/p_1} \left(\sum_{i=1}^{d} |y_i|^{p_2}\right)^{1/p_2}. \blacksquare$$

Proof. **1°)** *Case* $p = 1$. By Young's inequality (Proposition 9.1),

$$\sum_{i=1}^{d} |x_i y_i| \leq \frac{1}{p_1} \sum_{i=1}^{d} |x_i|^{p_1} + \frac{1}{p_2} \sum_{i=1}^{d} |y_i|^{p_2}.$$

Consider x/s and y/t, where $s = (\sum_{i=1}^{d} |x_i|^{p_1})^{1/p_1}$ and $t = (\sum_{i=1}^{d} |y_i|^{p_2})^{1/p_2}$. From this inequality, we obtain

$$\frac{1}{st} \sum_{i=1}^{d} |x_i y_i| \leq \frac{1}{p_1 s^{p_1}} \sum_{i=1}^{d} |x_i|^{p_1} + \frac{1}{p_2 t^{p_2}} \sum_{i=1}^{d} |y_i|^{p_2} = \frac{1}{p_1} + \frac{1}{p_2} = 1.$$

From which, if s and t are non-zero, we get

$$\sum_{i=1}^{d} |x_i y_i| \leq \left(\sum_{i=1}^{d} |x_i|^{p_1}\right)^{1/p_1} \left(\sum_{i=1}^{d} |y_i|^{p_2}\right)^{1/p_2}. \tag{9.1}$$

2 **Historical background of Hölder's inequality.** Otto HÖLDER established the inequality of Proposition 9.2 in 1889, for $p = 1$ [39, p. 44]; it is also said to appear in his thesis (1884).

If $s = 0$, then $x = 0$ and the inequality is immediate. The same argument holds if $t = 0$.

2°) Case $p > 1$. Inequality (9.1) with respect to the exponents p_1/p and p_2/p gives

$$\sum_{i=1}^{d} |x_i|^p |y_i|^p \leq \left(\sum_{i=1}^{d} (|x_i|^p)^{p_1/p}\right)^{p/p_1} \left(\sum_{i=1}^{d} (|y_i|^p)^{p_2/p}\right)^{p/p_2}.$$

By raising to the power $1/p$, we obtain the stated inequality. □

Let us compare some sums that involve distinct powers.

Proposition 9.3. — *Let* $1 \leq p \leq q < \infty$. *Then, for all* $x \in \mathbb{R}^d$,

$$\left(\sum_{i=1}^{d} |x_i|^q\right)^{1/q} \leq \left(\sum_{i=1}^{d} |x_i|^p\right)^{1/p} \leq d^{1/p-1/q} \left(\sum_{i=1}^{d} |x_i|^q\right)^{1/q}. \blacksquare$$

Proof. ***1°) First inequality.*** First of all, observe that if $t \geq 0$, then

$$1 + t^{q/p} \leq (1+t)^{q/p}.$$

Indeed, the function defined by $f(t) \stackrel{\text{def}}{=} 1 + t^{q/p} - (1+t)^{q/p}$ vanishes when $t = 0$ and, (Proposition A.80 (b)) since $f'(t) = (q/p)(t^{q/p-1} - (1+t)^{q/p-1}) \leq 0$ (Proposition A.82), it is decreasing.

When $d = 2$, the first inequality follows by taking $t = (|x_2|/|x_1|)^p$, provided $x_1 \neq 0$, multiplying by $|x_1|^q$, and then raising by the power $1/q$. If $x_1 = 0$, then the inequality is immediate.

The general inequality follows by induction on the dimension d. Indeed, if it is true for $d-1$ then, denoting $a_q = (\sum_{i=1}^{d-1} |x_i|^q)^{1/q}$, we have $a_q \leq a_p$, and

$$\left(\sum_{i=1}^{d} |x_i|^q\right)^{1/q} = (a_q^q + |x_d|^q)^{1/q} \leq (a_q^p + |x_d|^p)^{1/p} \leq$$
$$\leq (a_p^p + |x_d|^p)^{1/p} = \left(\sum_{i=1}^{d} |x_i|^p\right)^{1/p}.$$

2°) Second inequality. If $p < q$, define $1 \leq r < \infty$ by $1/r + 1/q = 1/p$ so that, using Hölder's inequality for finite sums (Proposition 9.2), we get

$$\left(\sum_{i=1}^{d} |x_i|^p\right)^{1/p} = \left(\sum_{i=1}^{d} (1|x_i|)^p\right)^{1/p} \leq \left(\sum_{i=1}^{d} 1^r\right)^{1/r} \left(\sum_{i=1}^{d} |x_i|^q\right)^{1/q}.$$

From which we obtain the stated inequality. If $p = q$, then instead we have an immediate equality. □

Now we also state **Minkowski's inequality**[3] for finite sums (the first one below), in addition to a variant.

Proposition 9.4. — *Let* $1 \leq p < \infty$, *and consider* x *and* y *in* $\mathbb{R}^d$, *and* t *in* $\mathbb{R}^d_+$. *Then,*

$$\left(\sum_{i=1}^{d} |x_i + y_i|^p\right)^{1/p} \leq \left(\sum_{i=1}^{d} |x_i|^p\right)^{1/p} + \left(\sum_{i=1}^{d} |y_i|^p\right)^{1/p},$$

$$\left(\sum_{i=1}^{d} |x_i + y_i|^p t_i\right)^{1/p} \leq \left(\sum_{i=1}^{d} |x_i|^p t_i\right)^{1/p} + \left(\sum_{i=1}^{d} |y_i|^p t_i\right)^{1/p}. \blacksquare$$

Proof. **1º)** *First inequality.* If $p = 1$, it is immediate. Otherwise,

$$\sum_{i=1}^{d} |x_i + y_i|^p \leq \sum_{i=1}^{d} |x_i|\,|x_i + y_i|^{p-1} + \sum_{i=1}^{d} |y_i|\,|x_i + y_i|^{p-1}.$$

Hölder's inequality (Proposition 9.2) relative to the coefficients p and $p/(p-1)$ applied to the sums on the right-hand side gives, since $1/p + (p-1)/p = 1$,

$$\sum_{i=1}^{d} |x_i{+}y_i|^p \leq \left(\sum_{i=1}^{d} |x_i|^p\right)^{\frac{1}{p}} \left(\sum_{i=1}^{d} |x_i{+}y_i|^p\right)^{\frac{p-1}{p}} + \left(\sum_{i=1}^{d} |y_i|^p\right)^{\frac{1}{p}} \left(\sum_{i=1}^{d} |x_i{+}y_i|^p\right)^{\frac{p-1}{p}}.$$

From which, after dividing by $(\sum_{i=1}^{d} |x_i + y_i|^p)^{(p-1)/p}$, we obtain the stated inequality.

2º) *Second inequality.* This is just the first inequality but with $x'_i = x_i t_i^{1/p}$ and $x'_i = x_i t_i^{1/p}$. □

9.2. Inequalities for continuous functions

Let us begin with **Hölder's inequality**[4] for uniformly continuous real-valued functions vanishing outside a bounded set (the general inequality for p-integrable functions will be given in Theorems 12.12 to 12.16, and supplemented by Theorem 13.4).

---

3 **Historical background of Minkowski's inequality.** This inequality was proven by Hermann MINKOWSKI, possibly in [52] or [53].

4 **Historical background of Hölder's inequality.** In 1910, Frigyes RIESZ stated and proved [63, p. 456] the inequality of Proposition 9.5 for the product of a function in $L^p(I)$ with a function in $L^{p'}(I)$, where I is a real interval. He named it after Otto HÖLDER, who had established the analogous inequality for finite sums, given here in Proposition 9.2 (see Note 2, p. 172).

Proposition 9.5. — *Let $f \in \mathcal{B}(\Omega)$ and $g \in \mathcal{B}(\Omega)$, where Ω is an open subset of $\mathbb{R}^d$, $1<p<\infty$, and let $1 < p' < \infty$ be the conjugate exponent defined by $1/p + 1/p' = 1$. Then,*

$$\int_\Omega |fg| \leq \Big(\int_\Omega |f|^p\Big)^{1/p}\Big(\int_\Omega |g|^{p'}\Big)^{1/p'}. \blacksquare$$

Proof. By Young's inequality (Proposition 9.1),

$$|fg| \leq \frac{|f|^p}{p} + \frac{|g|^{p'}}{p'}.$$

Since the Cauchy integral is increasing (Proposition 1.18 (a)) and linear, and since the functions $|fg|$, $|f|^p$, and $|g|^{p'}$ are continuous and vanish outside a bounded set, we have

$$\int_\Omega |fg| \leq \frac{1}{p}\int_\Omega |f|^p + \frac{1}{p'}\int_\Omega |g|^{p'}.$$

Applying this inequality to f/s and g/t, where $s = (\int_\Omega |f|^p)^{1/p}$ and $t = (\int_\Omega |g|^{p'})^{1/p'}$, gives

$$\frac{1}{st}\int_\Omega |fg| \leq \frac{1}{ps^p}\int_\Omega |f|^p + \frac{1}{p't^{p'}}\int_\Omega |g|^{p'} = \frac{1}{p} + \frac{1}{p'} = 1.$$

From which, we obtain the stated inequality, provided that s and t are non-zero.

If $s = 0$, Proposition 1.18 (c) shows that $|f|^p = 0$, so f and fg are zero and the inequality follows immediately. The same argument holds if $t = 0$. □

From this, for the same functions, we obtain **Hölder's generalized inequality**.

Proposition 9.6. — *Let $f_1, f_2, \ldots, f_n$ be functions in $\mathcal{B}(\Omega)$, where Ω is an open subset of $\mathbb{R}^d$, $0 < p < \infty$, and $0 < p_1 < \infty, \ldots, 0 < p_n < \infty$, such that*

$$\frac{1}{p} = \frac{1}{p_1} + \frac{1}{p_2} + \cdots + \frac{1}{p_n}.$$

Then,

$$\Big(\int_\Omega |f_1 f_2 \cdots f_n|^p\Big)^{1/p} \leq \Big(\int_\Omega |f_1|^{p_1}\Big)^{1/p_1}\Big(\int_\Omega |f_2|^{p_2}\Big)^{1/p_2} \cdots \Big(\int_\Omega |f_n|^{p_n}\Big)^{1/p_n}. \blacksquare$$

Proof. ***1º) Two-function case.*** Hölder's inequality (Proposition 9.5) with respect to the functions $|f_1|^p$ and $|f_2|^p$ and the exponents p_1/p and p_2/p (they are conjugate and strictly greater than 1), gives

$$\int_\Omega |f_1 f_2|^p \le \Big(\int_\Omega |f_1|^{p_1}\Big)^{p/p_1} \Big(\int_\Omega |f_2|^{p_2}\Big)^{p/p_2}.$$

This case is finished by raising this inequality to the power $1/p$.

2º) General case. Let us deduce this case from that with two functions by induction on n. Thus, assume that the stated inequality holds for $n-1$ functions, and define q by

$$\frac{1}{q} = \frac{1}{p_{n-1}} + \frac{1}{p_n}.$$

Then the inequality relative to $n-1$ functions gives

$$\Big(\int_\Omega |f_1 f_2 \cdots f_n|^p\Big)^{1/p} \le \Big(\int_\Omega |f_1|^{p_1}\Big)^{1/p_1} \cdots \Big(\int_\Omega |f_{n-2}|^{p_{n-2}}\Big)^{1/p_{n-2}} \Big(\int_\Omega |f_{n-1} f_n|^q\Big)^{1/q}.$$

The general inequality then follows by bounding the last term using the inequality from the two-function case. □

We now come to **Minkowski's inequality**[5] (the inequality for a p-integrable measure is implicit in Definition 10.2 of $L^p(\Omega; E)$ since $\|\ \|_{L^p(\Omega;E);\nu}$ is a semi-norm, and it is made explicit with $\|f\|_{L^p(\Omega;E);\nu} = (\int_\Omega \|f\|^p_{E;\nu})^{1/p}$, see Theorem 13.4).

Proposition 9.7. — *Let $f \in \mathcal{B}(\Omega; E)$ and $g \in \mathcal{B}(\Omega; E)$, where Ω is an open subset of $\mathbb{R}^d$, E is a Neumann space, $\|\ \|_{E;\nu}$ is a semi-norm of E, and $1 \le p < \infty$. Then,*

$$\Big(\int_\Omega \|f+g\|^p_{E;\nu}\Big)^{1/p} \le \Big(\int_\Omega \|f\|^p_{E;\nu}\Big)^{1/p} + \Big(\int_\Omega \|g\|^p_{E;\nu}\Big)^{1/p}. \blacksquare$$

Proof. Omitting the index ${}_{E;\nu}$, observe that

$$\|f+g\|^p = \|f+g\|\|f+g\|^{p-1} \le \|f\|\|f+g\|^{p-1} + \|g\|\|f+g\|^{p-1}.$$

Furthermore, $\|f\|$, $\|g\|$, and $\|f+g\|$ belong to $\mathcal{B}(\Omega)$ (Proposition 1.17) so, since the integral is increasing (Proposition 1.18 (a)) and linear,

$$\int_\Omega \|f+g\|^p \le \int_\Omega \|f\|\|f+g\|^{p-1} + \int_\Omega \|g\|\|f+g\|^{p-1}. \tag{9.2}$$

5 **Historical background of Minkowski's inequality.** Frigyes RIESZ stated and proved [63] the inequality of Proposition 9.7 for functions in $L^p(I)$ in 1910, where I is a real interval. He named it after Hermann MINKOWSKI, who had established the analogous inequality for finite sums.

By bounding the second integral using Hölder's inequality (Proposition 9.5) with respect to the functions $\|f\|$ and $\|f+g\|^{p-1}$, since $p' = p/(p-1)$, it becomes

$$\int_\Omega \|f\|\|f+g\|^{p-1} \leq \Big(\int_\Omega \|f\|^p\Big)^{1/p}\Big(\int_\Omega \|f+g\|^p\Big)^{(p-1)/p}.$$

Then, by bounding the third integral of (9.2) in an analogous way, we obtain

$$\int_\Omega \|f+g\|^p \leq \bigg(\Big(\int_\Omega \|f\|^p\Big)^{1/p} + \Big(\int_\Omega \|g\|^p\Big)^{1/p}\bigg)\Big(\int_\Omega \|f+g\|^p\Big)^{1-1/p}.$$

Which provides us with the stated inequality. □

9.3. Young's convolution inequality

We are going to establish a variant of Young's inequality that is related to the weighting operation, which differs from the convolution by a sign (for a reason explained in the comment *Weighting versus convolution* at the end of Section 7.5 of Volume 3 [74]).

Thus, let us define the weighting operation for a continuous function on $\mathbb{R}^d$ by a real-valued function with compact support.

Definition 9.8. — *Let $f \in \mathcal{C}(\mathbb{R}^d; E)$ and $\mu \in \mathcal{C}_D(\mathbb{R}^d)$, where E is a Neumann space and D is a compact subset of $\mathbb{R}^d$.*

*We say that the **weighting** of f by the weight μ is the function $f \diamond \mu$ defined on $\mathbb{R}^d$ by*

$$(f \diamond \mu)(x) \overset{\text{def}}{=} \int_{\mathbb{R}^d} f(x+y)\mu(y)\,dy. \blacksquare$$

Having defined weightings, let us establish the **Young's weighting inequality**.

Proposition 9.9. — *Let $f \in \mathcal{K}(\mathbb{R}^d; E)$ and $\mu \in \mathcal{C}_D(\mathbb{R}^d)$, where E is a Neumann space, D is a compact subset of $\mathbb{R}^d$, $1 \leq p < \infty$, $1 \leq r < \infty$, and $1 \leq q < \infty$, such that*

$$\frac{1}{q} = \frac{1}{p} + \frac{1}{r} - 1.$$

Then, $f \diamond \mu \in \mathcal{K}(\mathbb{R}^d; E)$ and, for every semi-norm $\| \;\|_{E;\nu}$ of E,

$$\Big(\int_{\mathbb{R}^d} \|f \diamond \mu\|_{E;\nu}^q\Big)^{1/q} \leq \Big(\int_{\mathbb{R}^d} \|f\|_{E;\nu}^p\Big)^{1/p}\Big(\int_{\mathbb{R}^d} |\mu|^r\Big)^{1/r}. \blacksquare$$

Proof. **1°)** *Uniform continuity of* $f \diamond \mu$. Since only the domain where a function does not vanish contributes to its Cauchy integral (Proposition 1.23),

$$(f \diamond \mu)(x) = \int_{\mathring{D}} f(x+y)\mu(y)\,\mathrm{d}y.$$

The bound for the semi-norms of the integral from Proposition 1.21 therefore gives, for every semi-norm of E, for all x and z in $\mathbb{R}^d$, and denoting $m = \sup_{y\in D} |\mu(y)|$,

$$\|(f \diamond \mu)(x) - (f \diamond \mu)(z)\|_{E;\nu} \leq |\mathring{D}| m \sup_{y\in D} \|f(x+y) - f(z+y)\|_{E;\nu}.$$

Uniform continuity (Definition 1.7) of f therefore implies that of $f \diamond \mu$.

2°) *Membership of* $\mathcal{K}(\mathbb{R}^d; E)$. Since $f \diamond \mu$ is uniformly continuous, it remains to verify that it vanishes outside a compact set, namely $S - D$, where $S = \operatorname{supp} f$. This is indeed the case since, if $x \notin S - D$, then the function $y \mapsto f(x+y)\mu(y)$ is zero, because:

— if $y \in D$, then $f(x+y) = 0_E$ as $x + y \notin S$ (if not, $x = (x+y) - y \in S - D$);

— otherwise, $y \notin D$ and thus $\mu(y) = 0$.

3°) *A pointwise bound.* Let $x \in \mathbb{R}^d$. The bound for the semi-norms of the integral from Proposition 1.17 gives, denoting $\| \ \|$ in the place of $\| \ \|_{E;\nu}$,

$$\begin{aligned}\|(f \diamond \mu)(x)\| &\leq \int_{\mathbb{R}^d} \|f(x+y)\| |\mu(y)|\,\mathrm{d}y = \\ &= \int_{\mathbb{R}^d} \Big(\|f(x+y)\|^{p/q} |\mu(y)|^{r/q}\Big) \|f(x+y)\|^{1-p/q} |\mu(y)|^{1-r/q}\,\mathrm{d}y.\end{aligned}$$

Hölder's inequality (Proposition 9.6) with $1/q + (1/p - 1/q) + (1/r - 1/q) = 1$ gets us

$$\begin{aligned}\|(f \diamond \mu)(x)\| \leq \Big(\int_{\mathbb{R}^d} &\|f(x+y)\|^p |\mu(y)|^r\,\mathrm{d}y\Big)^{1/q} \times \\ &\times \Big(\int_{\mathbb{R}^d} \|f(x+y)\|^p\,\mathrm{d}y\Big)^{1/p-1/q} \Big(\int_{\mathbb{R}^d} |\mu(y)|^r\,\mathrm{d}y\Big)^{1/r-1/q}.\end{aligned}$$

(If $r = 1$, then $p = q$ and the second integral disappears; if $p = 1$, then $r = q$ and the third integral disappears.) Raising it to the power q, since the integral is invariant under translation (Proposition 1.27), we get

$$\begin{aligned}&\|(f \diamond \mu)(x)\|^q \leq \\ &\leq \Big(\int_{\mathbb{R}^d} \|f(x+y)\|^p |\mu(y)|^r\,\mathrm{d}y\Big) \Big(\int_{\mathbb{R}^d} \|f(y)\|^p\,\mathrm{d}y\Big)^{q/p-1} \Big(\int_{\mathbb{R}^d} |\mu(y)|^r\,\mathrm{d}y\Big)^{q/r-1}.\end{aligned}$$

4°) Integral bound. By integrating with respect to x, since $x \mapsto \|(f \diamond \mu)(x)\|_{E;\nu}^q$ is uniformly continuous and vanishes outside a bounded set, it becomes

$$\int_{\mathbb{R}^d} \|(f \diamond \mu)(x)\|^q \, dx \le \Big(\int_{\mathbb{R}^d} dx \int_{\mathbb{R}^d} dy \, \|f(x+y)\|^p |\mu(y)|^r \Big) \times$$
$$\times \Big(\int_{\mathbb{R}^d} \|f(y)\|^p \, dy \Big)^{q/p-1} \Big(\int_{\mathbb{R}^d} |\mu(y)|^r \, dy \Big)^{q/r-1}.$$

By permuting the integrals (Proposition 1.28) then, together with the linearity of the integral (Proposition 1.16) and its translation invariance (Proposition 1.27), we get

$$\int_{\mathbb{R}^d} dx \int_{\mathbb{R}^d} dy \, \|f(x+y)\|^p |\mu(y)|^r = \int_{\mathbb{R}^d} dy \int_{\mathbb{R}^d} dx \, \|f(x+y)\|^p |\mu(y)|^r =$$
$$= \int_{\mathbb{R}^d} dy \, |\mu(y)|^r \int_{\mathbb{R}^d} dz \, \|f(z)\|^p.$$

The previous inequality therefore gives

$$\int_{\mathbb{R}^d} \|(f \diamond \mu)(x)\|^q \, dx \le \Big(\int_{\mathbb{R}^d} \|f(z)\|^p \, dz \Big)^{q/p} \Big(\int_{\mathbb{R}^d} |\mu(y)|^r \, dy \Big)^{q/r}.$$

By raising this inequality by the power $1/q$, we obtain Young's inequality. □

Young's convolution inequality. We call the **convolution** of $f \in \mathcal{C}(\mathbb{R}^d; E)$ by $\mu \in \mathcal{C}_D(\mathbb{R}^d)$ to be the function $f \star \mu$, defined on $\mathbb{R}^d$ by

$$(f \star \mu)(x) \overset{\text{def}}{=} \int_{\mathbb{R}^d} f(x-y) \, \mu(y) \, dy.$$

That is, denoting $\check{\mu}(y) \overset{\text{def}}{=} \mu(-y)$,

$$f \star \mu = f \diamond \check{\mu}.$$

Proposition 9.9 therefore gives, for $f \in \mathcal{K}(\mathbb{R}^d; E)$ and $1/q = 1/p + 1/r - 1$, **Young's convolution inequality**,

$$\Big(\int_{\mathbb{R}^d} \|f \star \mu\|_{E;\nu}^q \Big)^{1/q} \le \Big(\int_{\mathbb{R}^d} \|f\|_{E;\nu}^p \Big)^{1/p} \Big(\int_{\mathbb{R}^d} |\mu|^r \Big)^{1/r}. \; \square$$

9.4. Properties of regularizations of continuous functions

Let us give some properties regarding the integrals of the semi-norms of regularizations for continuous functions, which play a key role when justifying the definition of the semi-norms of $L^p(\Omega; E)$.

Proposition 9.10. — *Let $f \in \mathcal{B}(\Omega; E)$, where Ω is an open subset of $\mathbb{R}^d$, E is a Neumann space, $(R_n f)_{n \in \mathbb{N}^*}$ are regularizations for f (Definition 3.17), $\| \; \|_{E;\nu}$ is a semi-norm of E, and $1 \le p < \infty$.*

(a) *For every* $n \in \mathbb{N}^*$, *we have* $R_n f \in \mathcal{K}(\Omega; E)$, *and*

$$\Big(\int_\Omega \|R_n f\|_{E;\nu}^p\Big)^{1/p} \leq \Big(\int_\Omega \|f\|_{E;\nu}^p\Big)^{1/p},$$

$$\sup_{x\in\Omega} \|R_n f(x)\|_{E;\nu} \leq \sup_{x\in\Omega} \|f(x)\|_{E;\nu}.$$

(b) *When* $n \to \infty$,

$$\Big(\int_\Omega \|R_n f - f\|_{E;\nu}^p\Big)^{1/p} \to 0,$$

$$\Big(\int_\Omega \|R_n f\|_{E;\nu}^p\Big)^{1/p} \to \Big(\int_\Omega \|f\|_{E;\nu}^p\Big)^{1/p}. \blacksquare$$

Proof. ***1°)*** $R_n f \in \mathcal{K}(\Omega; E)$. This was established in Theorem 3.18.

2°) *First inequality of* (a). The expression for the regularizations of a continuous function from Proposition 3.21 and Definition 9.8 for the weighting give, for every $x \in \Omega$,

$$R_n f(x) = \int_{\mathbb{R}^d} (\widetilde{\alpha_n f})(x + y)\rho_n(y)\,\mathrm{d}y = (\widetilde{\alpha_n f} \diamond \rho_n)(x), \tag{9.3}$$

where $\widetilde{\ }$ is the extension by 0_E, $\alpha_n \in \mathcal{K}(\Omega)$ thus $\widetilde{\alpha_n f} \in \mathcal{K}(\mathbb{R}^d; E)$, and $\rho_n \in \mathcal{K}(\mathbb{R}^d)$.

As the Cauchy integral of a positive function increases with the domain (Proposition 1.18 (b)), by using Young's weighting inequality (Proposition 9.9 with $r = 1$, so that $q = p$), we get

$$\Big(\int_\Omega \|R_n f\|_{E;\nu}^p\Big)^{1/p} \leq \Big(\int_{\mathbb{R}^d} \|\widetilde{\alpha_n f} \diamond \rho_n\|_{E;\nu}^p\Big)^{1/p} \leq \Big(\int_{\mathbb{R}^d} \|\widetilde{\alpha_n f}\|_{E;\nu}^p\Big)^{1/p} \int_{\mathbb{R}^d} |\rho_n|.$$

This gives the stated inequality, since $|\alpha_n| \leq 1$ and $\int_{\mathbb{R}^d} |\rho_n| = 1$ (Definition 3.16).

3°) *Second inequality of* (a). By using the first equality in (9.3), together with the monotonicity of the real-valued integral and its translation invariance (Propositions 1.18 (a) and 1.27), we get

$$\|R_n f(x)\|_{E;\nu} \leq \sup_{z\in\mathbb{R}^d} \|\widetilde{\alpha_n f}(z)\|_{E;\nu} \int_{\mathbb{R}^d} \rho_n(y)\,\mathrm{d}y.$$

This implies the stated inequality, since again $|\alpha_n| \leq 1$ and $\int_{\mathbb{R}^d} \rho_n = 1$.

4°) *First convergence of* (b). We shall obtain it by approximating f by a function $\beta_k f$ with compact support in 4.a, and then by bounding the integral of $R_n(\beta_k f - f)$ using inequality (a) in 4.b.

4.a. *Localization of f.* Let $\epsilon > 0$ and $\mathcal{O} \stackrel{\text{def}}{=} \{x \in \Omega : f(x) \neq 0_E\}$. This is a bounded (since f vanishes outside a bounded set) open (since f is continuous) set. Let β_k be a localization function on $\mathcal{O}$ (Definition 3.16, extended by 0 outside $\mathcal{O}$) such that

$$\beta_k \in \mathcal{K}(\Omega), \quad 0 \leq \beta \leq 1, \quad \beta_k = \begin{cases} 1 & \text{on } \mathcal{O}^k_{3/k}, \\ 0 & \text{outside } \mathcal{O}^k_{2/k}. \end{cases}$$

Each set $\omega_k \stackrel{\text{def}}{=} \mathcal{O} \setminus \overline{\mathcal{O}^k_{3/k}}$ is open, bounded, decreases as k increases, and has an empty intersection for $k \in \mathbb{N}^*$, so (Theorem 2.17)

$$|\omega_k| \to 0 \ \text{ when } k \to \infty.$$

Let $b \stackrel{\text{def}}{=} \sup_{x\in\Omega} \|f(x)\|_{E;\nu}$, which is finite since any function in $\mathcal{B}(\Omega; E)$ is bounded (Proposition 1.13), and consider k such that

$$|\omega_k| \leq \Big(\frac{\epsilon}{3b}\Big)^p.$$

Then $\beta_k f = f$ outside ω_k (since $f = 0_E$ outside $\mathcal{O}$ and $\beta_k = 1$ on $\overline{\mathcal{O}^k_{3/k}}$) so, since the domain where a function vanishes does not contribute to its integral (Proposition 1.23), the bound for the Cauchy integral from Proposition 1.21 gives

$$\int_\Omega \|\beta_k f - f\|^p_{E;\nu} = \int_{\omega_k} \|\beta_k f - f\|^p_{E;\nu} \leq |\omega_k| \sup_{x\in\Omega} \|(\beta_k f - f)(x)\|^p_{E;\nu} \leq \frac{\epsilon^p}{3^p}. \tag{9.4}$$

4.b. *Estimate of $R_n f - f$.* Having fixed k, and denoting β in the place of β_k, decompose

$$R_n f - f = R_n(f - \beta f) + (R_n(\beta f) - \beta f) + (\beta f - f). \tag{9.5}$$

We now bound the integral of the semi-norm $\| \ \|_{E;\nu}$ of the three terms on the right-hand side.

For the first and third terms, observe, using the first bound from part (a), that inequality (9.4) implies

$$\Big(\int_\Omega \|R_n(f - \beta f)\|^p_{E;\nu}\Big)^{1/p} \leq \Big(\int_\Omega \|f - \beta f\|^p_{E;\nu}\Big)^{1/p} \leq \frac{\epsilon}{3}. \tag{9.6}$$

For the second term, observe that the support of βf is compact (because it is included in $\overline{\mathcal{O}^k_{2/k}}$, which is itself a compact set included in Ω), which implies (Proposition 3.22) that all the $R_n(\beta f)$ and βf are supported in the same compact set $Q \subset \Omega$. Since the domain where a function vanishes does not contribute to its integral (Proposition 1.23, again), the bound for the integral from Proposition 1.21 gives

$$\int_\Omega \|R_n(\beta f) - \beta f\|^p_{E;\nu} = \int_{\mathring{Q}} \|R_n(\beta f) - \beta f\|^p_{E;\nu} \leq |\mathring{Q}| \sup_{x\in\Omega} \|(R_n(\beta f) - \beta f)(x)\|^p_{E;\nu}.$$

However, when $n \to \infty$, $R_n(\beta f)$ converges (Theorem 3.20) to βf in $\mathrm{C_b}(\Omega; E)$. Therefore, for n large enough,

$$\Big(\int_\Omega \|R_n(\beta f) - \beta f\|^p_{E;\nu}\Big)^{1/p} \le |\mathring{Q}|^{1/p} \sup_{x\in\Omega} \|(R_n(\beta f) - \beta f)(x)\|_{E;\nu} \le \frac{\epsilon}{3}.$$

The decomposition (9.5) thus gives, together with (9.6) when n is large enough (depending on ϵ),

$$\Big(\int_\Omega \|R_n f - f\|^p_{E;\nu}\Big)^{1/p} \le \epsilon.$$

This proves that, as $n \to \infty$,

$$\Big(\int_\Omega \|R_n f - f\|^p_{E;\nu}\Big)^{1/p} \to 0. \tag{9.7}$$

5º) *Second convergence of* (b). The mapping $f \mapsto (\int_\Omega \|f\|^p_{E;\nu})^{1/p}$ is a semi-norm on $\mathcal{B}(\Omega; E)$ according to Minkowski's inequality (Proposition 9.7). It is therefore contracting, as is the case for any semi-norm (Proposition A.11), *i.e.*

$$\Big|\Big(\int_\Omega \|R_n f\|^p_{E;\nu}\Big)^{1/p} - \Big(\int_\Omega \|f\|^p_{E;\nu}\Big)^{1/p}\Big| \le \Big(\int_\Omega \|R_n f - f\|^p_{E;\nu}\Big)^{1/p}.$$

Therefore, the second convergence of (b) follows from (9.7). □

Chapter 10

$L^p(\Omega; E)$ Spaces

In this chapter, we construct the space $L^p(\Omega; E)$ and show that it is sequentially complete.

By definition, it is the set of the limits f in $\mathcal{M}(\Omega; E)$ of uniformly continuous functions with bounded support f_n such that $\sup_{m\geq n}(\int_\Omega \|f_m - f_n\|_{E;\nu}^p)^{1/p} \to 0$ as $n \to \infty$, when $1 \leq p < \infty$. It is endowed with the semi-norms $\|f\|_{L^p(\Omega;E);\nu} = \lim (\|f_n\|_{E;\nu}^p)^{1/p}$ (Definitions 10.1 and 10.2).

If $p = \infty$, the condition on f_n is $\sup_{n,x} \|f_n(x)\|_{E;\nu} < \infty$ and $\sup_{m\geq n} \int_\omega \|f_m - f_n\|_{E;\nu} \to 0$ as $n \to \infty$ for every bounded open set ω, and $\|f\|_{L^\infty(\Omega;E);\nu} = \inf \sup_{n,x} \|f_n(x)\|_{E;\nu}$, where the infimum is taken over all possible approximating sequences f_n.

After defining it (§ 10.1), we show that it is separated (§ 10.2). We give some properties of $L^p(\Omega; E)$ (§ 10.3), of $L^\infty(\Omega; E)$ (§ 10.4), and of approximations via regularizations (§ 10.5), prior to showing that it is Neumann (§ 10.6).

We conclude with some remarks regarding the methods used to construct $L^p(\Omega; E)$ (§ 10.7).

10.1. Definition of $L^p(\Omega; E)$

Let us define a p-integrable measure with values in a Neumann space.

Definition 10.1. — *A measure f, on an open subset Ω of $\mathbb{R}^d$ with values in a Neumann space E, is said to be p-**integrable**, where $1 \leq p \leq \infty$, if there exists a sequence $(f_n)_{n\in\mathbb{N}}$ in $\mathcal{B}(\Omega; E)$, that is, uniformly continuous functions that vanish outside a bounded set, such that, as $n \to \infty$,*

$$f_n \to f \text{ in } \mathcal{M}(\Omega; E), \tag{10.1}$$

and, for every semi-norm $\| \ \|_{E;\nu}$ of E:

— if $p < \infty$,

$$\sup_{m\geq n} \left(\int_\Omega \|f_m - f_n\|_{E;\nu}^p \right)^{1/p} \to 0; \tag{10.2}$$

— *if* $p = \infty$,

$$\sup_{n\in\mathbb{N},\, x\in\Omega} \|f_n(x)\|_{E;\nu} < \infty \tag{10.3}$$

and, for every bounded open subset ω *contained in* Ω,

$$\sup_{m\geq n} \int_\omega \|f_m - f_n\|_{E;\nu} \to 0. \tag{10.4}$$

We say that $(f_n)_{n\in\mathbb{N}}$ *is a* ***p-integrabilizing sequence*** *for* f. ▮

Now, let us turn to the **Lebesgue space**[1], $L^p(\Omega; E)$.

Definition 10.2. — *Let* Ω *be an open subset of* $\mathbb{R}^d$, E *is a Neumann space whose family of semi-norms is denoted* $\{\|\ \|_{E;\nu} : \nu \in \mathcal{N}_E\}$, *and* $1 \leq p \leq \infty$.

We denote by $L^p(\Omega; E)$ *the vector space of* p*-integrable measures, endowed with the semi-norms indexed by* $\nu \in \mathcal{N}_E$ *and defined by:*

— *if* $p < \infty$,

$$\|f\|_{L^p(\Omega;E);\nu} \stackrel{\text{def}}{=} \lim_{n\to\infty} \Big(\int_\Omega \|f_n\|^p_{E;\nu} \Big)^{1/p}, \tag{10.5}$$

where $(f_n)_{n\in\mathbb{N}}$ *is any* p*-integrabilizing sequence for* f;

— *if* $p = \infty$,

$$\|f\|_{L^\infty(\Omega;E);\nu} \stackrel{\text{def}}{=} \inf_{(f_n)_{n\in\mathbb{N}}} \sup_{n\in\mathbb{N},\, x\in\Omega} \|f_n(x)\|_{E;\nu}, \tag{10.6}$$

where the infimum is taken over the set of ∞*-integrabilizing sequences for* f. ▮

In particular, denote

$$L^p(\Omega) \stackrel{\text{def}}{=} L^p(\Omega; \mathbb{R})$$

and let $\|\ \|_{L^p(\Omega)}$ be its unique semi-norm (which is in fact a norm, since $L^p(\Omega; E)$ is separated, as we shall see in Theorem 10.5).

Remark. Belonging to $L^\infty(\Omega; E)$ is equivalent to belonging to $L^1(\omega; E)$ for any bounded open subset contained in Ω (by (10.4)), and being "bounded" (by (10.3)). This will be formalized in Theorem 10.11. □

---

1 **Historical background of the Lebesgue space.** Frigyes RIESZ introduced in 1910 [63] *the set of measurable functions on* $(0, 1)$ *for which* $|f|^p$ *is integrable*, which he called the *Lebesgue space* and denoted it by $[L^p]$. He showed that $(\int |f|^p)^{1/p}$ is a norm on this space.
In 1938, Bill PETTIS constructed [57] the space of *classes of functions* $L^p(\Omega; E)$ for a Banach space E.
A similar construction via completion to the one presented here was carried out by Nicolas BOURBAKI [8], see the comment *Scalar-integrable functions*, p. 206, when E is a Banach space.
Jacques SIMON constructed in 2009 the space of *measures* $L^p(\Omega; E)$ for a Neumann space E, in the form used here [71, définition 4.1].

Definition 10.2 is justified by the following result.

Theorem 10.3. — *The set of p-integrable measures is a vector space;*

— *in (10.5), the limit exists and does not depend on the choice of f_n;*

— *the equality (10.5), or (10.6) if $p = \infty$, defines a semi-norm on $L^p(\Omega; E)$.* ▮

To prove the uniqueness of the limit, we will use the following property[2].

Lemma 10.4. — *Let $(f_n)_{n\in\mathbb{N}}$ be a sequence in $\mathcal{B}(\Omega; E)$, where Ω is an open subset of $\mathbb{R}^d$, E is a Neumann space, $\| \, \|_{E;\nu}$ is a semi-norm of E, and $1 \leq p < \infty$, such that, as $n \to \infty$,*

$$f_n \to 0 \ \textit{in}\ \mathcal{M}(\Omega; E), \quad \sup_{m\geq n} \Big(\int_\Omega \|f_m - f_n\|^p_{E;\nu} \Big)^{1/p} \to 0.$$

Then,

$$\Big(\int_\Omega \|f_n\|^p_{E;\nu} \Big)^{1/p} \to 0. \ ▮$$

Proof. Given $\epsilon > 0$, let us bound the four terms successively in the following decomposition, where $m \geq n \geq \ell$ and R_k is the regularization operator (Definition 3.17),

$$f_n = (f_n - f_\ell) + (f_\ell - R_k f_\ell) + R_k(f_\ell - f_m) + R_k f_m. \tag{10.7}$$

1°) First term. By hypothesis, there exists ℓ such that, for all $n \geq \ell$ and $m \geq n$,

$$\Big(\int_\Omega \|f_m - f_n\|^p_{E;\nu} \Big)^{1/p} \leq \frac{\epsilon}{4}. \tag{10.8}$$

In particular,

$$\Big(\int_\Omega \|f_n - f_\ell\|^p_{E;\nu} \Big)^{1/p} \leq \frac{\epsilon}{4}. \tag{10.9}$$

2°) Second term. By the convergence of the regularizations (Proposition 9.10 (b)), there exists k such that

$$\Big(\int_\Omega \|f_\ell - R_k f_\ell\|^p_{E;\nu} \Big)^{1/p} \leq \frac{\epsilon}{4}. \tag{10.10}$$

2 **Historical background of Lemma 10.4.** Jacques SIMON established this result in 2009 [71, lemme 4.4]. It is the analogue for measures of the following result for classes of functions equal almost everywhere: a *real-valued positive function has a vanishing upper integral if and only if it is zero almost everywhere* (for example, see [BOURBAKI, 8, chap. IV, § 2, th. 1, p. 118]).

3º) Third term. The bound on the regularizations (Proposition 9.10 (a)), and also the bound (10.8) with $n = \ell$, yield

$$\Big(\int_\Omega \|R_k(f_\ell - f_m)\|^p_{E;\nu}\Big)^{1/p} \le \Big(\int_\Omega \|f_\ell - f_m\|^p_{E;\nu}\Big)^{1/p} \le \frac{\epsilon}{4}. \qquad (10.11)$$

4º) Fourth term. The support of $R_k f_m$ is contained in the compact set $\overline{\Omega^k_{1/k}}$ (Theorem 3.18), and also in the bounded open set $\Omega^k_{1/2k}$. Since the domain where a function vanishes does not contribute to its integral (Proposition 1.23), the bound on the semi-norms of the integral from Proposition 1.21 gives

$$\Big(\int_\Omega \|R_k f_m\|^p_{E;\nu}\Big)^{1/p} = \Big(\int_{\Omega^k_{1/2k}} \|R_k f_m\|^p_{E;\nu}\Big)^{1/p} \le |\Omega^k_{1/2k}|^{1/p} \sup_{x\in\Omega} \|R_k f_m(x)\|_{E;\nu}.$$

As $m \to \infty$, then $f_m \to 0$ in $\mathcal{M}(\Omega; E)$ by hypothesis whence, since R_k is sequentially continuous from $\mathcal{M}(\Omega; E)$ into $\mathrm{C_b}(\Omega; E)$ (Theorem 3.20), we have $R_k f_m \to 0$ in the latter space. That is, $\sup_{x\in\Omega} \|R_k f_m(x)\|_{E;\nu} \to 0$ by Definition 1.11. Thus, for m sufficiently large,

$$\Big(\int_\Omega \|R_k f_m\|^p_{E;\nu}\Big)^{1/p} \le \frac{\epsilon}{4}. \qquad (10.12)$$

5º) Conclusion. Decomposition (10.7), together with the bounds (10.9) to (10.12), as well as Minkowski's inequality (Proposition 9.7) give, for all $n \ge \ell$,

$$\Big(\int_\Omega \|f_n\|^p_{E;\nu}\Big)^{1/p} \le \epsilon. \ \Box$$

Let us now justify the definition of $L^p(\Omega; E)$.

Proof of Theorem 10.3. ***1º) Obtaining a vector space.*** Measures with a p-integrabilizing sequence, *i.e.* approximations satisfying (10.1) and (10.2), or (10.3) and (10.4) if $p = \infty$, satisfy the properties characterizing a vector subspace (Definition A.8 (b)), namely of $\mathcal{M}(\Omega; E)$. More precisely:

— their set is non-empty, since the zero measure has a p-integrabilizing sequence (the sequence of zero functions);

— if two measures f and g have p-integrabilizing sequences $(f_n)_{n\in\mathbb{N}}$ and $(g_n)_{n\in\mathbb{N}}$, and if $t \in \mathbb{R}$, then tf and $f+g$ have p-integrabilizing sequences, namely $(tf_n)_{n\in\mathbb{N}}$ and $(f_n + g_n)_{n\in\mathbb{N}}$. (They satisfy conditions (10.2), (10.3), and (10.4) by Minkowski's inequality for continuous functions from Proposition 9.7).

2º) Existence of a limit for $(\int_\Omega \|f_n\|^p_{E;\nu})^{1/p}$ when $p < \infty$. Again, by Minkowski's inequality (Proposition 9.7), the mapping $f \mapsto (\int_\Omega \|f\|^p_{E;\nu})^{1/p}$ is a semi-norm on

$\mathcal{B}(\Omega; E)$. It is therefore contracting (Proposition A.11), which can be written here as

$$\left|\Big(\int_\Omega \|f_n\|^p_{E;\nu}\Big)^{1/p} - \Big(\int_\Omega \|f_m\|^p_{E;\nu}\Big)^{1/p}\right| \le \Big(\int_\Omega \|f_n - f_m\|^p_{E;\nu}\Big)^{1/p}. \qquad (10.13)$$

According to criterion (10.2) of Definition 10.1 of an p-integrabilizing sequence, this tends to 0. Hence, $(\int_\Omega \|f_n\|^p_{E;\nu})^{1/p}$ is a Cauchy sequence in $\mathbb{R}$, and therefore has a limit.

3°*) Uniqueness of the limit when* $p < \infty$. Let $(g_n)_{n\in\mathbb{N}}$ be another p-integrabilizing sequence for f. Then, $(f_n - g_n)_{n\in\mathbb{N}}$ is a p-integrabilizing sequence for the zero function, so Lemma 10.4 yields

$$\Big(\int_\Omega \|f_n - g_n\|^p_{E;\nu}\Big)^{1/p} \to 0.$$

Inequality (10.13) with g_n in the place of f_m then shows that $(\int_\Omega \|g_n\|^p_{E;\nu})^{1/p}$ has the same limit as $(\int_\Omega \|f_n\|^p_{E;\nu})^{1/p}$.

4°*) Semi-norm when* $p < \infty$. Let $g \in L^p(\Omega; E)$, and let $(g_n)_{n\in\mathbb{N}}$ be a p-integrabilizing sequence for g. Minkowski's inequality with respect to f_n and g_n gives, in the limit, according to their Definition (10.5),

$$\|f+g\|_{L^p(\Omega;E);\nu} \le \|f\|_{L^p(\Omega;E);\nu} + \|g\|_{L^p(\Omega;E);\nu}. \qquad (10.14)$$

Similarly, $\|f\|_{L^p(\Omega;E);\nu} \ge 0$ and $\|tf\|_{L^p(\Omega;E);\nu} = |t|\,\|f\|_{L^p(\Omega;E);\nu}$. This proves that the mapping $f \mapsto \|f\|_{L^p(\Omega;E);\nu}$ is a semi-norm (Definition A.9) on $L^p(\Omega; E)$.

5°*) Semi-norm when* $p = \infty$. Given an ∞-integrabilizing sequence $(f_n)_{n\in\mathbb{N}}$ for f, denote

$$|||(f_n)_{n\in\mathbb{N}}||| \overset{\text{def}}{=} \sup_{n\in\mathbb{N},\, x\in\Omega} \|f_n(x)\|_{E;\nu}.$$

Its infimum taken over all sequences $(f_n)_{n\in\mathbb{N}}$ is finite by (10.3).

Let $g \in L^\infty(\Omega; E)$, and let $(g_n)_{n\in\mathbb{N}}$ be an ∞-integrabilizing sequence for g. Then $f_n + g_n$ is an ∞-integrabilizing sequence for $f+g$, so Definition (10.6) gives

$$\|f+g\|_{L^\infty(\Omega;E);\nu} \le |||(f_n+g_n)_{n\in\mathbb{N}}||| \le |||(f_n)_{n\in\mathbb{N}}||| + |||(g_n)_{n\in\mathbb{N}}|||.$$

By taking the infimum over all sequences $(f_n)_{n\in\mathbb{N}}$ and then over $(g_n)_{n\in\mathbb{N}}$, we have

$$\|f+g\|_{L^\infty(\Omega;E);\nu} \le \|f\|_{L^\infty(\Omega;E);\nu} + \|g\|_{L^\infty(\Omega;E);\nu}.$$

Similarly, $\|f\|_{L^\infty(\Omega;E);\nu} \ge 0$ and $\|tf\|_{L^\infty(\Omega;E);\nu} = |t|\,\|f\|_{L^\infty(\Omega;E);\nu}$. This proves that the mapping $f \mapsto \|f\|_{L^\infty(\Omega;E);\nu}$ is a semi-norm on $L^\infty(\Omega; E)$. □

Non-p-integrable measures. An example of a measure that is not p-integrable is the Dirac mass.

Note that this is a measure that cannot be represented by a class of p-integrable functions equal almost everywhere, see Theorems 17.31 and 17.33. □

10.2. Separability of $L^p(\Omega; E)$

Let us show that $L^p(\Omega; E)$ is a separated semi-normed space (we will see in Theorem 10.16 that it is a Neumann space).

Theorem 10.5. — *The space $L^p(\Omega; E)$, where Ω is an open subset of $\mathbb{R}^d$, E is a Neumann space, and $1 \le p \le \infty$, is a separated semi-normed space.* ▮

The proof will use the following inequality. It involves the conjugate exponent, whose definition given on p. 171 can be extended to cases involving $p = 1$ and $p = \infty$ by setting

$$\frac{1}{\infty} \stackrel{\text{def}}{=} 0,$$

so that $1' \stackrel{\text{def}}{=} \infty$ and $\infty' \stackrel{\text{def}}{=} 1$.

Lemma 10.6. — *Let*

$$f \in L^p(\Omega; E) \text{ and } \varphi \in \mathcal{K}(\Omega),$$

where Ω is an open subset of $\mathbb{R}^d$, E is a Neumann space, $1 \le p \le \infty$, and $\nu \in \mathcal{N}_E$ (the index set of the semi-norms of E), and let p' be the conjugate exponent of p defined by $1/p' + 1/p = 1$. Then,

$$\|\langle f, \varphi\rangle\|_{E;\nu} \le \|f\|_{L^p(\Omega;E);\nu} \|\varphi\|_{L^{p'}(\Omega)}. \text{ ▮}$$

Proof. Let $(f_n)_{n\in\mathbb{N}}$ be a p-integrabilizing sequence for f (Definition 10.1). Using the identification of a continuous function f_n with a measure (Theorem 3.15), and the bound on the semi-norms of the Cauchy integral (Proposition 1.17), we obtain

$$\|\langle f_n, \varphi\rangle\|_{E;\nu} = \Big\|\int_\Omega f_n \varphi\Big\|_{E;\nu} \le \int_\Omega \|f_n\|_{E;\nu} |\varphi|. \tag{10.15}$$

1°) Case $p < \infty$. Hölder's inequality from Proposition 9.5 gives

$$\|\langle f_n, \varphi\rangle\|_{E;\nu} \le \Big(\int_\Omega \|f_n\|^p_{E;\nu}\Big)^{1/p} \Big(\int_\Omega |\varphi|^{p'}\Big)^{1/p'}.$$

The stated inequality then follows by taking the limit as $n \to \infty$ and by using (10.1) and (10.5), since $\|\varphi\|_{L^{p'}(\Omega)} = (\int_\Omega |\varphi|^{p'})^{1/p'}$ (by Definition 10.2 with $f = f_n = \varphi$) .

2°) Case $p = \infty$. Then $p' = 1$, and we bound the last term of (10.15), using the monotonicity and linearity of the real-valued integral (Propositions 1.18 (a) and 1.16), by

$$\|\langle f_n, \varphi\rangle\|_{E;\nu} \le \sup_{x\in\mathbb{N},\, x\in\Omega} \|f_n(x)\|_{E;\nu} \int_\Omega |\varphi|.$$

The stated inequality then follows by taking the limit as $n \to \infty$ and with (10.1), and then by taking the infimum over all ∞-regularizing sequences (by Definition (10.6) of $\|f\|_{L^\infty(\Omega;E);\nu}$). □

Proof of Theorem 10.5. Let $f \in L^p(\Omega; E)$ such that, for every $\nu \in \mathcal{N}_E$,

$$\|f\|_{L^p(\Omega;E);\nu} = 0.$$

The inequality from Lemma 10.6 then shows, for every $\varphi \in \mathcal{K}(\Omega)$ and $\nu \in \mathcal{N}_E$, that

$$\|\langle f, \varphi\rangle\|_{E;\nu} = 0.$$

Since E is separated, as is any Neumann space (Definition 1.5), it follows then that $\langle f, \varphi\rangle = 0$ for every $\varphi \in \mathcal{K}(\Omega)$, hence

$$f = 0.$$

Therefore, $L^p(\Omega; E)$ also satisfies Definition 1.1 of a separated space, and it is additionally semi-normed by Definition 10.2. □

10.3. Some properties of $L^p(\Omega; E)$

Let us observe that functions in $\mathcal{B}(\Omega; E)$ are p-integrable: this is therefore the case for functions in p-integrabilizing sequences (the integral expression of semi-norms will be extended to any p-integrable measure in Theorem 13.4).

Proposition 10.7. — *Every function f in $\mathcal{B}(\Omega; E)$, where Ω is an open subset of $\mathbb{R}^d$ and E is a Neuman space, belongs to $L^p(\Omega; E)$ for $1 \leq p \leq \infty$ and, for every $\nu \in \mathcal{N}_E$ (the index set for the semi-norms of E),*

$$\|f\|_{L^p(\Omega;E);\nu} = \begin{cases} \left(\int_\Omega \|f\|^p_{E;\nu}\right)^{1/p} & \text{if } p \text{ is finite,} \\ \sup_{x\in\Omega} \|f(x)\|_{E;\nu} & \text{if } p = \infty. \end{cases}$$ ∎

Proof. ***1°)*** *Case $p < \infty$.* This is obtained by taking $f_n = f$ in both Definitions 10.1 and 10.2.

2°) *Case $p = \infty$.* Definition 10.1 with $f_n = f$ gives $f \in L^\infty(\Omega; E)$.

2.a. *Bound for the norm.* Definition 10.2 with $f_n = f$ gives

$$\|f\|_{L^\infty(\Omega;E);\nu} \leq \sup_{x\in\Omega} \|f(x)\|_{E;\nu}.$$

2.b. *Lower bound for the norm.* More generally, suppose that

$$f \in \mathcal{C}(\Omega; E) \cap L^\infty(\Omega; E).$$

Let $x \in \Omega$ and $\varphi_n(z) = \rho_n(z - x)$, where ρ_n is a regularizing function (Definition 3.16) and where n is large enough such that $B(x, 1/n) \subset \Omega$; then

$$\varphi_n \in \mathcal{K}(\Omega), \quad \operatorname{supp} \varphi_n \subset B(x, 1/n), \quad \int_\Omega \varphi_n = 1.$$

So, for every $\nu \in \mathcal{N}_E$,

$$\left\| \left(\int_\Omega f\varphi_n \right) - f(x) \right\|_{E;\nu} = \left\| \int_\Omega (f(x) - f)\varphi_n \right\|_{E;\nu} \leq \sup_{|z-x|<1/n} \|f(x) - f(z)\|_{E;\nu}.$$

This tends to 0 as $n \to \infty$, hence

$$\int_\Omega f\varphi_n \to f(x).$$

However (Theorem 3.15 and Lemma 10.6 with $p = \infty$, so that $p' = 1$),

$$\left\| \int_\Omega f\varphi_n \right\|_{E;\nu} = \|\langle f, \varphi_n \rangle\|_{E;\nu} \leq \|f\|_{L^\infty(\Omega;E);\nu} \left| \int_\Omega \varphi_n \right|.$$

Since $\int_\Omega \varphi_n = 1$, in the limit it follows that

$$\|f(x)\|_{E;\nu} \leq \|f\|_{L^\infty(\Omega;E);\nu}.$$

This holds for all $x \in \Omega$, and therefore

$$\sup_{x\in\Omega} \|f(x)\|_{E;\nu} \leq \|f\|_{L^\infty(\Omega;E);\nu}. \tag{10.16}$$ □

Let us compare $L^p(\Omega; E)$ to various other spaces.

Theorem 10.8. — *Let Ω be an open subset of $\mathbb{R}^d$, K be a compact set contained in Ω, E be a Neumann space, and $1 \leq p \leq \infty$. Then:*

(a) $$\mathbf{C}_K(\Omega; E) \subsetneq \mathcal{K}(\Omega; E) \subsetneq L^p(\Omega; E) \subsetneq \mathcal{M}(\Omega; E);$$

(b) $$\mathcal{C}_b(\Omega; E) \subsetneq L^\infty(\Omega; E) \subsetneq \mathcal{M}(\Omega; E).$$ ▮

Proof. Let $\{\| \ \|_{E;\nu} : \nu \in \mathcal{N}_E\}$ denote the semi-norms of E.

1°) Inclusions (a). **1.a.** $\mathbf{C}_K(\Omega; E) \subsetneq \mathcal{K}(\Omega; E)$. This inclusion was established in Volume 2 [73, Theorem 1.19].

1.b. $\mathcal{K}(\Omega; E) \underset{\rightarrow}{\subseteq} L^p(\Omega; E)$. Let $f \in \mathcal{K}(\Omega; E)$. The bound for the semi-norms of the integral from Proposition 1.17, gives, since the domain where a function vanishes does not contribute to its integral (Proposition 1.23) and $\int_{\overset{\circ}{\text{supp}} f} 1/(2 + |x|)^{d+2}\,dx \leq 2^d$ (Lemma 3.11), where supp f is the support of f,

$$\int_\Omega \|f\|_{E;\nu}^p = \int_{\overset{\circ}{\text{supp}} f} \frac{1}{(2 + |x|)^{d+2}} (2 + |x|)^{d+2} \|f\|_{E;\nu}^p \leq$$

$$\leq 2^d \sup_{x \in \Omega} (2 + |x|)^{d+2} \|f(x)\|_{E;\nu}^p.$$

The integral expression for the semi-norms $L^p(\Omega; E)$ of a function in $\mathcal{B}(\Omega; E)$ (Proposition 10.7) together with Definition 1.12 of the semi-norms of $\mathcal{K}(\Omega; E)$ therefore give, denoting $c = 2^{d/p}$ and $q(x) = (2 + |x|)^{(d+2)/p}$ (so $q \in \mathcal{C}^+(\Omega)$),

$$\|f\|_{L^p(\Omega;E);\nu} = \Big(\int_\Omega \|f\|_{E;\nu}^p\Big)^{1/p} \leq c \sup_{x \in \Omega} q(x) \|f(x)\|_{E;\nu} = c\|f\|_{\mathcal{K}(\Omega;E);q,\nu}.$$

This proves the topological inclusion (Definition 1.4 (a)) $\mathcal{K}(\Omega; E) \underset{\rightarrow}{\subseteq} L^p(\Omega; E)$.

1.c. $L^p(\Omega; E) \underset{\rightarrow}{\subseteq} \mathcal{M}(\Omega; E)$. Definition 3.2 of the semi-norms of $\mathcal{M}(\Omega; E)$ and the inequality from Lemma 10.6 give, for every $f \in L^p(\Omega; E)$, $\varphi \in \mathcal{K}(\Omega)$, and $\nu \in \mathcal{N}_E$,

$$\|f\|_{\mathcal{M}(\Omega;E);\varphi,\nu} = \|\langle f, \varphi \rangle\|_{E;\nu} \leq \|\varphi\|_{L^{p'}(\Omega)} \|f\|_{L^p(\Omega;E)}.$$

Which proves that $L^p(\Omega; E) \underset{\rightarrow}{\subseteq} \mathcal{M}(\Omega; E)$.

***2º) Inclusions (b).* 2.a.** $\mathcal{C}_b(\Omega; E) \underset{\rightarrow}{\subseteq} L^\infty(\Omega; E)$. Suppose that f is continuous and bounded, and consider a localizing sequence $(\alpha_n)_{n \in \mathbb{N}}$, *i.e.* (Definition 3.16)

$$\alpha_n \in \mathcal{K}(\Omega), \quad 0 \leq \alpha_n \leq 1, \quad \alpha_n = 1 \text{ on } \Omega_{3/n}^n,$$

where $\Omega_r^n = \{x \in \Omega : |x| < n,\ B(x, r) \subset \Omega\}$. With this, the $(\alpha_n f)_{n \in \mathbb{N}}$ form an ∞-integrabilizing sequence for f (Definition 10.1), since $\alpha_n f$ belongs to $\mathcal{K}(\Omega; E)$ and hence to $\mathcal{B}(\Omega; E)$. Given an open subset $\omega \Subset \Omega$, we have $\alpha_n f = f$ in ω once n is large enough, whence, as $n \to \infty$,

$$\alpha_n f \to f \text{ in } \mathcal{M}(\Omega; E), \quad \sup_{m \geq n} \int_\omega \|\alpha_m f - \alpha_n f\|_{E;\nu} = 0.$$

Thus (Definition 10.1), $f \in L^\infty(\Omega; E)$, and (Definition 10.2) we have

$$\|f\|_{L^\infty(\Omega;E);\nu} \leq \sup_{n \in \mathbb{N},\, x \in \Omega} \|\alpha_n f(x)\|_{E;\nu} = \sup_{x \in \Omega} \|f(x)\|_{E;\nu}. \tag{10.17}$$

Which proves that $\mathcal{C}_b(\Omega; E) \underset{\rightarrow}{\subseteq} L^\infty(\Omega; E)$.

2.b. $L^\infty(\Omega; E) \underset{\rightarrow}{\subseteq} \mathcal{M}(\Omega; E)$. This is 1.c when $p = \infty$. □

Let us show that each p-integrabilizing sequence converges in $L^p(\Omega; E)$ when p is finite (they were introduced for this purpose in Definition 10.1).

Theorem 10.9. — *Let $f \in L^p(\Omega; E)$ and $(f_n)_{n\in\mathbb{N}} \subset \mathcal{B}(\Omega; E)$, where Ω is an open subset of $\mathbb{R}^d$, E is a Neumann space, and $1 \le p < \infty$. Then:*

$$(f_n)_{n\in\mathbb{N}} \text{ is a } p\text{-integrabilizing sequence for } f \;\Leftrightarrow\; f_n \to f \text{ in } L^p(\Omega; E). \;\blacksquare$$

Warning. Unlike in the case when p is finite, ∞-integrabilizing sequences do not necessarily converge in $L^\infty(\Omega; E)$ (on the other hand, they do converge in $L^1_{\text{loc}}(\Omega; E)$, see Theorem 11.12). □

Proof of Theorem 10.9. **1°)** *Forward implication.* If the $(f_n)_{n\in\mathbb{N}}$ form a p-integrabilizing sequence for f (Definition 10.1), then the $(f_m - f_n)_{m\in\mathbb{N}}$ form a p-integrabilizing sequence for $f - f_n$, and therefore, for every semi-norm $\| \ \|_{E;\nu}$ of E,

$$\|f - f_n\|_{L^p(\Omega;E);\nu} = \lim_{m\to\infty} \Big(\int_\Omega \|f_m - f_n\|^p_{E;\nu} \Big)^{1/p}.$$

The right-hand side tends to 0 as $n \to \infty$, so we get $f_n \to f$ in $L^p(\Omega; E)$.

2°) *Backward implication.* If $f_n \to f$ in $L^p(\Omega; E)$, the integral expression for the $L^p(\Omega; E)$ semi-norms of a function in $\mathcal{B}(\Omega; E)$ (Proposition 10.7) yields

$$\Big(\int_\Omega \|f_m - f_n\|^p_{E;\nu} \Big)^{1/p} = \|f_m - f_n\|_{L^p(\Omega;E);\nu} \le$$
$$\le \|f - f_m\|_{L^p(\Omega;E);\nu} + \|f - f_n\|_{L^p(\Omega;E);\nu}$$

whose upper bound, for $m \ge n$, tends to 0 as $n \to \infty$; furthermore (Theorem 10.8), $f_n \to f$ in $\mathcal{M}(\Omega; E)$.

Thus, the $(f_n)_{n\in\mathbb{N}}$ form a p-integrabilizing sequence for f. □

10.4. Properties of $L^\infty(\Omega; E)$

Let us show that a continuous function belongs to $L^\infty(\Omega; E)$ if and only if it is bounded.

Theorem 10.10. — *Let*

$$f \in \mathcal{C}(\Omega; E)$$

where Ω is an open subset of $\mathbb{R}^d$ and E is a Neumann space.

(a) $$f \in L^\infty(\Omega; E) \quad \Leftrightarrow \quad f \text{ is bounded.}$$

(b) *If f is bounded, for every $\nu \in \mathcal{N}_E$ (the index set of the semi-norms of E),*

$$\|f\|_{L^\infty(\Omega;E);\nu} = \sup_{x\in\Omega} \|f(x)\|_{E;\nu} = \|f\|_{\mathcal{C}_b(\Omega;E);\nu}. \blacksquare$$

WARNING. To say that f is bounded does not mean that $\sup_{\nu\in\mathcal{N}_E} \sup_{x\in\Omega} \|f(x)\|_{E;\nu}$ is finite, but rather that $\sup_{x\in\Omega} \|f(x)\|_{E;\nu}$ is finite for each $\nu \in \mathcal{N}_E$. □

Proof of Theorem 10.10. ***1º)*** *Property* (a). **1.a.** *Bounded* $\Rightarrow L^\infty(\Omega; E)$. This was established in Theorem 10.8 (b).

1.b. $L^\infty(\Omega; E) \Rightarrow$ *bounded.* This was proven in (10.16), in 2.b of the proof of Proposition 10.7.

2º) *Property* (b). The first equality of the norms follows from (10.17) and (10.16). The second follows from Definition 1.11 (a) of $\|f\|_{\mathcal{C}_b(\Omega;E);\nu}$. □

Equivalent formulation of Theorem 10.10 (a). This property is equivalent to:

$$\mathcal{C}_b(\Omega; E) = \mathcal{C}(\Omega; E) \cap L^\infty(\Omega; E). \ \square$$

Now, let us show that $L^\infty(\Omega; E)$ measures are the same as $L^1_{\text{loc}}(\Omega; E)$ measures whose approximations are bounded on their set[3].

Theorem 10.11. — (a) *We have*

$$f \in L^\infty(\Omega; E),$$

where Ω is an open subset of $\mathbb{R}^d$ and E is a Neumann space, if and only if: for every open subset $\omega \Subset \Omega$, we have

$$f|_\omega \in L^1(\omega; E)$$

and the regularizations (Definition 3.17) satisfy, for every semi-norm $\| \ \|_{E;\nu}$ of E,

$$\sup_{n\in\mathbb{N}^*,\, x\in\Omega} \|R_n f(x)\|_{E;\nu} < \infty. \tag{10.18}$$

(b) *Then,*

$$\sup_{n\in\mathbb{N}^*,\, x\in\Omega} \|R_n f(x)\|_{E;\nu} = \|f\|_{L^\infty(\Omega;E);\nu}. \blacksquare$$

3 **Historical background of Theorem 10.11.** This result was established by Jacques SIMON in 2009 [71, Proposition 5.6].

Proof. **1º)** *Forward implication of (a).* Suppose $f \in L^\infty(\Omega; E)$, and let $(f_m)_{m\in\mathbb{N}}$ be an ∞-integrabilizing sequence for f (Definition 10.1). We need to verify the two stated criteria.

1.a. On the one hand, given an open subset $\omega \Subset \Omega$,

$$\text{the } (f_m|_\omega)_{m\in\mathbb{N}} \text{ form a 1-integrabilizing sequence for } f|_\omega. \tag{10.19}$$

Indeed, $f_m|_\omega \in \mathcal{B}(\omega; E)$ (this is clear) and, as $m \to \infty$, we have $f_m|_\omega \to f|_\omega$ in $\mathcal{M}(\omega; E)$ (Theorem 6.17) and $\sup_{m'\geq m} \int_\omega \|f_{m'} - f_m\|_{E;\nu} \to 0$ (Definition 10.1 and Proposition 1.18 (b)). Hence (Definition 10.1),

$$f|_\omega \in L^1(\omega; E). \tag{10.20}$$

1.b. On the other hand, given $\nu \in \mathcal{N}_E$, we have (the bound from Proposition 9.10 (a))

$$\sup_{x\in\Omega} \|R_n f_m(x)\|_{E;\nu} \leq \sup_{x\in\Omega} \|f_m(x)\|_{E;\nu}. \tag{10.21}$$

When $m \to \infty$, then $f_m \to f$ in $\mathcal{M}(\Omega; E)$ (Definition 10.1) so, as R_n is sequentially continuous from $\mathcal{M}(\Omega; E)$ into $\mathbf{C}_{K_n}(\Omega; E)$ (Theorem 3.20), $R_n f_m \to R_n f$ in $\mathbf{C}_{K_n}(\Omega; E)$. Thus (Theorem 10.8 (a)),

$$R_n f_m \to R_n f \ \text{ in } \mathcal{K}(\Omega; E).$$

By bounding the right-hand side of (10.21) by its supremum over m, and then taking the limit on the left-hand side, we obtain

$$\sup_{x\in\Omega} \|R_n f(x)\|_{E;\nu} \leq \sup_{m\in\mathbb{N},\, x\in\Omega} \|f_m(x)\|_{E;\nu}.$$

Then by the infimum over all ∞-integrabilizing sequences $(f_m)_{m\in\mathbb{N}}$ (Definition 10.1), it follows that (Definition 10.2)

$$\sup_{n\in\mathbb{N}^*,\, x\in\Omega} \|R_n f(x)\|_{E;\nu} \leq \|f\|_{L^\infty(\Omega;E);\nu} < \infty. \tag{10.22}$$

2º) *Converse of (a).* Suppose (10.18) and (10.20) are satisfied. Now we need to verify that

$$\text{the } (R_n f)_{n\in\mathbb{N}} \text{ form an } \infty\text{-integrabilizing sequence for } f. \tag{10.23}$$

First, $R_n f \in \mathcal{B}(\Omega; E)$ (Theorem 3.18);
— then, $R_n f \to f$ in $\mathcal{M}(\Omega; E)$ (Theorem 3.19);
— and, $\sup_{n\in\mathbb{N}^*,\, x\in\Omega} \|R_n f(x)\|_{E;\nu} < \infty$ for every $\nu \in \mathcal{N}_E$ (hypothesis (10.18));
— finally, given $\omega \Subset \Omega$, it remains to verify that

$$\sup_{m\geq n} \int_\omega \|R_m f - R_n f\|_{E;\nu} \to 0. \tag{10.24}$$

Let ω' be an open subset such that $\omega \Subset \omega' \Subset \Omega$ (Proposition A.20). From (10.20), we have $f \in L^1(\omega'; E)$, and so (Theorem 10.13 (c) on the convergence of $R_n f$ with $p = 1$) $R_n^{\omega'} f \to f$ in $L^1(\omega'; E)$. It then follows, using the integral expression of the semi-norms in $L^p(\Omega; E)$ of a function in $\mathcal{B}(\Omega; E)$ (Proposition 10.7), that

$$\sup_{m \geq n} \int_{\omega'} \|R_m^{\omega'} f - R_n^{\omega'} f\|_{E;\nu} \to 0.$$

This implies (10.24), since the integral over ω' bounds the integral over ω (Proposition 1.18 (b)), and $R_n^{\omega'} f = R_n^{\Omega} f$ on ω as soon as n is sufficiently large (so that the set $\omega' + B(0, 1/n)$ is included in Ω).

This proves (10.23), and therefore,

$$f \in L^\infty(\Omega; E).$$

3o) Property (b). By Definition (10.6), p. 184, of the semi-norms of $L^\infty(\Omega; E)$, it follows from (10.23) that

$$\|f\|_{L^\infty(\Omega;E);\nu} \leq \sup_{n \in \mathbb{N}^*,\, x \in \Omega} \|R_n f(x)\|_{E;\nu}.$$

Inequality (10.22) is therefore an equality. □

Let us complete Theorem 10.11 with an estimate that applies to restrictions (a more general estimate will be given in Theorem 11.1, completed by Theorem 11.5).

Proposition 10.12. — *Let $f \in L^\infty(\Omega; E)$, where Ω is an open subset of $\mathbb{R}^d$, E is a Neumann space, ω is an open set such that $\omega \Subset \Omega$, and $\nu \in \mathcal{N}_E$ (the index set of the semi-norms of E). Then,*

$$\|f|_\omega\|_{L^1(\omega;E);\nu} \leq |\omega| \|f\|_{L^\infty(\Omega;E);\nu}. \blacksquare$$

Proof. Let $(f_n)_{n\in\mathbb{N}}$ be an ∞-integrabilizing sequence for f (Definition 10.1). As established in (10.19), the $(f_n|_\omega)_{n\in\mathbb{N}}$ form a 1-integrabilizing sequence for $f|_\omega$. Thus (Definition 10.2 of the semi-norms of $L^1(\omega; E)$ and the bound from Proposition 1.21),

$$\|f|_\omega\|_{L^1(\omega;E);\nu} = \lim_{n\to\infty} \int_\omega \|f_n\|_{E;\nu} \leq |\omega| \sup_{n\in\mathbb{N},\, x\in\Omega} \|f_n(x)\|_{E;\nu}.$$

The stated inequality then follows by taking (Definition 10.2 of $L^\infty(\omega; E)$) the infimum over all sequences $(f_n)_{n\in\mathbb{N}}$. □

10.5. Approximation via regularizations and density

Let us state some approximation properties of regularizations[4].

Theorem 10.13. — *Let*

$$f \in L^p(\Omega; E), \quad 1 \leq p \leq \infty,$$

where Ω is an open subset of $\mathbb{R}^d$, E is a Neumann space, $(R_n f)_{n \in \mathbb{N}^*}$ are regularizations in $\mathcal{K}(\Omega; E)$ (Definition 3.17), and $\nu \in \mathcal{N}_E$ (index set of the semi-norms of E).

(a) *For all $n \in \mathbb{N}^*$,*

$$\|R_n f\|_{L^p(\Omega;E);\nu} \leq \|f\|_{L^p(\Omega;E);\nu}.$$

(b) *When $n \to \infty$,*

$$\|R_n f\|_{L^p(\Omega;E);\nu} \to \|f\|_{L^p(\Omega;E);\nu}.$$

(c) *If $p < \infty$, as $n \to \infty$,*

$$R_n f \to f \text{ in } L^p(\Omega; E). \blacksquare$$

Proof. *1o) Bound* (a). ***1.a.*** *Case $p < \infty$.* Let $(f_m)_{m \in \mathbb{N}}$ be a p-integrabilizing sequence for f (Definition 10.1). Then we have $f_m \in \mathbf{C}_{K_m}(\Omega; E)$ (Theorem 3.18) and (Propositions 10.7 and 9.10 (a))

$$\|R_n f_m\|_{L^p(\Omega;E);\nu} = \Big(\int_\Omega \|R_n f_m\|^p_{E;\nu}\Big)^{1/p} \leq \Big(\int_\Omega \|f_m\|^p_{E;\nu}\Big)^{1/p} = \|f_m\|_{L^p(\Omega;E);\nu}.$$

When $m \to \infty$, $f_m \to f$ in $\mathcal{M}(\Omega; E)$ (Definition 10.1), from which, since R_n is sequentially continuous (Theorem 3.20), we get $R_n f_m \to R_n f$ in $\mathbf{C}_{K_m}(\Omega; E)$ and therefore (Theorem 10.8)

$$R_n f_m \to R_n f \text{ in } L^p(\Omega; E).$$

Since (Theorem 10.9) $f_m \to f$ in $L^p(\Omega; E)$, by taking the limit, the above inequality gives

$$\|R_n f\|_{L^p(\Omega;E);\nu} \leq \|f\|_{L^p(\Omega;E);\nu}.$$

1.b. *Case $p = \infty$.* The bound (10.22) yields, together with Theorem 10.10,

$$\|R_n f\|_{L^\infty(\Omega;E);\nu} = \sup_{x \in \Omega} \|(R_n f)(x)\|_{E;\nu} \leq \|f\|_{L^\infty(\Omega;E);\nu}.$$

4 **Historical background of approximation via regularizations in $L^p(\Omega; E)$.** Jean LERAY showed in 1934 [46, lemme 2, p. 206] that $f \star \rho_n \to f$ in $L^2(\mathbb{R}^3)$ for $f \in L^2(\mathbb{R}^3)$.

2°) Convergence (c). Decompose

$$R_n f - f = R_n(f - f_m) + (R_n f_m - f_m) + (f_m - f) \tag{10.25}$$

and let $\epsilon > 0$. Since (Theorem 10.9, again) $f_m \to f$ in $L^p(\Omega; E)$, there exists $m \in \mathbb{N}$ such that

$$\|f_m - f\|_{L^p(\Omega;E);\nu} \leq \epsilon.$$

Using the bound from (a), it follows that

$$\|R_n(f_m - f)\|_{L^p(\Omega;E);\nu} \leq \|f_m - f\|_{L^p(\Omega;E);\nu} \leq \epsilon.$$

Finally (Propositions 10.7 and 9.10 (b)), $R_n f_m \to f_m$ in $L^p(\Omega; E)$, thus there exists n_m such that, for all $n \geq n_m$,

$$\|R_n f_m - f_m\|_{L^p(\Omega;E);\nu} \leq \epsilon.$$

These three inequalities imply, together with the decomposition (10.25),

$$\|R_n f - f\|_{L^p(\Omega;E);\nu} \leq 3\epsilon$$

which proves that

$$R_n f \to f \ \text{ in } L^p(\Omega; E). \tag{10.26}$$

3°) Convergence (b). **3.a.** *Case* $p < \infty$. The convergence of the norms follows from the convergence above.

3.b. *Case* $p = \infty$. If this convergence was not satisfied, then the bound (a) would imply the existence of $\epsilon > 0$ and a subsequence $(R_n f)_{n \in I}$ such that

$$\|R_n f\|_{L^\infty(\Omega;E);\nu} \leq \|f\|_{L^\infty(\Omega;E);\nu} - \epsilon. \tag{10.27}$$

However, as we established in (10.23), the $(R_n f)_{n \in \mathbb{N}^*}$ form an ∞-integrabilizing sequence for f. The subsequence would itself be an ∞-integrabilizing sequence for f, and therefore (Definition 10.2 and Theorem 10.10 (b))

$$\|f\|_{L^\infty(\Omega;E);\nu} \leq \sup_{n \in I,\, x \in \Omega} \|R_n f(x)\|_{E;\nu} = \sup_{n \in I} \|R_n f\|_{L^\infty(\Omega;E);\nu}.$$

This would contradict (10.27), hence (c) is satisfied. □

Theorem 10.13 (c) implies the following density property when p is finite.

Theorem 10.14. — *If* $1 \leq p < \infty$, *the spaces* $\mathcal{B}(\Omega; E)$, $\mathcal{K}(\Omega; E)$, *and* $\mathcal{K}^\infty(\Omega; E)$, *where* Ω *is an open subset of* $\mathbb{R}^d$ *and* E *is a Neumann space, are dense and sequentially dense in* $L^p(\Omega; E)$.

The positive functions in $\mathcal{B}(\Omega)$, $\mathcal{K}(\Omega)$, *and* $\mathcal{K}^\infty(\Omega)$ *are dense and sequentially dense in the subset of positive functions of* $L^p(\Omega)$. ▮

Proof. **1°)** *Sequential density of* $\mathcal{K}(\Omega; E)$. According to Theorem 10.13 (c), every integrable measure f is the limit in $L^p(\Omega; E)$ of its regularizations $R_n f$. Since these belong to $\mathcal{K}(\Omega; E)$ by Theorem 3.18, this proves that the space is sequentially dense in $L^p(\Omega; E)$.

2°) *Sequential density of* $\mathcal{B}(\Omega; E)$. This follows from 1°, since $\mathcal{K}(\Omega; E) \subset \mathcal{B}(\Omega; E)$.

3°) *Sequential density of* $\mathcal{K}^\infty(\Omega; E)$. This is obtained as in 1°, noting that if one chooses the localizing and regularizing functions in Definition 3.16 to be infinitely differentiable, then the regularizations $R_n f$ are also infinitely differentiable. This is established in Volume 3 [74, Theorem 8.15] for any distribution, which also includes every measure [74, Theorem 3.18].

4°) *Density.* Sequential density always implies density (Proposition A.16).

5°) *Positive functions.* This follows from the fact that, if $f \geq 0$, then $R_n f \geq 0$ (Definition 3.17, since α_n and ρ_n are positive). □

Case $p = \infty$. The spaces $\mathcal{B}(\Omega; E)$, $\mathcal{K}(\Omega; E)$, and $\mathcal{K}^\infty(\Omega; E)$ are **neither dense nor sequentially dense** in $L^\infty(\Omega; E)$ if $|\Omega| \neq 0$ and $E \neq \{0_E\}$.

Indeed, $L^\infty(\Omega; E)$ would then contain piecewise constant functions (Theorem 13.9) which are discontinuous, whereas the closures of $\mathcal{B}(\Omega; E)$, $\mathcal{K}(\Omega; E)$, and $\mathcal{K}^\infty(\Omega; E)$ consist solely of continuous functions. □

Let us characterize the measures of $L^p(\Omega; E)$ using the properties of their regularizations.

Theorem 10.15. — *Let* $f \in \mathcal{M}(\Omega; E)$, *where* Ω *is an open subset of* $\mathbb{R}^d$, E *is a Neumann space,* $R_n f \in \mathcal{K}(\Omega; E)$ *are regularizations (Definition 3.17), and* $1 \leq p \leq \infty$. *Then,*

$$f \in L^p(\Omega; E) \quad \Leftrightarrow \quad \textit{Definition 10.1 is satisfied with } f_n = R_n f. \ \blacksquare$$

Proof. **1°)** *Case* $p < \infty$. If $f \in L^p(\Omega; E)$, then (Theorem 10.13 (c)) $R_n f \to f$, so Definition 10.1 is satisfied with $f_n = R_n f$. The converse is clear.

2°) *Case* $p = \infty$. If $f \in L^\infty(\Omega; E)$, then Definition 10.1 is satisfied with $f_n = R_n f$ according to (10.23). The converse is clear. □

10.6. Completeness of $L^p(\Omega; E)$

Let us show that $L^p(\Omega; E)$ is sequentially complete, which corresponds, for real-valued measures, to the **Riesz–Fischer completeness theorem**[5].

Theorem 10.16. — *The space $L^p(\Omega; E)$, where Ω is an open subset of $\mathbb{R}^d$, E is a Neumann space, and $1 \leq p \leq \infty$, is a Neumann space.* ▌

Proof. Let $(f_k)_{k\in\mathbb{N}}$ be a Cauchy sequence in $L^p(\Omega; E)$.

1°) Construction of the limit. Since $L^p(\Omega; E) \subsetneq \mathcal{M}(\Omega; E)$ (Theorem 10.8), the sequence $(f_k)_{k\in\mathbb{N}}$ is also Cauchy (Proposition A.22) in $\mathcal{M}(\Omega; E)$, which is sequentially complete (Theorem 3.8), so it has a limit f in it, *i.e.*

$$f_k \to f \text{ in } \mathcal{M}(\Omega; E). \tag{10.28}$$

2°) $f \in L^p(\Omega; E)$ when $p < \infty$. By Definitions 10.1 and 10.2 of $L^p(\Omega; E)$, it suffices to show that the regularizations $R_n f$ form a p-integrabilizing sequence for f.

As $R_n f \in \mathcal{K}(\Omega; E)$ (Theorem 3.18) and $R_n f \to f$ in $\mathcal{M}(\Omega; E)$ (Theorem 3.19), it remains to verify criterion (10.2), which means here that, for any semi-norm $\| \; \|_{E;\nu}$ of E, when $n \to \infty$,

$$\sup_{m\geq n} \Big(\int_\Omega \|R_m f - R_n f\|_{E;\nu}^p \Big)^{1/p} \to 0. \tag{10.29}$$

For this, decompose, for m, n, and ℓ integers,

$$R_m f - R_n f = R_m(f - f_\ell) + (R_m f_\ell - R_n f_\ell) + R_n(f_\ell - f). \tag{10.30}$$

Given $\epsilon > 0$, we will now estimate the terms on the right-hand side.

2.a. *First and third terms.* By Definition 1.2 of a Cauchy sequence, there exists ℓ such that

$$\sup_{k\geq\ell} \|f_k - f_\ell\|_{L^p(\Omega;E);\nu} \leq \frac{\epsilon}{3}. \tag{10.31}$$

5 **Historical background of the Riesz–Fischer Theorem 10.16.** The completeness of $L^2(0,1)$ was established independently by Ernst FISCHER [27] and Frigyes RIESZ [62], and published in the same issue of the *Comptes rendus de l'Académie des Sciences*, Paris, in 1907.
We have not found any reference regarding the origin of the well-known completeness of $L^p(\Omega; E)$ when E is a Banach space.
Jacques SIMON extended this result to Neumann spaces E in 2009 [71, théorèmes 4.12 and 5.9], and he provided the proof used here (the classical proof for a normed E is not suitable, see the comment *Difficulty when E is not metrizable*, p. 205).

Using the integral expression for the semi-norms $L^p(\Omega; E)$ of a function in $\mathcal{B}(\Omega; E)$ (Proposition 10.7) with the bound for the regularizations (Theorem 10.13 (a)), it then follows, for all m and $k \geq \ell$, that

$$\Big(\int_\Omega \|R_m(f_k - f_\ell)\|_{E;\nu}^p\Big)^{1/p} = \|R_m(f_k - f_\ell)\|_{L^p(\Omega;E);\nu} \leq \|f_k - f_\ell\|_{L^p(\Omega;E);\nu} \leq \frac{\epsilon}{3}.$$

Since the mapping R_m is sequentially continuous from $\mathcal{M}(\Omega; E)$ into $\mathbf{C}_{K_m}(\Omega; E)$ (Theorem 3.20), the convergence (10.28) implies, when $k \to \infty$, that $R_m f_k \to R_m f$ in $\mathbf{C}_{K_m}(\Omega; E)$, and therefore

$$\|R_m(f_k - f_\ell)\|_{E;\nu}^p \to \|R_m(f - f_\ell)\|_{E;\nu}^p \text{ in } \mathbf{C}_{K_m}(\Omega).$$

Since the Cauchy integral is continuous from $\mathbf{C}_{K_m}(\Omega)$ into $\mathbb{R}$ (Proposition 1.26), it follows, for all m that, in the limit,

$$\Big(\int_\Omega \|R_m(f - f_\ell)\|_{E;\nu}^p\Big)^{1/p} \leq \frac{\epsilon}{3}. \tag{10.32}$$

2.b. *Second term.* Since the regularizations converge in $L^p(\Omega; E)$ (Theorem 10.13, part (c)), $R_n f_\ell - f_\ell$ converges to 0 as $n \to \infty$, so there exists n_1 such that, for all $n \geq n_1$,

$$\|R_n f_\ell - f_\ell\|_{L^p(\Omega;E);\nu} \leq \frac{\epsilon}{6}.$$

Again, by the expression for the semi-norms from Proposition 10.7, if $m \geq n \geq n_1$, then

$$\Big(\int_\Omega \|R_m f_\ell - R_n f_\ell\|_{E;\nu}^p\Big)^{1/p} = \|R_m f_\ell - R_n f_\ell\|_{L^p(\Omega;E);\nu} \leq \frac{\epsilon}{3}. \tag{10.33}$$

2.c. *p-integrability.* The decomposition (10.30), together with Minkowski's inequality (Proposition 9.7), the bound (10.32) with respect to m and n (in the place of m), and the bound (10.33), give, for all $m \geq n \geq n_1$,

$$\Big(\int_\Omega \|R_m f - R_n f\|_{E;\nu}^p\Big)^{1/p} \leq \epsilon. \tag{10.34}$$

This proves (10.29), and thus the $R_n f$ form a p-integrabilizing sequence for f, and

$$f \in L^p(\Omega; E). \tag{10.35}$$

3º) *$f_k \to f$ in $L^p(\Omega; E)$ when $p < \infty$.* Using Proposition 10.7 again, inequality (10.32) can be written as

$$\|R_m(f - f_\ell)\|_{L^p(\Omega;E);\nu} \leq \frac{\epsilon}{3}.$$

Since the regularizations converge in $L^p(\Omega; E)$ (Theorem 10.13 (c), again), it follows in the limit that, when $m \to \infty$,

$$\|f - f_\ell\|_{L^p(\Omega;E);\nu} \leq \frac{\epsilon}{3}.$$

From which, together with (10.31), we get, for all $k \geq \ell$,

$$\|f_k - f\|_{L^p(\Omega;E);\nu} \leq \epsilon.$$

Thus,

$$f_k \to f \text{ in } L^p(\Omega; E).$$

***4º) Conclusion for** $p < \infty$.* This proves that $L^p(\Omega; E)$ is sequentially complete, *i.e.* it is a Neumann space, since it is semi-normed and separated (Theorem 10.5).

***5º)** ∞-integrability.* Let us verify the two criteria of Theorem 10.11 (a) when $p = \infty$.

5.a. *First criterion.* On the one hand, for any open subset $\omega \Subset \Omega$,

$$f|_\omega \in L^1(\omega; E). \tag{10.36}$$

Indeed, $(f_k|_\omega)_{k\in\mathbb{N}}$ is a Cauchy sequence in $L^1(\omega; E)$ according to the inequality from Proposition 10.12, hence the case when $p = 1$ shows that $f_k|_\omega$ converges in it.

Moreover, it converges to $f|_\omega$ in $\mathcal{M}(\omega; E)$ by (10.28) and the continuity of the restriction from $\mathcal{M}(\Omega; E)$ into $\mathcal{M}(\omega; E)$ (Theorem 6.17).

5.b. *Second criterion.* Let $\epsilon > 0$. For ℓ sufficiently large, say $\ell \geq \ell_0$, we have

$$\sup_{k\geq\ell} \|f_k - f_\ell\|_{L^\infty(\Omega;E);\nu} \leq \epsilon.$$

From the expressions for the semi-norms of $L^\infty(\Omega; E)$ in Theorem 10.10 (b), and by the bound on the regularizations from Theorem 10.13 (a), for all m and $k \geq \ell \geq \ell_0$, we get

$$\sup_{x\in\Omega} \|R_m(f_k - f_\ell)(x)\|_{E;\nu} = \|R_m(f_k - f_\ell)\|_{L^\infty(\Omega;E);\nu} \leq \|f_k - f_\ell\|_{L^\infty(\Omega;E);\nu} \leq \epsilon.$$

However, $f_k \to f$ in $\mathcal{M}(\Omega; E)$ (by (10.28)), and R_m is sequentially continuous from $\mathcal{M}(\Omega; E)$ into $\mathbf{C}_b(\Omega; E)$ (Theorem 3.20), hence $R_m f_k \to R_m f$ in $\mathbf{C}_b(\Omega; E)$. In the limit,

$$\sup_{x\in\Omega} \|R_m(f - f_\ell)(x)\|_{E;\nu} \leq \epsilon. \tag{10.37}$$

From which it follows, together with the bound on the regularizations from Theorem 10.13 (a), that

$$\sup_{m\in\mathbb{N}^*,\, x\in\Omega} \|R_m f(x)\|_{E;\nu} \leq \sup_{m\in\mathbb{N}^*,\, \ell\geq\ell_0,\, x\in\Omega} \|R_m f_\ell(x)\|_{E;\nu} + \epsilon \leq$$

$$\leq \sup_{\ell\geq\ell_0,\, x\in\Omega} \|f_\ell\|_{L^\infty(\Omega;E);\nu} + \epsilon.$$

Since $(f_\ell)_{\ell\in\mathbb{N}}$ is bounded in $L^\infty(\Omega;E)$ (as is any Cauchy sequence, see Proposition A.12),

$$\sup_{m\in\mathbb{N}^*,\, x\in\Omega} \|R_m f(x)\|_{E;\nu} < \infty. \tag{10.38}$$

5.c. *∞-integrability.* From (10.36) and (10.38), we have (using the characterization of $L^\infty(\Omega;E)$ from Theorem 10.11 (a)),

$$f \in L^\infty(\Omega;E).$$

6º) $f_\ell \to f$ in $L^\infty(\Omega;E)$. Inequality (10.37) can be written, together with the expression for the semi-norms of $L^\infty(\Omega;E)$ from Theorem 10.10 (b), as

$$\|R_m(f - f_\ell)\|_{L^\infty(\Omega;E);\nu} \leq \epsilon.$$

The convergence of the semi-norms of the regularizations (Theorem 10.13 (b)) gives us, in the limit,

$$\|f - f_\ell\|_{L^\infty(\Omega;E);\nu} \leq \epsilon.$$

Whence,

$$f_\ell \to f \ \text{ in } L^\infty(\Omega;E).$$

7º) Conclusion for $p=\infty$. This proves that $L^\infty(\Omega;E)$ is sequentially complete, *i.e.* it is a Neumann space, since it is separated (Theorem 10.5, again). □

Alternative proof of Theorem 10.16 when E is normed. When E is normed, then so is $L^p(\Omega;E)$, given a Cauchy sequence $(f_n)_{n\in\mathbb{N}}$ in $L^p(\Omega;E)$, we can extract a subsequence from it such that

$$\|f_{n+1} - f_n\|_{L^p(\Omega;E)} \leq 2^{-n}.$$

By choosing an approximation $g_n \in \mathcal{B}(\Omega;E)$ such that $\|g_n - f_n\|_{L^p(\Omega;E)} \leq 2^{-n}$, we obtain a sequence $(g_n)_{n\in\mathbb{N}}$ which satisfies $\sup_{m\geq n}\int_\Omega \|g_m - g_n\|_E^p \to 0$ and converges in $\mathcal{M}(\Omega;E)$ to the limit f of the f_n in $\mathcal{M}(\Omega;E)$.

This proves (Definition 10.2) that $f \in L^p(\Omega;E)$, from which it follows that $f_n \to f$ in $L^p(\Omega;E)$. □

Let us show that $L^p(\Omega;E)$ is metrizable or normed, provided that E is too.

Theorem 10.17. — *Let Ω be an open subset of $\mathbb{R}^d$ and $1 \leq p \leq \infty$.*

If E is a Fréchet space, then $L^p(\Omega;E)$ is a Fréchet space.

If E is a Banach space, then $L^p(\Omega;E)$ is a Banach space; in particular, $L^p(\Omega)$ is a Banach space. ▌

Proof. ***1°) Fréchet space*** $L^p(\Omega; E)$. If E is a Fréchet space, then its family of semi-norms is equivalent to a countable family (Definitions 1.5 and A.24). The same holds for the family of semi-norms of $L^p(\Omega; E)$ (Definition 10.2), which is therefore metrizable. Since this space is sequentially complete (Theorem 10.16), it is Fréchet.

2°) Banach space $L^p(\Omega; E)$. If E is a Banach space, then it is normed, so the family of semi-norms of $L^p(\Omega; E)$ reduces to a norm. Since this space is sequentially complete, it is Banach.

3°) Space $L^p(\Omega)$. Since $\mathbb{R}$ is a Banach space (Proposition A.35), $L^p(\Omega) = L^p(\Omega; \mathbb{R})$ is Banach by 2°. □

We shall see in Theorem 12.17 that $L^2(\Omega; E)$ is a Hilbert space, provided that E is Hilbert too.

10.7. Remarks on methods of construction

One may define integrable distributions by requiring that the p-integrabilizing sequence converges only in the sense of distributions, rather than in the sense of measures, as follows.

Definition 10.18. — *A **p-integrable distribution** is a distribution $f \in \mathcal{D}'(\Omega; E)$, where Ω is an open subset of $\mathbb{R}^d$, E is a Neumann space, and $1 \le p \le \infty$, such that there exists a sequence $(f_n)_{n\in\mathbb{N}}$ in $\mathcal{B}(\Omega; E)$ such that:*

$$f_n \to f \ \textit{in}\ \mathcal{D}'(\Omega; E) \tag{10.39}$$

and, for every semi-norm $\| \ \|_{E;\nu}$ of E, $(f_n)_{n\in\mathbb{N}}$ satisfies (10.2) if $p < \infty$, or (10.3) and (10.4) if $p = \infty$. ▮

Let us show that each integrable distribution coincides with an integrable measure (after identifying each measure f with the distribution $\overline{f}$ defined by $\langle \overline{f}, \varphi\rangle \overset{\text{def}}{=} \langle f, \varphi\rangle$, where $\varphi \in \mathcal{D}(\Omega)$).

Theorem 10.19. — *A distribution is p-integrable if and only if it is a p-integrable measure.* ▮

Proof. Conditions (10.2), (10.3), and (10.4) depend only on the semi-norms of E, so it suffices to compare the two notions of convergence for the f_n.

1º) Measure implies distribution. The convergence $f_n \to f$ in the sense of measures implies that in the sense of distributions [Vol. 3, Theorem 3.18].

2º) Distribution implies measure. **2.a.** Choose the functions ρ_n in the regularizing sequence so that they belong to $\mathcal{D}(\mathbb{R}^d)$. Then they satisfy Definition 8.1 of Volume 3, so [Vol. 3, Theorem 8.23] R_n is sequentially continuous from $\mathcal{D}'(\Omega; E)$ into $\mathbf{C}_b(\Omega; E)$.

Lemma 10.4 then holds under the hypothesis that $f_n \to f$ in $\mathcal{D}'(\Omega; E)$, instead of in $\mathcal{M}(\Omega; E)$ (only point 4º is modified in its proof, where now the sequential continuity of R_k is used).

Hence, the unchanged proof of Theorem 10.3 justifies Definition 10.2 of the space of p-integrable distributions, which we denote by $L^p_*(\Omega; E)$.

2.b. If the $(f_n)_{n\in\mathbb{N}}$ form a $*$-p-integrabilizing sequence (*i.e.* it satisfies hypothesis (10.39) instead of (10.1)), then the $(f_m - f_n)_{m\in\mathbb{N}}$ form a $*$-p-integrabilizing sequence for $f - f_n$. Hence, for every semi-norm $\| \ \|_{E;\nu}$ of E,

$$\|f - f_n\|_{L^p_*(\Omega;E);\nu} = \lim_{m\to\infty} \Big(\int_\Omega \|f_m - f_n\|^p_{E;\nu} \Big)^{1/p}.$$

The right-hand side tends to 0 by assumption as $n \to \infty$, hence

$$f_n \to f \ \text{ in } L^p_*(\Omega; E). \tag{10.40}$$

Let us revisit the proof of Lemma 10.6, now though with a $*$-p-integrabilizing sequence for f. Then, for every $f \in L^p_*(\Omega; E)$, $\varphi \in \mathcal{D}(\Omega)$, and $\nu \in \mathcal{N}_E$, we obtain

$$\|\langle f, \varphi\rangle\|_{E;\nu} \le \|f\|_{L^p_*(\Omega;E);\nu} \|\varphi\|_{L^{p'}(\Omega)}. \tag{10.41}$$

Let $\varphi_0 \in \mathcal{K}(\Omega)$ and $\varphi = R_m\varphi_0 - R_q\varphi_0$, where $q \ge m$. When $m \to \infty$, we have $R_m\varphi_0 \to \varphi_0$ in $L^{p'}(\Omega)$ if $p' < \infty$ (Theorem 10.13 (c)), and thus if $p > 1$. Then $(\langle f, R_m\varphi_0\rangle)_{m\in\mathbb{N}}$ is a Cauchy sequence in E, which is sequentially complete, and hence it converges.

Thus, for every $f \in L^p_*(\Omega; E)$ with $p > 1$, the mapping $\varphi \mapsto \langle f, \varphi\rangle$ admits a continuous linear extension to $\varphi \in \mathcal{K}(\Omega)$; it satisfies (10.41) for such a φ. Hence, for $\varphi \in \mathcal{K}(\Omega)$,

$$\|\langle f_n - f, \varphi\rangle\|_{E;\nu} \le \|f_n - f\|_{L^p_*(\Omega;E);\nu} \|\varphi\|_{L^{p'}(\Omega)}.$$

The convergence (10.40) therefore implies

$$f_n \to f \ \text{ in } \mathcal{M}(\Omega; E).$$

Whence, f is an integrable measure and $L^p_*(\Omega; E) \underset{\leftrightarrow}{\equiv} L^p(\Omega; E)$ when $p > 1$.

2.c. This is true when $p = 1$ because, if $\varphi_0 \in \mathcal{K}(\Omega)$, we have $R_m\varphi_0 \to \varphi_0$ in $L^\infty(\Omega)$, since φ_0 is continuous. □

Let us describe some various other methods for constructing $L^p(\Omega; E)$. Up to identification, all of them yield the same space characterized by sequential completeness and the sequential density of $\mathcal{K}(\Omega; E)$.

Construction of $L^p(\Omega; E)$ by completion. One can define $L^p(\Omega; E)$, for example when E is a Banach space, as a completion of $\mathcal{B}(\Omega; E)$ or $\mathcal{K}(\Omega; E)$ endowed with the norm $(\int_\Omega \|f\|_E^p)^{1/p}$. This leads to a non-separated space, whose separation is $L^p(\Omega; E)$. Since there exist an infinite number of completions, it is necessary to resort to identifications to compare the space $L^p(\Omega; E)$ with $L^q(\Omega; E)$, $L^q_{\text{loc}}(\Omega; E)$, $\mathcal{C}(\Omega; E)$, or others. This is avoided by "completing" within a specific space.

That is why [Bourbaki, 8, déf. 2, p. 129] "completes" $\mathcal{K}(\Omega; E)$ in the space $\mathcal{F}^p(\Omega; E)$ of functions f from Ω into E, for which $\|f\| = (\int_\Omega^* \|f\|_E^p)^{1/p}$ is finite and endowed with this semi-norm ($\int^*$ is the upper integral of a positive function). □

Advantage of "completing" in $\mathcal{M}(\Omega; E)$. "Completing" in the space of measures (our Definition 10.2) is simpler (because the completion is separated) and better suited to PDEs (because the derivatives are defined in $\mathcal{M}(\Omega; E)$, which can be identified with a subspace of $\mathcal{D}'(\Omega; E)$ [Vol. 3, § 3.5, Identification (3.15), p. 61], whereas they are not defined in $\mathcal{F}^p(\Omega; E)$).

The difficulty with the completion in $\mathcal{M}(\Omega; E)$ lies in the verification that the limit of $\int_\Omega \|f_n\|_E^p$ in (10.5) does not depend on the choice of approximations f_n, which is the purpose of Lemma 10.4. This is analogous to the following property [Bourbaki, 8, chap. IV, § 2, th. 1 p. 118], which underlies the completion in $\mathcal{F}^p(\Omega; E)$: *a positive real-valued function has a zero upper integral if and only if it is zero almost everywhere.* □

Difficulty when E is not normed. The construction of $L^p(\Omega; E)$ when E is a non-normed space relies upon using the semi-norms of E, which one can raise by a power p, unlike an open set! A metric d is also unsuitable, since $d(tu, 0)$ is not proportional to t.

This is why, if one only uses topologies, metrics, or norms, one constructs $L^p(\Omega; E)$ only for values in a Banach space, *e.g.* [Bourbaki, 8, chap. IV, § 3–6, p. 123–215] or [Edwards, 22, § 8.18–8.20, p. 586–607]. □

Difficulty when E is not metrizable. The main difficulty here lies in proving sequential completeness (Theorems 4.12 and 10.16), as the classical proof, which consists of extracting an absolutely convergent subsequence from a Cauchy sequence (see the comment *Alternative proof...* p. 202), is not suitable for a non-metrizable space E. Indeed, extracting an absolutely convergent subsequence for each semi-norm no longer yields a common subsequence, since the family of semi-norms of E is no longer countable.

This difficulty can be circumvented by using the regularization operators R_m, which provide an approximation process in $L^p(\Omega; E)$ that is independent of the semi-norm. □

Completion of $\mathcal{B}(\Omega; E)$ or $\mathcal{K}(\Omega; E)$. We would obtain the same $L^p(\Omega; E)$ space by "completing" $\mathcal{K}(\Omega; E)$ as we would by completing $\mathcal{B}(\Omega; E)$, as was done in Section 10.1, since $\mathcal{K}(\Omega; E)$ is also sequentially dense in $L^p(\Omega; E)$ (Theorem 10.14).

This would simplify certain proofs but complicate others, including:
— the integration of constant functions, and thus the relation between the integral and the measure of an open set (Proposition 1.25);
— the restriction of an integrable function f, because the restrictions $(f_n|_\omega)_{n\in\mathbb{N}}$ in an integrabilizing sequence would not form an integrabilizing sequence for $f|_\omega$. □

Construction via measurability. The $L^p(\Omega; E)$ space is usually defined as the set of *classes of measurable functions f equal almost everywhere such that $\|f\|_E^p$ is integrable*. This approach will be followed in Chapter 17. It has the advantage of being valid for any measurable set Ω, but has the drawback of being limited to Fréchet spaces E. □

Advantage of construction via completion. Construction via completion avoids the study of measurability, which is not always straightforward and is of no use to us: we rely only on integrability, which is much stronger than measurability; it involves only the Cauchy integral of a uniformly continuous function with bounded support. Once the measures are known, and they are indispensable to use, they provide a framework that is very well-suited to studying integrability (Definition 5.1). □

Scalar-valued integrable functions. One can integrate a scalar-valued integrable function taking values in an arbitrary separated semi-normed space E, but whose integral belongs to E'^*, which is larger than E [BOURBAKI, 8, chap. VI, § 1, définition 1, p. 8]. To obtain $\int_\Omega f \in E$, BOURBAKI assumes [*loc. cit.*, proposition 8, p. 14] that f is locally bounded and E is semi-reflexive, which is more restrictive than E being a Neumann space [Vol. 1, Theorem 17.7].

Moreover, scalar integrability is not suitable to construct $L^p(\Omega; E)$. This is why [BOURBAKI, 8] restricts its construction to the case where E is a Banach space. □

Case where Ω has finite measure. If $|\Omega| < \infty$, the space $L^\infty(\Omega; E)$ remains unchanged when ω is replaced by Ω in Property (10.4) of Definition 10.1. □

Case where Ω is no longer open. One could still construct the space $L^p(\Omega; E)$ if Ω were only measurable. It would no longer be a space of measures, but rather a quotient space of measures.

More precisely, a *quasi-measure* is a continuous linear mapping from $\mathcal{B}(\Omega)$ into E. A quasi-measure f is *negligible* if, for every $\nu \in \mathcal{N}_E$, its extensions $\widetilde{f}$ satisfy $\inf \|\widetilde{f}\|_{L^p(\mathbb{R}^d; E); \nu} = 0$. Then, $L^p(\Omega; E)$ would be the quotient space of $L^p(\mathbb{R}^d; E)$ by the subspace of negligible quasi-measures. It would hence be a quotient space of quasi-measures.

This is of no use for PDEs, which are defined on open sets, and would take us too far afield. □

Chapter 11

Dependence on p and Ω, Local Spaces

We show that $L^p(\Omega; E)$ increases as p decreases provided that the measure of Ω is finite (Theorem 11.1), and also state the Riesz convexity theorem (Theorem 11.2).

Then, we define local spaces $L^p_{\text{loc}}(\Omega; E)$, and present some of their properties (§ 11.2).

Finally, we define localization–extension (§ 11.3), provide some gluing properties for a finite (§ 11.4) or infinite (§ 11.5) number of domains, and lastly study the continuity in p (Theorem 11.20).

11.1. Dependence on p

Let us begin with the dependence on p for a domain of finite measure.

Theorem 11.1. — *Let Ω be an open subset of $\mathbb{R}^d$, E be a Neumann space, and p and q such that*

$$|\Omega| < \infty, \quad 1 \leq q \leq p \leq \infty.$$

Then:

(a) $$L^\infty(\Omega; E) \subsetneq L^p(\Omega; E) \subsetneq L^q(\Omega; E) \subsetneq L^1(\Omega; E);$$

(b) *for every $f \in L^p(\Omega; E)$ and $\nu \in \mathcal{N}_E$ (the index set for the semi-norms of E),*

$$\|f\|_{L^q(\Omega;E);\nu} \leq |\Omega|^{1/q-1/p} \|f\|_{L^p(\Omega;E);\nu}. \;\blacksquare$$

Proof. **1°)** *Inequality* (b) *when* $p < \infty$. Let $(f_n)_{n\in\mathbb{N}}$ be a p-integrabilizing sequence for f (Definition 10.1) and ω_n be a bounded set included in Ω, outside which f_n

vanishes. Hölder's inequality (Proposition 9.6 with $1/q = 1/p + (1/q - 1/p)$) yields (with Propositions 1.23, 1.25, and 2.2, whilst omitting the index $_{E;\nu}$)

$$\Big(\int_\Omega \|f_n\|^q\Big)^{1/q} = \Big(\int_{\omega_n} (1\|f_n\|)^q\Big)^{1/q} \leq \Big(\int_{\omega_n} 1\Big)^{1/q-1/p}\Big(\int_{\omega_n} \|f_n\|^p\Big)^{1/p} \leq$$
$$\leq |\Omega|^{1/q-1/p}\Big(\int_\Omega \|f_n\|^p\Big)^{1/p}. \quad (11.1)$$

Applying this inequality to $f_m - f_n$ gives

$$\sup_{m\geq n}\Big(\int_\Omega \|f_m - f_n\|^q\Big)^{1/q} \leq |\Omega|^{1/q-1/p} \sup_{m\geq n}\Big(\int_\Omega \|f_m - f_n\|^p\Big)^{1/p}.$$

By Definition 10.1 of a p-integrable measure, this tends to 0 as $n \to \infty$, whence

$$f \in L^q(\Omega; E).$$

By Definition 10.2 of the semi-norms of $L^p(\Omega; E)$, after taking the limit, we obtain inequality (b) from inequality (11.1).

2º) Inequality (b) *when* $p = \infty$. Now, $(f_n)_{n\in\mathbb{N}}$ is an ∞-integrabilizing sequence for f (Definition 10.1), $\nu \in \mathcal{N}_E$, $\epsilon > 0$, and $b = 2\sup_{n\in\mathbb{N},\, x\in\Omega} \|f_n(x)\|_{E;\nu}$. Since we have $|\Omega| < \infty$, there exists a compact set $K \subset \Omega$ such that (Theorem 2.12)

$$|\Omega \setminus K| \leq \frac{\epsilon}{b^q}.$$

Let ω be an open set such that $K \subset \omega \Subset \Omega$ (Proposition A.20, since $K \Subset \Omega$ as K is bounded and $\overline{K} = K \subset \Omega = \mathring{\Omega}$). Then, $\omega \cup (\Omega \setminus K) = \Omega$, and therefore (Theorems 6.3 and Proposition 1.18 (a), whilst omitting the index $_{E;\nu}$)

$$\int_\Omega \|f_n\|^q \leq \int_{\Omega\setminus K} \|f_n\|^q + \int_\omega \|f_n\|^q \leq b^q|\Omega \setminus K| + b^{q-1}\int_\omega \|f_n\| \leq$$
$$\leq \epsilon + b^{q-1}\int_\omega \|f_n\|. \quad (11.2)$$

With respect to $f_m - f_n$ (we have $\|f_m - f_n\| \leq b$), this inequality gives

$$\sup_{m\geq n}\int_\Omega \|f_m - f_n\|^q \leq \epsilon + b^{q-1}\sup_{m\geq n}\int_\omega \|f_m - f_n\|. \quad (11.3)$$

By Definition 10.1 of an ∞-integrable measure, this is bounded by 2ϵ once n is sufficiently large, so (Definition 10.1 of a q-integrable measure)

$$f \in L^q(\Omega; E).$$

According to inequality (11.3), the f_n form a q-integrabilizing sequence for f. Furthermore, the open set $\omega_n = \{x \in \Omega : f_n(x) \neq 0\}$ is bounded and (Propositions 1.23 and 1.21),

$$\Big(\int_\Omega \|f_n\|^q\Big)^{1/q} = \Big(\int_{\omega_n} \|f_n\|^q\Big)^{1/q} \leq |\omega_n|^{1/q} \sup_{n\in\mathbb{N},\, x\in\Omega} \|f_n(x)\|_{E;\nu}.$$

In the limit, by Definition 10.2 of the semi-norms of $L^q(\Omega; E)$, and since the measure of the open sets is increasing (Proposition 2.2),

$$\|f\|_{L^q(\Omega;E);\nu} \leq |\Omega|^{1/q} \sup_{n\in\mathbb{N},\, x\in\Omega} \|f_n(x)\|_{E;\nu}.$$

By taking the infimum over all sequences of f_n, this gives, by Definition 10.2 of the semi-norms of $L^\infty(\Omega; E)$,

$$\|f\|_{L^q(\Omega;E);\nu} \leq |\Omega|^{1/q} \|f\|_{L^\infty(\Omega;E);\nu}.$$

3o) *Inclusion* $L^p(\Omega; E) \subsetneq L^q(\Omega; E)$. This follows from inequality (b) and by Definition 1.4 of a topological inclusion.

4o) *Other inclusions.* These are just particular cases of the previous inclusion. □

Let us now state the (Marcel) **Riesz convexity theorem**[1] on the dependence in p for an arbitrary domain. (Continuity in p is established in Theorems 11.20 and 18.9.)

Theorem 11.2. — *Let Ω be an open subset of $\mathbb{R}^d$, E be a Neumann space, and*

$$1 \leq p_0 \leq p \leq p_1 \leq \infty.$$

Then,

$$L^{p_0}(\Omega; E) \cap L^{p_1}(\Omega; E) \subsetneq L^p(\Omega; E)$$

and, for every $f \in L^{p_0}(\Omega; E) \cap L^{p_1}(\Omega; E)$ and $\nu \in \mathcal{N}_E$ (the index set for the semi-norms of E),

$$\|f\|_{L^p(\Omega;E);\nu} \leq \|f\|^{1-\theta}_{L^{p_0}(\Omega;E);\nu} \|f\|^{\theta}_{L^{p_1}(\Omega;E);\nu}$$

where

$$\theta = \frac{1/p_0 - 1/p}{1/p_0 - 1/p_1}, \quad \textit{that is,} \quad \frac{1}{p} = \frac{1-\theta}{p_0} + \frac{\theta}{p_1}$$

(thus $0 \leq \theta \leq 1$, and θ increases from 0 to 1 as p increases from p_0 to p_1). ▮

---

1 **Historical background of the Riesz convexity Theorem 11.2.** This result is attributed to Marcel RIESZ.

Proof. ***1º) $f \in L^p(\Omega; E)$ when $p_1 < \infty$.*** For now, suppose $f \in \mathcal{B}(\Omega; E)$. Hölder's generalized inequality (Proposition 9.6 with $1/p = (1-\theta)/p_0 + \theta/p_1$) then yields

$$\Big(\int_\Omega \|f\|_{E;\nu}^p\Big)^{1/p} = \Big(\int_\Omega (\|f\|_{E;\nu}^{1-\theta}\|f\|_{E;\nu}^{\theta})^p\Big)^{1/p} \leq$$

$$\leq \Big(\int_\Omega \|f\|_{E;\nu}^{p_0}\Big)^{(1-\theta)/p_0}\Big(\int_\Omega \|f\|_{E;\nu}^{p_1}\Big)^{\theta/p_1}. \qquad (11.4)$$

Let us return to $f \in L^{p_0}(\Omega; E) \cap L^{p_1}(\Omega; E)$. Its regularizations $(R_n f)_{n\in\mathbb{N}^*}$ given by Definition 3.17 satisfy

$$\Big(\int_\Omega \|R_m f - R_n f\|_{E;\nu}^p\Big)^{1/p} \leq$$

$$\leq \Big(\int_\Omega \|R_m f - R_n f\|_{E;\nu}^{p_0}\Big)^{(1-\theta)/p_0}\Big(\int_\Omega \|R_m f - R_n f\|_{E;\nu}^{p_1}\Big)^{\theta/p_1}.$$

This tends to 0 when $m \geq n$ and $n \to \infty$ (Theorem 10.13 (c)). Moreover (Theorem 3.19), $R_n f \to f$ in $\mathcal{M}(\Omega; E)$, hence (Definition 10.2)

$$f \in L^p(\Omega; E).$$

2º) $f \in L^p(\Omega; E)$ when $p_1 = \infty$. Suppose $p_0 < \infty$, since otherwise $p = \infty$ and the property is immediate. For $f \in \mathcal{B}(\Omega; E)$, by the monotonicity and linearity of the real-valued integral (Propositions 1.18 (a) and 1.16), we have the bound

$$\int_\Omega \|f\|_{E;\nu}^p = \int_\Omega \|f\|_{E;\nu}^{p_0}\|f\|_{E;\nu}^{p-p_0} \leq \Big(\int_\Omega \|f\|_{E;\nu}^{p_0}\Big) \sup_{x\in\Omega} \|f(x)\|_{E;\nu}^{p-p_0}.$$

Here, since $\theta = 1 - p_0/p$ and $\sup_{x\in\Omega}\|f(x)\|_{E;\nu} = \|f\|_{L^\infty(\Omega;E);\nu}$ (Theorem 10.10), we have

$$\Big(\int_\Omega \|f\|_{E;\nu}^p\Big)^{1/p} \leq \Big(\int_\Omega \|f\|_{E;\nu}^{p_0}\Big)^{(1-\theta)/p_0} \|f\|_{L^\infty(\Omega;E);\nu}^{\theta}. \qquad (11.5)$$

The regularizations for $f \in L^{p_0}(\Omega; E) \cap L^\infty(\Omega; E)$ therefore satisfy

$$\Big(\int_\Omega \|R_m f - R_n f\|_{E;\nu}^p\Big)^{1/p} \leq \Big(\int_\Omega \|R_m f - R_n f\|_{E;\nu}^{p_0}\Big)^{(1-\theta)/p_0} \big(2\|f\|_{L^\infty(\Omega;E);\nu}\big)^{\theta},$$

because (Theorem 10.13 (a)) $\|R_n f\|_{L^\infty(\Omega;E);\nu} \leq \|f\|_{L^\infty(\Omega;E);\nu}$.

The right-hand side tends to 0, from which it follows, as in the case $p_1 < \infty$, that

$$f \in L^p(\Omega; E).$$

3º) Inequality of the norms. Inequality (11.4) or (11.5) with respect to $R_n f$ gives, in the limit, by Definition 10.2 of the norm of $L^p(\Omega; E)$, the inequality of the norms.

4º) Inclusion. This follows from the inequality of the norms, by Definition 1.4 of a topological inclusion. □

11.2. $L^p_{\text{loc}}(\Omega; E)$ spaces

Let us define local Lebesgue spaces $L^p_{\text{loc}}(\Omega; E)$.

Definition 11.3. — *Let Ω be an open subset of $\mathbb{R}^d$, E be a Neumann space with semi-norms $\{\| \ \|_{E;\nu} : \nu \in \mathcal{N}_E\}$, and $1 \leq p \leq \infty$. We denote*

$$L^p_{\text{loc}}(\Omega; E) \stackrel{\text{def}}{=} \{f \in \mathcal{M}(\Omega; E) : f|_\omega \in L^p(\omega; E),\ \forall \text{ open } \omega \Subset \Omega\}$$

endowed with the semi-norms, indexed by $\nu \in \mathcal{N}_E$ and open sets $\omega \Subset \Omega$,

$$\|f\|_{L^p_{\text{loc}}(\Omega;E);\omega,\nu} \stackrel{\text{def}}{=} \|f|_\omega\|_{L^p(\omega;E);\nu}. \ \blacksquare$$

Recall that the **compact inclusion** $\omega \Subset \Omega$ of open subsets of $\mathbb{R}^d$ indicates:

$$\omega \text{ is bounded and } \overline{\omega} \subset \mathring{\Omega}.$$

Now, let us characterize $L^p_{\text{loc}}(\Omega; E)$, with the help of a sequence of open sets $\omega_n \Subset \Omega$.

Theorem 11.4. — *Let Ω be an open subset of $\mathbb{R}^d$, E be a Neumann space with semi-norms $\{\| \ \|_{E;\nu} : \nu \in \mathcal{N}_E\}$, and $1 \leq p \leq \infty$. Let us denote*

$$\omega_n \stackrel{\text{def}}{=} \{x \in \Omega : |x| < n,\ B(x, 1/n) \subset \Omega\},$$

where $B(x, r)$ is the closed ball centered at x with radius r (so $\omega_n = \Omega^n_{1/n}$). Then,

$$L^p_{\text{loc}}(\Omega; E) = \{f \in \mathcal{M}(\Omega; E) : f|_{\omega_n} \in L^p(\omega_n; E),\ \forall n \in \mathbb{N}^*\},$$

and its topology is defined by the semi-norms, indexed by $n \in \mathbb{N}^*$ and $\nu \in \mathcal{N}_E$,

$$\|f\|_{L^p_{\text{loc}}(\Omega;E);n,\nu} \stackrel{\text{def}}{=} \|f|_{\omega_n}\|_{L^p(\omega_n;E);\nu}. \ \blacksquare$$

Proof. Let X be the topological space obtained by replacing every open set $\omega \Subset \Omega$ with ω_n. The space $L^p_{\text{loc}}(\Omega; E)$ is, of course, topologically included in X.

Conversely, let $\omega \Subset \Omega$. Then $\overline{\omega}$ is a compact set included in Ω, and hence in one of the ω_n, which form an increasing open cover of Ω. Thus, $X \subsetneqq L^p_{\text{loc}}(\Omega; E)$. $\square$

Let us show that restriction within the spaces $L^p(\Omega; E)$ is continuous.

Theorem 11.5. — *Let*

$$f \in L^p(\Omega; E),$$

and let $f|_\omega$ be its restriction (Definition 6.16), where Ω and ω are open subsets of $\mathbb{R}^d$ such that $\omega \subset \Omega$, E is a Neumann space, and $1 \leq p \leq \infty$. Then:

(a) $$f|_\omega \in L^p(\omega; E);$$

(b) *for every $\nu \in \mathcal{N}_E$ (the index set of the semi-norms of E),*

$$\|f|_\omega\|_{L^p(\omega;E);\nu} \leq \|f\|_{L^p(\Omega;E);\nu};$$

(c) *the mapping $f \mapsto f|_\omega$ is continuous linear, and hence sequentially continuous, from $L^p(\Omega; E)$ into $L^p(\omega; E)$.* ▮

Proof. **1°)** Property (a). Suppose that $(f_n)_{n\in\mathbb{N}}$ is a p-integrabilizing sequence for f (Definition 10.1). Then the $f_n|_\omega$ form a p-integrabilizing sequence for $f|_\omega$ in $L^p(\omega; E)$ (the convergence (10.1) in $\mathcal{M}(\omega; E)$ is given by Theorem 6.17). Whence,

$$f|_\omega \in L^p(\omega; E).$$

2°) *Inequality (b).* The bound on the semi-norms follows from Definition 10.2.

3°) *Property (c).* Continuity follows from the bound in (b), according to the characterization of continuous linear mappings from Proposition 1.9.

Any continuous mapping is also sequentially continuous (Proposition 1.8). □

Let us now show that $L^p_{\text{loc}}(\Omega; E)$ is a Neumann space and provide some bounds for it.

Theorem 11.6. — *The space $L^p_{\text{loc}}(\Omega; E)$, where Ω is an open subset of $\mathbb{R}^d$, E is a Neumann space, and $1 \leq p \leq \infty$, is a Neumann space, and*

$$L^p(\Omega; E) \subsetneq L^p_{\text{loc}}(\Omega; E) \subsetneq \mathcal{M}(\Omega; E). \; ▮$$

Proof of Theorem 11.6. **1°)** Semi-norms. The space $L^p_{\text{loc}}(\Omega; E)$ is semi-normed by Definition 11.3.

2°) Separation. Suppose that $\|f\|_{L^p_{\text{loc}}(\Omega;E);\omega,\nu} = 0$ for every $\nu \in \mathcal{N}_E$ and $\omega \Subset \Omega$.

Then, $f|_\omega$ vanishes for every $\omega \Subset \Omega$. Let $\varphi \in \mathcal{K}(\Omega)$ whose support is denoted K, and let ω be an open set such that $K \subset \omega \Subset \Omega$ (Proposition A.20). By Definition 6.16 of the restriction,

$$\langle f, \varphi \rangle_\Omega = \langle f|_\omega, \varphi|_\omega \rangle_\omega = 0_E.$$

Therefore, f vanishes. Hence, $L^p_{\text{loc}}(\Omega; E)$ is separated (Definition 1.1).

3º) *First topological inclusion.* Let $f \in L^p(\Omega; E)$ and $\omega \Subset \Omega$. Its restriction satisfies (Theorem 11.5) $f|_\omega \in L^p(\omega; E)$, thus $f \in L^p_{\text{loc}}(\Omega; E)$, and

$$\|f\|_{L^p_{\text{loc}}(\Omega;E);\omega,\nu} = \|f|_\omega\|_{L^p(\omega;E);\nu} \le \|f\|_{L^p(\Omega;E);\nu}.$$

Which provides the topological inclusion (Definition 1.4) $L^p(\Omega; E) \subsetneqq L^p_{\text{loc}}(\Omega; E)$.

4º) *Second topological inclusion.* Let $f \in L^p_{\text{loc}}(\Omega; E)$, $\varphi \in \mathcal{K}(\Omega)$ whose support is denoted K, and let ω be an open set such that $K \subset \omega \Subset \Omega$. Definitions 3.2 and 6.16 of the semi-norms of $\mathcal{M}(\Omega; E)$ and the restriction, together with the inequality from Lemma 10.6 with $c_\varphi = \|\varphi\|_{L^{p'}(\Omega}$, yield

$$\|f\|_{\mathcal{M}(\Omega;E);\varphi,\nu} = \|\langle f, \varphi \rangle_\Omega\|_{E;\nu} = \|\langle f|_\omega, \varphi|_\omega \rangle_\omega\|_{E;\nu} \le c_\varphi \|f|_\omega\|_{L^p(\omega;E);\nu}.$$

Using Definition 11.3 of $L^p_{\text{loc}}(\Omega; E)$, this can be written as

$$\|f\|_{\mathcal{M}(\Omega;E);\varphi,\nu} \le c_\varphi \|f\|_{L^p_{\text{loc}}(\Omega;E);\omega,\nu}.$$

Which provides the topological inclusion $L^p_{\text{loc}}(\Omega; E) \subsetneqq \mathcal{M}(\Omega; E)$.

5º) *Sequential completion.* Let $(f_n)_{n\in\mathbb{N}}$ be a Cauchy sequence in $L^p_{\text{loc}}(\Omega; E)$.

Since $L^p_{\text{loc}}(\Omega; E) \subsetneqq \mathcal{M}(\Omega; E)$ by 4º, $(f_n)_{n\in\mathbb{N}}$ is a Cauchy sequence in $\mathcal{M}(\Omega; E)$ (Proposition A.22), which is sequentially complete (Theorem 3.8). It therefore has a limit f, namely

$$f_n \to f \text{ in } \mathcal{M}(\Omega; E).$$

Let $\omega \Subset \Omega$ be an open subset. According to Definition 11.3 of the semi-norms of $L^p_{\text{loc}}(\Omega; E)$, $(f_n|_\omega)_{n\in\mathbb{N}}$ is a Cauchy sequence in $L^p(\omega; E)$, which is sequentially complete (Theorem 10.16). It therefore has a limit, which must be $f|_\omega$, and so,

$$f_n \to f \text{ in } L^p_{\text{loc}}(\Omega; E).$$

Hence, $L^p_{\text{loc}}(\Omega; E)$ is sequentially complete, *i.e.* it is a Neumann space. □

Let us state some properties of metrizability.

Theorem 11.7. — *If E is a Fréchet space, $1 \leq p \leq \infty$, and Ω is an open subset of $\mathbb{R}^d$, then $L^p_{\text{loc}}(\Omega; E)$ is a Fréchet space.*

In particular, $L^p_{\text{loc}}(\Omega)$ is a Fréchet space. ▮

Proof. *1°) $L^p_{\text{loc}}(\Omega; E)$ space.* If E is a Fréchet space, its family of semi-norms is equivalent to a family indexed by a countable set $\mathcal{N}_E$. The family of semi-norms of $L^p_{\text{loc}}(\Omega; E)$ is therefore equivalent to a family indexed by $(\omega_n)_{n\in\mathbb{N}^*} \times \mathcal{N}_E$ (Theorem 11.4), which is countable (Proposition A.2 (c)). Hence, the space $L^p_{\text{loc}}(\Omega; E)$ is metrizable; since it is sequentially complete (Theorem 11.6), it is a Fréchet space.

2°) $L^p_{\text{loc}}(\Omega)$ space. Since the space $\mathbb{R}$ is a Banach space (Proposition A.35), it is also a Fréchet space (Proposition A.32). Hence, $L^p_{\text{loc}}(\Omega) = L^p_{\text{loc}}(\Omega; \mathbb{R})$ is a Fréchet space. □

Let us show that $L^p_{\text{loc}}(\Omega; E)$ increases when p decreases.

Theorem 11.8. — *Let $1 \leq q \leq p \leq \infty$, E be a Neumann space, and Ω be an open subset of $\mathbb{R}^d$. Then,*

$$L^\infty_{\text{loc}}(\Omega; E) \subsetneqq L^p_{\text{loc}}(\Omega; E) \subsetneqq L^q_{\text{loc}}(\Omega; E) \subsetneqq L^1_{\text{loc}}(\Omega; E).$$ ▮

Proof. This follows from the fact that $L^p(\omega; E)$ increases as p decreases and that the measure $|\omega|$ is finite (Theorem 11.1). □

Let us compare $\mathcal{C}(\Omega; E)$ with $L^p_{\text{loc}}(\Omega; E)$.

Theorem 11.9. — *Let Ω be an open subset of $\mathbb{R}^d$, E be a Neumann space, and $1 \leq p \leq \infty$. Then,*

$$\mathcal{C}(\Omega; E) \subsetneqq L^p_{\text{loc}}(\Omega; E).$$

For every $f \in \mathcal{C}(\Omega; E)$, every open subset $\omega \Subset \Omega$, and every semi-norm $\| \ \|_{E;\nu}$ of E,

$$\|f\|_{L^p_{\text{loc}}(\Omega;E);\omega,\nu} = \begin{cases} \left(\int_\omega \|f\|^p_{E;\nu} \right)^{1/p} & \text{if } p < \infty, \\ \sup_{x\in\omega} \|f(x)\|_{E;\nu} & \text{if } p = \infty. \end{cases}$$ ▮

Proof. These equalities follow from Definition 11.3 of $L^p_{\text{loc}}(\Omega; E)$ and the integral expression for the $L^p(\omega; E)$ semi-norms of a function in $\mathcal{B}(\omega; E)$ (Proposition 10.7) for $f|_\omega$.

Proposition 10.7 applies here since $f|_\omega \in \mathcal{B}(\omega; E)$ because $\overline{\omega}$ is compact (Proposition A.36 (a)), and hence f is uniformly continuous (Proposition A.41) and supp $f|_\omega$ is bounded (as it is contained in $\overline{\omega}$). □

Let us show that $R_n f$ are approximations for f in $L^p_{\text{loc}}(\Omega; E)$ when p is finite.

Theorem 11.10. — *Let $f \in L^p_{\text{loc}}(\Omega; E)$, where Ω is an open subset of $\mathbb{R}^d$, E is a Neumann space, and $(R_n f)_{n\in\mathbb{N}^*}$ are regularizations (Definition 3.17).*

Then, when $1 \le p < \infty$ and $n \to \infty$,

$$R_n f \to f \ \text{ in } L^p_{\text{loc}}(\Omega; E). \ \blacksquare$$

Proof. Let $\omega \Subset \Omega$ be an open subset. Choose an open subset σ such that $\omega \Subset \sigma \Subset \Omega$ and an m such that

$$\omega \subset \sigma^m_{4/m},$$

where $\sigma^m_r = \{x \in \sigma : |x| < m,\ B(x, r) \subset \sigma\}$. For example, this is realized if we choose $\sigma = \Omega^m_{4/m}$ and m is sufficiently large enough so that $\omega \subset \Omega^m_{8/m}$.

Thus, $f \in L^p(\sigma; E)$ and (Theorem 10.13 (c))

$$R^\sigma_n f \to f \ \text{ in } L^p(\sigma; E). \tag{11.6}$$

Furthermore, as soon as $n \ge m$,

$$R^\sigma_n f = R^\Omega_n f \ \text{ in } \omega.$$

Indeed, $\omega \subset \sigma^m_{4/m} \subset \sigma^n_{4/n} \subset \Omega^n_{4/n}$, hence we have $\alpha^\sigma_n = \alpha^\Omega_n = 1$ on ω by Definition 3.16 of α_n. Since the restriction is continuous (Theorem 11.5 (c)), (11.6) implies that

$$R^\Omega_n f|_\omega \to f|_\omega \ \text{ in } L^p(\omega; E).$$

Therefore, by Definition 11.3 of the semi-norms of $L^p_{\text{loc}}(\Omega; E)$,

$$R^\Omega_n f \to f \ \text{ in } L^p_{\text{loc}}(\Omega; E). \ \square$$

Theorem 11.10 implies the following density property.

Theorem 11.11. — *The space $\mathcal{K}(\Omega; E)$ is sequentially dense, and hence dense, in $L^p_{\text{loc}}(\Omega; E)$, where Ω is an open subset of $\mathbb{R}^d$, E is a Neumann space, and $1 \leq p < \infty$.* ▮

Proof. Sequential density follows from Theorem 11.10, since we have (Theorem 3.18) $R_n f \in \mathcal{K}(\Omega; E)$. Density then always follows from this (Proposition A.16). □

Let us characterize the elements of $L^\infty(\Omega; E)$ as well as the ∞-integrabilizing sequences (this is the analogue of Theorem 10.9 when $p = \infty$).

Theorem 11.12. —*Let Ω be an open subset of $\mathbb{R}^d$, $1 \leq p < \infty$, and E be a Neumann space.*

(a) *Then,*

$$f \in L^\infty(\Omega; E)$$

if and only if there exists a sequence $(f_n)_{n\in\mathbb{N}}$ in $\mathcal{B}(\Omega; E)$ such that

$$(f_n)_{n\in\mathbb{N}} \text{ is bounded in } \mathcal{C}_b(\Omega; E), \quad f_n \to f \text{ in } L^p_{\text{loc}}(\Omega; E). \tag{11.7}$$

(b) *The two conditions (11.7) are equivalent, for a sequence $(f_n)_{n\in\mathbb{N}}$ in $\mathcal{B}(\Omega; E)$, to*

$$(f_n)_{n\in\mathbb{N}} \text{ is an } \infty\text{-integrabilizing sequence for } f. \text{ ▮}$$

Proof. ***1º)*** Forward implication of (a). If

$$f \in L^\infty(\Omega; E):$$

— by Theorem 3.18, $(R_n f)_{n\in\mathbb{N}^*}$ belongs to $\mathcal{K}(\Omega; E)$, and hence to $\mathcal{B}(\Omega; E)$;

— by Theorem 10.13 (b), $(R_n f)_{n\in\mathbb{N}}$ is bounded in $L^\infty(\Omega; E)$ and thus, since each $R_n f$ is continuous, we have (Theorem 10.10 (b))

$$(R_n f)_{n\in\mathbb{N}} \text{ is bounded in } \mathcal{C}_b(\Omega; E);$$

— we have $f \in L^\infty_{\text{loc}}(\Omega; E)$ (Theorem 11.6), hence $f \in L^p_{\text{loc}}(\Omega; E)$ (Theorem 11.8), from which it follows (Theorem 11.10) that

$$R_n f \to f \text{ in } L^p_{\text{loc}}(\Omega; E).$$

2º) Converse. If (11.7) is satisfied, then we have the three properties required to define $L^\infty(\Omega; E)$:

— (10.1), as $L^p_{\text{loc}}(\Omega; E) \subsetneq \mathcal{M}(\Omega; E)$ (Theorem 11.6), whence $f_n \to f$ in $\mathcal{M}(\Omega; E)$;

— (10.3), which is equivalent (Theorem 10.10) to $(f_n)_{n\in\mathbb{N}}$ being bounded in $\mathcal{C}_b(\Omega; E)$, since f_n is continuous;

— (10.4), since $\sup_{m\geq n} \int_\omega \|f_m - f_n\|_{E;\nu} \to 0$ for every open set $\omega \Subset \Omega$ by Theorem 11.9, and $L^p_{\text{loc}}(\Omega; E) \subsetneq L^1_{\text{loc}}(\Omega; E)$ (Theorem 11.8).

Thus, $f \in L^\infty(\Omega; E)$; furthermore, $(f_n)_{n\in\mathbb{N}}$ is an ∞-integrabilizinge sequence.

3º) Converse of (b). If $(f_n)_{n\in\mathbb{N}}$ is an ∞-integrabilizing sequence and $\omega \Subset \Omega$, then:

— by (10.3), $(f_n)_{n\in\mathbb{N}}$ is bounded in $\mathcal{C}_b(\Omega; E)$;

— from Definition 10.1, $(f_n|_\omega)_{n\in\mathbb{N}}$ is a 1-integrabilizing sequence for $f|_\omega$, hence $f_n \to f$ in $L^1(\omega; E)$ and, by Riesz inequality (Theorem 11.2 with $p_0 = 1, p_1 = \infty$),

$$\|f_n - f\|_{L^p(\omega;E);\nu} \leq \|f_n - f\|^{1/p}_{L^1(\omega;E);\nu} \|f_n - f\|^{1-1/p}_{L^\infty(\omega;E);\nu}.$$

According to (10.3) and (10.4), $f_n \to f$ in $L^p(\omega; E)$ and, *a fortiori* (Theorem 11.6), in $L^p_{\text{loc}}(\Omega; E)$. □

11.3. Localization–extension

Let us define the *localized–extension* of a measure.

Theorem 11.13. — *Let*

$$f \in \mathcal{M}(\omega; E) \text{ and } \alpha \in \mathcal{C}(\Omega) \text{ such that } \operatorname{supp}\alpha \subset \omega \subset \Omega,$$

where Ω and ω are two open subsets of $\mathbb{R}^d$, and E is a Neumann space.

We define the ***localized–extension*** $\widetilde{\alpha f} \in \mathcal{M}(\Omega; E)$ *by: for every $\varphi \in \mathcal{K}(\Omega)$,*

$$\langle \widetilde{\alpha f}, \varphi \rangle_\Omega \stackrel{\text{def}}{=} \langle f, (\alpha\varphi)|_\omega \rangle_\omega.$$

It coincides with αf on ω and 0_E on $\Omega \setminus \operatorname{supp}\alpha$, and the mapping $f \mapsto \widetilde{\alpha f}$ is continuous linear, and hence sequentially continuous, from $\mathcal{M}(\omega; E)$ into $\mathcal{M}(\Omega; E)$. ▌

Hypotheses on α**.** The hypotheses $\alpha \in \mathcal{C}(\Omega)$ and $\operatorname{supp}\alpha \subset \omega$ can be replaced by $\alpha \in \mathcal{C}(\overline{\omega})$ and $\alpha = 0$ on $\partial\omega$. Alternatively, one may assume that $\alpha \in \mathcal{C}(\omega)$ admits a (unique) continuous extension that vanishes on $\partial\omega$. □

Proof of Theorem 11.13. **1o)** *Membership of* $\mathcal{M}(\Omega; E)$. The right-hand side satisfies the characterization of a measure on Ω from Theorem 3.3 since, for every semi-norm $\| \ \|_{E;\nu}$ of E, there exist $q \in \mathcal{C}^+(\omega)$ and $c \in \mathbb{R}$ such that

$$\|\langle f, (\alpha\varphi)|_\omega\rangle_\omega\|_{E;\nu} \leq c \sup_{x \in \omega} q(x)|(\alpha\varphi)(x)| = c \sup_{x \in \Omega} (\alpha\widetilde{q})(x)|\varphi(x)|,$$

where $\alpha\widetilde{q} \in \mathcal{C}^+(\Omega)$. Indeed, $\alpha\widetilde{q}$ is continuous on the open subsets ω (where it equals αq) and $\Omega \setminus \operatorname{supp}\alpha$ (where it vanishes), and therefore also on their union Ω.

2o) *Local equalities.* By Definition 6.16 of the restriction, for every $\varphi \in \mathcal{K}(\omega)$,

$$\langle \widetilde{\alpha f}|_\omega, \varphi\rangle_\omega = \langle \widetilde{\alpha f}, \widetilde{\varphi}\rangle_\Omega = \langle \alpha f, \varphi\rangle_\omega.$$

Thus,

$$\widetilde{\alpha\varphi} = \alpha\varphi \ \text{ on } \omega.$$

Similarly, $\alpha = 0$ on $\Omega \setminus \operatorname{supp}\alpha$, hence

$$\widetilde{\alpha\varphi} = 0_E \ \text{ on } \Omega \setminus \operatorname{supp}\alpha.$$

3o) *Continuity.* For every $\varphi \in \mathcal{K}(\Omega)$, by Definition 3.2 of the semi-norms of a measure,

$$\|\widetilde{\alpha f}\|_{\mathcal{M}(\Omega;E);\varphi,\nu} = \|\langle \widetilde{\alpha f}, \varphi\rangle\|_{E;\nu} = \|\langle f, (\alpha\varphi)|_\omega\rangle\|_{E;\nu} = \|f\|_{\mathcal{M}(\omega;E);(\alpha\varphi)|_\omega,\nu}.$$

Which shows that the mapping $f \mapsto \widetilde{\alpha\varphi}$ from $\mathcal{M}(\omega; E)$ into $\mathcal{M}(\Omega; E)$ is continuous, according to the characterization of continuous linear mappings from Proposition 1.9.

4o) *Sequential continuity.* This follows from continuity (Proposition 1.8). □

Let us define the *localized–extension* of a p-integrable measure (we shall see in Theorem 13.6 that any measure in $L^p(\Omega; E)$ can be extended by 0_E without compactification).

Theorem 11.14. — *Let*

$$f \in L^p(\omega; E), \quad \alpha \in \mathcal{C}_b(\omega), \quad \omega \subset \Omega, \quad \operatorname{supp}\alpha \textit{ is closed},$$

where ω and Ω are two open subsets of $\mathbb{R}^d$, E is a Neumann space, $1 \leq p \leq \infty$, and $\widetilde{\alpha f}$ is the extension of αf by 0_E (Theorem 11.13).

Then, $\widetilde{\alpha f} \in L^p(\Omega; E)$ and for every $\nu \in \mathcal{N}_E$ (the index set of the semi-norms of E),

$$\|\widetilde{\alpha f}\|_{L^p(\Omega;E);\nu} \leq \sup_{x \in \omega} |\alpha(x)| \, \|f\|_{L^p(\omega;E);\nu}. \ \blacksquare$$

Proof. **1°)** *Case where* $p < \infty$. **1.a.** *Membership of* $L^p(\Omega; E)$. Let $(f_n)_{n\in\mathbb{N}}$ be a p-integrabilizing sequence for f (Definition 10.1). We will show that the $(\widetilde{\alpha f_n})_{n\in\mathbb{N}}$ form a p-integrabilizing sequence for $\widetilde{\alpha f}$:

— on the one hand, $\widetilde{\alpha f_n} \in \mathcal{B}(\Omega; E)$;

— on the other hand, $\widetilde{\alpha f_n} \to \widetilde{\alpha f}$ in $\mathcal{M}(\Omega; E)$, since $f_n \to f$ in $\mathcal{M}(\omega; E)$ and the mapping $f \mapsto \widetilde{\alpha f}$ is sequentially continuous from $\mathcal{M}(\omega; E)$ into $\mathcal{M}(\Omega; E)$ (Theorem 11.13);

— finally, as $\widetilde{\alpha f_n}$ vanishes on $\Omega \setminus \omega$, and since the domain where a function vanishes does not contribute to its integral, together with the monotonicity of the real-valued integral (Propositions 1.23 and 1.18 (a)), denoting $b = \sup_{x\in\omega} |\alpha(x)|$ (which is finite since $\alpha \in \mathcal{C}_b(\Omega)$), when $n \to \infty$, we get

$$\sup_{m\geq n} \int_\Omega \|\widetilde{\alpha f_m} - \widetilde{\alpha f_n}\|^p_{E;\nu} = \sup_{m\geq n} \int_\omega \|\alpha(f_m - f_n)\|^p_{E;\nu} \leq$$
$$\leq b \sup_{m\geq n} \int_\omega \|f_m - f_n\|^p_{E;\nu} \to 0.$$

Therefore (Definition 10.2),

$$\widetilde{\alpha f} \in L^p(\Omega; E).$$

1.b. *Inequality.* Again, Propositions 1.23 and 1.18 (a) give

$$\Big(\int_\Omega \|\widetilde{\alpha f_n}\|^p_{E;\nu}\Big)^{1/p} = \Big(\int_\omega \|\alpha f_n\|^p_{E;\nu}\Big)^{1/p} \leq \sup_{x\in\omega} |\alpha(x)| \Big(\int_\omega \|f_n\|^p_{E;\nu}\Big)^{1/p}.$$

From which, in the limit (Definition 10.2), we obtain the stated bound, since $\widetilde{\alpha f_n}$ and f_n are p-integrabilizing sequences for $\widetilde{\alpha f}$ and f, respectively.

2°) *Case where* $p = \infty$. **2.a.** *Membership of* $L^\infty(\Omega; E)$. Now let $(f_n)_{n\in\mathbb{N}}$ be an ∞-integrabilizing sequence for f (Definition 10.1). We will now show that the $(\widetilde{\alpha f_n})_{n\in\mathbb{N}}$ form an ∞-integrabilizing sequence for $\widetilde{\alpha f}$:

— on the one hand, $\widetilde{\alpha_n f} \in \mathcal{B}(\Omega; E)$ and $\widetilde{\alpha f_n} \to \widetilde{\alpha f}$ in $\mathcal{M}(\Omega; E)$ when $p < \infty$;

— on the other hand,

$$\sup_{n\in\mathbb{N},\, x\in\Omega} \|(\widetilde{\alpha f_n})(x)\|_{E;\nu} \leq \sup_{x\in\omega} |\alpha(x)| \sup_{n\in\mathbb{N},\, x\in\omega} \|f_n(x)\|_{E;\nu} < \infty; \tag{11.8}$$

— finally, let λ and σ be open sets such that $\lambda \Subset \Omega$ and $\operatorname{supp} \alpha \subset \sigma \Subset \omega$; since $\widetilde{\alpha f_n}$ vanishes on $\lambda \setminus (\lambda \cap \sigma)$ and $\lambda \cap \sigma \Subset \omega$, as $n \to \infty$, we have

$$\sup_{m\geq n} \int_\lambda \|\widetilde{\alpha f_m} - \widetilde{\alpha f_n}\|_{E;\nu} = \sup_{m\geq n} \int_{\lambda\cap\sigma} \|\alpha(f_m - f_n)\|_{E;\nu} \leq$$
$$\leq b \sup_{m\geq n} \int_{\lambda\cap\sigma} \|f_m - f_n\|_{E;\nu} \to 0.$$

Therefore (Definition 10.1),

$$\widetilde{\alpha f} \in L^\infty(\Omega; E).$$

2.b. *Inequality.* By Definition 10.2 of the semi-norms of $L^\infty(\Omega; E)$, for every ∞-integrabilizing sequence for f, together with (11.8) and $b = \sup_{x\in\omega} |\alpha(x)|$, we have

$$\|\widetilde{\alpha f}\|_{L^\infty(\Omega;E);\nu} \leq \inf_{\{(f_n)_{n\in\mathbb{N}}\}} \sup_{n\in\mathbb{N},\, x\in\Omega} \|(\widetilde{\alpha f_n})(x)\|_{E;\nu} \leq$$
$$\leq b \inf_{\{(f_n)_{n\in\mathbb{N}}\}} \sup_{n\in\mathbb{N},\, x\in\omega} \|f_n(x)\|_{E;\nu} = b\|f\|_{L^\infty(\omega;E);\nu}.$$

Which is the stated bound. □

11.4. Dependence on Ω

Let us show that p-integrable measures possess the finite gluing property.

Theorem 11.15. — *Let $f_1 \in L^p(\Omega_1; E)$ and $f_2 \in L^p(\Omega_2; E)$, where Ω_1 and Ω_2 are open subsets of $\mathbb{R}^d$, E is a Neumann space, and $1 \leq p \leq \infty$, such that*

$$f_1 = f_2 \ \textit{in}\ \Omega_1 \cap \Omega_2. \tag{11.9}$$

Then, there exists a unique $f \in L^p(\Omega_1 \cup \Omega_2; E)$ such that

$$f = f_1 \ \textit{in}\ \Omega_1, \quad f = f_2 \ \textit{in}\ \Omega_2.$$ ▮

Proof. ***1o)*** *Existence.* Let (α_1, α_2) be a partition of unity subordinate to the covering of Ω by Ω_1 and Ω_2 (Proposition A.51) and, denoting by $\widetilde{\alpha_i f_i}$ the extension of $\alpha_i f_i$ to Ω by 0_E (Theorems 11.13 and 11.14), let $f \in L^p(\Omega; E)$ be given by

$$f = \widetilde{\alpha_1 f_1} + \widetilde{\alpha_2 f_2}.$$

Given $\varphi \in \mathcal{K}(\Omega_1)$, we have $\alpha_2\varphi \in \mathcal{K}(\Omega_1 \cap \Omega_2)$. Again denoting by φ the extension of φ by 0, we have (definition of $\tilde{}$ in Theorem 11.13 and hypothesis (11.9))

$$\langle \widetilde{\alpha_2 f_2}, \varphi\rangle_\Omega = \langle f_2, \alpha_2\varphi\rangle_{\Omega_1\cap\Omega_2} = \langle f_1, \alpha_2\varphi\rangle_{\Omega_1\cap\Omega_2} = \langle f_1, \alpha_2\varphi\rangle_{\Omega_1}.$$

Therefore,

$$\langle f, \varphi\rangle_\Omega = \langle \widetilde{\alpha_1 f_1}, \varphi\rangle_\Omega + \langle \widetilde{\alpha_2 f_2}, \varphi\rangle_\Omega = \langle f_1, \alpha_1\varphi\rangle_{\Omega_1} + \langle f_1, \alpha_2\varphi\rangle_{\Omega_1} = \langle f_1, \varphi\rangle_{\Omega_1},$$

which proves (Definition 6.16) that

$$f = f_1 \ \text{on}\ \Omega_1.$$

Similarly, $f = f_2$ on Ω_2.

2o) *Uniqueness.* As the measure f is unique on Ω_1 and Ω_2, it is therefore unique on their union. □

Let us state some additivity properties of the semi-norms relative to the domain.

Theorem 11.16. — *Let $f \in L^p(\Omega_1 \cup \Omega_2; E)$, where Ω_1 and Ω_2 are open subsets of $\mathbb{R}^d$, E is a Neumann space, $1 \leq p < \infty$, and $\nu \in \mathcal{N}_E$ (the index set of the semi-norms of E).*

Then, denoting $\|f\|_{L^p(\omega;E);\nu}$ in the place of $\|f|_\omega\|_{L^p(\omega;E);\nu}$:

(a) $$\|f\|^p_{L^p(\Omega_1\cup\Omega_2;E);\nu} + \|f\|^p_{L^p(\Omega_1\cap\Omega_2;E);\nu} = \|f\|^p_{L^p(\Omega_1;E);\nu} + \|f\|^p_{L^p(\Omega_2;E);\nu};$$

(b) $$\|f\|^p_{L^p(\Omega_1\cup\Omega_2;E);\nu} \leq \|f\|^p_{L^p(\Omega_1;E);\nu} + \|f\|^p_{L^p(\Omega_2;E);\nu};$$

(c) *if $\Omega_1 \cap \Omega_2 = \emptyset$,*

$$\|f\|^p_{L^p(\Omega_1\cup\Omega_2;E);\nu} = \|f\|^p_{L^p(\Omega_1;E);\nu} + \|f\|^p_{L^p(\Omega_2;E);\nu};$$

(d) *if $p = \infty$,*

$$\|f\|_{L^\infty(\Omega_1\cup\Omega_2;E);\nu} = \sup(\|f\|_{L^\infty(\Omega_1;E);\nu}, \|f\|_{L^\infty(\Omega_2;E);\nu}). \blacksquare$$

The proof will use the following extension of an equality via continuity [Vol. 1, Theorem 10.2].

Proposition 11.17. — *Let E and F be separated semi-normed spaces, $U \subset X \subset E$, and let T_1 and T_2 be mappings from X into F such that $T_1 = T_2$ on U. Then,*

$$T_1 = T_2 \ \textit{on}\ X:$$

(a) *if T_1 and T_2 are continuous from X into F, and U is dense in X;*

(b) *or if T_1 and T_2 are sequentially continuous from X into F, and U is sequentially dense in X.* $\blacksquare$

We can now prove the aforementioned additivity properties.

Proof of Theorem 11.16. **1°)** Equality (a). Let $f \in \mathcal{B}(\Omega_1 \cup \Omega_2; E)$. Then, $\|f\|_{E;\nu}$ belongs to $\mathcal{B}(\Omega_1 \cup \Omega_2)$ (Proposition 1.17) and (Theorem 6.1 (a))

$$\int_{\Omega_1\cup\Omega_2} \|f\|^p_{E;\nu} + \int_{\Omega_1\cap\Omega_2} \|f\|^p_{E;\nu} = \int_{\Omega_1} \|f\|^p_{E;\nu} + \int_{\Omega_2} \|f\|^p_{E;\nu},$$

from which inequality (a) follows, since then $\int_\omega \|f\|^p_{E;\nu} = \|f\|^p_{L^p(\omega;E);\nu}$ (Proposition 10.7 for $f|_\omega$) for every open subset $\omega \subset \Omega_1 \cup \Omega_2$ and, in particular, for $\Omega_1 \cup \Omega_2$, $\Omega_1 \cap \Omega_2$, Ω_1, and Ω_2.

The stated equality extends to $f \in L^p(\Omega_1 \cup \Omega_2; E)$ via continuity by Proposition 11.17 (b) (or by (a)), whose hypotheses are satisfied because:

— $\mathcal{B}(\Omega_1 \cup \Omega_2; E)$ is sequentially dense in $L^p(\Omega_1 \cup \Omega_2; E)$ (Theorem 10.14);

— the mapping $f \mapsto \|f\|^p_{L^p(\omega;E);\nu}$ is continuous from $L^p(\Omega_1 \cup \Omega_2; E)$ into $\mathbb{R}$ (as it is the composition of the continuous mappings: $|_\omega$, $\| \; \|_{L^p(\omega;E);\nu}$, and $t \mapsto t^p$, see Proposition A.39, Theorem 11.5, and also Propositions A.44 and A.82).

2o) Properties (b) *and* (c). These follow from equality (a).

3o) Property (d). **3.a.** *Notation.* Denote $\Omega = \Omega_1 \cup \Omega_2$, $f_1 = f|_{\Omega_1}$, $f_2 = f|_{\Omega_2}$, and

$$a = \sup(\|f_1\|_{L^\infty(\Omega_1;E);\nu}, \|f_2\|_{L^\infty(\Omega_2;E);\nu}).$$

3.b. *An inequality.* For $i = 1$ or 2, f_i belongs to $L^\infty(\Omega_i; E)$ (Theorem 11.5 (b)) and $\|f_i\|_{L^\infty(\Omega_i;E);\nu} \leq \|f\|_{L^\infty(\Omega;E);\nu}$. Therefore,

$$a \leq \|f\|_{L^\infty(\Omega;E);\nu}. \tag{11.10}$$

3.c. *Converse inequality.* For all $n \in \mathbb{N}^*$, the regularized (Definition 3.17) $R_n^{\Omega_i} f_i$ of the f_i belong to $\mathcal{K}(\Omega_i; E)$ (Theorems 10.10 (b) and 10.13 (a)), and

$$\sup_{x \in \Omega_i} \|R_n^{\Omega_i} f_i(x)\|_{E;\nu} = \|R_n^{\Omega_i} f_i\|_{L^\infty(\Omega_i;E);\nu} \leq \|f_i\|_{L^\infty(\Omega_i;E);\nu}. \tag{11.11}$$

Let (α_1, α_2) be a partition of unity subordinate to the cover of Ω by Ω_1 and Ω_2 given by Proposition A.51. As we will verify below, where $\tilde{}$ denotes the extension to Ω by 0_E:

$$\text{the } (\widetilde{\alpha_i R_n^{\Omega_i} f_i})_{n \in \mathbb{N}^*} \text{ form a } p\text{-integrabilizing sequence for } \alpha_i f. \tag{11.12}$$

Thus, $f_n = \widetilde{\alpha_1 R_n^{\Omega_1} f_1} + \widetilde{\alpha_2 R_n^{\Omega_2} f_2}$ is a p-integrabilizing sequence for $\alpha_1 f + \alpha_2 f$, i.e. for f since $\alpha_1 + \alpha_2 = 1$.

For all $n \in \mathbb{N}$ and $x \in \Omega$, (11.11) implies

$$\|f_n(x)\|_{E;\nu} \leq \alpha_1(x) \|\widetilde{R_n^{\Omega_1} f_1(x)}\|_{E;\nu} + \alpha_2(x) \|\widetilde{R_n^{\Omega_2} f_2(x)}\|_{E;\nu} \leq a.$$

From which it follows (Definition 10.1) that

$$\|f\|_{L^\infty(\Omega;E);\nu} \leq a.$$

3.d. *Verification of (11.12).* If suffices to verify the three criteria of Definition 10.1 for an ∞-integrabilizing sequence. For every $\varphi \in \mathcal{K}(\Omega)$, we have $\alpha_i \varphi \in \mathcal{K}(\Omega_i)$, and:

$$\langle \widetilde{\alpha_i R_n^{\Omega_i} f_i}, \varphi \rangle = \int_\Omega \widetilde{\alpha_i R_n^{\Omega_i} f_i} \varphi = \int_{\Omega_i} \alpha_i R_n^{\Omega_i} f_i \varphi \to \int_{\Omega_i} \alpha_i f_i \varphi;$$

$$\sup_{n \in \mathbb{N}^*,\, x \in \Omega} \|\widetilde{\alpha_i R_n^{\Omega_i} f_i(x)}\|_{E;\nu} = \sup_{n \in \mathbb{N}^*,\, x \in \Omega_i} \|\alpha_i R_n^{\Omega_i} f_i(x)\|_{E;\nu} < \infty;$$

$$\sup_{m \geq n} \int_\Omega \left\| \alpha_i (\widetilde{R_m^{\Omega_i} f_i} - \widetilde{R_n^{\Omega_i} f_i}) \right\|_{E;\nu} \leq \sup_{m \geq n} \int_{\Omega_i} \left\| \alpha_i (R_m^{\Omega_i} f_i - R_n^{\Omega_i} f_i) \right\|_{E;\nu} \to 0. \;\square$$

11.5. Infinite gluing on Ω and continuity in p

Let us state an infinite-gluing property for an increasing sequence of open sets.

Theorem 11.18. — *For all $n \in \mathbb{N}$, let $f_n \in L^p(\omega_n; E)$, where ω_n is an open subset of $\mathbb{R}^d$ increasing with n, E is a Neumann space, and $1 \leq p \leq \infty$, such that*

$$f_{n+1} = f_n \ \text{ on } \omega_n$$

and, for every $\nu \in \mathcal{N}_E$ (the index set of the semi-norms of E),

$$\sup_{n \in \mathbb{N}} \|f_n\|_{L^p(\omega_n;E);\nu} < \infty. \tag{11.13}$$

Then, there exists $f \in L^p(\Omega; E)$, where $\Omega \stackrel{\text{def}}{=} \bigcup_{n\in\mathbb{N}} \omega_n$, such that

$$f = f_n \ \text{ on } \omega_n.$$

Furthermore, for every $\nu \in \mathcal{N}_E$,

$$\|f\|_{L^p(\Omega;E);\nu} = \lim_{n\to\infty} \|f_n\|_{L^p(\omega_n;E);\nu}. \ \blacksquare$$

Proof of Theorem 11.18. ***1°) Membership of $L^p(\Omega; E)$.*** **1.a.** Preliminary. Consider the open sets

$$\lambda_n = \Omega^n_{1/n} = \{x \in \Omega : B(x, 1/n) \subset \Omega, \ |x| < n\},$$

where $\overline{B(x,r)} = \{y \in \mathbb{R}^d : |y - x| \leq r\}$. For m sufficiently large, we have $\lambda_n \subset \omega_m$ (since $\overline{\lambda_n}$ is a compact set included in Ω). Let g_n be the restriction of f_m to λ_n, then (Theorem 11.5),

$$\|g_n\|_{L^p(\lambda_n;E);\nu} \leq \|f_m\|_{L^p(\omega_m;E);\nu}. \tag{11.14}$$

The sequence $(\|g_n\|_{L^p(\lambda_n;E);\nu})_{n\in\mathbb{N}}$ is increasing and bounded by hypothesis (11.13), hence (Proposition A.5) it has a limit ℓ_ν as $n \to \infty$, i.e.

$$\|g_n\|_{L^p(\lambda_n;E);\nu} \to \ell_\nu. \tag{11.15}$$

***1.b.** Case $p < \infty$.* For $m \geq n$, let

$$C_{n,m} = \lambda_m \setminus \overline{\lambda_n}.$$

Since $\lambda_n \cup C_{n,m} \subset \lambda_m$ and $\lambda_n \cap C_{n,m} = \emptyset$, we have (Theorem 11.16 (c) and 11.5 (b))

$$\|g_m\|^p_{L^p(\lambda_n;E);\nu} + \|g_m\|^p_{L^p(C_{n,m};E);\nu} = \|g_m\|^p_{L^p(\lambda_n \cup C_{n,m};E);\nu} \leq \|g_m\|^p_{L^p(\lambda_m;E);\nu}.$$

Therefore,

$$\sup_{m\geq n} \|g_m\|^p_{L^p(C_{n,m};E);\nu} \leq \ell^p_\nu - \|g_m\|^p_{L^p(\lambda_n;E);\nu}. \tag{11.16}$$

Since $\overline{\lambda_{n-1}}$ is a compact set included in the open set λ_n, there exists a function γ_n (for example, the function given by (3.17), p. 67, relative to $\Omega = \lambda_n$) such that

$$\gamma_n \in \mathcal{K}(\lambda_n), \quad \gamma_n = 1 \text{ on } \lambda_{n-1}, \quad 0 \leq \gamma_n \leq 1 \text{ on } \lambda_n.$$

By extending $\gamma_n g_n$ by 0_E, we obtain $\widetilde{\gamma_n g_n} \in L^p(\Omega; E)$ (Theorem 11.14), and

$$\|\widetilde{\gamma_m g_m} - \widetilde{\gamma_n g_n}\|_{L^p(\Omega;E);\nu} \leq \|g_m\|_{L^p(C_{n-1,m};E);\nu}$$

(since $\operatorname{supp}(\gamma_n - \gamma_m) \subset C_{n-1,m}$ and $|\gamma_n - \gamma_m| \leq 1$). From which it follows, by (11.16) and (11.15), when $n \to \infty$,

$$\sup_{m\geq n} \|\widetilde{\gamma_m g_m} - \widetilde{\gamma_n g_n}\|_{L^p(\Omega;E);\nu} \to 0.$$

Therefore, $(\widetilde{\gamma_n g_n})_{n\in\mathbb{N}}$ is a Cauchy sequence in $L^p(\Omega; E)$, which is sequentially complete (Theorem 10.16). It therefore has a limit

$$f \in L^p(\Omega; E).$$

For all m, it satisfies

$$f = g_m \text{ on } \lambda_{m-1}.$$

This guarantees the uniqueness of f on all of λ_m, hence also on their union Ω, and $f = f_n$ on ω_n.

1.c. *Case* $p = \infty$. Let us verify the two criteria of Theorem 10.11 (a).

— On the one hand, let $\omega \Subset \Omega$; for n sufficiently large such that $\omega_n \supset \omega$, we have (Propositions 10.12 and 2.2 on the monotonicity of the measure)

$$\|f_n|_\omega\|_{L^1(\omega;E);\nu} \leq |\omega| \|f_n\|_{L^\infty(\omega_n;E);\nu}$$

which is bounded independently of n. Thus, from the case when $p = 1$,

$$f|_\omega \in L^1(\omega; E).$$

— On the one hand, if n is large enough such that $x \in (\omega_n)^n_{4/n}$, then we have $\alpha^\Omega_n(x) = \alpha^{\omega_n}_n = 1$, so

$$R_n f(x) = R^{\omega_n}_n f(x) = R^{\omega_n}_n f_n(x).$$

From this (Theorem 10.10 (b) and 10.13 (a) on the semi-norms of $L^\infty(\Omega; E)$ and on the regularizations), we get

$$\|R_n f(x)\|_{E;\nu} = \|R_n^{\omega_n} f_n(x)\|_{E;\nu} \leq \|f_n\|_{L^\infty(\omega_n;E);\nu} \leq \sup_{m\in\mathbb{N}} \|f_m\|_{L^\infty(\omega_m;E);\nu},$$

which is bounded independently of both n and x.

Therefore (Theorem 10.11 (a)),

$$f \in L^\infty(\Omega; E).$$

***2º) Inequality.* 2.a.** *Case* $p < \infty$. By Theorem 11.14 (since $\operatorname{supp} \gamma_n \subset \lambda_n$ and $|\gamma_n| \leq 1$) and (11.14),

$$\|\widetilde{\gamma_n g_n}\|_{L^p(\Omega;E);\nu} \leq \|g_n\|_{L^p(\lambda_n;E);\nu} \leq \|f_m\|_{L^p(\omega_m;E);\nu}$$

where m is such that $\omega_m \supset \lambda_n$. In the limit, since $\widetilde{\gamma_n g_n} \to f$ in $L^p(\Omega; E)$ from 1.b, we get

$$\|f\|_{L^p(\Omega;E);\nu} \leq \lim_{m\to\infty} \|f_m\|_{L^p(\omega_m;E);\nu}.$$

As the inverse inequality follows from Theorem 11.5 (b), this is in fact an equality.

2.b. *Case* $p = \infty$. Observe that, for $\epsilon > 0$ and $\nu \in \mathcal{N}_E$, there exists an open subset $\omega \Subset \Omega$ such that

$$\|f\|_{L^\infty(\omega;E)} \geq \|f\|_{L^\infty(\Omega;E)} - \epsilon. \tag{11.17}$$

If not, then, for every $\omega \Subset \Omega$, we would have $\|f\|_{L^\infty(\omega;E)} \leq \|f\|_{L^\infty(\Omega;E)} - \epsilon$; since $R_n f$ has compact support (Theorem 3.18), for a particular $\omega_n \Subset \Omega$, we would have

$$\|R_n f\|_{L^\infty(\Omega;E);\nu} = \|f\|_{L^\infty(\omega_n;E);\nu} \leq \|f\|_{L^\infty(\Omega;E)} - \epsilon;$$

which, in the limit as $n \to \infty$ (Theorem 10.13 (b)), would give

$$\|f\|_{L^\infty(\Omega;E);\nu} \leq \|f\|_{L^\infty(\Omega;E);\nu} - \epsilon,$$

which is absurd. Hence, for $\epsilon > 0$, (11.17) implies that there exist $\omega \Subset \Omega$ and $n \in \mathbb{N}$ such that

$$\|f\|_{L^\infty(\Omega;E)} \leq \|f\|_{L^\infty(\omega;E)} + \epsilon \leq \|f_n\|_{L^\infty(\omega_m;E)} + \epsilon \leq \sup_{n\in\mathbb{N}} \|f_n\|_{L^\infty(\omega_n;E)} + \epsilon.$$

As this is true for all positive ϵ, it is also true for $\epsilon = 0$, so

$$\|f\|_{L^\infty(\Omega;E)} \leq \sup_{n\in\mathbb{N}} \|f_n\|_{L^\infty(\omega_n;E)}.$$

As the inverse inequality follows from Theorem 11.5 (b), this is in fact an equality. □

Let us now consider infinite gluing on arbitrary open sets.

Theorem 11.19. — *For all $n \in \mathbb{N}$, let $f_n \in L^p(\omega_n; E)$, where ω_n is an open subset of $\mathbb{R}^d$, E is a Neumann space, and $1 \leq p \leq \infty$, such that, for all n and m in $\mathbb{N}$,*

$$f_n = f_m \ \text{ on } \omega_n \cap \omega_m$$

and, for every $\nu \in \mathcal{N}_E$ (the index set for the semi-norms of E),

$$\sum_{m \in \mathbb{N}} \|f_m\|^p_{L^p(\omega_m;E);\nu} < \infty \quad \text{if } p < \infty,$$

$$\sup_{m \in \mathbb{N}} \|f_m\|_{L^\infty(\omega_m;E);\nu} < \infty \quad \text{if } p = \infty.$$

Then, there exists a unique $f \in L^p(\Omega; E)$, where $\Omega = \bigcup_{n \in \mathbb{N}} \omega_n$, such that

$$f = f_n \ \text{ on } \omega_n.$$

Furthermore, for every $\nu \in \mathcal{N}_E$,

$$\|f\|_{L^p(\Omega;E);\nu} \leq \Big(\sum_{m \in \mathbb{N}} \|f_m\|^p_{L^p(\omega_m;E);\nu}\Big)^{1/p} \quad \text{if } p < \infty,$$

$$\|f\|_{L^\infty(\Omega;E);\nu} = \sup_{m \in \mathbb{N}} \|f_m\|_{L^\infty(\omega_m;E);\nu} \quad \text{if } p = \infty. \ \blacksquare$$

Proof. Here, ω_n does not necessarily increase with n, but we can reduce to that case by considering

$$\Omega_n \stackrel{\text{def}}{=} \omega_1 \cup \omega_2 \cup \ldots \cup \omega_n.$$

By applying the finite gluing property $n-1$ times (Theorem 11.15), we obtain an F_n in $L^p(\Omega_n; E)$ such that, for all $m \leq n$, we have $F_n = f_m$ on ω_m. Then, if $p < \infty$, by applying the subadditivity property $n-1$ times (Theorem 11.16 (b)), we obtain the bound

$$\|F_n\|^p_{L^p(\Omega_n;E);\nu} \leq \|F_n\|^p_{L^p(\omega_1;E);\nu} + \cdots + \|F_n\|^p_{L^p(\omega_n;E);\nu} \leq \sum_{m \in \mathbb{N}} \|f_m\|^p_{L^p(\omega_m;E);\nu}.$$

Whereas, if $p = \infty$, we obtain (Theorem 11.16 (d)) the bound

$$\|F_n\|_{L^\infty(\Omega_n;E);\nu} = \sup\{\|F_n\|_{L^\infty(\omega_1;E);\nu}, \ldots, \|F_n\|_{L^\infty(\omega_n;E);\nu}\} = \\ = \sup_{m \leq n} \|f_m\|_{L^\infty(\omega_m;E);\nu}.$$

By gluing over an increasing sequence (Theorem 11.18), this implies the existence of a unique $f \in L^p(\Omega; E)$ such that $f = F_n$ on Ω_n, and hence $f = f_n$ on ω_n. Moreover, f satisfies the stated bound. □

Let us state some properties regarding continuity in p. (We will generalize (b) to an arbitrary Banach space E in Theorem 18.9.)

Theorem 11.20. — *Let $f \in L^p(\Omega; E)$, where Ω is an open subset of $\mathbb{R}^d$ and E is a Neumann space, for every $p \in (p_0, p_1)$ with $1 \leq p_0 < p_1 < \infty$ and, for every $\nu \in \mathcal{N}_E$ (the index set of the semi-norms of E) and $c_\nu \in \mathbb{R}$ such that*

$$\|f\|_{L^p(\Omega;E);\nu} \leq c_\nu.$$

(a) *Then, $f \in L^{p_0}(\Omega; E)$ and, for every $\nu \in \mathcal{N}_E$,*

$$\|f\|_{L^{p_0}(\Omega;E);\nu} \leq c_\nu.$$

(b) *If E is a reflexive Banach space and $\|f\|_{L^p(\Omega;E)} \leq c$ for every $p < p_1$, then $f \in L^{p_1}(\Omega; E)$, and*

$$\|f\|_{L^{p_1}(\Omega;E)} \leq c. \blacksquare$$

Proof. ***1o)*** *Property* (a). For all $n \in \mathbb{N}$, let

$$\omega_n \stackrel{\text{def}}{=} \{x \in \Omega : |x| < n\}.$$

According to Theorem 11.1 regarding the dependence of a domain of finite measure on p, the restriction f_n of f to ω_n satisfies, for every $p \in (p_0, p_1)$ and $\nu \in \mathcal{N}_E$,

$$\|f_n\|_{L^{p_0}(\omega_n;E);\nu} \leq |\omega_n|^{1/p_0 - 1/p} \|f_n\|_{L^p(\omega_n;E);\nu} \leq c_\nu |\omega_n|^{1/p_0 - 1/p}.$$

In the limit as $p \to p_0$,

$$\|f_n\|_{L^{p_0}(\omega_n;E);\nu} \leq c_\nu.$$

Since $\Omega = \bigcup_{n\in\mathbb{N}} \omega_n$, Theorem 11.18 on the infinite gluing property implies that $f \in L^{p_0}(\Omega; E)$, and

$$\|f\|_{L^{p_0}(\Omega;E);\nu} \leq c_\nu.$$

2o) *Property* (b). Here, E is a reflexive Banach space, and we shall make an exception here (to avoid postponing this result until after Chapter 16) by using Theorem 16.2:

$$L^p(\Omega; E) \text{ is a reflexive Banach space for } 1 < p < \infty. \tag{11.18}$$

For all $n \in \mathbb{N}^*$ and $p < p_1$, the regularizations satisfy (Theorem 10.13 (a))

$$\|R_n f\|_{L^p(\Omega;E)} \leq \|f\|_{L^p(\Omega;E)} \leq c.$$

Since $R_n f \in \mathcal{K}(\Omega; E)$ (Theorem 3.18), we have $\|R_n f\|^p \to \|R_n f\|^{p_1}$ uniformly as $p \to p_1$. Using the integral expression for the norm (Proposition 10.7) and the convergence of the Cauchy integral (Theorem 1.26),

$$\|R_n f\|_{L^p(\Omega;E)} = \Big(\int_\Omega \|R_n f\|^p\Big)^{1/p} \to \Big(\int_\Omega \|R_n f\|^{p_1}\Big)^{1/p_1} = \|R_n f\|_{L^{p_1}(\Omega;E)}.$$

Thus, for all $n \in \mathbb{N}^*$,

$$\|R_n f\|_{L^{p_1}(\Omega;E)} \leq c. \tag{11.19}$$

As this sequence is bounded, it therefore has (Proposition A.77) a subsequence such that

$$R_{\sigma(n)} f \to g \ \text{ in } L^{p_1}(\Omega; E)\text{-weak}. \tag{11.20}$$

Since $L^{p_1}(\Omega; E) \underset{\to}{\subseteq} \mathcal{M}(\Omega; E)$ (Theorem 10.8 (a)), we then have (Proposition A.70) $L^{p_1}(\Omega; E)$-weak $\underset{\to}{\subseteq} \mathcal{M}(\Omega; E)$-weak, from which it follows that

$$R_{\sigma(n)} f \to g \ \text{ in } \mathcal{M}(\Omega; E)\text{-weak}.$$

However, $R_{\sigma(n)} f \to f$ in $\mathcal{M}(\Omega; E)$ (Theorem 3.19), and therefore in $\mathcal{M}(\Omega; E)$-weak as well (Proposition A.68). Hence,

$$g = f.$$

Since F-weak $= F$ (Definition A.67), (11.20) implies

$$f \in L^{p_1}(\Omega; E).$$

The bound on the semi-norms of a weak limit (Proposition A.71, with $p(f)$ equal to $\|f\|_{L^{p_1}(\Omega;E)}$) gives, together with (11.19), (11.20), and the fact that $g = f$,

$$\|f\|_{L^{p_1}(\Omega;E)} \leq c. \ \square$$

Chapter 12

Image Under a Linear Mapping

We are first interested in the image Lf of $f \in L^p(\Omega; E)$ under a linear mapping L from E into F (Theorem 12.1), and in its application to the dependence on E.

We then move on to multilinear mappings. After reviewing these mappings (§ 12.2), we define the image $T(f_1, \ldots, f_I)$ under a continuous multilinear mapping T from $E_1 \times \ldots \times E_d$ into F (Definition 12.10), and we show that the mapping $(f_1, \ldots, f_I) \mapsto T(f_1, \ldots, f_I)$ is continuous from $L^{p_1}(\Omega; E_1) \times \ldots \times L^{p_I}(\Omega; E_I)$ into $L^p(\Omega; F)$ (Theorem 12.12), where $1/p = 1/p_1 + \cdots + 1/p_I$.

We then present some particular cases when the E_i are Banach spaces (§ 12.3), and more generally define the image on the $L^p_{\text{loc}}(\Omega; E_i)$ spaces (§ 12.4).

12.1. Image under a linear mapping and dependence on E

Let us present some properties of the image under a linear mapping.

Theorem 12.1. — *Let*

$$L \in \mathcal{L}(E; F)$$

where E and F are Neumann spaces, Ω is an open subset of $\mathbb{R}^d$, and $1 \le p \le \infty$.

(a) *The mapping $f \mapsto Lf$ (Definition 6.9) is continuous linear, and thus sequentially continuous, from $L^p(\Omega; E)$ into $L^p(\Omega; F)$.*

(b) *If L is an isomorphism, then so is the mapping $f \mapsto Lf$.*

(c) *Suppose that, for every $u \in E$, there exist a finite set N in $\mathcal{N}_E$ and $c \ge 0$, such that*

$$\|Lu\|_{F;\mu} \le c \sup_{\nu \in N} \|u\|_{E;\nu}, \tag{12.1}$$

where $\{\| \; \|_{E;\nu} : \nu \in \mathcal{N}_E\}$ and $\{\| \; \|_{F;\mu} : \mu \in \mathcal{F}\}$ are the semi-norms of E and F.

Then, for every $f \in L^p(\Omega; E)$,

$$\|Lf\|_{L^p(\Omega;F);\mu} \leq c \sup_{\nu \in N} \|f\|_{L^p(\Omega;E);\nu}.$$

(d) *The mapping* $f \mapsto Lf$ *is continuous linear, and hence sequentially continuous, from* $L^p_{\text{loc}}(\Omega; E)$ *into* $L^p_{\text{loc}}(\Omega; F)$. ▮

Proof. **1º)** *Preliminaries.* Since L is continuous, for every $\mu \in \mathcal{N}_F$, there exist (Proposition 1.9) $c \in \mathbb{R}$ and a finite subset N of $\mathcal{N}_E$ satisfying (12.1). Then, for every $u \in E$,

$$\|Lu\|_{F;\mu} \leq c \sum_{\nu \in N} \|u\|_{E;\nu}.$$

By the monotonicity of the integral (Proposition 1.18 (a)) and Minkowski's inequality (Proposition 9.7), it follows then, for every $f \in \mathcal{B}(\Omega; E)$, that

$$\Big(\int_\Omega \|Lf\|^p_{F;\mu}\Big)^{1/p} \leq c\Big(\int_\Omega \Big(\sum_{\nu \in N} \|f\|_{E;\nu}\Big)^p\Big)^{1/p} \leq c \sum_{\nu \in N} \Big(\int_\Omega \|f\|^p_{E;\nu}\Big)^{1/p}. \quad (12.2)$$

2º) *Inequality* (c). **2.a.** *Case* $p < \infty$. Let $f \in L^p(\Omega; E)$ and $(f_n)_{n \in \mathbb{N}} \subset \mathcal{B}(\Omega; E)$ be a p-integrabilizing sequence for f (Definition 10.1). We will show that the Lf_n form a p-integrabilizing sequence for Lf:

— first, Lf_n belongs to $\mathcal{B}(\Omega; F)$ since it is uniformly continuous (as L is by Proposition 1.9, and hence Lf_n is too by Proposition A.39), and its support is bounded as it is contained in the support of the f_n;

— next, $Lf_n \to Lf$ in $\mathcal{M}(\Omega; F)$ (Theorem 6.10 on the continuity of $f \mapsto Lf$);

— finally, with (12.2), when $n \to \infty$,

$$\sup_{m \geq n} \Big(\int_\Omega \|Lf_m - Lf_n\|^p_{F;\mu}\Big)^{1/p} \leq c \sup_{\nu \in N} \sup_{m \geq n} \Big(\int_\Omega \|f_m - f_n\|^p_{E;\nu}\Big)^{1/p} \to 0. \quad (12.3)$$

From which (Definition 10.2) we get,

$$Lf \in L^p(\Omega; F).$$

Moreover, Definition 10.2 of the norm of $L^p(\Omega; E)$ and inequality (12.2) relative to f_n imply

$$\begin{aligned} \|Lf\|_{L^p(\Omega;F);\mu} &= \lim_{n \to \infty} \Big(\int_\Omega \|Lf_n\|^p_{F;\mu}\Big)^{1/p} \leq \\ &\leq \lim_{n \to \infty} c \sum_{\nu \in N} \Big(\int_\Omega \|f_n\|^p_{E;\nu}\Big)^{1/p} = c \sum_{\nu \in N} \|f\|_{L^p(\Omega;E);\nu}. \end{aligned} \quad (12.4)$$

2.b. *Case* $p = \infty$. We proceed similarly as before, albeit now we replace inequality (12.3) with the same one, but now with $p = 1$ and in a bounded open set ω included in Ω, and with the inequality

$$\sup_{n\in\mathbb{N},\, x\in\Omega} \|Lf_n(x)\|_{F;\mu} \leq c \sup_{\nu\in N,\, n\in\mathbb{N},\, x\in\Omega} \|f_n(x)\|_{E;\nu} < \infty.$$

Moreover, by taking the infimum over all ∞-integrabilizing sequences for f, Definition 10.2 of the norm of $L^\infty(\Omega; E)$ gives

$$\begin{aligned}\|Lf\|_{L^\infty(\Omega;F);\mu} &\leq \inf_{(f_n)_{n\in\mathbb{N}}} \sup_{n\in\mathbb{N},\, x\in\Omega} \|Lf_n(x)\|_{F;\mu} \leq \\ &\leq c \inf_{(f_n)_{n\in\mathbb{N}}} \sup_{n\in\mathbb{N},\, x\in\Omega} \sup_{\nu\in N} \|f_n(x)\|_{E;\nu} = c \sup_{\nu\in N} \|f\|_{L^\infty(\Omega;E);\nu}.\end{aligned} \tag{12.5}$$

3º) Continuity. Inequalities (12.4) or (12.5) (depending on the value of p) implies (Proposition 1.9) that

the mapping $f \mapsto Lf$ is continuous from $L^p(\Omega; E)$ into $L^p(\Omega; F)$.

4º) Sequential continuity. This always follows from continuity (Proposition 1.8).

5º) Isomorphism. If L is an isomorphism, then L is continuous linear from E into F, and L^{-1} is similarly from F into E. Thus, the mapping $f \mapsto Lf$ is continuous linear from $L^p(\Omega; E)$ into $L^p(\Omega; F)$, and $g \mapsto L^{-1}g$ is likewise from $L^p(\Omega; F)$ into $L^p(\Omega; E)$, and is the inverse of $f \mapsto Lf$.

6º) Property (d). Now, let

$$f \in L^p_{\text{loc}}(\Omega; E).$$

For an open subset $\omega \Subset \Omega$, we have $f|_\omega \in L^p(\omega; E)$, and hence $L(f|_\omega) \in L^p(\omega; F)$ from (a). Since $L(f|_\omega) = (Lf)|_\omega$ (this is a simple consequence of Definitions 6.16 and 6.9 of the restriction), it follows that

$$Lf \in L^p_{\text{loc}}(\Omega; F).$$

Moreover, by (12.4) or (12.5) (depending on the value of p),

$$\|Lf\|_{L^p_{\text{loc}}(\Omega;F);\omega,\mu} = \|Lf\|_{L^p(\omega;F);\mu} \leq c \sup_{\nu\in N} \|f\|_{L^p(\omega;E);\nu} = c \sup_{\nu\in N} \|f\|_{L^p_{\text{loc}}(\omega;E);\omega,\nu}.$$

Thus (Proposition 1.9),

the mapping $f \mapsto Lf$ is continuous from $L^p_{\text{loc}}(\Omega; E)$ into $L^p_{\text{loc}}(\Omega; F)$. □

Let us estimate the norms in the Banach spaces. Recall that, if E and F are such spaces, then

$$|||L|||_{\mathcal{L}(E;F)} \stackrel{\text{def}}{=} \sup_{u\in E,\, u\neq 0} \frac{\|Lu\|_F}{\|u\|_E}. \tag{12.6}$$

Theorem 12.2. — *Let $f \in L^p(\Omega; E)$ and $L \in \mathcal{L}(E; F)$, where $1 \leq p \leq \infty$, Ω is an open subset of $\mathbb{R}^d$,*

$$E \text{ and } F \text{ are Banach spaces.}$$

Then, for the norm of L defined by (12.6),

$$\|Lf\|_{L^p(\Omega;F)} \leq |||L|||_{\mathcal{L}(E;F)} \|f\|_{L^p(\Omega;E)}. \blacksquare$$

Proof. This is a particular case of Theorem 12.1, since

$$\|Lu\|_F \leq |||L|||_{\mathcal{L}(E;F)} \|u\|_E. \square$$

Norm estimates in Banach spaces. If E and F are Banach spaces such that $E \subsetneq F$, then, denoting $c = \sup_{v \in E, v \neq 0} \|v\|_F / \|v\|_E$ and applying Theorem 12.2 to the identity mapping, we get: for every $f \in L^p(\Omega; E)$,

$$\|f\|_{L^p(\Omega;F)} \leq c \|f\|_{L^p(\Omega;E)}. \quad (12.7) \ \square$$

Let us show that $L^p(\Omega; E)$ and $L^p_{\text{loc}}(\Omega; E)$ both increase with E.

Theorem 12.3. — *Let*

$$E \subsetneq F,$$

where E and F are Neumann spaces, Ω is an open subset of $\mathbb{R}^d$, and $1 \leq p \leq \infty$. Then,

$$L^p(\Omega; E) \subsetneq L^p(\Omega; F),$$

$$L^p_{\text{loc}}(\Omega; E) \subsetneq L^p_{\text{loc}}(\Omega; F). \blacksquare$$

Proof. This follows from applying Theorem 12.1 to the identity mapping. $\square$

The density of $L^p(\Omega; E)$ in $L^p(\Omega; F)$ will be examined in Theorem 13.10. Next, let us observe that topological equivalences preserve $L^p(\Omega; E)$ and $L^p_{\text{loc}}(\Omega; E)$.

Theorem 12.4. — *The spaces $L^p(\Omega; E)$ and $L^p_{\text{loc}}(\Omega; E)$, where Ω is an open subset of $\mathbb{R}^d$, E is a Neumann space, and $1 \leq p \leq \infty$, are preserved when the family of semi-norms of E is replaced by an equivalent family.* $\blacksquare$

Proof. This follows from applying Theorem 12.1 to the identity mapping. $\square$

Let us observe that the semi-norms of $L^p(\Omega; E)$ increase as those of E.

Theorem 12.5. — *Let E be a Neumann space endowed with two semi-norms such that $\| \ \|_{E;\nu} \leq \| \ \|_{E;\mu}$. Then, for every open subset Ω of $\mathbb{R}^d$ and $1 \leq p \leq \infty$,*

$$\| \ \|_{L^p(\Omega;E);\nu} \leq \| \ \|_{L^p(\Omega;E);\mu}. \blacksquare$$

Proof. We have the inequality $\|f\|_{L^p(\Omega;E);\nu} \leq \|f\|_{L^p(\Omega;E);\mu}$ for every f in $\mathcal{B}(\Omega; E)$ (Definition 1.14), and hence for every f in $L^p(\Omega; E)$ by Definition 10.2. □

12.2. Image under a multilinear mapping

We shall begin with some preliminaries. First, we shall characterize continuous multilinear mappings [Vol. 1, Theorem 7.20].

Proposition 12.6. — *A multilinear mapping T from a product $E_1 \times \cdots \times E_I$ of separated semi-normed spaces into a separated semi-normed space F is continuous if and only if, for every semi-norm $\| \ \|_{F;\mu}$ of F, there exist I finite sets N_i of semi-norms $\| \ \|_{E_i;\nu_i}$ of E_i and $c \geq 0$ such that: for all $(u_1, \ldots, u_I) \in E_1 \times \cdots \times E_I$,*

$$\|T(u_1, \ldots, u_I)\|_{F;\mu} \leq c \sup_{\nu_1 \in N_1} \|u_1\|_{E_1;\nu_1} \cdots \sup_{\nu_I \in N_I} \|u_d\|_{E_I;\nu_I}. \blacksquare$$

Let us now define the space of continuous multilinear mappings.

Definition 12.7. — *Let $E_1, \ldots, E_I$, and F be separated semi-normed spaces.*

Denote by $\mathcal{L}^I(E_1 \times \cdots \times E_I; F)$ the space of continuous multilinear mappings from $E_1 \times \cdots \times E_I$ into F, endowed with semi-norms indexed by the bounded subsets B_1 of $E_1, \ldots, B_I$ of E_I, and by $\mu \in \mathcal{N}_F$,

$$\|T\|_{\mathcal{L}^I(E_1 \times \cdots \times E_I;F);B_1,\ldots,B_I,\mu} \stackrel{\text{def}}{=} \sup_{u_1 \in B_1,\ldots,u_I \in B_I} \|T(u_1, \ldots, u_I)\|_{F;\mu}. \blacksquare$$

When the E_i are normed, the topology of $\mathcal{L}^I(E_1 \times \cdots \times E_I; F)$ is equivalently generated by the following family of semi-norms [Vol. 1, Theorems 12.19 and 7.2].

Proposition 12.8. — *Let $E_1, \ldots, E_I$ be normed spaces and F be a separated semi-normed space.*

(a) *The topology of* $\mathcal{L}^I(E_1 \times \cdots \times E_I; F)$ *is generated by the family of semi-norms indexed by* $\mu \in \mathcal{N}_F$,

$$|||T|||_{\mathcal{L}^I(E_1\times\cdots\times E_I;F);\mu} \stackrel{\text{def}}{=} \sup_{u_i\in E_i,\, u_i\neq 0,\, 1\leq i\leq I} \frac{\|T(u_1,\ldots,u_I)\|_{F;\mu}}{\|u_1\|_{E_1}\cdots\|u_I\|_{E_I}}.$$

(b) $$\|T(u_1,\ldots,u_I)\|_{F;\mu} \leq |||T|||_{\mathcal{L}^I(E_1\times\cdots\times E_I;F);\mu}\|u_1\|_{E_1}\cdots\|u_I\|_{E_I}.$$

(c) *The mapping* $(T, u_1, \ldots, u_I) \mapsto T(u_1, \ldots, u_I)$ *is continuous multilinear, and therefore sequentially continuous, from* $\mathcal{L}^I(E_1 \times \cdots \times E_I; F) \times E_1 \times \cdots \times E_I$ *into* F. ∎

We shall make use of the following **extension theorem** [Vol. 1, Theorem 10.3].

Proposition 12.9. — *Let* T *be a uniformly continuous mapping from* U *into a Neumann space* F*, where* U *is sequentially dense in a subset* X *of a separated semi-normed space* E.

(a) *For every* $u \in X$*, there exists a unique* $\overline{T}(u) \in F$ *such that* $\overline{T}(u) = \lim T(u_n)$*, where* $(u_n)_{n\in\mathbb{N}}$ *is any sequence in* U *such that* $u_n \to u$ *in* E.

(b) *The mapping* $\overline{T}$ *is the unique sequentially continuous extension of* T *to* X*. It is continuous and uniformly continuous.* ∎

Let us finally turn to the definition of the image under a multilinear mapping of $L^p(\Omega; E_i)$ measures. The definition becomes more complicated for $L^\infty(\Omega; E_i)$ measures, since they cannot be approximated in $L^\infty(\Omega; E_i)$ using continuous functions.

Definition 12.10. — *Let*

$$T \in \mathcal{L}^I(E_1 \times \cdots \times E_I; F),$$

$$f_i \in L^{p_i}(\Omega; E_i), \quad 1 \leq p_i \leq \infty, \quad \frac{1}{p_1} + \cdots + \frac{1}{p_I} = \frac{1}{p}, \quad 1 \leq p \leq \infty,$$

where $E_1, \ldots, E_I$*, and* F *are Neumann spaces, and* Ω *is an open subset of* $\mathbb{R}^d$.

We define

$$T(f_1, \ldots, f_I) \in L^p(\Omega; F)$$

by

$$T(f_1, \ldots, f_I) \stackrel{\text{def}}{=} \lim_{n\to\infty}\big(\lim_{m\to\infty} T(f_1^{n,m}, \ldots, f_I^{n,m})\big),$$

where

$$f_i^{n,m} \stackrel{\text{def}}{=} \begin{cases} f_i^n & \text{if } p_i < \infty, \\ f_i^m & \text{if } p_i = \infty, \end{cases}$$

and $(f_i^n)_{n\in\mathbb{N}}$ *and* $(f_i^m)_{m\in\mathbb{N}}$ *are any* p_i*-integrabilizing sequences (Definition 10.1) for* f_i*. That is,* f_i^n *and* f_i^m *belong to* $\mathcal{B}(\Omega; E_i)$*, and*

$$f_i^n \to f_i \ \text{ in } L^{p_i}(\Omega; E_i),$$

$$f_i^m \to f_i \ \text{ in } L^1_{\text{loc}}(\Omega; E_i), \quad (f_i^m)_{m\in\mathbb{N}} \text{ is bounded in } \mathcal{C}_b(\Omega; E_i),$$

and $\big(T(f_1^{n,m}, \ldots, f_I^{n,m})\big)(x) \stackrel{\text{def}}{=} T\big(f_1^{n,m}(x), \ldots, f_I^{n,m}(x)\big)$*, for every* $x \in \Omega$*.* ▮

This definition is justified by the following result.

Lemma 12.11. — *In Definition 12.10:*

— $T(f_1^{n,m}, \ldots, f_I^{n,m})$ *has a limit relative to* m *in* $L^1_{\text{loc}}(\Omega;F)$*, and this limit also belongs to* $L^\infty(\Omega; F)$*;*

— *this limit relative to* m *also has a limit relative to* n *that belongs to* $L^p(\Omega; F)$*;*

— *neither limit depends on the choice of* f_i^n *and* f_i^m*, nor on that of* p *and* p_i *(when several suitable values exist).* ▮

Proof. ***1o) Preliminary.*** To simplify notation, endow (Proposition A.23) each E_i with a filtering family of semi-norms $\{\| \ \|_{E_i;\nu_i} : \nu_i \in \mathcal{N}_{E_i}\}$ that generate its topology. This does not affect (Theorem 12.4) either $L^p(\Omega; E_i)$ nor $L^1_{\text{loc}}(\Omega; E_i)$. By Definition 1.6 of a filtering family, for every finite subset $N_i \subset \mathcal{N}_{E_i}$, there exists $\nu_i \in \mathcal{N}_{E_i}$ such that

$$\sup_{\nu\in N_i} \|u_i\|_{E_i;\nu} \le \|u_i\|_{E_i;\nu_i}.$$

The characterization of continuous multilinear mappings from Proposition 12.6 then implies that, for every semi-norm $\| \ \|_{F;\mu}$ of F, there exist $\nu_i \in \mathcal{N}_{E_i}$ and $c \ge 0$ such that

$$\|T(u_1, \ldots, u_I)\|_{F;\mu} \le c\|u_1\|_{E_1;\nu_1} \cdots \|u_I\|_{E_I;\nu_I}. \tag{12.8}$$

2o) Case where each p_i ***is finite.*** In this case, $p < \infty$, and the limit

$$T(f_1, \ldots, f_I) \stackrel{\text{def}}{=} \lim_{n\to\infty} T(f_1^n, \ldots, f_I^n) \tag{12.9}$$

exists in $L^p(\Omega; F)$ independently of the choices for $(f_i^n)_{n\in\mathbb{N}}$ by Proposition 12.9 with respect to the sets $U = \mathcal{B}(\Omega; E_1) \times \cdots \times \mathcal{B}(\Omega; E_I)$, $F = L^p(\Omega; F)$, and $X = L^{p_1}(\Omega; E_1) \times \cdots \times L^{p_I}(\Omega; E_I)$. Indeed:

— $L^p(\Omega; F)$ is a Neumann space (Theorem 10.16);

— $\mathcal{B}(\Omega; E_1)\times\cdots\times\mathcal{B}(\Omega; E_I)$ is sequentially dense in $L^{p_1}(\Omega; E_1)\times\cdots\times L^{p_I}(\Omega; E_I)$, because each $\mathcal{B}(\Omega; E_i)$ is sequentially dense in $L^{p_i}(\Omega; E_i)$ (Theorem 10.14);

— the mapping $(f_1,\dots,f_I)\mapsto T(f_1,\dots,f_I)$ is continuous multilinear from the space $\mathcal{B}(\Omega;E_1)\times\cdots\times\mathcal{B}(\Omega;E_I)$ endowed with the topology of $L^{p_1}(\Omega; E_1)\times\cdots\times L^{p_I}(\Omega; E_I)$ into $L^p(\Omega; F)$ since, for every $f_i \in \mathcal{B}(\Omega; E_i)$, inequality (12.8) and Hölder's inequality for continuous functions (Proposition 9.6) imply

$$\Big(\int_\Omega \|T(f_1,\dots,f_I)\|^p_{F;\mu}\Big)^{1/p} \le c\Big(\int_\Omega (\|f_1\|_{E_1;\nu_1}\cdots\|f_I\|_{E_I;\nu_I})^p\Big)^{1/p} \le$$

$$\le c\Big(\int_\Omega \|f_1\|^{p_1}_{E_1;\nu_1}\Big)^{1/p_1}\cdots\Big(\int_\Omega \|f_I\|^{p_I}_{E_I;\nu_I}\Big)^{1/p_I};$$

i.e. using the integral expression for the $L^p(\Omega; E)$ semi-norms of functions in $\mathcal{B}(\Omega; E)$ (Proposition 10.7),

$$\|T(f_1,\dots,f_I)\|_{L^p(\Omega;F);\mu} \le c\|f_1\|_{L^{p_1}(\Omega;E_1);\nu_1}\cdots\|f_I\|_{L^{p_I}(\Omega;E_I);\nu_I}; \qquad (12.10)$$

— the mapping $(f_1,\dots,f_I)\mapsto T(f_1,\dots,f_I)$ is therefore uniformly continuous on any bounded subset of X (Proposition A.59).

3°) Case where each p_i is infinite. When the $(f_i^m)_{m\in\mathbb{N}}$ are bounded in $\mathcal{C}_b(\Omega; E)$, the hypothesis $f_i^m \to f_i$ in $L^1_{\text{loc}}(\Omega; E_i)$ shows, after applying Theorem 11.12 (b) twice, that, for every open set $\omega \Subset \Omega$, $(f_i^m)_{m\in\mathbb{N}}$ is an ∞-integrabilizing sequence for f. Thus,

$$f_i^m \to f_i \text{ in } L^I(\omega; E_i).$$

The case where each p_i is finite (and equal to I) then shows that the limit

$$T(f_1,\dots,f_I) \overset{\text{def}}{=} \lim_{m\to\infty} T(f_1^m,\dots,f_I^m) \qquad (12.11)$$

exists in $L^1(\omega; F)$ (since $1/I+\cdots+1/I = 1$) and is independent of the choice of the f_i^m.

There exists a unique $T(f_1,\dots,f_I)\in L^1_{\text{loc}}(\Omega; F)$, because the limits relative to the open sets ω and ω' coincide on $\omega\cap\omega'$ by Theorem 11.15 on gluing. The convergence (12.11) therefore takes place (Definition 11.3) in this $L^1_{\text{loc}}(\Omega; F)$ space.

Moreover, by (12.8),

$$\sup_{x\in\Omega}\|T(f_1^m,\dots,f_I^m)(x)\|_{F;\mu} \le c\sup_{x\in\Omega}\|f_1^m(x)\|_{E_1;\nu_1}\cdots\sup_{x\in\Omega}\|f_I^m(x)\|_{E_I;\nu_I}, \qquad (12.12)$$

which is bounded independently of m by hypothesis.

Therefore, by Definition 10.2 of $L^\infty(\Omega; F)$,

$$T(f_1, \dots, f_I) \in L^\infty(\Omega; F).$$

4º) Case where some p_i are finite but not all. To simplify the notation, we can reduce to the case where $p_i < \infty$ for $i \leq I$ and $p_i = \infty$ otherwise, by permuting variables. The definition then reads as

$$T(f_1, \dots, f_I) \stackrel{\text{def}}{=} \lim_{n\to\infty} \big(\lim_{m\to\infty} T(f_1^n, \dots, f_J^n, f_{J+1}^m, \dots, f_I^m)\big).$$

We shall proceed in three steps.

4.1. *Passing to the limit relative to m.* The case where each p_i is infinite shows, since $f_i^n \in \mathcal{B}(\Omega; E_i) \subset L^\infty(\Omega; E_i)$ (Theorem 10.8 (b)), that the limit

$$T(f_1^n, \dots, f_J^n, f_{J+1}, \dots, f_I) \stackrel{\text{def}}{=} \lim_{m\to\infty} T(f_1^n, \dots, f_J^n, f_{J+1}^m, \dots, f_I^m) \qquad (12.13)$$

exists in $L^1_{\text{loc}}(\Omega; F)$, belongs to $L^\infty(\Omega; F)$, and is independent of the choice of the measures $(f_i^m)_{m\in\mathbb{N}}$ for $i \geq J+1$.

4.2. *Property of the limit in m.* By (12.8),

$$\|T(f_1^n, \dots, f_J^n, f_{J+1}^m, \dots, f_I^m)\|_{F;\mu} \leq cP\|f_1^n\|_{E_1;\nu_1} \cdots \|f_J^n\|_{E_J;\nu_J},$$

where

$$P = \prod_{i=J+1}^{I} \sup_{m\in\mathbb{N},\, x\in\Omega} \|f_i^m(x)\|_{E_i;\nu_i},$$

which is finite by hypothesis. For every open set $\omega \Subset \Omega$, Hölder's inequality for continuous functions (Proposition 9.6) states that

$$\Big(\int_\omega \|T(f_1^n, \dots, f_J^n, f_{J+1}^m, \dots, f_I^m)\|^p_{F;\mu}\Big)^{1/p} \leq cP\prod_{i=1}^{J}\Big(\int_\omega \|f_i^n\|^{p_i}_{E_i;\nu_i}\Big)^{1/p_i}.$$

By bounding the integral on the right-hand side by the integral over Ω (Proposition 1.18 (b)), it becomes, again using the integral expression for the $L^p(\Omega; E)$ semi-norms of functions in $\mathcal{B}(\Omega; E)$ (Proposition 10.7),

$$\|T(f_1^n, \dots, f_J^n, f_{J+1}^m, \dots, f_I^m)\|_{L^p(\omega;F);\mu} \leq cP\prod_{i=1}^{J} \|f_i^n\|_{L^{p_i}(\Omega;E_i);\nu_i}.$$

As the convergence (12.13) takes place in $L^p(\omega; F)$ then, according to the characterization of ∞-integrabilizing sequences (Theorem 11.12 (b)), in the limit, we get

$$\|T(f_1^n, \dots, f_J^n, f_{J+1}, \dots, f_I)\|_{L^p(\omega;F);\mu} \leq cP\prod_{i=1}^{J} \|f_i^n\|_{L^{p_i}(\Omega;E_i);\nu_i}.$$

By gluing (Theorem 11.18 for an increasing sequence of ω whose union is Ω), this implies

$$T(f_1^n, \ldots, f_J^n, f_{J+1}, \ldots, f_I) \in L^p(\Omega; F)$$

and

$$\|T(f_1^n, \ldots, f_J^n, f_{J+1}, \ldots, f_I)\|_{L^p(\Omega;F);\mu} \leq cP \prod_{i=1}^{J} \|f_i^n\|_{L^{p_i}(\Omega;E_i);\nu_i}. \qquad (12.14)$$

4.3. *Passing to the limit relative to n.* By using the extension theorem (Proposition 12.9), as was done in the case where all the p_i are finite, this inequality implies that the limit

$$T(f_1, \ldots, f_I) \overset{\text{def}}{=} \lim_{n \to \infty} T(f_1^n, \ldots, f_J^n, f_{J+1}, \ldots, f_I)) \qquad (12.15)$$

exists in $L^p(\Omega; F)$ and is independent on the choice of $(f_i^n)_{m \in \mathbb{N}}$ for $i \leq J$.

5º) *Influence of the choice of p_i.* For a fixed i, suppose that the hypotheses are satisfied for $p_{i1} < p_{i2} < \infty$, and let the two corresponding approximations be

$$f_{i1}^n \to f_i \text{ in } L^{p_{i1}}(\Omega; E_i), \quad f_{i2}^n \to f_i \text{ in } L^{p_{i2}}(\Omega; E_i).$$

Let $\omega \Subset \Omega$ be an open set. Then, $f_{i2}^n \to f_i$ in $L^{p_{i1}}(\omega; E_i)$ (Theorem 11.1 (a) on the dependence on p), thus f_{i1}^n and f_{i2}^n are two approximations for f_i in $L^{p_{i1}}(\omega; E_i)$. The corresponding limits therefore coincide on ω and, futhermore, by gluing (Theorem 11.15), they coincide on the entirety of Ω.

If $p_{i2} = \infty$, we proceed in the same way by observing that the corresponding approximation converges in $L^{p_{i1}}(\omega; E_i)$ by the characterization of ∞-integrabilizing sequences (Theorem 11.12 (b), again). □

Let us give the continuity of the image under T, and **Hölder's general inequality**.

Theorem 12.12. — *Suppose that the hypotheses of Definition 12.10 are satisfied.*

(a) *The mapping $(f_1, \ldots, f_I) \mapsto T(f_1, \ldots, f_I)$ is continuous multilinear, and thus sequentially continuous, from $L^{p_1}(\Omega; E_1) \times \cdots \times L^{p_I}(\Omega; E_I)$ into $L^p(\Omega; F)$.*

(b) *Let $\| \|_{E_i;\nu_i}$ and $\| \|_{F;\mu}$ be semi-norms of E_i and F, and let $N_1, \ldots, N_I$ be the respective finite subsets of $\mathcal{N}_{E_1}, \ldots, \mathcal{N}_{E_I}$, and $c \geq 0$ (given by Proposition 12.6) such that: for every $u_1 \in E_1, \ldots, u_I \in E_I$,*

$$\|T(u_1, \ldots, u_I)\|_{F;\mu} \leq c \sup_{\nu_1 \in N_1} \|u_1\|_{E_1;\nu_1} \cdots \sup_{\nu_I \in N_I} \|u_I\|_{E_I;\nu_I}.$$

Then,

$$\|T(f_1, \ldots, f_I)\|_{L^p(\Omega;F);\mu} \leq c \sup_{\nu_1 \in N_1} \|f_1\|_{L^{p_1}(\Omega;E_1);\nu_1} \cdots \sup_{\nu_I \in N_I} \|f_I\|_{L^{p_I}(\Omega;E_I);\nu_I}. \blacksquare$$

Proof. ***1º)*** *Bound* (b). To simplify the notation, we can reduce it to the case where p_i is finite for $i \leq I$ and infinite otherwise, by permuting variables. Hölder's inequality for continuous functions (Proposition 9.6) therefore reads, for all f_i^n and f_i^m in $\mathcal{B}(\Omega; E_i)$, as

$$\Big(\int_\Omega \|T(f_1^n, \dots, f_I^n, f_{I+1}^m, \dots f_d^m)\|_{F;\mu}^p\Big)^{1/p} \leq$$
$$\leq c \sup_{\nu_1 \in N_1} \Big(\int_\Omega \|f_1^n\|_{E_1;\nu_1}^{p_1}\Big)^{1/p_1} \cdots \Big(\int_\Omega \|f_I^n\|_{E_I;\nu_I}^{p_I}\Big)^{1/p_I} \times$$
$$\times \sup_{x\in\Omega,\, \nu_{I+1} \in \mathcal{N}_{E_{I+1}}} \|f_{I+1}^m(x)\|_{E_{I+1};\nu_{I+1}} \cdots \sup_{x\in\Omega,\, \nu_d \in \mathcal{N}_{E_d}} \|f_{I+1}^m(x)\|_{E_d;\nu_d}.$$

Taking the infimum over all ∞-integrabilizing sequences f_j^m, we obtain (Definitions 10.2 and 12.10)

$$\Big(\int_\Omega \|T(f_1^n, \dots, f_I^n, f_{I+1}, \dots f_d)\|_{F;\mu}^p\Big)^{1/p} \leq$$
$$\leq c \sup_{\nu_1 \in N_1} \Big(\int_\Omega \|f_1^n\|_{E_1;\nu_1}^{p_1}\Big)^{1/p_1} \cdots \Big(\int_\Omega \|f_I^n\|_{E_I;\nu_I}^{p_I}\Big)^{1/p_I} \times$$
$$\times \sup_{\nu_{I+1} \in \mathcal{N}_{E_{I+1}}} \|f_{I+1}\|_{L^\infty(\Omega;E_{I+1});\nu_{I+1}} \cdots \sup_{\nu_d \in \mathcal{N}_{E_d}} \|f_{I+1}\|_{L^\infty(\Omega;E_d);\nu_d}.$$

Lastly, by passing to the limit as $n \to \infty$, we obtain (Definitions 10.2 and 12.10, again) the stated inequality.

2º) *Continuity* (a). The bound implies continuity (Proposition 12.6), and sequential continuity always follows from continuity (Proposition 1.8). □

12.3. Images in Banach and Hilbert spaces

Let us first observe that $\mathcal{L}^I(E_1 \times \cdots \times E_I; F)$ is normable (Definition A.24) provided that the E_i and F are normed [Vol. 1, Theorem 12.20].

Proposition 12.13. — *If $E_1, E_2, \dots, E_I$, and F are normed spaces, then the topology of $\mathcal{L}^I(E_1 \times \cdots \times E_I; F)$ is generated by the norm*

$$|||T|||_{\mathcal{L}^I(E_1\times\cdots\times E_I;F)} \stackrel{\text{def}}{=} \sup_{u_i\in E_i,\, u_i\neq 0,\, 1\leq i\leq I} \frac{\|T(u_1, \dots, u_I)\|_F}{\|u_1\|_{E_1} \cdots \|u_I\|_{E_I}}. \;\blacksquare$$

Now, let us state **Hölder's inequality** [1] for Banach spaces.

---

1 **Historical background of Hölder's inequality.** In 1910, Frigyes RIESZ stated and proved inequality (b) of Theorem 12.15 in the case of two real-valued functions on an interval with $\omega = \Omega$ [63, p. 456]. He named it after Otto HÖLDER, who had established the analogous inequality for finite sums in 1889 [39, p. 44]; it would also appear in his thesis (1884).

Theorem 12.14. — *Let Ω be an open subset of $\mathbb{R}^d$, and let T, E_i, F, and f_i be such that:*

$$T \in \mathcal{L}^I(E_1 \times \cdots \times E_I; F), \quad E_1, \ldots, E_I, \text{ and } F \text{ are Banach spaces,}$$

$$f_i \in L^{p_i}(\Omega; E_i), \quad 1 \le p_i \le \infty, \quad \frac{1}{p_1} + \cdots + \frac{1}{p_I} = \frac{1}{p}, \quad 1 \le p \le \infty.$$

(a) *Then, for the norm of T defined in Proposition 12.13:*

$$\|T(f_1, \ldots, f_I)\|_{L^p(\Omega;F)} \le |||T|||_{\mathcal{L}^I(E_1 \times \cdots \times E_I;F)} \|f_1\|_{L^{p_1}(\Omega;E_1)} \cdots \|f_I\|_{L^{p_I}(\Omega;E_I)};$$

the mapping $(T, f_1, \ldots, f_I) \mapsto T(f_1, \ldots, f_I)$ is continuous multilinear, and sequentially continuous, from $\mathcal{L}^I(E_1 \times \ldots \times E_I; F) \times L^{p_1}(\Omega; E_1) \times \cdots \times L^{p_I}(\Omega; E_I)$ into $L^p(\Omega; F)$.

(b) *If $p = 1$, i.e. if $1/p_1 + \cdots + 1/p_I = 1$, then, for every open set $\omega \subset \Omega$:*

$$\Big\| \int_\omega T(f_1, \ldots, f_I) \Big\|_F \le |||T|||_{\mathcal{L}^I(E_1 \times \cdots \times E_I;F)} \|f_1\|_{L^{p_1}(\Omega;E_1)} \cdots \|f_I\|_{L^{p_I}(\Omega;E_I)};$$

the mapping $(T, f_1, \ldots, f_I) \mapsto \int_\omega T(f_1, \ldots, f_I)$ is continuous multilinear, and sequentially continuous, from $\mathcal{L}^I(E_1 \times \cdots \times E_I; F) \times L^{p_1}(\Omega; E_1) \times \cdots \times L^{p_I}(\Omega; E_I)$ into F. ▮

Proof. ***1º)*** *Property* (a). The inequality from Theorem 12.12 (b) can be written, together with the definition of the norm for T from Proposition 12.8 (a), as

$$\|T(u_1, \ldots, u_I)\|_F \le |||T|||_{\mathcal{L}^d(E_1 \times \cdots \times E_I;F)} \|u_1\|_{E_1} \cdots \|u_I\|_{E_d}.$$

This implies (Proposition 12.8 (b) and (c)) the desired inequality and continuity.

2º) *Property* (b). The norm inequality follows from the inequality in (a) and from the bound on the semi-norms of the integral from Theorem 5.4.

Multilinearity follows from Theorems 12.12 (a) and 5.3. Continuity follows from the norm inequality, according to the characterization of continuous multilinear mappings (Proposition 12.6). Finally, and as always, sequential continuity follows from continuity (Proposition 1.8). □

We now come to Hölder's real-valued inequality.

Theorem 12.15. — *Let*

$$f_i \in L^{p_i}(\Omega), \quad 1 \le p_i \le \infty, \quad \frac{1}{p_1} + \cdots + \frac{1}{p_I} = \frac{1}{p}, \quad 1 \le p \le \infty,$$

where Ω is an open subset of $\mathbb{R}^d$. Then:

(a) *the product is continuous multilinear from* $L^{p_1}(\Omega) \times \cdots \times L^{p_I}(\Omega)$ *into* $L^p(\Omega)$, *and*

$$\|f_1 f_2 \dots f_I\|_{L^p(\Omega)} \leq \|f_1\|_{L^{p_1}(\Omega)} \cdots \|f_I\|_{L^{p_I}(\Omega)};$$

(b) *if* $p = 1$, *i.e. if* $1/p_1 + \cdots + 1/p_I = 1$ *then, for every open subset* $\omega \subset \Omega$,

$$\left| \int_\omega f_1 f_2 \dots f_I \right| \leq \|f_1\|_{L^{p_1}(\Omega)} \cdots \|f_I\|_{L^{p_I}(\Omega)}. \blacksquare$$

Proof. These are simply some particular cases of Theorems 12.12 and 12.14. □

Now we consider the instance where a multilinear mapping varies pointwise.

Theorem 12.16. — *Let*

$$T \in L^{p_0}(\Omega; \mathcal{L}^I(E_1 \times \cdots \times E_I; F)), \quad 1 \leq p_0 \leq \infty, \quad E_1, \dots, E_I \text{ be Banach,}$$

$$f_i \in L^{p_i}(\Omega; E_i)), \quad 1 \leq p_i \leq \infty, \quad \frac{1}{p_0} + \frac{1}{p_1} + \cdots + \frac{1}{p_I} = \frac{1}{p}, \quad 1 \leq p \leq \infty,$$

where Ω *is an open subset of* $\mathbb{R}^d$, *and* F *is a Neumann space.*

Then, $T(f_1, \dots, f_I)$ *belongs to* $L^p(\Omega; F)$ *and, for the semi-norms of* T *defined in Proposition 12.8 (a), for every* $\mu \in \mathcal{N}_F$,

$$\|T(f_1, \dots, f_I)\|_{L^p(\Omega;F);\mu} \leq$$
$$\leq |||T|||_{L^{p_0}(\Omega;\mathcal{L}^I(E_1 \times \cdots \times E_I;F));\mu} \|f_1\|_{L^{p_1}(\Omega;E_1)} \cdots \|f_I\|_{L^{p_I}(\Omega;E_I)}.$$

The mapping $(T, f_1, \dots, f_I) \mapsto T(f_1, \dots, f_I)$ *is continuous multilinear, and sequentially continuous, from* $L^{p_0}(\Omega; \mathcal{L}^I(E_1 \times \cdots \times E_I; F)) \times L^{p_1}(\Omega; E_1) \times \cdots \times L^{p_I}(\Omega; E_I)$ *into* $L^p(\Omega; F)$. ▮

Proof. Let Φ be the mapping $(T, u_1, \dots, u_I) \mapsto T(u_1, \dots, u_I)$, which is continuous multilinear from $\mathcal{L}^I(E_1 \times \cdots \times E_I; F) \times E_1 \times \cdots \times E_I$ into $L^p(\Omega; F)$ (Proposition 12.8) and, for every $\mu \in \mathcal{N}_F$,

$$\|T(u_1, \dots, u_I)\|_{F;\mu} \leq |||T|||_{\mathcal{L}^I(E_1 \times \cdots \times E_I;F);\mu} \|u_1\|_{E_1} \cdots \|u_I\|_{E_I}.$$

Definition 12.10 and Theorem 12.12 applied to Φ then give the stated results. □

Let us show that $L^2(\Omega; E)$ is a Hilbert space provided that E is a Hilbert space.

Theorem 12.17. — *The space $L^2(\Omega; E)$, where E is a Hilbert space and Ω is an open subset of $\mathbb{R}^d$, is a Hilbert space, and*

$$(f, g)_{L^2(\Omega;E)} \stackrel{\text{def}}{=} \int_\Omega (f, g)_E.$$

In particular, $L^2(\Omega)$ is a Hilbert space, and

$$(f, g)_{L^2(\Omega)} \stackrel{\text{def}}{=} \int_\Omega fg. \ \blacksquare$$

Proof. ***1º) First property.*** The product $(\ ,\ )_E$ is (Definition A.30) continuous bilinear from $E \times E$ into $\mathbb{R}$, hence Hölder's inequality (Theorem 12.14 (b) with $\omega = \Omega$ and $p_1 = p_2 = 2$) shows that the bilinear mapping

$$(f, g) \mapsto \int_\Omega (f, g)_E \text{ is continuous from } L^2(\Omega; E) \times L^2(\Omega; E) \text{ into } \mathbb{R}. \tag{12.16}$$

Furthermore, for every f and g in $L^2(\Omega; E)$,

$$\int_\Omega (f, g)_E = \int_\Omega (g, f)_E, \quad \int_\Omega (f, f)_E = \|f\|_{L^2(\Omega;E)}. \tag{12.17}$$

Indeed, these properties are satisfied for f and g in $\mathcal{B}(\Omega; E)$ (the first is because $(f(x), g(x)) = (g(x), f(x))$, and the second is from $(f(x), f(x)) = \|f(x)\|_E^2$ and Proposition 10.7); they extend by continuity (Proposition 11.17 (a)) thanks to (12.16), since $\mathcal{B}(\Omega; E)$ is dense in $L^2(\Omega; E)$ (Theorem 10.14).

The mapping $(f, g) \mapsto \int_\Omega (f, g)_E$ is a scalar product on $L^2(\Omega; E)$ and its norm follows from (12.16) and (12.17). Hence, $L^2(\Omega; E)$ is a pre-Hilbert space. Since it is sequentially complete (Theorem 10.16), it is moreover a Hilbert space (Definition A.30).

2º) Second property. This is just a special case of the first property. □

12.4. Images in local spaces

Let us define the image under a multilinear mapping of $L^p_{\text{loc}}(\Omega; E_i)$ measures.

Theorem 12.18. — *Let*

$$T \in \mathcal{L}^I(E_1 \times \cdots \times E_I; F),$$

$$f_i \in L^{p_i}_{\text{loc}}(\Omega; E_i), \quad 1 \le p_i \le \infty, \quad \frac{1}{p_1} + \cdots + \frac{1}{p_I} = \frac{1}{p}, \quad 1 \le p \le \infty,$$

where $E_1, \ldots, E_I$, and F are Neumann spaces and Ω is an open subset of $\mathbb{R}^d$.

(a) *We define*

$$T(f_1, \ldots, f_I) \in L^p_{\text{loc}}(\Omega; F)$$

by: for every open set $\omega \Subset \Omega$,

$$T(f_1, \ldots, f_I) \stackrel{\text{def}}{=} T(f_1|_\omega, \ldots, f_I|_\omega) \text{ in } \omega.$$

(b) *The mapping* $(f_1, \ldots, f_I) \mapsto T(f_1, \ldots, f_I)$ *is continuous multilinear, and hence sequentially continuous, from* $L^{p_1}_{\text{loc}}(\Omega; E_1) \times \cdots \times L^{p_I}_{\text{loc}}(\Omega; E_I)$ *into* $L^p_{\text{loc}}(\Omega; F)$. ▮

Proof. ***1°)*** *Definition of* $T(f_1, \ldots, f_I)$. We have $f_i \in L^{p_i}(\omega; E_i)$, and therefore (Definition 12.10) $g_\omega = T(f_1|_\omega, \ldots, f_I|_\omega)$ belongs to $L^p(\omega; F)$.

As $g_\omega = g_{\omega'}$ on $\omega \cap \omega'$ (since $f_i|_\omega$ equals $f_i|_{\omega'}$ there), the g_ω admits a unique gluing (Theorem 11.15), denoted $g \in \mathcal{M}(\Omega; F)$. Therefore, it satisfies (Definition 11.3) $g \in L^p_{\text{loc}}(\Omega; F)$.

2°) *Continuity.* Let us endow (Proposition A.23) E_i with a filtering family of semi-norms $\{\| \ \|_{E_i;\nu_i} : \nu_i \in \mathcal{N}_{E_i}\}$ which generates its topology and does not change the space $L^p_{\text{loc}}(\Omega; E_i)$ (Theorem 12.4). Hence, see (12.8), p. 235, for every semi-norm $\| \ \|_{F;\mu}$ of F, there exist $\nu_i \in \mathcal{N}_{E_i}$ and $c \geq 0$ such that

$$\|T(u_1, \ldots, u_I)\|_{F;\mu} \leq c\|u_1\|_{E_1;\nu_1} \cdots \|u_I\|_{E_I;\nu_I}.$$

For every open set $\omega \Subset \Omega$, the inequality from Theorem 12.12 (b) is therefore satisfied on ω and, by Definition 11.3 of $L^p_{\text{loc}}(\Omega; E)$, it can be written as

$$\|T(f_1, \ldots, f_I)\|_{L^p_{\text{loc}}(\Omega;F);\omega,\mu} \leq c\|f_1\|_{L^{p_1}_{\text{loc}}(\omega;E_1);\omega,\nu_1} \cdots \|f_I\|_{L^{p_I}_{\text{loc}}(\omega;E_I);\omega,\nu_I}.$$

Continuity of $(f_1, \ldots, f_I) \mapsto T(f_1, \ldots, f_I)$ then follows from Proposition 12.6.

3°) *Sequential continuity.* As always, sequential continuity follows from continuity (Proposition 1.8). □

Let us show that the value of a locally p-integrable measure on a test function equals the integral of their product.

Theorem 12.19. — *Let*

$$f \in L^p(\Omega; E) \text{ or } f \in L^p_{\text{loc}}(\Omega; E), \quad \varphi \in \mathcal{K}(\Omega),$$

where Ω *is an open subset of* $\mathbb{R}^d$, E *is a Neumann space, and* $1 \leq p \leq \infty$.

Then, $f\varphi \in L^1(\Omega; E)$ *and*

$$\int_\Omega f\varphi = \langle f, \varphi \rangle_{\mathcal{M}(\Omega;E)\times\mathcal{K}(\Omega)}.$$

The mapping $(f,\varphi) \mapsto \int_\Omega f\varphi$ *is sequentially continuous from* $L^p(\Omega; E) \times \mathcal{K}(\Omega)$ *and from* $L^p_{\text{loc}}(\Omega; E) \times \mathcal{K}(\Omega)$ *into* E. ▌

Proof. **1o)** $f\varphi \in L^1(\Omega; E)$. Since $f \in L^1_{\text{loc}}(\Omega; E)$ (Theorems 11.6 and 11.8) and $\varphi \in L^\infty(\Omega)$ (Theorem 10.8), then (Definition 12.18 with the product mapping for T)

$$f\varphi \in L^1_{\text{loc}}(\Omega; E).$$

Let ω be an open set such that

$$\operatorname{supp} \varphi \subset \omega \Subset \Omega. \tag{12.18}$$

Then, $f \in L^1(\omega; E)$ hence (Definition 12.10 with the product mapping for T)

$$f\varphi \in L^1(\omega; E).$$

Moreover, $\varphi = 0$ on the open set $\omega' = \Omega \setminus \operatorname{supp} \varphi$, and so (Definition 12.18) $f\varphi$ vanishes there, whence

$$f\varphi \in L^1(\omega'; E).$$

Since $\Omega = \omega \cap \omega'$, by gluing (Theorem 11.15),

$$f\varphi \in L^1(\Omega; E).$$

2o) *Equality.* Let ω be an open set such that $\operatorname{supp} \varphi \Subset \omega \Subset \Omega$. Then, $f\varphi$ vanishes outside ω hence, as the domain where a function vanishes does not contribute to its integral, and by the definition of the integral (Theorem 6.19 and Definition 5.1),

$$\int_\Omega f\varphi = \int_\omega f\varphi = \int_\omega f|_\omega \varphi|_\omega.$$

Moreover, by the definition of the restriction (Definition 6.16, since $\varphi|_\omega \in \mathcal{K}(\omega)$),

$$\langle f, \varphi \rangle_{\mathcal{M}(\Omega;E)\times\mathcal{K}(\Omega)} = \langle f|_\omega, \varphi|_\omega \rangle_{\mathcal{M}(\omega;E)\times\mathcal{K}(\omega)}.$$

Let $g_n \in \mathcal{B}(\omega; E)$ be a p-integrabilizing sequence for $f|_\omega$ (Definition 10.1). The stated equality is satisfied for g_n (Theorem 3.15) from which, in the limit as $n \to \infty$, we get

$$\int_\omega f|_\omega \varphi|_\omega = \langle f|_\omega, \varphi|_\omega \rangle_{\mathcal{M}(\omega;E)\times\mathcal{K}(\omega)}.$$

Thus,

$$\int_\Omega f\varphi = \langle f, \varphi \rangle_{\mathcal{M}(\Omega;E)\times\mathcal{K}(\Omega)}.$$

3o) *Sequential continuity.* The stated continuity follows from the fact that the mapping $(f,\varphi) \mapsto \langle f, \varphi \rangle$ is sequentially continuous from $\mathcal{M}(\Omega; E) \times \mathcal{K}(\Omega)$ into E (Theorem 3.9), from the inclusion $L^p_{\text{loc}}(\Omega; E) \subsetneq \mathcal{M}(\Omega; E)$ (Theorem 11.6), and from the above equality. □

Chapter 13

Various Operations

First, we consider the image $\|f\|_{E;\nu}$ of f under the semi-norms of E (§ 13.1), and the image $\|f\|^s_{E;\nu}$ of this value raised by some power s (§ 13.2). Then, we define its extension $\widetilde{f}$ by 0 (§ 13.3), and present some properties of step functions (§ 13.4).

Also, if E is dense in F, then we show that $L^p(\Omega; E)$ is dense in $L^p(\Omega; F)$, and is sequentially dense if F is additionally a Fréchet space. Moreover, $L^p(\Omega; E)$ is separable if E is separable, and sequentially separable if E is additionally a Fréchet space (§ 13.5).

Finally, we provide (§ 13.6) some conditions under which a weak limit of measures in $L^\infty(\Omega; E)$ is itself in $L^\infty(\Omega; E)$.

13.1. Image under a semi-norm of E

Let us define the image of a $L^1_{\rm loc}(\Omega; E)$ measure under a semi-norm of E.

Theorem 13.1. — *Let $f \in L^1_{\rm loc}(\Omega; E)$, where Ω is an open subset of $\mathbb{R}^d$, E is a Neumann space, and $\|\ \|_{E;\nu}$ is one of its semi-norms.*

(a) *We define $\|f\|_{E;\nu} \in L^1_{\rm loc}(\Omega)$ by*

$$\|f\|_{E;\nu} \overset{\text{def}}{=} \lim_{n\to\infty} \|f_n\|_{E;\nu} \ \text{ in } L^1_{\rm loc}(\Omega),$$

where $(f_n)_{n\in\mathbb{N}}$ is any sequence in $\mathcal{C}(\Omega; E)$ such that $f_n \to f$ in $L^1_{\rm loc}(\Omega; E)$.

(b) *The mapping $f \mapsto \|f\|_{E;\nu}$ is uniformly continuous, and hence continuous and sequentially continuous, from $L^1_{\rm loc}(\Omega; E)$ into $L^1_{\rm loc}(\Omega)$.* ∎

Proof. The limit exists, does not depend on the choice of the f_n, and defines a uniformly continuous mapping by the extension theorem (Proposition 12.9). Indeed:

— $L^1_{\text{loc}}(\Omega)$ is a Neumann space (Theorem 11.6);

— $\mathcal{C}(\Omega; E)$ is sequentially dense in $L^1_{\text{loc}}(\Omega; E)$ (Theorem 11.11);

— the mapping $f \mapsto \|f\|_{E;\nu}$ is uniformly continuous from $\mathcal{C}(\Omega; E)$ endowed with the topology of $L^1_{\text{loc}}(\Omega; E)$ into $L^1_{\text{loc}}(\Omega)$ since, for every f and g in $\mathcal{C}(\Omega; E)$, and for every open set $\omega \Subset \Omega$, we have

$$\big\|\|f\|_{E;\nu} - \|g\|_{E;\nu}\big\|_{L^1_{\text{loc}}(\Omega);\omega} = \int_\omega \big|\|f\|_{E;\nu} - \|g\|_{E;\nu}\big| \leq$$

$$\leq \int_\omega \|f - g\|_{E;\nu} = \|f - g\|_{L^1_{\text{loc}}(\Omega;E);\omega,\nu}.$$

These equalities follow from the integral expression for the $L^1_{\text{loc}}(\Omega; E)$ semi-norms of a continuous function (Theorem 11.9); the inequality follows from the monotonicity of the real-valued integral (Proposition 1.18 (a)), since (Proposition A.11)

$$\big|\|f(x)\|_{E;\nu} - \|g(x)\|_{E;\nu}\big| \leq \|f(x) - g(x)\|_{E;\nu}. \quad \square$$

Let us state some properties of the image under a semi-norm of E.

Theorem 13.2. — *Let Ω be an open subset of $\mathbb{R}^d$, E be a Neumann space, $\|\ \|_{E;\nu}$ be one of its semi-norms, and $1 \leq p \leq \infty$.*

(a) *If $f \in L^1_{\text{loc}}(\Omega; E)$ and $g \in L^1_{\text{loc}}(\Omega; E)$,*

$$\|f\|_{E;\nu} \geq 0,$$

$$\big|\|f\|_{E;\nu} - \|g\|_{E;\nu}\big| \leq \|f - g\|_{E;\nu}.$$

(b) *For every $f \in L^1_{\text{loc}}(\Omega; E)$ and $\varphi \in \mathcal{K}(\Omega)$ such that $\varphi \geq 0$,*

$$\|\langle f, \varphi\rangle\|_{E;\nu} \leq \langle \|f\|_{E;\nu}, \varphi\rangle.$$

(c) *The mapping $f \mapsto \|f\|_{E;\nu}$ is uniformly continuous, and hence continuous and sequentially continuous, from $L^p(\Omega; E)$ into $L^p(\Omega)$ and from $L^p_{\text{loc}}(\Omega; E)$ into $L^p_{\text{loc}}(\Omega)$.*

(d) *If $f \in L^p(\Omega; E)$,*

$$\|f\|_{L^p(\Omega;E);\nu} = \big\|\|f\|_{E;\nu}\big\|_{L^p(\Omega)}.$$

(e) *If $f \in L^1(\Omega; E)$,*

$$\|f\|_{L^1(\Omega;E);\nu} = \int_\Omega \|f\|_{E;\nu}. \ ■$$

Proof. ***1º)*** *Properties* (a). ***1.a.*** *Second inequality.* Let $(f_n)_{n\in\mathbb{N}}$ and $(g_n)_{n\in\mathbb{N}}$ be sequences in $\mathcal{C}(\Omega; E)$ such that $f_n \to f$ and $g_n \to g$ in $L^1_{\rm loc}(\Omega; E)$. For all $x \in \Omega$ (Proposition A.11),

$$\big| \|f_n(x)\|_{E;\nu} - \|g_n(x)\|_{E;\nu} \big| \le \|f_n(x) - g_n(x)\|_{E;\nu}.$$

For every $\varphi \in \mathcal{K}(\Omega)$ such that $\varphi \ge 0$,

$$\int_\Omega \big| \|f_n\|_{E;\nu} - \|g_n\|_{E;\nu} \big| \varphi \le \int_\Omega \|f_n - g_n\|_{E;\nu} \varphi.$$

In the limit,

$$\int_\Omega \big| \|f\|_{E;\nu} - \|g\|_{E;\nu} \big| \varphi \le \int_\Omega \|f - g\|_{E;\nu} \varphi.$$

This proves (Definition 5.7) that

$$\big| \|f\|_{E;\nu} - \|g\|_{E;\nu} \big| \le \|f - g\|_{E;\nu}. \tag{13.1}$$

1.b. *First inequality.* This follows from the second inequality with $g = 0$.

2º) *Inequality* (b). According to the identification of a continuous function with a measure (Theorem 3.15), the bound for the semi-norms of the integral of a continuous and bounded function (Proposition 1.17), and using Theorem 3.15 again, we have

$$\|\langle f_n, \varphi\rangle\|_{E;\nu} = \Big\| \int_\Omega f_n \varphi \Big\|_{E;\nu} \le \int_\Omega \|f_n\|_{E;\nu}\, \varphi = \langle \|f_n\|_{E;\nu}, \varphi\rangle.$$

Since convergence in $L^1_{\rm loc}(\Omega; E)$ implies that in $\mathcal{M}(\Omega; E)$ (Theorem 11.6), we have $\langle f_n, \varphi\rangle \to \langle f, \varphi\rangle$ as $n \to \infty$; also, we have $\|f_n\|_{E,\nu} \to \|f\|_{E,\nu}$ in $L^1_{\rm loc}(\Omega)$ (Theorem 13.1 (b)), and hence in $\mathcal{M}(\Omega)$ (Theorem 11.6). Whence

$$\|\langle f, \varphi\rangle\|_{E;\nu} \le \langle \|f\|_{E;\nu}, \varphi\rangle.$$

3º) *Uniform continuity* (c) *in* $L^p(\Omega; E)$ *when* $p < \infty$. Now, let $(f_n)_{n\in\mathbb{N}}$ be a sequence in $\mathcal{B}(\Omega; E)$ such that $f_n \to f$ in $L^p(\Omega; E)$. Define

$$\|\dot{f}\|_{E;\nu} \stackrel{\rm def}{=} \lim_{n\to\infty} \|f_n\|_{E;\nu} \ \text{ in } L^p(\Omega). \tag{13.2}$$

The extension theorem (Proposition 12.9) shows that $\|\dot{f}\|_{E;\nu}$ exists and does not depend on the sequence $(f_n)_{n\in\mathbb{N}}$, and that $f \mapsto \|\dot{f}\|_{E;\nu}$ is a uniformly continuous mapping from $L^p(\Omega; E)$ into $L^p(\Omega)$, since:

— $L^p(\Omega; E)$ is a Neumann space (Theorem 10.16);

— $\mathcal{C}(\Omega; E)$ is sequentially dense in $L^p(\Omega; E)$ (Theorem 10.14);

— the mapping $f \mapsto \|\dot{f}\|_{E;\nu}$ is uniformly continuous from $\mathcal{C}(\Omega; E)$ endowed with the topology of $L^p(\Omega; E)$ into $L^p(\Omega)$ since, by the integral expression for the $L^p(\Omega; E)$ semi-norms of the functions in $\mathcal{B}(\Omega; E)$ (Proposition 10.7), for f_n and g_n in $\mathcal{C}(\Omega; E)$,

$$\Big\| \|f_n\|_{E;\nu} - \|g_n\|_{E;\nu} \Big\|_{L^p(\Omega)} = \Big(\int_\Omega \big| \|f_n\|_{E;\nu} - \|g_n\|_{E;\nu} \big|^p \Big)^{1/p} \leq$$
$$\leq \Big(\int_\Omega \|f_n - g_n\|^p_{E;\nu} \Big)^{1/p} = \|f_n - g_n\|_{L^p(\Omega;E);\nu}.$$

The mapping $\|\dot{f}\|_{E;\nu}$ coincides with $\|f\|_{E;\nu}$, since it is its unique sequentially continuous extension with values in $L^1_{\text{loc}}(\Omega)$ (Proposition 12.9) and, *a fortiori*, in $L^p(\Omega)$ (this is included in $L^1_{\text{loc}}(\Omega)$ by Theorems 11.6 and 11.8).

4o) Equality (d) *when* $p < \infty$. Let $(f_n)_{n \in \mathbb{N}}$ be a p-integrabilizing sequence for f. Then (Proposition 10.7, again),

$$\|f_n\|_{L^p(\Omega;E)} = \Big(\int_\Omega \|f_n\|^p_{E;\nu} \Big)^{1/p} = \Big\| \|f_n\|_{E;\nu} \Big\|_{L^p(\Omega)}.$$

In the limit (Definition 10.2 of the norm of $L^p(\Omega; E)$ and (13.2)), we get

$$\|f\|_{L^p(\Omega;E)} = \Big\| \|f\|_{E;\nu} \Big\|_{L^p(\Omega)}.$$

5o) Equality (d) *when* $p = \infty$. Given $(f_n)_{n \in \mathbb{N}}$ an ∞-integrabilizing sequence for f (Definition 10.1), $(\|f_n\|_{E;\nu})_{n \in \mathbb{N}}$ is an ∞-integrabilizing sequence for $\|f\|_{E;\nu}$. Indeed, it satisfies conditions (10.3) and (10.4), and $\|f_n\|_{E;\nu} \to \|f\|_{E;\nu}$ in $\mathcal{M}(\Omega)$ since, for every $\varphi \in \mathcal{K}(\Omega)$,

$$\Big| \int_\Omega \|f_m\|_{E;\nu} \varphi - \|f_n\|_{E;\nu} \varphi \Big| \leq \int_\Omega \|f_m - f_n\|_{E;\nu} (\varphi^+ - \varphi^-),$$

where $\varphi^+(x) = \sup(\varphi(x), 0)$ and $\varphi^-(x) = \inf(\varphi(x), 0)$, which tends to 0 for $m \geq n$ and $n \to \infty$ since φ^+ and φ^- also belong to $\mathcal{K}(\Omega)$; hence $\int_\Omega \|f_n\|_{E;\nu} \varphi$, being a Cauchy sequence, has a limit; which is $\int_\Omega \|f\|_{E;\nu} \varphi$ (as in 3.b).

Thus,

$$\Big\| \|f\|_{E;\nu} \Big\|_{L^\infty(\Omega)} \leq \inf_{(f_n)_{n \in \mathbb{N}}} \sup_{n \in \mathbb{N},\, x \in \Omega} \|f_n(x)\| = \|f\|_{L^\infty(\Omega;E);\nu}.$$

Conversely, if $f \in \mathcal{C}(\Omega; E)$, the expression for the regularizations from Proposition 3.21 gives, since α_n and ρ_n are positive, for all $x \in \Omega$,

$$\|(R_n f)(x)\|_{E;\nu} = \Big\| \int_{\mathring{B}(0,1/n)} \widetilde{\alpha_n f}(x+y) \rho_n(y)\, dy \Big\|_{E;\nu} \leq$$
$$\leq \int_{\mathring{B}(0,1/n)} \widetilde{\alpha_n \|f\|_{E;\nu}}(x+y) \rho_n(y)\, dy = R_n(\|f\|_{E;\nu}).$$

Therefore (Theorem 10.10),

$$\|R_n f\|_{L^\infty(\Omega;E);\nu} = \sup_{x\in\Omega} \|(R_n f)(x)\|_{E;\nu} \leq \|\|R_n f\|_{E;\nu}\|_{L^\infty(\Omega)}.$$

In the limit (Theorem 10.13 (b), since $(\|R_n f\|_{E;\nu})_{n\in\mathbb{N}^*}$ is an ∞-integrabilizing sequence for $\|f\|_{E;\nu}$),

$$\|f\|_{L^\infty(\Omega;E);\nu} \leq \big\|\|f\|_{E;\nu}\big\|_{L^\infty(\Omega)}.$$

6º) Uniform continuity (c) *in* $L^\infty(\Omega;E)$. This follows from the inequality

$$\big\|\|f\|_{E;\nu} - \|g\|_{E;\nu}\big\|_{L^\infty(\Omega)} \leq \big\|\|f-g\|_{E;\nu}\big\|_{L^\infty(\Omega)} = \|f-g\|_{L^\infty(\Omega;E);\nu}.$$

7º) Uniform continuity (c) *in* $L^p_{\text{loc}}(\Omega;E)$. For every f and g in $L^p_{\text{loc}}(\Omega;E)$, and for every open set $\omega \Subset \Omega$, together with inequality (13.1) and equality (d), we have

$$\big\|\|f\|_{E;\nu} - \|g\|_{E;\nu}\big\|_{L^p_{\text{loc}}(\Omega);\omega} = \big\|\|f\|_{E;\nu} - \|g\|_{E;\nu}\big\|_{L^p(\omega)} \leq$$
$$\leq \big\|\|f-g\|_{E;\nu}\big\|_{L^p(\omega)} = \|f-g\|_{L^p(\omega;E);\nu} = \|f-g\|_{L^p_{\text{loc}}(\Omega;E);\omega,\nu}.$$

8º) Continuity (d). Uniform continuity implies continuity (Definition 1.7), which itself implies sequential continuity (Proposition 1.8).

9º) Equality (e). Proposition 10.7 gives, for every 1-integrabilizing sequence for f,

$$\|f_n\|_{L^1(\Omega;E);\nu} = \int_\Omega \|f_n\|_{E;\nu}.$$

In the limit, according to Definition 10.2 of the norm of $L^1(\Omega;E)$ and the continuity of the integral (Theorem 5.5),

$$\|f\|_{L^1(\Omega;E);\nu} = \int_\Omega \|f\|_{E;\nu}. \ \square$$

13.2. Powers

Let us now consider the powers of a locally p-integrable real-valued measure.

Theorem 13.3. — *Let Ω be an open subset of $\mathbb{R}^d$, and*

$$f \in L^p_{\text{loc}}(\Omega), \quad f \geq 0, \quad 1 \leq p \leq \infty, \quad 0 < s < \infty, \quad s \leq p.$$

(a) *We define $f^s \in L^{p/s}_{\text{loc}}(\Omega)$ by*

$$f^s \stackrel{\text{def}}{=} \lim_{n\to\infty} f_n^s,$$

where $(f_n)_{n\in\mathbb{N}}$ is any sequence in $\mathcal{C}^+(\Omega)$ such that $f_n \to f$ in $L^p_{\text{loc}}(\Omega)$.

(b) *For every positive f and g in $L^p(\Omega)$,*

$$\|f^s\|_{L^{p/s}(\Omega)} = \|f\|^s_{L^p(\Omega)}, \tag{13.3}$$

$$\|f^s - g^s\|_{L^{p/s}(\Omega)} \leq \begin{cases} \|f-g\|^s_{L^p(\Omega)} & \text{if } s \leq 1, \\ s\|f-g\|_{L^p(\Omega)}\|f+g\|^{s-1}_{L^p(\Omega)} & \text{if } s \geq 1. \end{cases} \tag{13.4}$$

(c) *The mapping $f \mapsto f^s$ is uniformly continuous, and thus continuous and sequentially continuous, from the subset of positive functions in $L^p_{\text{loc}}(\Omega)$ into $L^{p/s}_{\text{loc}}(\Omega)$, and from $L^p(\Omega)$ into $L^{p/s}(\Omega)$.* ∎

Proof. ***1º) Real-valued inequality.*** For every u and v in $\mathbb{R}^+$, we have

$$|u^s - v^s| \leq \begin{cases} |u-v|^s & \text{if } s \leq 1, \\ s|u-v|(u+v)^{s-1} & \text{if } s \geq 1. \end{cases} \tag{13.5}$$

Indeed, if u and v are zero, then it is immediate. Otherwise, suppose that $u \geq v$ (this is always possible by exchanging u and v if required), and divide through by u. The situation thus reduces to $0 \leq v \leq 1$ and

$$|1 - v^s| \leq \begin{cases} |1-v|^s & \text{if } s \leq 1, \\ s|1-v|(1+v)^{s-1} & \text{if } s \geq 1. \end{cases}$$

The case when $s \leq 1$ follows from $v^s \geq v$ and hence $|1-v^s| \leq |1-v| \leq |1-v|^s$.

When $s \geq 1$, observe that $|1-v^s| \leq s|1-v|$, which is sufficient. Indeed, if $F(v) = 1 - v^s - s(1-v)$, then $F'(v) = -sv^{s-1} + s$ and $F''(v) = -s(s-1)v^{s-2}$; since $F(0) = 1 - s \leq 0$, $F(1) = 0$, and $F''(v) \leq 0$, we obtain $F(v) \leq 0$.

2º) Inequality (13.4) in $\mathcal{B}(\Omega)$. For every positive function f and g in $\mathcal{B}(\Omega)$, we get (13.4) from (13.5): if $s \leq 1$, then it is immediate; if $s \geq 1$, we use Hölder's inequality (Proposition 9.6, since $s \leq p$) with $s/p = 1/p + (s-1)/p$; if $p = \infty$, this follows from (Theorem 10.10) $\|f\|_{L^\infty(\Omega)} = \sup_{x\in\Omega} |f(x)|$.

3º) Continuity in $L^p_{\text{loc}}(\Omega)$ when $p < \infty$. The limit exists, is independent of the choice of f_n, and the mapping $f \mapsto f^s$ is uniformly continuous, continuous, and sequentially continuous on the subset of positive functions in $L^p_{\text{loc}}(\Omega)$ by the extension theorem (Proposition 12.9). Indeed:

— $L^{p/s}_{\text{loc}}(\Omega)$ is a Neumann space (Theorem 11.6);

— $\mathcal{C}^+(\Omega)$ is sequentially dense in the subset of positive functions of $L^p_{\text{loc}}(\Omega)$, since the regularizations for f satisfy $R_n f \in \mathcal{C}(\Omega)$ (Theorem 3.18), $R_n f \geq 0$ (see (5.15), p. 99), and $R_n f \to f$ in $L^p_{\text{loc}}(\Omega)$ (Theorem 11.10);

— the mapping $f \mapsto f^s$ is uniformly continuous from $\mathcal{C}^+(\Omega)$ endowed with the topology of $L^p_{\text{loc}}(\Omega)$ into $L^{p/s}_{\text{loc}}(\Omega)$ since, for every f and g in $\mathcal{C}^+(\Omega)$, and every open set $\omega \Subset \Omega$, we have inequality (13.4) with $\Omega = \omega$.

4º) Continuity in $L^p(\Omega)$ when $p < \infty$. This is the same proof but on Ω.

5º) Inequalities when $p < \infty$. Since the mapping $f \mapsto f^s$ is continuous from the subset of positive functions of $L^p(\Omega)$ into $L^{p/s}(\Omega)$, inequalities (13.3) and (13.4) that are satisfied by the positive functions in $\mathcal{B}(\Omega)$ are also satisfied for those in $L^p(\Omega)$, since the positive functions in $\mathcal{B}(\Omega)$ are dense there (Theorem 10.14).

6º) Continuity and inequalities when $p = \infty$. To pass to the limit in equalities (13.3) and (13.4), which hold for the positive functions in $\mathcal{B}(\Omega)$ by 2º, it suffices to observe that (Theorem 10.13 (b))

$$\|R_n f\|_{L^\infty(\Omega)} \to \|f\|_{L^\infty(\Omega)}.$$

This provides these inequalities for positive functions in $L^\infty(\Omega)$, from which the continuity properties follow. The same reasoning can also be applied to $L^\infty_{\text{loc}}(\Omega)$. □

Let us express the semi-norms of $L^p(\Omega; E)$ in terms of $\|f\|_{E;\nu}$.

Theorem 13.4. — *Let $f \in L^p(\Omega; E)$, where Ω is an open subset of $\mathbb{R}^d$, E is a Neumann space, $1 \leq p < \infty$, and $\|\ \|_{E;\nu}$ is a semi-norm of E.*

Then, $\|f\|^p_{E;\nu} \in L^1(\Omega)$ and

$$\|f\|_{L^p(\Omega;E);\nu} = \Big(\int_\Omega \|f\|^p_{E;\nu}\Big)^{1/p}.$$

If $E = \mathbb{R}$,

$$\|f\|_{L^p(\Omega)} = \Big(\int_\Omega |f|^p\Big)^{1/p}. \blacksquare$$

Proof. *1º) Belonging to $L^1(\Omega)$.* Let $(f_n)_{n\in\mathbb{N}}$ be a p-integrabilizing sequence for f. Using equality (13.2) from 2.a in the proof of Theorem 13.2,

$$\|f\|_{E;\nu} = \lim_{n\to\infty} \|f_n\|_{E;\nu} \text{ in } L^p(\Omega). \tag{13.6}$$

From this it follows, with Theorem 13.3 (c), that

$$\|f\|^p_{E;\nu} = \lim_{n\to\infty} \|f_n\|^p_{E;\nu} \text{ in } L^1(\Omega).$$

2º) Semi-norms. As the integral is sequentially continuous on $L^1(\Omega)$ (Theorem 5.5), we have, together with Definition 10.2,

$$\int_\Omega \|f\|^p_{E;\nu} = \lim_{n\to\infty} \int_\Omega \|f_n\|^p_{E;\nu} = \|f\|^p_{L^p(\Omega;E);\nu}. \quad \square$$

We may use this result to bound the integral.

Theorem 13.5. — *Let $f \in L^1(\Omega; E)$, where Ω is an open subset of $\mathbb{R}^d$, E is a Neumann space, and ω is a measurable subset of Ω. For every semi-norm $\|\ \|_{E;\nu}$ of E,*

$$\Big\| \int_\omega f \Big\|_{E;\nu} \leq \int_\omega \|f\|_{E;\nu}. \quad \blacksquare$$

Proof. The approximations in Definition 5.1 of the integral satisfy (Theorem 1.17)

$$\Big\| \int_{\omega_n} f_n \Big\|_{E;\nu} \leq \int_{\omega_n} \|f_n\|_{E;\nu}.$$

The stated equality is obtained by taking the limit as $n \to \infty$ in Definition 5.1 of the integral. Indeed, $\|f_n\|_{E;\nu}$ is an integrabilizing sequence for $\|f\|_{E;\nu}$ by the characterization given in Theorem 4.10, since it converges in $L^1(\Omega)$ by (13.6). $\square$

13.3. Extensions

Let us show that every measure in $L^p(\omega; E)$ can be extended by 0_E.

Theorem 13.6. — *Let $f \in L^p(\omega; E)$, where ω is an open subset of $\mathbb{R}^d$, E is a Neumann space, $1 \leq p \leq \infty$, and Ω is an open set containing ω.*

(a) *We define the **extension** $\widetilde{f} \in L^p(\Omega; E)$ by: for every $\varphi \in \mathcal{K}(\Omega)$,*

$$\langle \widetilde{f}, \varphi \rangle_\Omega \stackrel{\text{def}}{=} \int_\omega f\varphi. \tag{13.7}$$

(b) *For every $\nu \in \mathcal{N}_E$ (the index set of the semi-norms of E),*

$$\|\widetilde{f}\|_{L^p(\Omega;E);\nu} = \|f\|_{L^p(\omega;E);\nu},$$

and the mapping $f \mapsto \widetilde{f}$ is continuous linear from $L^p(\omega; E)$ into $L^p(\Omega; E)$.

(c) $$\widetilde{f} = f \text{ on } \omega, \quad \widetilde{f} = 0_E \text{ on } \Omega \setminus \overline{\omega}. \quad \blacksquare$$

Proof. **1º)** *Property* (a) *when* $p < \infty$. Let $f \in \mathcal{K}(\omega; E)$. Its extension $\widetilde{f}$ by 0_E belongs to $\mathcal{K}(\Omega; E)$ and, according to the integral expression for the semi-norms of $L^p(\Omega; E)$ for a function in $\mathcal{B}(\Omega; E)$ (Proposition 10.7), and since the domain where a function vanishes does not contribute to its integral (Proposition 1.23),

$$\|\widetilde{f}\|_{L^p(\Omega;E);\nu} = \Big(\int_\Omega \|\widetilde{f}\|_{E;\nu}\Big)^{1/p} = \Big(\int_\omega \|f\|_{E;\nu}\Big)^{1/p} = \|f\|_{L^p(\omega;E);\nu}. \qquad (13.8)$$

From this it follows, together with the extension theorem (Proposition 12.9), that $f \mapsto \widetilde{f}$ has a unique continuous extension from $L^p(\omega; E)$ into $L^p(\Omega; E)$. Indeed:

— $L^p(\Omega; E)$ is a Neumann space (Theorem 10.16);

— $\mathcal{K}(\omega; E)$ is sequentially dense in $L^p(\omega; E)$ (Theorem 10.14);

— the mapping $\widetilde{\ }$ is continuous linear, and thus (Proposition 1.9) uniformly continuous, from the subset $\mathcal{K}(\omega; E)$ of $L^p(\omega; E)$ into $L^p(\Omega; E)$ by (13.8).

For every $f \in \mathcal{K}(\omega; E)$ and $\varphi \in \mathcal{K}(\Omega)$, the identification of f with a measure (Theorem 3.15) gives, since the domain where a function vanishes does not contribute to its integral (Proposition 1.23),

$$\langle \widetilde{f}, \varphi\rangle_\Omega = \int_\Omega \widetilde{f}\varphi = \int_\omega f\varphi.$$

This equality extends by continuity (Proposition 11.17 (b)) to $f \in L^p(\omega; E)$, since these terms depend continuously on f from $L^p(\omega; E)$ into E (for the first term, since $L^p(\Omega; E) \subsetneq \mathcal{M}(\Omega; E)$, see Theorem 10.8 (a)). It therefore indeed defines the measure $\widetilde{f}$ in $L^p(\Omega; E)$.

2º) *Property* (a) *when* $p = \infty$. Here, we shall verify the characterization of Theorem 10.11 (a):

— on the one hand, f belongs to $L^\infty(\omega; E)$, which is included in $L^\infty_{\text{loc}}(\omega; E)$ and hence in $L^1_{\text{loc}}(\omega; E)$ (Theorems 11.6 and 11.8), and the proof presented in 1º then gives

$$\widetilde{f} \in L^1_{\text{loc}}(\Omega; E);$$

— on the other hand, using (13.7), Definition 3.17 of a regularization can be written, for every $x \in \Omega$, as

$$(R_n\widetilde{f})(x) = \langle \widetilde{f}, \alpha_n \tau_x \rho_n\rangle_\Omega = \int_\omega \alpha_n \tau_x \rho_n f$$

and, since (Definition 3.16) $|\alpha_n| \leq 1$ and $\int_{\mathbb{R}^d} |\rho_n| = 1$ for all $n \in \mathbb{N}$, Hölder's inequality (Theorem 12.14 (b)) gives

$$\sup_{x\in\Omega} \|R_n\widetilde{f}(x)\|_{E;\nu} \leq \|f\|_{L^\infty(\omega;E);\nu}\|\alpha_n\|_{L^\infty(\omega)}\int_{\mathbb{R}^d} |\rho_n| \leq \|f\|_{L^\infty(\omega;E);\nu}. \qquad (13.9)$$

Therefore, Theorem 10.11 (a) gives $\widetilde{f} \in L^\infty(\Omega; E)$.

3º) Property (c). From Definition 6.16 of the restriction and the definition in (13.7), for $\varphi \in \mathcal{K}(\omega)$, we get $\langle \widetilde{f}|_\omega, \varphi\rangle_\omega = \int_\omega f\varphi$, from which $\widetilde{f}|_\omega = f$.

For $\varphi \in \mathcal{K}(\Omega \setminus \overline{\omega})$, the definition in (13.7) gives $\widetilde{f}|_{\Omega\setminus\overline{\omega}} = 0_E$.

4º) Equality (b) *when* $p < \infty$. This equality is established in (13.8) for $f \in \mathcal{K}(\omega; E)$. By continuity, it extends (Proposition 11.17 (b)) to $f \in L^p(\omega; E)$, since $\mathcal{K}(\omega; E)$ is sequentially dense in $L^p(\omega; E)$ (Theorem 10.14).

The continuity of the mapping $f \mapsto \widetilde{f}$ follows from the characterization of continuous linear mappings of Proposition 1.9.

5º) Equality (b) *when* $p = \infty$. We can rewrite inequality (13.9) in the form (Theorem 10.10 (b)) $\|R_n\widetilde{f}\|_{L^\infty(\Omega;E);\nu} \leq \|f\|_{L^\infty(\omega;E);\nu}$, from which, when $n \to \infty$ (Theorem 10.13 (b)), we obtain

$$\|\widetilde{f}\|_{L^\infty(\Omega;E);\nu} \leq \|f\|_{L^\infty(\Omega;E);\nu}.$$

The reverse inequality follows from the fact that $f = \widetilde{f}|_\Omega$, along with (c). □

Non-extendable measures. The restriction is defined for any measure (Definition 6.16). By contrast, a (non-integrable) measure is not always extendable. Such is the case for the measure associated with the continuous function $f \in \mathcal{C}(0, 1)$ defined by

$$f(x) = \mathrm{e}^{1/x},$$

as seen in the comment *Example of a non-extendable distribution*, in Volume 3, § 16.7, p. 342. □

13.4. Step measures

Let us define and study the **characteristic measure** 1_Ω of an open set.

Theorem 13.7. — *Let ω and Ω be open subsets of $\mathbb{R}^d$ such that $\omega \subset \Omega$.*

(a) *We define*

$$1_\omega \in L^\infty(\Omega)$$

by: for every $\varphi \in \mathcal{K}(\Omega)$,

$$\langle 1_\omega, \varphi\rangle_\Omega \overset{\text{def}}{=} \int_\omega \varphi.$$

(b) *If* χ_ω *is the function equal to* 1 *on* ω, *and if* $\widetilde{\ }$ *is the extension from Theorem 13.6,*

$$1_\omega = \widetilde{\chi_\omega}.$$

(c) *If* $\omega \neq \emptyset$,

$$\|1_\omega\|_{L^\infty(\Omega)} = 1.$$

(d) *If* $|\omega| < \infty$ *and* $1 \leq p < \infty$, *then* $1_\omega \in L^p(\Omega)$, *and*

$$\|1_\omega\|_{L^p(\Omega)} = |\omega|^{1/p}. \ ▮$$

Proof. ***1º)*** *Properties* (a) *and* (b). The function χ_ω belongs to $L^\infty(\omega)$ by Theorem 10.10. It therefore has an extension $\widetilde{\chi_\omega} \in L^\infty(\Omega)$ (Theorem 13.6 (a)), namely 1_ω.

2º) *Property* (c). We have ((b), Theorems 13.6 (b) and 10.10)

$$\|1_\omega\|_{L^\infty(\Omega)} = \|\widetilde{\chi_\omega}\|_{L^\infty(\Omega)} = \|\chi_\omega\|_{L^\infty(\omega)} = 1.$$

3º) *Property* (d). When $|\omega| < \infty$, we have $1_\omega \in L^p(\omega)$ since $L^\infty(\omega) \subset L^p(\omega)$ (Theorem 11.1 (a)). Furthermore ((b), Theorem 13.6 (b), Proposition 10.7 and Proposition 1.25),

$$\|1_\omega\|_{L^p(\Omega)} = \|\widetilde{\chi_\omega}\|_{L^p(\Omega)} = \|\chi_\omega\|_{L^p(\omega)} = \Big(\int_\omega 1\Big)^{1/p} = |\omega|^{1/p}. \ \square$$

Now let us define and study finite **piecewise constant measures.**

Theorem 13.8. — *Let*

$$f \stackrel{\text{def}}{=} \sum_{1\leq i\leq I} 1_{\omega_i} e_i,$$

where $\omega_1, \ldots, \omega_I$, *and* Ω *are open subsets of* $\mathbb{R}^d$ *such that, for all* i *and* j *in* $[\![1, I]\!]$, $i \neq j$,

$$\omega_i \subset \Omega, \quad \omega_i \text{ is bounded and non-empty}, \quad \omega_i \cap \omega_j = \emptyset,$$

$e_1, \ldots, e_I$ *are elements of a separated semi-normed space* E, *and* $1 \leq p < \infty$.

Then, $f \in L^p(\Omega; E)$ *and, for every* $\nu \in \mathcal{N}_E$ *(the index set for the semi-norms of* E*),*

$$\|f\|_{L^p(\Omega;E);\nu} = \begin{cases} \Big(\sum_{1\leq i\leq I} |\omega_i| \|e_i\|^p_{E;\nu}\Big)^{1/p} & \text{if } p < \infty, \\ \sup_{1\leq i\leq I} \|e_i\|_{E;\nu} & \text{if } p = \infty, \end{cases}$$

and

$$\int_\Omega f = \sum_{1\leq i\leq I} |\omega_i|\, e_i. \ ▮$$

Proof. ***1º) Belonging to*** $L^p(\Omega; E)$. The measure $1_{\omega_i} e_i$ is the image (Definition 6.9) of $1_{\omega_i} \in L^p(\Omega; \mathbb{R})$ (Theorem 13.7 (d)) under the continuous linear mapping $t \mapsto te_i$ from $\mathbb{R}$ into E. Hence it belongs to $L^p(\Omega; E)$ (Theorem 12.1 (a)), as does their sum f.

2º) Semi-norm when $p < \infty$. Since the domain where a function vanishes does not contribute to its integral, and by the additivity with respect to the domain (Theorems 6.19 and 11.16 (c)), we get

$$\|f\|^p_{L^p(\Omega;E);\nu} = \|f\|^p_{L^p(\bigcup_i \omega_i;E);\nu} = \sum_{1 \le i \le I} \|1_{\omega_i} e_i\|^p_{L^p(\omega_i;E);\nu}.$$

Using the image under the mapping $t \mapsto te_i$ (Theorem 12.1) and the calculation of the $L^p(\Omega)$ norm for 1_ω (Theorem 13.7 (d)), it becomes

$$\|f\|^p_{L^p(\Omega;E);\nu} = \sum_{1 \le i \le I} \|1_{\omega_i}\|^p_{L^p(\omega_i)} \|e_i\|^p_{E;\nu} = \sum_{1 \le i \le I} |\omega_i| \|e_i\|^p_{E;\nu}.$$

3º) Semi-norm when $p = \infty$. Again, by using Theorem 6.19, the supremum relative to the domains (Theorem 11.16 (d)), the image under the mapping $t \mapsto te_i$ (Theorem 12.1), and the calculation of the $L^\infty(\omega)$ norm for 1_ω (Theorem 13.7 (c)), it becomes

$$\|f\|_{L^\infty(\Omega;E);\nu} = \|f\|_{L^\infty(\bigcup_i \omega_i;E);\nu} = \sup_{1 \le i \le I} \|1_{\omega_i} e_i\|_{L^\infty(\omega_i;E);\nu} = \sup_{1 \le i \le I} \|e_i\|_{E;\nu}.$$

4º) Integral. Using the linearity of the integral (Theorem 5.3), its image under the linear mapping $t \mapsto te_i$ (Theorem 6.13), and the calculation of the $L^1(\Omega)$ norm for 1_ω (Theorem 13.7 (d)), it becomes

$$\int_\Omega f = \sum_{1 \le i \le I} \int_\Omega 1_{\omega_i} e_i = \sum_{1 \le i \le I} \Big(\int_\Omega 1_{\omega_i} \Big) e_i = \sum_{1 \le i \le I} |\omega_i| \, e_i. \quad \square$$

Now, let us construct **piecewise constant approximations**.

Theorem 13.9. — *Let $f \in L^p(\Omega; E)$, where Ω is an open subset of $\mathbb{R}^d$, E is a separated semi-normed space, and $1 \le p < \infty$. Let $n \in \mathbb{N}$, and consider*

$$f_n = \sum_{s \in \mathcal{S}_n} 1_{s,n} f_{s,n}$$

where, denoting by $\Delta_{s,n}$ the closed cube of side length 2^{-n} centered at $2^{-n}s$,

$$\mathcal{S}_n = \{s \in \mathbb{Z}^d : \Delta_{s,n} \subset \Omega \cap B(0,n)\}, \quad 1_{s,n} = 1_{\mathring{\Delta}_{s,n}}, \quad f_{s,n} = 2^{nd} \int_{\mathring{\Delta}_{s,n}} f.$$

Then, $f_n \in L^p(\Omega; E)$ and, for every $\nu \in \mathcal{N}_E$ (the index set for the norms of E),

$$\|f_n\|_{L^p(\Omega;E);\nu} = 2^{-nd}\Big(\sum_{s\in S_n} \|f_{s,n}\|^p_{E;\nu}\Big)^{1/p}, \tag{13.10}$$

$$\|f_n\|_{L^p(\Omega;E);\nu} \leq \|f\|_{L^p(\Omega;E);\nu}, \tag{13.11}$$

and, when $n \to \infty$,

$$f_n \to f \;\; in \; L^p(\Omega; E). \;\blacksquare \tag{13.12}$$

Proof. **1º)** *Belonging to $L^p(\Omega; E)$ and equality (13.10).* This is provided by Theorem 13.8 since $|\mathring{\Delta}_{s,n}| = 2^{-nd}$ (Proposition 2.1) and $f_{s,n} \in E$.

2º) *Inequality (13.11).* Since the domain where a function vanishes does not contribute to its integral, and using the monotonicity relative to the domain (Theorems 6.19 and 11.16 (c)), as well as (Propositions 1.25 and 2.1) $\int_{\mathring{\Delta}_{s}n} 1 = |\mathring{\Delta}_{s,n}| = 2^{-nd}$, we obtain

$$\|f_n\|^p_{L^p(\Omega;E);\nu} = \sum_{s\in S_n} \int_{\mathring{\Delta}_{s,n}} 1_{s,n} \Big\| 2^{nd} \int_{\mathring{\Delta}_{s,n}} f \Big\|^p_{E;\nu} = \sum_{s\in S_n} 2^{nd(p-1)} \Big\| \int_{\mathring{\Delta}_{s,n}} f \Big\|^p_{E;\nu}.$$

Using the bound on the semi-norms of E of the integral by the semi-norms of $L^1(\Omega; E)$, and the dependence of the latter on p (Theorems 5.4 and 11.1(b)), we get

$$\Big\| \int_{\mathring{\Delta}_{s,n}} f \Big\|_{E;\nu} \leq \|f\|_{L^1(\mathring{\Delta}_{s,n};E);\nu} \leq 2^{-nd(1-1/p)} \|f\|_{L^p(\mathring{\Delta}_{s,n};E);\nu}.$$

Again, using Theorems 11.16 (c) and 6.19, now with f, we obtain

$$\|f_n\|^p_{L^p(\Omega;E);\nu} \leq \sum_{s\in S_n} \|f\|^p_{L^p(\mathring{\Delta}_{s,n};E);\nu} = \|f\|^p_{L^p(\bigcup_{s\in S_n} \mathring{\Delta}_{s,n};E);\nu} = \|f\|^p_{L^p(\Omega;E);\nu}.$$

3º) *Convergence (13.12).* The space $\mathcal{K}(\Omega; E)$ is sequentially dense in $L^p(\Omega; E)$ (Theorem 10.14) so, for every $\nu \in \mathcal{N}_E$ and $\epsilon > 0$, there exists $g \in \mathcal{K}(\Omega; E)$ such that

$$\|f - g\|_{L^p(\Omega;E);\nu} \leq \frac{\epsilon}{3}. \tag{13.13}$$

We have the bound

$$\begin{aligned}&\|f - f_n\|_{L^p(\Omega;E);\nu} \leq \\ &\quad \leq \|f - g\|_{L^p(\Omega;E);\nu} + \|g - g_n\|_{L^p(\Omega;E);\nu} + \|g_n - f_n\|_{L^p(\Omega;E);\nu}.\end{aligned} \tag{13.14}$$

Inequality (13.11) then gives

$$\|g_n - f_n\|_{L^p(\Omega;E);\nu} \leq \|g - f\|_{L^p(\Omega;E);\nu} \leq \frac{\epsilon}{3}. \tag{13.15}$$

On the other hand, since the domain where a function vanishes does not contribute to its integral, with the dependence on p, and the calculation of the $L^\infty(\Omega; E)$ norm (Theorems 6.19, 11.1 and 10.10), we have

$$\|g - g_n\|_{L^p(\Omega;E);\nu} \leq |\operatorname{supp} g + B(0,1)|^{1/p} \|g - g_n\|_{L^\infty(\Omega;E)} \leq$$
$$\leq |\operatorname{supp} g + B(0,1)|^{1/p} \sup_{|x-y| \leq \sqrt{d}2^{-nd}} \|g(x) - g(y)\|_{E;\nu}.$$

Finally, for n sufficiently large, this is less than $\epsilon/3$ since g is uniformly continuous so (13.14) gives, together with (13.13) and (13.15),

$$\|f - f_n\|_{L^p(\Omega;E);\nu} \leq \epsilon,$$

which proves (13.12). □

13.5. Density and separability

Let us present some results on the density of $L^p(\Omega; E)$ in $L^p(\Omega; F)$.

Theorem 13.10. — *Let E and F be Neumann spaces such that $E \subsetneq F$, Ω be an open subset of $\mathbb{R}^d$, and $1 \leq p < \infty$.*

(a) *If E is dense in F, then $L^p(\Omega; E)$ is dense in $L^p(\Omega; F)$.*

(b) *If E is sequentially dense (and thus dense) in a Fréchet space F, then $L^p(\Omega; E)$ is sequentially dense in $L^p(\Omega; F)$.* ▌

Proof. **1°)** *Density.* Suppose that E is dense in F, let $\{\| \|_{F;\mu} : \mu \in \mathcal{N}_F\}$ be the semi-norms of F, and $f \in L^p(\Omega; F)$.

The piecewise constant approximations $f_n = \sum_{s\in\mathcal{S}_n} 1_{s,n} f_{s,n}$ converge to f by (13.12), thus, by denoting the closure in $L^p(\Omega; F)$ by $^-$,

$$\overline{(f_n)_{n\in\mathbb{N}}} \ni f.$$

For each n, there is a finite number of values $f_{s,n} \in F$, hence there exist $e_{s,n} \in E$ such that $g \overset{\text{def}}{=} \sum_{s\in\mathcal{S}_n} 1_{s,n} e_{s,n}$ satisfies (see (13.10)), for every $\mu \in \mathcal{N}_F$ (the index set for the semi-norms of F) and $\epsilon > 0$,

$$\|g - f_n\|_{L^p(\Omega;F);\mu} = 2^{-nd}\Big(\sum_{s\in\mathcal{S}_n} \|e_{s,n} - f_{s,n}\|^p_{F;\mu}\Big)^{1/p} \leq \epsilon.$$

Let

$$\mathcal{F} \stackrel{\text{def}}{=} \Big\{ g = \sum_{s \in \mathcal{S}_n} 1_{s,n} e_{s,n} : e_{s,n} \in E,\ s \in \mathcal{S}_n,\ n \in \mathbb{N} \Big\}. \tag{13.16}$$

We have $\overline{\mathcal{F}} \supset (f_n)_{n\in\mathbb{N}}$, so $\overline{\mathcal{F}}$ contains $\overline{(f_n)_{n\in\mathbb{N}}}$, which itself contains f. Therefore, $L^p(\Omega; E)$ is dense in $L^p(\Omega; F)$.

2º) Sequential density. If F is a Fréchet space, then $L^p(\Omega; E)$ is also a Fréchet space (Theorems 10.17) so, in this case, density is equivalent to sequential density (Proposition A.25 (b)). Property (b) therefore follows from Property (a). □

Let us define separable sets.

Definition 13.11. — (a) *A separated semi-normed space E is said to be* ***separable*** *if it contains a countable dense subset V. That is, if $\overline{V} = E$.*

(b) *We say that E is* ***sequentially separable*** *if it contains a countable sequentially dense subset V. That is, if every $u \in E$ is the limit of a sequence in V.* ▮

Let us present some results on the separability of $L^p(\Omega; E)$.

Theorem 13.12. — *Let E be a Neumann space and Ω be an open subset of $\mathbb{R}^d$.*

(a) *If E is separable and $1 \le p < \infty$, then $L^p(\Omega; E)$ is separable.*

(b) *If E is a sequentially separable Fréchet space and $1 \le p < \infty$, then $L^p(\Omega; E)$ is sequentially separable.* ▮

Proof. 1º) Separability. Let X be a countable dense subset of E and $f \in L^p(\Omega; E)$.

The piecewise constant approximations $f_n = \sum_{s\in\mathcal{S}_n} 1_{s,n} f_{s,n}$ converge to f by (13.12), thus, by denoting the closure in $L^p(\Omega; E)$ by $\overline{}$,

$$\overline{(f_n)_{n\in\mathbb{N}}} \ni f.$$

For each n, there is a finite number of values $f_{s,n} \in E$, so there exist $e_{s,n} \in X$ such that $g \stackrel{\text{def}}{=} \sum_{s\in\mathcal{S}_n} 1_{s,n} e_{s,n}$ satisfies, for every semi-norm $\| \; \|_{E;\nu}$ of E and $\epsilon > 0$,

$$\|g - f_n\|_{L^p(\Omega;E);\nu} = 2^{-nd} \Big(\sum_{s\in\mathcal{S}_n} \|e_{s,n} - f_{s,n}\|^p_{E;\nu} \Big)^{1/p} \le \epsilon.$$

Let $\mathcal{F} \stackrel{\text{def}}{=} \{g = \sum_{s \in \mathcal{S}_n} 1_{s,n} e_{s,n} : e_{s,n} \in X,\ s \in \mathcal{S}_n,\ n \in \mathbb{N}\}$. We have $\overline{\mathcal{F}} \supset (f_n)_{n\in\mathbb{N}}$, so $\overline{\mathcal{F}}$ contains $\overline{(f_n)_{n\in\mathbb{N}}}$, which itself contains f. Therefore, $\mathcal{F}$ is dense in $L^p(\Omega; E)$.

Being a countable union of finite sets, $\mathcal{F}$ is countable (Proposition A.2 (d)). Therefore, $L^p(\Omega; E)$ is separable.

*2*º*) Sequential separability.* If E is a Fréchet space, then $L^p(\Omega; E)$ (Theorem 10.17) is also a Fréchet space so, in this case, separability is equivalent to sequential separability (Proposition A.25 (c)). Property (b) then follows from (a). □

Observe that $L^\infty(\Omega; E)$ is not a separable space.

Theorem 13.13. — *The space $L^\infty(\Omega; E)$ is not separable if Ω is a non-empty open subset of $\mathbb{R}^d$, and E is a Neumann space that cannot be reduced to 0_E.*

In particular, $L^\infty(\Omega)$ is not separable if Ω is a non-empty open set. ▌

The proof is based upon the following lemma.

Lemma 13.14. — *A semi-normed space Y is not separable if it contains an uncountable family $\{\mathcal{O}_i\}_{i\in I}$ of non-empty pairwise disjoint open subsets.* ▌

Proof. Suppose that Y contains an uncountable family $\{\mathcal{O}_i\}_{i\in I}$ of non-empty pairwise disjoint open subsets, and that there exists a dense sequence $(y_n)_{n\in\mathbb{N}}$ in Y.

For each $i \in I$, the set $\mathcal{O}_i \cap (y_n)_{n\in\mathbb{N}}$ would be non-empty, and we could choose $n(i)$ such that $y_{n(i)} \in \mathcal{O}_i$. The mapping $i \to n(i)$ would be injective since, otherwise, if $n(i) = n(j)$, then $i = j$ as the $\mathcal{O}_i$ are pairwise disjoint. Thus, I would be countable.

Hence, there is no dense sequence in Y: it is not separable. □

Proof of Theorem 13.13. Let e be a non-zero point in E and, for every $x \in \Omega$, let $r_x > 0$ such that $B(x, 2r_x) \subset \Omega$, and

$$\mathcal{O}_x = \Big\{ f \in L^\infty(\Omega; E) : \|f - 1_{B(x,r_x)} e\|_{L^\infty(\Omega;E)} < \frac{1}{2} \Big\}.$$

The family $\{\mathcal{O}_x\}_{x\in\Omega}$ is not countable since it is indexed by $x \in \Omega$ (Proposition A.3). It is composed of non-empty open sets that are pairwise disjoint since, if $x \neq y$, then $B(x, r_x)$ is distinct from $B(y, r_y)$, so there exists at least one ball belonging to one but not the other. Hence, on this ball, all the f on one subset are of the form te with $t < 1/2$, and all the f on the other subset are of the form te with $t > 1/2$.

The $\mathcal{O}_x$ form an uncountable family of pairwise disjoint open subsets contained in $L^\infty(\Omega; E)$. Therefore, this space is not separable by Lemma 13.14. □

13.6. Limit of a bounded sequence in $L^\infty(\Omega; E)$

Let us show that the weak limit of a bounded sequence of measures in $L^\infty(\Omega; E)$ itself belongs to $L^\infty(\Omega; E)$, provided it is locally integrable[1].

Theorem 13.15. — *Let f and $(f_n)_{n\in\mathbb{N}}$ be such that, given an open subset Ω of $\mathbb{R}^d$, E a Neumann space, $\mathcal{N}_E$ the index set for its semi-norms, and $(c_\nu)_{\nu\in\mathcal{N}_E} \subset \mathbb{R}$:*

$$f_n \in L^\infty(\Omega; E), \quad \|f_n\|_{L^\infty(\Omega;E);\nu} \leq c_\nu, \ \forall n \in \mathbb{N}, \ \forall \nu \in \mathcal{N}_E,$$

$$f_n \to f \ \text{ in } \mathcal{M}(\Omega; E)\text{-weak}, \tag{13.17}$$

$$f \in L^1_{\text{loc}}(\Omega; E). \tag{13.18}$$

Then,

$$f \in L^\infty(\Omega; E)$$

and, for every $\nu \in \mathcal{N}_E$,

$$\|f\|_{L^\infty(\Omega;E);\nu} \leq c_\nu. \ ∎$$

*Proof. **1o)** Belonging to* $L^\infty(\Omega; E)$. Let $m \in \mathbb{N}$ and $x \in \Omega$. The regularizations $R_m f_n$ (Definition 3.17) satisfy, according to the expression for the $L^\infty(\Omega; E)$ semi-norms of a continuous function and their bound (Theorems 10.10 (b) and 10.13 (a)), for all $n \in \mathbb{N}^*$,

$$\|(R_m f_n)(x)\|_{E;\nu} \leq \|R_m f_n\|_{L^\infty(\Omega;E);\nu} \leq \|f_n\|_{L^\infty(\Omega;E);\nu} \leq c_\nu.$$

Since the mapping $f \mapsto R_m f(x)$ is continuous linear from $\mathcal{M}(\Omega; E)$ into E (Theorem 3.20), by the Banach–Dieudonné theorem (Proposition A.69), it is continuous from $\mathcal{M}(\Omega; E)$-weak into E-weak (Definition A.67). It is *a fortiori* sequentially continuous (Proposition 1.8), so

$$R_m f_n(x) \to R_m f(x) \ \text{ in } E\text{-weak}.$$

From which, together with the bound on the weak limits from Proposition A.71, we get

$$\|(R_m f)(x)\|_{E;\nu} \leq c_\nu. \tag{13.19}$$

Since $f \in L^1_{\text{loc}}(\Omega; E)$, by the characterization of $L^\infty(\Omega; E)$ from Theorem 10.11 (a),

$$f \in L^\infty(\Omega; E).$$

1 **Historical background of Theorem 13.15.** Jacques SIMON established this result for a Banach space E and when $d = 1$ in 2007 [70, lemma 15, p. 54]. He established the general result in 2009 [71, théorème 5.10].

If E is a separable reflexive Banach space, or more generally the dual of a separable Banach space, then this is a classical consequence of the Banach–Bourbaki theorem, see the supplement that follows the proof.

2º) Bound on the semi-norms. According to Theorem 10.13 (a) and (b) (for the first equality), Theorem 10.10 (b) (for the second), and (13.19),

$$\|f\|_{L^\infty(\Omega;E);\nu} = \sup_{m\in\mathbb{N}^*} \|R_m f\|_{L^\infty(\Omega;E);\nu} = \sup_{m\in\mathbb{N}^*,\, x\in\Omega} \|(R_m f)(x)\|_{E;\nu} \leq c_\nu. \ \square$$

Let us state a particular case of Theorem 13.15.

Theorem 13.16. — *Properties (13.17) and (13.18) are satisfied if there exists p such that* $1 \leq p \leq \infty$, *and*

$$f_n \to f \text{ in } L^p(\Omega; E)\text{-weak.} \ \blacksquare$$

Proof. Property (13.17) is satisfied since $L^p(\Omega; E) \subsetneq \mathcal{M}(\Omega; E)$ (Theorem 10.8 (a)), from which (Proposition A.70) we get $L^p(\Omega; E)$-weak $\subsetneq \mathcal{M}(\Omega; E)$-weak.

Property (13.18) follows from (Definition A.67 and Theorems 11.6 and 11.8)

$$L^p(\Omega; E)\text{-weak} = L^p(\Omega; E) \subset L^p_{\text{loc}}(\Omega; E) \subset L^1_{\text{loc}}(\Omega; E). \ \square$$

Supplement to Theorem 13.15. If E is the dual of a separable Fréchet space F (so, in particular, when E is a separable reflexive Banach space), we can omit the hypothesis $f \in L^1_{\text{loc}}(\Omega; E)$ in Theorem 13.15.

Indeed, $L^\infty(\Omega; E)$ is then isomorphic to the dual of $L^1(\Omega; F)$ which is separable and, therefore, by the Banach–Bourbaki theorem (Proposition A.76):

$$\textit{every bounded sequence in } L^\infty(\Omega; E) \textit{ has a convergent subsequence in } L^\infty(\Omega; E)\text{-}*\text{weak} \quad (13.20)$$

(when $L^\infty(\Omega; E)$ is identified with $(L^1(\Omega; F))'$). In other words, for any bounded sequence $(f_n)_{n\in\mathbb{N}}$ in $L^\infty(\Omega; E)$, there exist $f \in L^\infty(\Omega; E)$ and a subsequence such that

$$\int_\Omega \langle f_n - f, g\rangle_{E\times F} \to 0, \quad \forall g \in L^1(\Omega; F). \ \square$$

Chapter 14

Change of Variable, Weightings

We first consider the image $f \circ T$ of a p-integrable measure f under a regular change of variable T from Λ into Ω (Theorem 14.1).

Then we separate the variables of a p-integrable measure (Theorem 14.2): this is an isomorphism from $L^p(\Omega_1 \times \Omega_2; E)$ onto $L^p(\Omega_1; L^p(\Omega_2; E))$ when p is finite, but not when it is infinite. We also permute the variables (Theorem 14.5): this is an isomorphism from $L^p(\Omega_1; L^q(\Omega_2; E))$ onto $L^q(\Omega_1; L^p(\Omega_2; E))$ when $q = p$, but not when they are distinct.

Finally, we define the weighting $f \diamond \mu$ on the space of measures (Definition 14.6), before showing that it is continuous bilinear from $L^p(\Omega; E) \times L^r_D(\Omega)$ into $L^q(\Omega_D; E)$ (Theorem 14.7) and that, if $\Omega = \mathbb{R}^d$, then the hypothesis on the compact support for μ can be dropped (Theorem 14.8).

14.1. Change of variable

Let us show that the $L^p(\Omega; E)$ spaces are preserved under a regular change of variables.

Theorem 14.1. — *Let $f \in L^p(\Omega; E)$, where Ω is an open subset of $\mathbb{R}^d$, E is a Neumann space, $1 \leq p \leq \infty$, $\mathcal{N}_E$ is the index set for the semi-norms of E, and*

$$T \in \mathcal{C}^1(\Lambda; \Omega), \quad T^{-1} \in \mathcal{C}^1(\Omega; \Lambda) \textit{ and } |\det[\nabla T^{-1}]| \textit{ is bounded.}$$

Then,

$$f \circ T \in L^p(\Lambda; E)$$

and, for every $\nu \in \mathcal{N}_E$ and $c \stackrel{\text{def}}{=} \sup_{x \in \Omega} |\det[\nabla T^{-1}(x)]|$,

$$\|f \circ T\|_{L^p(\Lambda;E);\nu} \leq c^{1/p} \|f\|_{L^p(\Omega;E);\nu} \quad \textit{if } p < \infty,$$
$$\|f \circ T\|_{L^\infty(\Lambda;E);\nu} = \|f\|_{L^\infty(\Omega;E);\nu} \quad \textit{if } p = \infty.$$

The mapping $f \mapsto f \circ T$ is continuous linear, and hence sequentially continuous, from $L^p(\Omega; E)$ into $L^p(\Lambda; E)$. ∎

Proof. **1o)** *Belonging to* $L^p(\Omega; E)$ *when* $p < \infty$. Let $(f_n)_{n\in\mathbb{N}}$ be a p-integrabilizing sequence for f such that (Theorem 10.15) $f_n \in \mathcal{K}(\Omega; E)$; recall (Definition 10.1), when $n \to \infty$, then $f_n \to f$ in $\mathcal{M}(\Omega; E)$ and, for every $\nu \in \mathcal{N}_E$, that

$$\sup_{m\geq n} \Big(\int_\Omega \|f_m - f_n\|_{E;\nu} \Big)^{1/p} \to 0. \tag{14.1}$$

Let us verify that $(f_n \circ T)_{n\in\mathbb{N}}$ is a p-integrabilizing sequence (Definition 10.1) for $f \circ T$:

— on the one hand, $f_n \circ T$ is continuous, being the composition of continuous mappings (Proposition A.39) and vanishes outside $T^{-1}(S_n)$, where S_n is the support of f_n, which is a compact subset of Λ (Proposition A.45 (b)), and thus a bounded set (Proposition A.36 (a));

— on the other hand, since the mapping $f \mapsto f \circ T$ is continuous from $\mathcal{M}(\Omega; E)$ into $\mathcal{M}(\Lambda; E)$ (Theorem 7.13), and hence sequentially continuous (Proposition 1.8), the convergence $f_n \to f$ in $\mathcal{M}(\Omega; E)$ gives

$$f_n \circ T \to f \circ T \ \text{ in } \mathcal{M}(\Lambda; E);$$

— finally, from Theorem 7.9 (b) on the change of variables in a Cauchy integral (since the f_n are continuous), we get

$$\int_\Lambda \|f_m \circ T - f_n \circ T\|^p_{E;\nu} = \int_\Omega \|f_m - f_n\|^p_{E;\nu} |\det[\nabla T^{-1}]| \leq c \int_\Omega \|f_m - f_n\|^p_{E;\nu},$$

which, together with convergence (14.1), implies

$$\sup_{m\geq n} \Big(\int_\Lambda \|f_m \circ T - f_n \circ T\|^p_{E;\nu} \Big)^{1/p} \to 0.$$

Therefore, $(f_n \circ T)_{n\in\mathbb{N}}$ is a p-integrabilizing sequence for $f \circ T$. Thus, by Definition 10.2,

$$f \circ T \in L^p(\Lambda; E).$$

2o) *Semi-norms when* $p < \infty$. By Definition 10.2 of the semi-norms of $L^p(\Omega; E)$,

$$\|f\|_{L^p(\Omega;E);\nu} = \lim_{n\to\infty} \Big(\int_\Omega \|f_n\|_{E;\nu} \Big)^{1/p}.$$

Again, by Theorem 7.9 (b) on the change of variables,

$$\int_\Lambda \|f_n \circ T\|^p_{F;\mu} = \int_\Omega \|f_n\|^p_{E;\nu} |\det[\nabla T^{-1}]| \leq c \int_\Omega \|f_n\|^p_{E;\nu}.$$

Since $(f_n \circ T)_{n\in\mathbb{N}}$ is a p-integrabilizing sequence for $f \circ T$, in the limit, with Definition 10.2 of the semi-norms, we obtain

$$\|f \circ T\|_{L^p(\Lambda;E);\nu} \leq c^{1/p} \|f\|_{L^p(\Omega;E);\nu}.$$

3°) Belonging to $L^\infty(\Omega; E)$. If $f \in L^\infty(\Omega; E)$, condition (14.1) is replaced (Definition 10.1) by

$$\sup_{n\in\mathbb{N},\, x\in\Omega} \|f_n(x)\|_{E;\nu} < \infty$$

and, for every bounded open subset ω of Ω, by

$$\sup_{m\geq n} \int_\omega \|f_m - f_n\|_{E;\nu} \to 0. \tag{14.2}$$

However,

$$\sup_{n\in\mathbb{N},\, y\in\Lambda} \|(f_n \circ T)(y)\|_{E;\nu} = \sup_{n\in\mathbb{N},\, x\in\Omega} \|f_n(x)\|_{E;\nu} < \infty.$$

Furthermore, for every bounded open subset λ of Λ,

$$\int_\lambda \|f_m \circ T - f_n \circ T\|_{E;\nu} = \int_{T(\lambda)} \|f_m - f_n\|_{E;\nu} |\det[\nabla T^{-1}]| \leq c \int_{T(\lambda)} \|f_m - f_n\|_{E;\nu},$$

which tends to 0 by hypothesis (14.2).

Therefore, $(f_n \circ T)_{n\in\mathbb{N}}$ is an ∞-integrabilizing sequence for $f \circ T$ and, by Definition 10.2,

$$f \circ T \in L^\infty(\Lambda; E).$$

4°) Semi-norms when $p = \infty$. Using Definition 10.2, by taking the infimum over the set of ∞-integrabilizing sequences for f, we get

$$\|f\|_{L^\infty(\Omega;E);\nu} = \inf_{(f_n)_{n\in\mathbb{N}}} \sup_{n\in\mathbb{N},\, x\in\Omega} \|f_n(x)\|_{E;\nu}.$$

Thus, by taking the infimum over the set of ∞-integrabilizing sequences $(g_n)_{n\in\mathbb{N}}$ for $f \circ T$, we get

$$\|f \circ T\|_{L^\infty(\Lambda;E);\nu} = \inf_{(g_n)_{n\in\mathbb{N}}} \sup_{n\in\mathbb{N},\, y\in\Lambda} \|g_n(y)\|_{E;\nu} =$$

$$= \inf_{(g_n \circ T^{-1})_{n\in\mathbb{N}}} \sup_{n\in\mathbb{N},\, x\in\Omega} \|(g_n \circ T^{-1})(x)\|_{E;\nu} = \|f\|_{L^\infty(\Omega;E);\nu},$$

since $g_n \circ T^{-1}$ is an ∞-integrabilizing sequence for f, and since, in this way, we obtain every ∞-integrabilizing sequence for f (here, we can omit the boundedness hypothesis on $|\det[\nabla T^{-1}]|$).

5°) Continuity. The continuity of $f \mapsto f \circ T$ follows from the above bounds on the semi-norms, according to the characterization of continuous linear mappings from Proposition 1.9.

6°) Sequential continuity. As always, this follows from continuity (Proposition 1.8).

□

14.2. Regrouping and separation of variables

Let us show that the **regrouping of variables**, and hence also their **separation**, preserves the space $L^p(\Omega; E)$ when p is finite, but that this is not the case when p is infinite[1].

Theorem 14.2. — *Let*

$$f \in L^p(\Omega_1; L^p(\Omega_2; E)),$$

where $1 \leq p \leq \infty$, Ω_1 *is an open subset of* $\mathbb{R}^{d_1}$, Ω_2 *is an open subset of* $\mathbb{R}^{d_2}$, *and* E *is a Neumann space with the semi-norms* $\{\|\ \|_{E;\nu} : \nu \in \mathcal{N}_E\}$.

(a) *Then, there exists a unique*

$$\overline{f} \in L^p(\Omega_1 \times \Omega_2; E)$$

such that, by separating the variables (Definition 8.9), we obtain f.

For every $\nu \in \mathcal{N}_E$,

$$\|\overline{f}\|_{L^p(\Omega_1 \times \Omega_2; E);\nu} = \|f\|_{L^p(\Omega_1; L^p(\Omega_2; E));\nu} \quad \text{if } p < \infty, \tag{14.3}$$

$$\|\overline{f}\|_{L^\infty(\Omega_1 \times \Omega_2; E);\nu} \leq \|f\|_{L^\infty(\Omega_1; L^\infty(\Omega_2; E));\nu} \quad \text{if } p = \infty. \tag{14.4}$$

(b) *If* $1 \leq p < \infty$, *the mapping* $f \mapsto \overline{f}$ *is an isomorphism from* $L^p(\Omega_1; L^p(\Omega_2; E))$ *onto* $L^p(\Omega_1 \times \Omega_2; E)$.

(c) *If* $p = \infty$, *the mapping* $f \mapsto \overline{f}$ *is continuous, linear, and injective from the space* $L^\infty(\Omega_1; L^\infty(\Omega_2; E))$ *into* $L^\infty(\Omega_1 \times \Omega_2; E)$.

If Ω_1 *and* Ω_2 *are non-empty sets and if* E *does not reduce to* 0_E, *then it is not surjective.* ∎

---

1 **Historical background of Theorem 14.2.** Guido FUBINI proved in 1907 [31] an analogous result for $E = \mathbb{R}$ and $p = 1$. More precisely, he showed that, when $f \in L^1(\Omega_1 \times \Omega_2)$, then $x_2 \mapsto f(x_1, x_2)$ belongs to $L^1(\Omega_2)$ for almost every $x_1 \in \Omega_1$ and $\int_{\Omega_1} dx_1 \int_{\Omega_1} dx_2 |f(x_1, x_2)| < \infty$.
Conversely, Leonida TONELLI showed in 1923 [77] an analogous result for surjectivity when $E = \mathbb{R}$ and $p = 1$. More precisely [DUNFORD–SCHWARTZ, 21, p. 194–195], he established a result which implied that, given $f(x_1) \in L^1(\Omega_2)$ for almost every x_1 such that one can define $\int_{\Omega_1} dx_1 \int_{\Omega_2} |f|\, dx_2$ and that it is finite, then $(x_1, x_2) \mapsto f(x_1)(x_2)$ belong to $L^1(\Omega_1 \times \Omega_2)$.
Alain HARAUX and Thierry CAZENAVE established in 1990 the non-surjectivity of the regrouping of variables in the spaces L^∞ by constructing the example used here. They do not appear to have published it.

Proof. ***1°)*** *Separation of variables for finite* p. Consider

$$g \in L^p(\Omega_1 \times \Omega_2; E).$$

Let $(g_n)_{n\in\mathbb{N}}$ be a p-integrabilizing sequence for g (Definition 10.1), *i.e.* g_n belongs to $\mathcal{B}(\Omega_1 \times \Omega_2; E)$ and, when $n \to \infty$, $g_n \to g$ in $\mathcal{M}(\Omega_1 \times \Omega_2; E)$ and, for every $\nu \in \mathcal{N}_E$,

$$\sup_{m\geq n} \int_{\Omega_1\times\Omega_2} \|g_m - g_n\|^p_{E;\nu} \to 0. \tag{14.5}$$

By separating the variables, we obtain (Theorem 8.11) the measures $\underline{g}$ and $\underline{g_n}$ such that

$$\underline{g_n} \to \underline{g} \ \text{ in } \mathcal{M}(\Omega_1; \mathcal{M}(\Omega_2; E)).$$

The function $\|g_n\|^p_{E;\nu}$ is uniformly continuous (being the composition of g_n with $\| \ \|_{E;\nu}$ and $t \mapsto t^p$, which is uniformly continuous on bounded sets, see Propositions A.41, A.44, and A.82) and its support is bounded (since it is included in the support of g_n). By separating its variables, we obtain (Proposition 8.6)

$$\underline{\|g_n\|^p_{E;\nu}} \in \mathbf{C}_{Q_1}(\Omega_1; \mathbf{C}_{Q_2}(\Omega_2)),$$

where Q_1 and Q_2 are compact sets (depending on n) whose product contains the support of g_n, and

$$\int_{\Omega_1\times\Omega_2} \|g_n\|^p_{E;\nu} = \int_{\Omega_1} \int_{\Omega_2} \underline{\|g_n\|^p_{E;\nu}}.$$

Since $\underline{\|g_n\|^p_{E;\nu}} = \|\underline{g_n}\|^p_{E;\nu}$, this gives, together with the integral expression for the $L^p(\Omega; E)$ semi-norms of functions in $\mathcal{B}(\Omega; E)$ (Proposition 10.7) applied twice,

$$\int_{\Omega_1\times\Omega_2} \|g_n\|^p_{E;\nu} = \int_{\Omega_1} \|\underline{g_n}\|^p_{L^p(\Omega_2;E);\nu} = \|\underline{g_n}\|^p_{L^p(\Omega_1;L^p(\Omega_2;E));\nu}. \tag{14.6}$$

With respect to $g_m - g_n$, this equality shows, together with (14.5), that the $\underline{g_n}$ form a Cauchy sequence in $L^p(\Omega_1; L^p(\Omega_2; E))$, which is sequentially complete (Theorem 10.16, applied twice). This sequence therefore has a limit there, which is also the limit in the $\mathcal{M}(\Omega_1; \mathcal{M}(\Omega_2; E))$ space, namely $\underline{g}$. Therefore,

$$\underline{g} \in L^p(\Omega_1; L^p(\Omega_2; E)).$$

2°) *Semi-norm for finite* p. According to (14.6) and Proposition 10.7 again,

$$\|g_n\|_{L^p(\Omega_1\times\Omega_2;E);\nu} = \|\underline{g_n}\|_{L^p(\Omega_1;L^p(\Omega_2;E));\nu}.$$

Whence, in the limit, equality (14.3) holds for all such g that admit a regrouping of variables.

3º) Isomorphism for finite p. The linear mapping $g \mapsto \underline{g}$ is continuous and injective by equality (14.3). Therefore, it remains to verify that it is surjective. To this end, let

$$f \in L^p(\Omega_1; L^p(\Omega_2; E)).$$

Further, let R_n^1 and R_n^2 be the respective regularization operators on Ω_1 and Ω_2 given by Definition 3.17, and

$$f_n = R_n^2 R_n^1 g.$$

As we shall verify in Lemma 14.3 (a) and (c), $f_n \in \mathbf{C}_{K_n^1}(\Omega_1; \mathbf{C}_{K_n^2}(\Omega_2; E))$, where $K_n^i = \overline{(\Omega_i)_{1/n}^n}$ is a compact set included in Ω_i, and

$$f_n \to f \text{ in } L^p(\Omega_1; L^p(\Omega_2; E)). \tag{14.7}$$

By regrouping the variables, we obtain (Proposition 8.7)

$$\overline{f_n} \in \mathbf{C}_{K_n^1 \times K_n^2}(\Omega_1 \times \Omega_2; E).$$

With equality (14.3) and the convergence (14.7),

$$\sup_{m \geq n} \|\overline{f_m} - \overline{f_n}\|_{L^p(\Omega_1 \times \Omega_2; E);\nu} = \sup_{m \geq n} \|f_m - f_n\|_{L^p(\Omega_1; L^p(\Omega_2; E));\nu} \to 0.$$

The sequence $(\overline{f_n})_{n \in \mathbb{N}}$ is therefore Cauchy in $L^p(\Omega_1 \times \Omega_2; E)$, which is sequentially complete (Theorem 10.16) and hence has a limit g, *i.e.*

$$\overline{f_n} \to g \text{ in } L^p(\Omega_1 \times \Omega_2; E).$$

By separating the variables, it follows that $f_n \to \underline{g}$, which, again with (14.7), gives

$$\underline{g} = f.$$

This proves surjectivity, once Lemma 14.3 has been verified.

4º) Existence of $\overline{f}$ when $p = \infty$. Let

$$f \in L^\infty(\Omega_1; L^\infty(\Omega_2; E)).$$

For bounded open sets $\omega_1 \subset \Omega_1$ and $\omega_2 \subset \Omega_2$, we have $f \in L^1(\omega_1; L^1(\omega_2; E))$ (Theorems 11.1 and 12.3). The case $p = 1$ then yields $\overline{f_{\omega_1 \times \omega_2}} \in L^1(\omega_1 \times \omega_2; E)$. These measures admit a gluing (Theorem 11.18) such that

$$\overline{f} \in L^1_{\text{loc}}(\Omega_1 \times \Omega_2; E). \tag{14.8}$$

Let R_n^1 and R_n^2 be regularization operators respectively on Ω_1 and Ω_2 given by Definition 3.17. Theorem 10.11 (b) gives

$$\sup_{x_1 \in \Omega_1} \|R_n^1 f(x_1)\|_{L^\infty(\Omega_2; E);\nu} \leq \|f\|_{L^\infty(\Omega_1; L^\infty(\Omega_2; E));\nu}. \tag{14.9}$$

Again by using Theorem 10.11 (b) (now with Ω_2 in the place of Ω_1), we obtain

$$\sup_{x_1\in\Omega_1,\, x_2\in\Omega_2} \|R_n^2(R_n^1 f(x_1))(x_2)\|_{E);\nu} \le \sup_{x_1\in\Omega_1} \|R_n^1 f(x_1)\|_{L^\infty(\Omega_2;E);\nu}. \tag{14.10}$$

Observe that $R_n = R_n^2 R_n^1$ is a regularization operator on $\Omega_1 \times \Omega_2$ with respect to $\alpha_n^1 \otimes \alpha_n^2$ and $\rho_n^1 \otimes \rho_n^2$. We have

$$\sup_{(x_1,\, x_2)\in\Omega_1\times\Omega_2} \|\overline{R_n f}(x_1, x_2)\|_{E;\nu} = \sup_{x_1\in\Omega_1,\, x_2\in\Omega_2} \|R_n^2(R_n^1 f(x_1))(x_2)\|_{E;\nu}. \tag{14.11}$$

With (14.10) and (14.9), it follows that

$$\sup_{(x_1,\, x_2)\in\Omega_1\times\Omega_2} \|\overline{R_n f}(x_1, x_2)\|_{E;\nu} < \infty.$$

Thus, Theorem 10.11 (a) on the characterization of the elements of $L^\infty(\Omega; E)$ gives

$$\overline{f} \in L^\infty(\Omega_1 \times \Omega_2; E).$$

5º) Semi-norm when $p = \infty$. In the limit, and by taking the infimum over all ∞-integrabilizing sequences for f with respect to x_2 and then with respect to x_1, inequality (14.11) yields (Definition 10.2) the equality (14.4).

6º) Uniqueness, continuity, and injectivity when $p = \infty$. These follow from the inequality for the semi-norms.

7º) Non-surjectivity when $p = \infty$. **7.a.** *A particular case.* We begin with the case where Ω_1 and Ω_2 are $(0, 1)$. Let

$$\omega \overset{\text{def}}{=} \{x \in \mathbb{R}^2 : 0 < x_1 < 1,\ 0 < x_2 < x_1\},$$

and consider the characteristic measure on ω defined in Theorem 13.7 (a),

$$1_\omega \in L^\infty((0, 1) \times (0, 1)).$$

If we have

$$\underline{1_\omega} \in L^\infty((0, 1); L^\infty(0, 1)),$$

then Theorem 11.1 would imply $\underline{1_\omega} \in L^1((0, 1); L^\infty(0, 1))$, and therefore we would have (Theorem 10.13 (c))

$$R_n \underline{1_\omega} \to \underline{1_\omega} \text{ in } L^1((0, 1); L^\infty((0, 1))). \tag{14.12}$$

However, $\underline{1_\omega}(x_1) = 1_{(0,x_1)}$ and (Definition 3.17), when $n \ge 12$ and $x_1 \in (1/4, 3/4)$,

$$(R_n \underline{1_\omega})(x_1) = \langle \underline{1_\omega}, \alpha_n \tau_{x_1} \rho_n \rangle = \int_{x_1 - 1/n}^{x_1 + 1/n} \underline{1_\omega}(y) \rho_n(y - x_1)\, dy.$$

Thus, $(R_n \underline{1_\omega})(x_1)$ is a continuous function in x_2, and

$$((R_n \underline{1_\omega})(x_1))(x_2) = \int_{x_1-1/n}^{x_2} \rho_n(y - x_1)\, dy.$$

In particular, $((R_n \underline{1_\omega})(x_1))(x_1) = 1/2$, and

$$\|(R_n \underline{1_\omega})(x_1) - 1_{(0,x_1)}\|_{L^\infty((0,1))} \geq \frac{1}{2}.$$

This holds for all $x_1 \in (1/4, 3/4)$, which would contradict (14.12). Therefore,

$$\underline{1_\omega} \notin L^\infty((0,1); L^\infty(0,1)).$$

7.b. *General case.* Let $a \in \Omega_1 \times \Omega_2$ and $\epsilon > 0$ such that

$$(a_1, a_1 + \epsilon) \times \ldots \times (a_d, a_d + \epsilon) \subset \Omega_1 \times \Omega_2.$$

Further, let $e \in E$, $e \neq 0_E$, and

$$\omega \stackrel{\text{def}}{=} \{a_1 < x_1 < a_1 + \epsilon, \ldots,\ a_{d-1} < x_{d-1} < a_{d-1} + \epsilon,\ a_d < x_d < a_d + x_1 - a_1\}.$$

Then,

$$\underline{1_\omega e} \notin L^\infty(\Omega_1; L^\infty(\Omega_2; E)).$$

Thus, $1_\omega e$, which belongs to $L^\infty(\Omega_1 \times \Omega_2; E)$, is not contained in the image after regouping the variables of an element in $L^\infty(\Omega_1; L^\infty(\Omega_2; E))$. □

It remains to establish the following properties.

Lemma 14.3. — *Let*

$$f \in L^p(\Omega_1; L^p(\Omega_2; E)),$$

where Ω_1 is an open subset of $\mathbb{R}^{d_1}$, Ω_2 is an open subset of $\mathbb{R}^{d_2}$, E is a Neumann space, and $1 \leq p < \infty$. Futhermore, let R_n^1 and R_n^2 be regularization operators on Ω_1 and Ω_2, respectively, given by Definition 3.17. Then, when $n \to \infty$ for (b) and (c):

(a) $$R_n^2 R_n^1 f \in \mathbf{C}_{K_n^1}(\Omega_1; \mathbf{C}_{K_n^2}(\Omega_2; E));$$

(b) $$R_n^2 f \to f \ \text{ in } L^p(\Omega_1; L^p(\Omega_2; E));$$

(c) $$R_n^2 R_n^1 f \to f \ \text{ in } L^p(\Omega_1; L^p(\Omega_2; E)).$$ ▮

Proof. ***1º)*** *Property* (a). First of all, observe that

$$R_n^2 \text{ is sequentially continuous from } L^p(\Omega_2; E) \text{ into } \mathbf{C}_{K_n^2}(\Omega_2; E). \tag{14.13}$$

Indeed, the mapping R_n^2 is linear and sequentially continuous from $\mathcal{M}(\Omega_2; E)$ into $\mathbf{C}_{K_n^2}(\Omega_2; E)$ (Theorem 3.20), and (Theorem 10.8 (a)) $L^p(\Omega_2; E)$ is topologically included in $\mathcal{M}(\Omega_2; E)$, and we apply Proposition A.40 (b) to it.

Since $R_n^1 f \in \mathbf{C}_{K_n^1}(\Omega_1; L^p(\Omega_2; E))$ (Theorem 3.18), it follows from (14.13) that

$$R_n^2 R_n^1 f \in \mathbf{C}_{K_n^1}(\Omega_1; \mathbf{C}_{K_n^2}(\Omega_2; E));$$

indeed, $R_n^2 R_n^1 f$ is sequentially continuous, and thus continuous (Proposition A.47); it furthermore vanishes outside the compact subset K_n^1 with values in $\mathbf{C}_{K_n^2}(\Omega_2; E)$, hence it is uniformly continuous by Heine's theorem (Proposition A.41).

2º) *Property* (b). If $R_n^2 f$ did not converge to f, then there would exist $\nu \in \mathcal{N}_E$ (index set for the semi-norms of E), $a > 0$, and a sequence of integers n such that

$$\|R_n^2 f - f\|_{L^p(\Omega_1; L^p(\Omega_2; E)); \nu} \geq a. \tag{14.14}$$

By the density result (Theorem 10.14), there would exist $h \in \mathcal{B}(\Omega_1; L^p(\Omega_2; E))$ such that

$$\|f - h\|_{L^p(\Omega_1; L^p(\Omega_2; E)); \nu} \leq \frac{a}{3}.$$

Then, since R_n^2 is contracting (Theorem 10.13 (a)),

$$\|R_n^2 (f - h)\|_{L^p(\Omega_1; L^p(\Omega_2; E)); \nu} \leq \frac{a}{3}.$$

Decomposing $R_n^2 f - f = R_n^2(f - h) + (R_n^2 h - h) + (h - f)$, (14.14) would then imply

$$a \leq \|R_n^2 f - f\|_{L^p(\Omega_1; L^p(\Omega_2; E)); \nu} \leq \|R_n^2 h - h\|_{L^p(\Omega_1; L^p(\Omega_2; E)); \nu} + \frac{2a}{3}. \tag{14.15}$$

As h, and thus $R_n^2 h$, vanishes outside a bounded subset of $\mathbb{R}^{d_1}$, then (by the integral expression for the $L^p(\Omega; E)$ semi-norms of a function in $\mathcal{B}(\Omega; E)$ of Proposition 10.7, since the set where a function vanishes does not contribute to its integral, see Proposition 1.23, and by the bound for the integral from Proposition 1.21), if we denote $\omega_1 = \{x_1 \in \Omega_1 : h(x_1) \neq 0\}$ (which is a bounded open set), we would have

$$\|R_n^2 h - h\|_{L^p(\Omega_1; L^p(\Omega_2; E)); \nu} = \Big(\int_{\omega_1} \|R_n^2 h - h\|^p_{L^p(\Omega_2; E); \nu} \Big)^{1/p} \leq$$

$$\leq |\omega_1|^{1/p} \sup_{x_1 \in \omega_1} \|(R_n^2 h - h)(x_1)\|_{L^p(\Omega_2; E); \nu}.$$

Thus, $|\omega_1| < \infty$ and (14.15) would imply the existence of a sequence of $x_1^n \in \omega_1$ such that

$$\|(R_n^2 h - h)(x_1^n)\|_{L^p(\Omega_2;E);\nu} \geq \frac{a}{3|\omega_1|^{1/p}} = b. \tag{14.16}$$

By the Bolzano–Weierstrass theorem (Proposition A.37), there would exist a subsequence x_1^n converging to a limit $x_1 \in \overline{\omega_1}$. Since h is uniformly continuous (Definition 1.12), $h(x_1^n)$ would have (Proposition 12.9 (a)) a limit ℓ in the space $L^p(\Omega_2; E)$. For all n larger than a particular n_0, we would have

$$\|h(x_1^n) - \ell\|_{L^p(\Omega_2;E);\nu} \leq \frac{b}{3}.$$

Then, since R_n^2 is contracting (Theorem 10.13 (a)), we would have

$$\|R_n^2(h(x_1^n) - \ell)\|_{L^p(\Omega_2;E);\nu} \leq \frac{b}{3}.$$

Decomposing $(R_n^2 h - h)(x_1^n) = R_n^2(h(x_1^n) - \ell) + (R_n^2 \ell - \ell) + (\ell - h(x_1^n))$, (14.16) would then imply

$$b \leq \|(R_n^2 h - h)(x_1^n)\|_{L^p(\Omega_2;E);\nu} \leq \|R_n^2 \ell - \ell\|_{L^p(\Omega_2;E);\nu} + \frac{2b}{3}.$$

Therefore,

$$\|R_n^2 \ell - \ell\|_{L^p(\Omega_2;E);\nu} \geq \frac{b}{3}.$$

However, this would contradict the convergence $R_n^2 \ell \to \ell$ established in Theorem 10.13 (c), therefore the convergence stated in (b) does indeed hold.

3o) Property (c). Decompose

$$R_n^2 R_n^1 f - f = R_n^2(R_n^1 f - f) + R_n^2 f - f.$$

Since (Theorem 10.13 (a)), for every $\nu \in \mathcal{N}_E$,

$$\|R_n^2(R_n^1 f - f)\|_{L^p(\Omega_1;L^p(\Omega_2;E));\nu} \leq \|R_n^1 f - f\|_{L^p(\Omega_1;L^p(\Omega_2;E));\nu}$$

and $R_n^1 f - f \to 0$ (Theorem 10.13 (c)), we therefore have $R_n^2(R_n^1 f - f) \to 0$ in $L^p(\Omega_1; L^p(\Omega_2; E))$. Furthermore, $R_n^2 f - f \to 0$ by Property (b). Hence,

$$R_n^2 R_n^1 f - f \to 0. \ \Box$$

Alternative proof of the non-surjectivity of the regrouping of variables when p is finite. Let us give another proof to part 7.a of the proof of Theorem 14.2. The mapping $x \mapsto 1_{(0,x)}$, namely $\underline{1_\omega}$, is discontinuous everywhere; Lusin's theorem (Theorem 14.4, below) shows that it does not belong to the space $L^1_{\text{loc}}((0,1); L^\infty(0,1))$ and, *a fortiori*, not to $L^\infty((0,1); L^\infty(0,1))$. And yet, the mapping 1_ω belongs to $L^\infty((0,1) \times (0,1))$. $\Box$

Let us state a variant of **Lusin's theorem**[2].

Theorem 14.4. — *Given $f \in L^p(\Omega; E)$, where Ω is an open subset of $\mathbb{R}^d$, $1 \le p < \infty$ and E is a Fréchet space, and $\epsilon > 0$, there exists a measurable set ω of measure less than ϵ such that $f \in \mathcal{C}(\Omega \setminus \omega; E)$.* □

Proof. Every $f \in L^p(\Omega; E)$ is the limit of a sequence $f_n \in \mathcal{B}(\Omega; E)$ (Theorem 10.14). Since f is a measurable function almost everywhere (Theorems 17.31), Egoroff's theorem (Theorem 17.3) implies that, for every bounded open subset $\Omega_n \subset \Omega$, there exists a measurable subset ω_n such that $|\omega_n| \le \epsilon 2^{-n-1}$ and $f_n \to f$ uniformly on $\Omega_n \setminus \omega_n$. Since the set of continuous functions is closed under uniform convergence [Vol. 2, Theorem 1.10], f is continuous on $\Omega_n \setminus \omega_n$. By covering Ω using a sequence $(\Omega_n)_{n\in\mathbb{N}}$, we find that f is continuous outside $\omega = \bigcup_{n\in\mathbb{N}} \omega_n$ with measure (Theorem 2.18 (a)) $|\omega| \le \sum_{n\in\mathbb{N}} \epsilon 2^{-n-1} = \epsilon$. □

Identifications. If one were to identify f with $\overline{f}$ in Theorem 14.2 (which we do not recommend), one would obtain: if $1 \le p < \infty$,

$$L^p(\Omega_1; L^p(\Omega_2; E)) \underset{\rightleftarrows}{\approx} L^p(\Omega_1 \times \Omega_2; E); \tag{14.17}$$

if $p = \infty$,

$$L^\infty(\Omega_1; L^\infty(\Omega_2; E) \underset{\not\rightleftarrows}{\subsetneq} L^\infty(\Omega_1 \times \Omega_2; E)), \tag{14.18}$$

$$L^\infty(\Omega_1; L^\infty(\Omega_2; E) \not\approx L^\infty(\Omega_1 \times \Omega_2; E)). \;\square \tag{14.19}$$

14.3. Permutation of variables

Let us permute the variables of a p-integrable measure of q-integrable measures.

Theorem 14.5. — *Let Ω_1 be an open subset of $\mathbb{R}^{d_1}$, Ω_2 be an open subset of $\mathbb{R}^{d_2}$, E be a Neumann space, and*

$$1 \le p \le q \le \infty, \quad p \ne \infty.$$

(a) *The permutation of variables $f \mapsto \breve{f}$ (Definition 8.2) is linear, continuous, sequentially continuous, and injective from $L^p(\Omega_1; L^q(\Omega_2; E))$ into $L^q(\Omega_2; L^p(\Omega_1; E))$. For every $f \in L^q(\Omega_1; L^p(\Omega_2; E))$ and $\nu \in \mathcal{N}_E$ (the index set of the semi-norms of E),*

$$\|\breve{f}\|_{L^q(\Omega_2; L^p(\Omega_1; E)); \nu} \le \|f\|_{L^p(\Omega_1; L^q(\Omega_2; E)); \nu}.$$

The mapping $f \mapsto \breve{f}$ is an isomorphism from the space $L^p(\Omega_1; L^p(\Omega_2; E))$ onto $L^p(\Omega_2; L^p(\Omega_1; E))$. For every $f \in L^p(\Omega_1; L^p(\Omega_2; E))$,

$$\|\breve{f}\|_{L^p(\Omega_2; L^p(\Omega_1; E)); \nu} = \|f\|_{L^p(\Omega_1; L^p(\Omega_2; E)); \nu}.$$

(b) *If $\Omega_1 \ne \emptyset$, $\Omega_2 \ne \emptyset$, $E \ne \{0_E\}$, and $p < q$, the mapping $f \mapsto \breve{f}$ does not map the space $L^q(\Omega_2; L^p(\Omega_1; E))$ into $L^p(\Omega_1; L^q(\Omega_2; E))$.* ▮

2 **Historical background of Lusin's theorem.** Theorem 17.4, in one dimension and for real-valued functions, was stated in 1903 by Henri LEBESGUE and proven in 1905 by Giuseppe VITALI [79], before being rediscovered in 1912 by Nikolai LUSIN [50] ([BOURBAKI, 10, p. 280]).

Proof. ***1°) Inequality for q finite.*** For $g \in L^1(\Omega_1; L^r(\Omega_2))$, $1 \le r \le \infty$, we have, with the bound on the integral from Theorem 13.5 with $E = L^r(\Omega)$,

$$\Big\| \int_{\Omega_1} g \Big\|_{L^r(\Omega_2)} \le \int_{\Omega_1} \|g\|_{L^r(\Omega_2)}. \tag{14.20}$$

Using the expression for the integral from Theorem 13.4, and if $r < \infty$, this gives

$$\Big(\int_{\Omega_2} \Big| \int_{\Omega_1} g \Big|^r \Big)^{1/r} \le \int_{\Omega_1} \Big(\int_{\Omega_2} |g|^r \Big)^{1/r}.$$

With $g = \|f\|^p_{E;\nu}$ and $r = q/p$ (which is ≥ 1), then by raising it to the power $1/p$, we get

$$\Big(\int_{\Omega_2} \Big| \int_{\Omega_1} \|f\|^p_{E;\nu} \Big|^{q/p} \Big)^{1/q} \le \Big(\int_{\Omega_1} \Big(\int_{\Omega_2} \|f\|^q_{E;\nu} \Big)^{p/q} \Big)^{1/p}.$$

Again using Theorem 13.4, this gives

$$\|\breve{f}\|_{L^q(\Omega_2; L^p(\Omega_1; E)); \nu} \le \|f\|_{L^p(\Omega_1; L^q(\Omega_2; E)); \nu}. \tag{14.21}$$

2°) Inequality for $q = \infty$. From inequality (14.20) with $g = \|f\|^p_{E;\nu}$ and $r = \infty$, together with inequality (13.3) from Theorem 13.3, we get

$$\Big\| \int_{\Omega_1} \|f\|^p_{E;\nu} \Big\|_{L^\infty(\Omega_2)} \le \int_{\Omega_1} \big\| \|f\|^p_{E;\nu} \big\|_{L^\infty(\Omega_2)} = \int_{\Omega_1} \big\| \|f\|_{E;\nu} \big\|^p_{L^\infty(\Omega_2)}.$$

Using Theorem 13.4, then by raising it to the power $1/p$, we get

$$\|\breve{f}\|_{L^\infty(\Omega_2; L^p(\Omega_1; E)); \nu} \le \|f\|_{L^p(\Omega_1; L^\infty(\Omega_2; E)); \nu}. \tag{14.22}$$

3°) Linearity and injectivity. These were established in Theorem 8.2 for any measure, and hence for those which are p-integrable.

4°) Continuity. Inequality (14.21) or (14.22), depending on the value of q, implies continuity, according to the characterization of continuous linear mappings (Proposition 1.9).

Continuity implies sequential continuity by Proposition 1.8.

5°) Isomorphism. If $p = q$, the inverse inequality to (14.21) is also true, and the mapping $\breve{f} \mapsto f$ is continuous linear; therefore, it is an isomorphism.

6°) Property (b) when $q < \infty$. 6.a. A particular case. Let $\Omega_1 = \Omega_2 = (0, 1)$ and $(f(x))(y) \stackrel{\text{def}}{=} (x - y)^{-1/q} 1_{\{y \le x\}}$. Then, since $p/q < 1$,

$$\|f\|_{L^q(\Omega_1; L^p(\Omega_2))} = \Big(\int_{(0,1)} \Big(\int_{(0,1)} |(f(x))(y)|^p \, dy \Big)^{q/p} dx \Big)^{1/q} =$$

$$= \Big(\int_0^1 \Big(\int_0^x \frac{dy}{(x - y)^{p/q}} \Big)^{q/p} dx \Big)^{1/q} < \infty.$$

Therefore, $f \in L^q(\Omega_1; L^p(\Omega_2))$. However,

$$\|\breve{f}\|_{L^p(\Omega_2;L^q(\Omega_1))} = \Big(\int_{(0,1)} \Big(\int_{(0,1)} |(f(x))(y)|^q \, dx\Big)^{p/q} dy\Big)^{1/p} =$$

$$= \Big(\int_0^1 \Big(\int_y^1 \frac{dx}{x-y}\Big)^{p/q} dy\Big)^{1/p} = \infty.$$

Hence, $\breve{f} \notin L^p(\Omega_2; L^q(\Omega_1))$, and so,

$$f \mapsto \breve{f} \text{ is not a mapping from } L^q(\Omega_2; L^p(\Omega_1)) \text{ into } L^p(\Omega_1; L^q(\Omega_2)).$$

6.b. *General case.* Let $a \in \Omega_1 \times \Omega_2$ and $\epsilon > 0$ such that

$$(a_1, a_1 + \epsilon) \times \ldots \times (a_d, a_d + \epsilon) \subset \Omega_1 \times \Omega_2.$$

Consider $e \in E$, $e \neq 0_E$,

$$\omega \stackrel{\text{def}}{=} \{a_1 < x_1 < a_1 + \epsilon, \ldots, \; a_{d-1} < x_{d-1} < a_{d-1} + \epsilon, \; a_d < x_d < a_d + x_1 - a_1\},$$

and

$$(f(x_1, \ldots, x_{d_1}))(x_{d_1+1}, \ldots, x_d) \stackrel{\text{def}}{=} \frac{1}{\Big(\dfrac{x_1 - a_1}{\epsilon} - \dfrac{x_d - a_d}{\epsilon}\Big)^{1/q}} 1_\omega e.$$

By 6.a, we have $f \in L^q(\Omega_1; L^p(\Omega_2))$ and $\breve{f} \notin L^p(\Omega_2; L^q(\Omega_1))$.

7°) *Property* (b) *when* $q = \infty$. Let $1_\omega e$ be the measure considered in part 7° of the proof of Theorem 14.2.

It belongs to $L^\infty(\Omega_1; L^p(\Omega_2; E))$ since it is continuous and bounded with values in $L^p(\Omega_2; E)$ for any finite p.

However, $\breve{1}_\omega$ does not belong to $L^p(\Omega_2; L^\infty(\Omega_1; E))$, since $R_n \breve{1}_\omega$ does not converge to $\breve{1}_\omega$ in $L^1(\Omega_1; L^\infty(\Omega_2; E))$ (as we saw in 7.a of the proof of Theorem 14.2 when proving that (14.12) was false). □

14.4. Weightings of measures

Let us define the weighting of two measures[3].

---

3 **Historical background of weightings.** Weightings were introduced by Jacques SIMON in 1993 [69, Definition 3, p. 205] for distributions with values in a Banach space E. Given an open subset Ω of $\mathbb{R}^d$, they generalize the definition of the convolution, up to a symmetry in the weight μ.

Definition 14.6. — *Let*

$$f \in \mathcal{M}(\Omega; E), \quad \mu \in \mathcal{M}_D(\mathbb{R}^d),$$

where Ω is an open subset of $\mathbb{R}^d$, D is a compact subset of $\mathbb{R}^d$, and E is a Neumann space.

We define the ***weighting***

$$f \diamond \mu \in \mathcal{M}(\Omega_D; E)$$

on the open set

$$\Omega_D \stackrel{\text{def}}{=} \{x \in \mathbb{R}^d : x + D \subset \Omega\}$$

by: for every $\varphi \in \mathcal{K}(\Omega_D)$,

$$\langle f \diamond \mu, \varphi \rangle_{\Omega_D} \stackrel{\text{def}}{=} \langle f, \mu \diamond \breve{\varphi} \rangle_\Omega,$$

where $\breve{\varphi} \in \mathcal{K}(\mathbb{R}^d)$ is defined by $\breve{\varphi}(x) \stackrel{\text{def}}{=} \varphi(-x)$ if $-x \in \Omega_D$ and $\breve{\varphi}(x) \stackrel{\text{def}}{=} 0$ otherwise, and $\mu \diamond \breve{\varphi} \in \mathcal{K}(\Omega)$ is given by

$$(\mu \diamond \breve{\varphi})(x) \stackrel{\text{def}}{=} \langle \mu, \tau_x \breve{\varphi} \rangle_{\mathbb{R}^d},$$

where $\tau_x \breve{\varphi} \in \mathcal{K}(\mathbb{R}^d)$ is defined by $(\tau_x \breve{\varphi})(y) \stackrel{\text{def}}{=} \breve{\varphi}(y - x)$. ▮

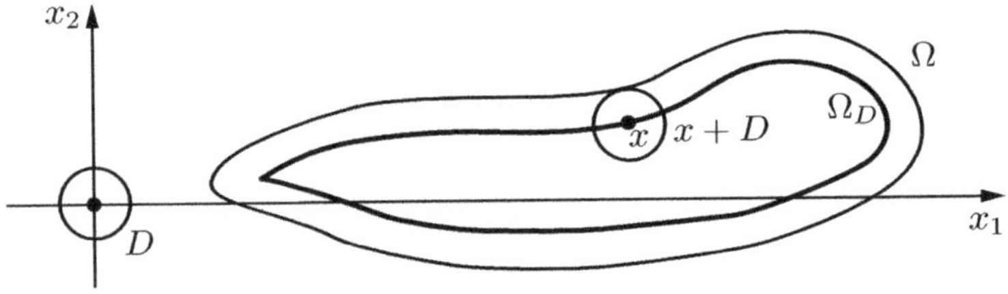

Figure 14.1. *Domain of definition Ω_D of the weighting $f \diamond \mu$*

Terminology. *Weighting* is a **new term**, so the reader should recall its definition when using it. We say that $f \diamond \mu$ is the weighting of f by μ since, when f is continuous, $(f \diamond \mu)(x)$ is the integral $\int_{\mathring{D}} f(x + y)\mu(y)\, dy$ of f over a neighborhood of x weighted by the weight μ [Vol. 3, Theorem 7.4]. □

Domain Ω_D. The domain of definition Ω_D of $f \diamond \mu$ is represented in Figure 14.1 above. It is "relatively" smaller than Ω, in the sense that it can be included inside it after a suitable translation. If $0 \in D$, then a translation is not needed since then $\Omega_D \subset \Omega$.

If Ω is bounded, Ω_D cannot strictly contain Ω, even after a translation; and it cannot be equal to it except when D is empty or can be reduced to a point, in which case $\mu = 0$. In contrast, if Ω is not bounded and $0 \notin D$, then Ω_D can contain Ω: for example, if $\Omega = (0, \infty)$ and $D = [1, 2]$, then $\Omega_D = (-1, \infty)$. □

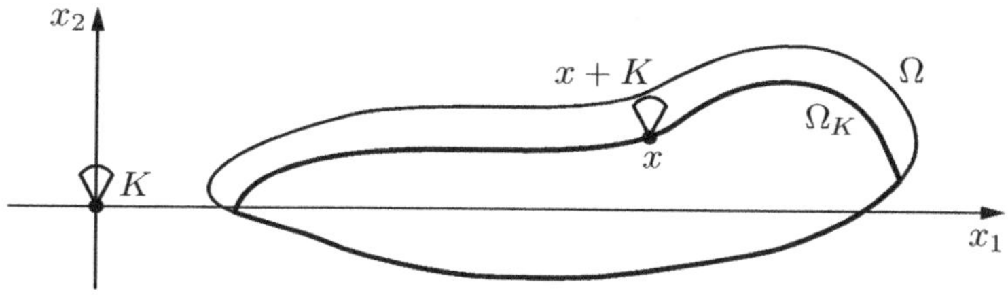

Figure 14.2. *Domain Ω_K going up to a section of the boundary. The boundary of Ω_K is shown in bold*

Choice of "domain" D for μ. To define the weighting in a neighborhood of the boundary, in the following volumes we shall use a conic section K, in which case Ω_K goes up to a section of the boundary if it is regular, as shown in Figure 14.2. □

Justification of Definition 14.6. ***1°)*** *Openness of Ω_D.* Let $x \in \Omega_D$. The set $x + D$ is a compact subset contained in the open set Ω, so (Proposition A.19) there exists a ball B centered at 0 such that $x + D + B \subset \Omega$. That is, $x + B \subset \Omega_D$, hence Ω_D is open.

2°) $\tau_x \breve{\varphi} \in \mathcal{K}(\mathbb{R}^d)$. The translated function $\tau_x \breve{\varphi}$ is continuous and its support is a compact subset of $\mathbb{R}^d$.

3°) $\mu \diamond \breve{\varphi} \in \mathcal{K}(\Omega)$. **3.a.** *Continuity.* The mappings $x \mapsto \tau_x \breve{\varphi}$ and μ are continuous, so the same can be said for their composition $\langle \mu, \tau_x \breve{\varphi} \rangle_{\mathbb{R}^d}$ (Proposition A.39).

3.b. *Support.* We shall verify below that

$$\operatorname{supp}(\mu \diamond \breve{\varphi}) \subset \operatorname{supp} \mu - \operatorname{supp} \breve{\varphi}. \tag{14.23}$$

Hence,

$$\operatorname{supp}(\mu \diamond \breve{\varphi}) \subset D + \Omega_D \subset \Omega.$$

Which proves $\mu \diamond \breve{\varphi} \in \mathcal{K}(\Omega)$, once (14.23) has been verified.

4°) *Verification of (14.23).* **4.a.** *A sufficient condition.* By Definition 6.18 of the support,

$$\operatorname{supp}(\mu \diamond \breve{\varphi}) = \mathbb{R}^d \setminus \mathcal{O},$$

where $\mathcal{O}$ is the largest open set on which $\mu \diamond \breve{\varphi}$ vanishes. The inclusion (14.23) is therefore equivalent to

$$\mathbb{R}^d \setminus (\operatorname{supp} \mu - \operatorname{supp} \breve{\varphi}) \subset \mathcal{O}.$$

Hence, after denoting by S_μ the support of μ and $S_{\breve{\varphi}}$ that of $\breve{\varphi}$, it suffices to show that

$$\mathbb{R}^d \setminus (S_\mu - S_{\breve{\varphi}}) \text{ is an open set on which } \mu \diamond \breve{\varphi} \text{ vanishes.} \tag{14.24}$$

4.b. *Vanishing of $\mu \diamond \breve{\varphi}$ on $\mathbb{R}^d \setminus (S_\mu - S_{\breve{\varphi}})$.* Let

$$x \in \mathbb{R}^d \setminus (S_\mu - S_{\breve{\varphi}}).$$

Then, $(x + S_{\breve{\varphi}}) \cap S_\mu = \emptyset$ (since $x \notin S_\mu - S_{\breve{\varphi}}$), so

$$x + S_{\breve{\varphi}} \subset \mathbb{R}^d \setminus S_\mu. \tag{14.25}$$

In other words, $\operatorname{supp} \tau_x \breve{\varphi}$ is contained in the largest open set on which μ vanishes. Therefore,

$$\langle \mu, \tau_x \breve{\varphi} \rangle_{\mathbb{R}^d} = 0_E.$$

That is,

$$(\mu \diamond \breve{\varphi})(x) = 0_E. \tag{14.26}$$

4.c. *Openness of $\mathbb{R}^d \setminus (S_\mu - S_{\breve{\varphi}})$.* Again, let

$$x \in \mathbb{R}^d \setminus (S_\mu - S_{\breve{\varphi}}).$$

In the inclusion (14.25), the set $x + S_{\breve{\varphi}}$ is compact (since $S_{\breve{\varphi}}$ is compact there) and $\mathbb{R}^d \setminus S_\mu$ is open. Hence, by the strong inclusion theorem (Proposition A.19), there exists a ball $B(0, \epsilon)$, where $\epsilon > 0$, such that

$$x + S_{\breve{\varphi}} + B(0, \epsilon) \subset \mathbb{R}^d \setminus S_\mu.$$

Thus, $B(x, \epsilon) \subset \mathbb{R}^d \setminus (S_\mu - S_{\breve{\varphi}})$, which proves that

$$\mathbb{R}^d \setminus (S_\mu - S_{\breve{\varphi}}) \text{ is open.}$$

Combining this with (14.26) finishes the proof of the sufficient condition (14.24). □

14.5. Weightings

Let us weight a p-integrable measure using an r-integrable weight with compact support D, where

$$L^r_D(\mathbb{R}^d) \stackrel{\text{def}}{=} \{\mu \in L^r(\mathbb{R}^d) : \operatorname{supp} \mu \subset D\}.$$

Theorem 14.7. — *Let*

$$f \in L^p(\Omega; E), \quad \mu \in L^r_D(\mathbb{R}^d), \quad \frac{1}{p} + \frac{1}{r} \geq 1,$$

where Ω is an open subset of $\mathbb{R}^d$, D is a compact subset of $\mathbb{R}^d$, E is a Neumann space, $1 \leq p \leq \infty$, $1 \leq r \leq \infty$, and $\nu \in \mathcal{N}_E$ (the index set for the semi-norms of E).

(a) *Then,* $f \diamond \mu \in L^q(\Omega_D; E)$, *where* $\dfrac{1}{q} = \dfrac{1}{p} + \dfrac{1}{r} - 1$, *and*

$$\|f \diamond \mu\|_{L^q(\Omega_D;E);\nu} \leq \|f\|_{L^p(\Omega;E);\nu} \|\mu\|_{L^r(\mathbb{R}^d)}.$$

(b) *The mapping* $\diamond$ *is continuous bilinear, and thus sequentially continuous, from* $L^p(\Omega; E) \times L^r_D(\mathbb{R}^d)$ *into* $L^q(\Omega_D; E)$. ∎

Proof. **1°)** *Property* (a) *when* $\Omega = \mathbb{R}^d$. **1.a.** *An inequality.* If we have $\mu \in \mathcal{B}(\mathbb{R}^d)$ and $f \in \mathcal{B}(\mathbb{R}^d; E)$, then Young's convolution inequality for continuous functions (Proposition 9.9) can be written, together with the integral expression for the $L^p(\Omega; E)$ semi-norms of a function in $\mathcal{B}(\Omega; E)$ (Proposition 10.7), and if p and r are finite, as

$$\|f \diamond \mu\|_{L^q(\mathbb{R}^d;E);\nu} \leq \|f\|_{L^p(\mathbb{R}^d;E);\nu} \|\mu\|_{L^r(\mathbb{R}^d)}. \tag{14.27}$$

If $p = \infty$, this holds since $r = 1$, $q = \infty$, and (Theorem 10.10, Definition 9.8, Theorems 5.3, and 5.10, and Theorems 10.10 and 4.8),

$$\|f \diamond \mu\|_{L^\infty(\mathbb{R}^d;E);\nu} = \sup_{x \in \mathbb{R}^d} \|(f \diamond \mu)(x)\|_{E;\nu} = \sup_{x \in \mathbb{R}^d} \left\| \int_{\mathbb{R}^d} f(x+y)\mu(y)\, dy \right\|_{E;\nu} \leq$$

$$\leq \sup_{x \in \mathbb{R}^d} \|f(x)\|_{E;\nu} \int_{\mathbb{R}^d} |\mu(y)|\, dy = \|f\|_{L^\infty(\mathbb{R}^d;E);\nu} \|\mu\|_{L^1(\mathbb{R}^d)}.$$

If $r = \infty$, this again holds since $p = 1$, $q = \infty$, and, similarly to the above along with Theorem 10.10,

$$\|f \diamond \mu\|_{L^\infty(\mathbb{R}^d;E);\nu} \leq \int_{\mathbb{R}^d} \|f(x)\|_{E;\nu}\, dx \sup_{y \in \mathbb{R}^d} |\mu(y)| = \|f\|_{L^1(\mathbb{R}^d;E);\nu} \|\mu\|_{L^\infty(\mathbb{R}^d)}.$$

1.b. *Belonging to* $L^q(\mathbb{R}^d; E)$. Let $R_n f$ be regularizations (Definition 3.17), and decompose

$$R_n f \diamond R_n \mu - R_m f \diamond R_m \mu = (R_n f - R_m f) \diamond R_n \mu + R_m f \diamond (R_n \mu - R_m \mu).$$

When $n \to \infty$, then $R_n f \to f$ in $L^p(\mathbb{R}^d; E)$ and $R_n \mu \to \mu$ in $L^r(\mathbb{R}^d)$ so, for all n, $\|R_n f\|_{L^p(\mathbb{R}^d;E);\nu} \leq \|f\|_{L^p(\mathbb{R}^d;E);\nu}$ and $\|R_n \mu\|_{L^r(\mathbb{R}^d)} \leq \|\mu\|_{L^r(\mathbb{R}^d)}$ (Theorem 10.13); therefore, inequality (14.27) implies, when p and r are either finite or infinite, that

$$(R_n f \diamond R_n \mu)_{n \in \mathbb{N}} \text{ is a Cauchy sequence in } L^q(\mathbb{R}^d; E).$$

Since this space is sequentially complete (Theorem 10.16), $R_n f \diamond R_n \mu$ converges to

$$f \diamond \mu \in L^q(\mathbb{R}^d; E).$$

1.c. *Inequality.* This is the limit of inequality (14.27) applied to $R_n f$ and $R_n \mu$, since $\|R_n f\|_{L^p(\Omega;E);\nu} \to \|f\|_{L^p(\Omega;E);\nu}$ (Theorem 10.13 (b)), when p is either finite or infinite.

2º) *Property* (a) *in the general case.* 2.a. *Preliminary.* First of all, let us show that

$$f \diamond \mu \text{ is the restriction of } \widetilde{f} \diamond \mu \text{ to } \Omega_D,$$

where $\widetilde{f} \in L^p(\mathbb{R}^d; E)$ is the extension by 0_E given by Theorem 13.6 (a).

Indeed, for every $\varphi \in \mathcal{K}(\Omega_D)$, on the one hand, by the definition of the weighting (Definition 14.6) and the integral of a product (Theorem 12.19, since $\mu \diamond \check{\varphi} \in \mathcal{K}(\Omega)$ by Definition 14.6), we have

$$\langle f \diamond \mu, \varphi \rangle_{\Omega_D} = \langle f, \mu \diamond \check{\varphi} \rangle_\Omega = \int_\Omega f(\mu \diamond \check{\varphi})$$

whereas, on the other hand, by the definition of the restriction and the weighting (Definitions 6.16 and 14.6), and extending by 0_E (Theorem 13.6 (a)),

$$\langle (\widetilde{f} \diamond \mu)|_{\Omega_D}, \varphi \rangle_{\Omega_D} = \langle \widetilde{f} \diamond \mu, \varphi \rangle_{\mathbb{R}^d} = \langle \widetilde{f}, \mu \diamond \check{\varphi} \rangle_{\mathbb{R}^d} = \int_\Omega f(\mu \diamond \check{\varphi}).$$

2.b. *Belonging to* $L^q(\Omega_D; E)$. The case $\Omega = \mathbb{R}^d$ provides $\widetilde{f} \diamond \mu \in L^q(\mathbb{R}^d; E)$, and its restriction satisfies (Theorem 11.5 (a))

$$f \diamond \mu \in L^q(\Omega_D; E).$$

2.c. *Inequality.* We have (Theorem 11.5 (b) for $\Omega = \mathbb{R}^d$, and Theorem 13.6 (b))

$$\|f \diamond \mu\|_{L^q(\Omega_D;E);\nu} \leq \|\widetilde{f} \diamond \mu\|_{L^q(\mathbb{R}^d;E);\nu} \leq$$
$$\leq \|\widetilde{f}\|_{L^p(\mathbb{R}^d;E);\nu} \|\mu\|_{L^r(\mathbb{R}^d)} = \|f\|_{L^p(\Omega;E);\nu} \|\mu\|_{L^r(\mathbb{R}^d)}.$$

3º) *Property* (b). Continuity follows (Proposition 12.6) from inequality (a).

Finally, sequential continuity follows from continuity (Proposition 1.8). □

Let us show that the compactness hypothesis on the support of μ can be dropped when $\Omega = \mathbb{R}^d$.

Theorem 14.8. — *Let*

$$f \in L^p(\mathbb{R}^d; E), \quad \mu \in L^r(\mathbb{R}^d), \quad \frac{1}{p} + \frac{1}{r} \geq 1,$$

where E is a Neumann space, $1 \leq p \leq \infty$, and $1 \leq r < \infty$.

(a) *We define*

$$f \diamond \mu \in L^q(\mathbb{R}^d; E), \quad \textit{where } \frac{1}{q} = \frac{1}{p} + \frac{1}{r} - 1,$$

by

$$f \diamond \mu \overset{\text{def}}{=} \lim_{n \to \infty} f \diamond \mu_n,$$

where μ_n vanishes outside a bounded set and $\mu_n \to \mu$ in $L^r(\mathbb{R}^d)$.

(b) *For every $\nu \in \mathcal{N}_E$ (the index set for the semi-norms of E),*

$$\|f \diamond \mu\|_{L^q(\mathbb{R}^d;E);\nu} \leq \|f\|_{L^p(\mathbb{R}^d;E);\nu} \|\mu\|_{L^r(\mathbb{R}^d)}.$$

(c) *The mapping $\diamond$ is continuous bilinear, and hence sequentially continuous, from $L^p(\mathbb{R}^d; E) \times L^r(\mathbb{R}^d)$ into $L^q(\mathbb{R}^d; E)$.* ▮

Proof. **1°)** *Belonging to $L^q(\mathbb{R}^d; E)$.* The limit exists and is independent of the choice of μ_n by the extension theorem (Proposition 12.9), which applies when $r < \infty$, because:

— $L^q(\mathbb{R}^d; E)$ is a Neumann space (Theorem 10.16);

— integrable measures vanishing outside a bounded set are sequentially dense in $L^r(\mathbb{R}^d)$, as this was already true for $\mathcal{B}(\mathbb{R}^d)$ (Theorem 10.14);

— the mapping $\mu \mapsto f \diamond \mu$ is continuous linear from $\mathcal{B}(\mathbb{R}^d)$ endowed with the $L^r(\mathbb{R}^d)$ topology (Theorem 14.7 (a)).

2°) *Inequality.* Inequality (b) follows by continuity from that of Theorem 14.7 (a).

3°) *Continuity.* Continuity follows (Proposition 12.6) from inequality (b).

Finally, sequential continuity follows from continuity (Proposition 1.8). □

Let us verify that we recover the weighting by a compactly supported weight, which justifies the use of the same notation.

Theorem 14.9. — *If the support of μ is compact, then the definition of $f \diamond \mu$ from Theorem 14.8 coincides with that from Definition 14.6.* ▮

Proof. If the support of μ is compact then, by choosing μ_n to be equal to a regularized $R_n\mu$ whose support is contained in the compact set $D = \operatorname{supp} \mu + B(0, 1)$, we recover the weighting from Definition 14.6 via the continuity property stated in Theorem 14.7 (b). □

Chapter 15

Compact Sets

We aim to characterize the compact subsets of $L^p(\Omega; E)$ and $L^p_{\mathrm{loc}}(\Omega; E)$.

It is necessary to distinguish between the notions of compactness and sequential compactness, which are neither stronger nor weaker than one another, and we also present some technical results (§ 15.1).

We characterize the relatively compact subsets $\mathcal{F}$ of $L^p(\Omega; E)$, which coincide with the relatively sequentially compact subsets when E is a Fréchet space (§ 15.2). When $E = \mathbb{R}^\ell$, we present some particular cases when $\mathcal{F}$ is a family of bounded subsets, or when Ω is an open set with a regular boundary (§ 15.3).

We then characterize the relatively compact subsets of $L^p_{\mathrm{loc}}(\Omega; E)$, which recovers the Fréchet–Kolmogorov theorem when $E = \mathbb{R}$ (§ 15.4).

We end this chapter with some compactness results for "intermediate" spaces (§ 15.5).

15.1. Preliminaries

Let us define some notions of compactness related to covering properties.

Definition 15.1. — *Let U be a subset of a separated semi-normed space E.*

(a) *We say that U is* ***compact*** *if every cover of U by open sets has a finite subcover.*

(b) *We say that U is* ***relatively compact*** *if its closure $\overline{U}$ is compact.*

(c) *We say that U is* ***precompact*** *if, for every semi-norm $\| \ \|_{E;\nu}$ of E and $\epsilon > 0$, there exists a finite number of elements $u \in E$ such that U can be covered by the semi-balls $B_{E;\nu}(u, \epsilon) = \{v \in E : \|v - u\|_{E;\nu} \leq \epsilon\}$.* ▮

Let us define some notions of compactness related to the convergence of subsequences.

Definition 15.2. — *Let U be a subset of a separated semi-normed space E.*

(a) *We say that U is* ***sequentially compact*** *if every sequence in U has a convergent subsequence whose limit belongs to U.*

(b) *We say that U is* ***relatively sequentially compact*** *if every sequence in U has a convergent subsequence (in E).* ▌

In a general separated semi-normed space, recall that sequential compactness is neither stronger nor weaker than compactness [Vol. 1, Properties (2.6) and (2.7), p. 27]. On the other hand, we do have the following equivalences [Vol. 1, Theorems 4.6 (b) and (c) and 4.14 (b)].

Proposition 15.3. — *In a Fréchet space, and hence in a Banach space, sequential compactness is equivalent to compactness, and relative sequential compactness is equivalent to relative compactness, which in turn is equivalent to precompactness.* ▌

Let us state some translation properties [1].

Theorem 15.4. — *Let $f \in L^p(\Omega; E)$, where Ω is an open subset of $\mathbb{R}^d$, E is a Neumann space, $1 \le p \le \infty$, $\nu \in \mathcal{N}_E$ (the index set for the semi-norms of E), and $x \in \mathbb{R}^d$.*

(a) *Then, $\tau_x f \in L^p(\Omega + x; E)$, and*

$$\|\tau_x f\|_{L^p(\Omega+x;E);\nu} = \|f\|_{L^p(\Omega;E);\nu}.$$

(b) *If $p < \infty$, when $x \to 0$,*

$$\|\tau_x f - f\|_{L^p(\Omega\cap(\Omega+x);E);\nu} \to 0$$

and, for every open subset ω and compact subset D of $\mathbb{R}^d$ such that $\omega + D \subset \Omega$,

$$[x \mapsto \tau_x f|_\omega] \in \mathbf{C}_b(D; L^p(\omega; E)).$$ ▌

Proof. **1°)** Property (a). The change of variables from Theorem 14.1 with $T = \tau_x$, and also the fact that $|\det[\nabla T^{-1}]| = |\det[\nabla \tau_{-x}]| = 1$, gives $\tau_x f \in L^p(\Omega + x; E)$ and

$$\|\tau_x f\|_{L^p(\Omega+x;E);\nu} \le \|f\|_{L^p(\Omega;E);\nu}.$$

1 **Historical background on translation properties.** The translation invariance of the integral, which forms the basis of Theorem 15.4, was one of the six conditions required by Henri LEBESGUE for his definition of the integral given in 1904 [44, p. 106 and 122], see Note 2 of Chapter 5.

This inequality is in fact an equality, since

$$\|f\|_{L^p(\Omega;E);\nu} = \|\tau_{-x}\tau_x f\|_{L^p(\Omega;E);\nu} \leq \|\tau_x f\|_{L^p(\Omega+x;E);\nu}.$$

2°) First property of (b). The space $\mathcal{K}(\Omega; E)$ is sequentially dense in $L^p(\Omega; E)$ (Theorem 10.14), *i.e.* given $\epsilon > 0$, there exists $g \in \mathcal{K}(\Omega; E)$ such that

$$\|f - g\|_{L^p(\Omega;E);\nu} \leq \frac{\epsilon}{3}.$$

For every $x \in \mathbb{R}^d$,

$$\|\tau_x f - f\|_{L^p((\Omega+x)\cap\Omega;E);\nu} \leq$$
$$\leq \|\tau_x(f-g)\|_{L^p(\Omega+x;E);\nu} + \|\tau_x g - g\|_{L^p((\Omega+x)\cap\Omega;E);\nu} + \|g - f\|_{L^p(\Omega;E);\nu}.$$

Translation invariance gives $\|\tau_x(f-g)\|_{L^p(\Omega+x;E);\nu} = \|f-g\|_{L^p(\Omega;E);\nu} \leq \epsilon/3$. The function g has support $\omega \Subset \Omega$, thus (Proposition A.20) there exists an open subset ω' such that $\omega \Subset \omega' \Subset \Omega$. When $x \to 0$, for x small enough,

$$\|\tau_x g - g\|_{L^p((\Omega+x)\cap\Omega;E);\nu} = \|\tau_x g - g\|_{L^p(\omega';E);\nu} \to 0$$

as the convergence takes place in $\mathcal{K}(\omega'; E)$ and therefore (Theorem 10.8) in $L^p(\omega'; E)$. Hence, there exists $\eta > 0$ such that it is bounded by $\epsilon/3$, and so

$$\|\tau_x f - f\|_{L^p((\Omega+x)\cap\Omega;E);\nu} \leq \epsilon.$$

3°) Second property of (b). If $\omega + D \subset \Omega$, then the function $x \mapsto \tau_x f|_\omega$ is continuous from D into $L^p(\omega; E)$. Since D is compact, it is uniformly continuous and bounded by Heine's theorem (Proposition A.41). □

We will also use the following variant of **Ascoli's theorem** [Vol. 2, Theorem 1.29].

Proposition 15.5. — *Let $\mathcal{F} \subset \mathbf{C}_b(\Omega; E)$, where Ω is a bounded subset of $\mathbb{R}^d$, E is a separated semi-normed space, and $\{\|\ \|_{E;\nu} : \nu \in \mathcal{N}_E\}$ is the family of semi-norms of E, such that:*

(a) *$\forall x \in \Omega$, every sequence in $\{f(x) : f \in \mathcal{F}\}$ has a convergent subsequence in E;*

(b) *$\mathcal{F}$ is uniformly equicontinuous, that is, for every $\nu \in \mathcal{N}_E$ and $\epsilon > 0$, there exists $\eta > 0$ such that $|y - y'| \leq \eta$ implies $\sup_{f\in\mathcal{F}} \|f(y) - f(y')\|_{E;\nu} \leq \epsilon$.*

Then every sequence in $\mathcal{F}$ has a convergent subsequence in $\mathbf{C}_b(\Omega; E)$. ∎

15.2. Compact subsets of $L^p(\Omega; E)$

Let us characterize the relatively compact subsets of $L^p(\Omega; E)$ when E is a Fréchet space. This is referred to as the **Riesz–Kolmogorov** theorem[2], although that name is perhaps more suited to Theorem 15.15.

Theorem 15.6. — *Let $\mathcal{F} \subset L^p(\Omega; E)$, where*

$$E \text{ is a Fréchet space}, \quad 1 \le p < \infty,$$

Ω is an open subset of $\mathbb{R}^d$ and $\{\| \, \|_{E;\nu} : \nu \in \mathcal{N}_E\}$ is the family of semi-norms of E.

Then, the three properties together:

(a) $\forall \nu \in \mathcal{N}_E$, $\forall \omega \Subset \Omega$, *when* $x \in \mathbb{R}^d$, $\omega + x \subset \Omega$, *and* $x \to 0$,

$$\sup_{f \in \mathcal{F}} \|\tau_x f - f\|_{L^p(\omega;E);\nu} \to 0;$$

(b) $\forall \nu \in \mathcal{N}_E$, $\forall \epsilon > 0$, $\exists$ *compact subset* $K \subset \Omega$ *such that,*

$$\sup_{f \in \mathcal{F}} \|f\|_{L^p(\Omega \setminus K;E);\nu} \le \epsilon;$$

(c) $\forall \omega \Subset \Omega$,

$$\textit{every sequence in } \Big\{ \int_\omega f : f \in \mathcal{F} \Big\} \textit{ has a convergent subsequence in } E;$$

are equivalent to:

$$\textit{every sequence in } \mathcal{F} \textit{ has a convergent subsequence in } L^p(\Omega; E),$$

which itself is equivalent to:

$$\mathcal{F} \textit{ is relatively compact in } L^p(\Omega; E). \ ∎$$

---

2 **Historical background of the Riesz–Kolmogorov theorem.** Marcel RIESZ proved in 1933 [64] that properties (a) and (b) characterize the compact subsets of $L^p(\mathbb{R})$, provided $\mathcal{F}$ is bounded.
Andrey KOLMOGOROV had shown in 1931 [40] that (a) characterizes the relatively sequentially compact subsets of $L^p(\mathbb{R})$ when $1 < p < \infty$, assuming all the f in $\mathcal{F}$ have their support contained in a common compact set.
This theorem is sometimes called the **Fréchet–Kolmogorov** theorem, perhaps because Maurice FRÉCHET proved in 1937 [30] that compactness in $L^p(X)$ is equivalent to *equisummability* and to *almost everywhere continuity* in X. For more details, see HANCHE-OLSEN and HOLDEN [33].
The general result presented here is new.

In (a), $\sup_{f\in\mathcal{F}} \|\tau_x f - f\|_{L^p(\omega;E)}$ is well-defined since $\tau_x f \in L^p(\Omega + x; E)$ (Theorem 15.4 (a)) and $\omega \subset \Omega + x$ provided x is small enough.

The proof will use the approximation M_r obtained by averaging over a ball of radius r. For $f \in L^p(\Omega; E)$, we define $M_r f$ on $\Omega_B = \{x \in \mathbb{R}^d : x + B \in \Omega\}$, where $B = \{x \in \mathbb{R}^d : |x| \leq r\}$, by: for every $x \in \Omega_B$,

$$M_r f(x) \stackrel{\text{def}}{=} \frac{1}{|\mathring{B}|} \int_{\mathring{B}(x,r)} f.$$

More precisely, we shall use the following properties.

Lemma 15.7. — *Let $f \in L^p(\Omega; E)$, where Ω is an open subset of $\mathbb{R}^d$, E is a semi-normed space, $1 \leq p < \infty$, ω is an open subset of Ω, and $r > 0$, such that $\omega + B \subset \Omega$. Then:*

(a) $$M_r f \in \mathcal{C}(\Omega_B; E);$$

(b) *for every $\eta > 0$ and x_1 and x_2 in ω_B such that $|x_1 - x_2| \leq \eta$,*

$$\|M_r f(x_2) - M_r f(x_1)\|_{E;\nu} \leq \frac{1}{|\mathring{B}|^{1/p}} \sup_{|x|\leq\eta} \|\tau_x f - f\|_{L^p(\omega;E);\nu};$$

(c) $$\|M_r f - f\|_{L^p(\omega;E);\nu} \leq \sup\nolimits_{|x|\leq r} \|\tau_x f - f\|_{L^p(\omega;E);\nu}. \blacksquare$$

Proof. ***1º)*** *Inequality* (b). We have

$$M_r f(x_2) - M_r f(x_1) = \frac{1}{|\mathring{B}|} \int_{\mathring{B}(x_1,r)} \tau_{x_1-x_2} f - f.$$

Since the integral is bounded by the semi-norm of $L^1(\mathring{B}; E)$, which in turn is bounded by the semi-norm of $L^p(\mathring{B}; E)$ up to powers of $|\mathring{B}|$ (Theorems 5.4 and 11.1 (b) with $q = 1$),

$$\begin{aligned} \|M_r f(x_2) - M_r f(x_1)\|_{E;\nu} &\leq \frac{1}{|\mathring{B}|} \|\tau_{x_1-x_2} f - f\|_{L^1(\mathring{B}(x_1,r);E);\nu} \leq \\ &\leq \frac{1}{|\mathring{B}|^{1/p}} \|\tau_{x_1-x_2} f - f\|_{L^p(\mathring{B}(x_1,r);E)\nu}. \end{aligned} \tag{15.1}$$

From which, we obtain inequality (b), because $\mathring{B}(x_1, r) \subset \omega$ since $x_1 \in \omega_B$.

2º) *Continuity* (a). Inequality (b) with $\omega = \Omega_B$ provides the continuity of M_f on Ω_B, since $\tau_{x_2-x_1} f \to f$ in $L^p(\mathring{B}(x_1, r); E)$ as $x_2 \to x_1$ from Theorem 15.4 (b).

3º) Inequality (c). Observe that

$$M_r f = \frac{1}{|\mathring{B}|} \int_{\mathring{B}} \tau_{-y} f \, dy,$$

where the integral takes values in $L^p(\omega; E)$, *i.e.* where we use the fact that (Theorem 15.4) $[x \mapsto \tau_x f] \in \mathbf{C}_b(\mathring{B}; L^p(\omega; E))$. Indeed, for $f \in \mathcal{B}(\Omega; E)$ and $x \in \omega$,

$$(M_r f)(x) = \frac{1}{|\mathring{B}|} \int_{\mathring{B}(x,r)} f(y) \, dy = \frac{1}{|\mathring{B}|} \int_{\mathring{B}} f(x+y) \, dy = \frac{1}{|\mathring{B}|} \int_{\mathring{B}} \tau_{-y} f(x) \, dy \, ;$$

we finish by taking the limit with continuous approximations for $f \in L^p(\Omega; E)$.

Thus,

$$M_r f - f = \frac{1}{|\mathring{B}|} \int_{\mathring{B}} \tau_{-y} f - f \, dy.$$

Finally, inequality (c) follows from the bound in Proposition 1.21 with values in $L^p(\omega; E)$, since $-y \in \mathring{B}$ is equivalent to $y \in \mathring{B}$. □

We are now able to prove our theorem.

Proof of Theorem 15.6. ***1º) Preliminary.*** Since $L^p(\Omega; E)$ is a Fréchet space (Theorem 10.17), relative sequentially compactness is equivalent to relative compactness, which itself is equivalent to precompactness (Proposition 15.3). Therefore, it suffices to show that (a), (b), and (c) together are equivalent to

$$\mathcal{F} \text{ is precompact in } L^p(\Omega; E). \tag{15.2}$$

That is (Definition 15.1 (c)): for every $\nu \in \mathcal{N}_E$ and $\epsilon > 0$, there exists a finite subset $\mathcal{G}$ of $L^p(\Omega; E)$ such that, for every $f \in \mathcal{F}$, there exists $g \in \mathcal{G}$ such that

$$\|g - f\|_{L^p(\Omega;E);\nu} \leq \epsilon.$$

2º) Forward implication. **2.a.** *A bound.* Let $\nu \in \mathcal{N}_E$ and $\epsilon > 0$, and consider a compact set $K \subset \Omega$ satisfying (b) and an open set ω such that $K \subset \omega \Subset \Omega$ (Proposition A.20). From (a), there exists $r > 0$ such that, for every $f \in \mathcal{F}$ and $x \in \mathbb{R}^d$ such that $\omega + x \subset \Omega$ and $|x| \leq r$, we have

$$\|\tau_x f - f\|_{L^p(\omega;E);\nu} \leq \epsilon.$$

Denote $B = B(0, r)$, and furthermore choose r small enough so that $K + B \subset \omega$ (Proposition A.19), namely $K \subset \omega_B$, and $\omega + B \subset \Omega$. Then (Lemma 15.7 (c))

$$\|M_r f - f\|_{L^p(\omega;E);\nu} \leq \sup_{|x| \leq r} \|\tau_x f - f\|_{L^p(\omega;E)} \leq \epsilon. \tag{15.3}$$

2.b. *Precompactness of $M_r\mathcal{F}$ in $\mathcal{C}_b(\omega_B; E)$.* Let us show that $\{M_r f|_{\omega_B} : f \in \mathcal{F}\}$ satisfies the two compactness criteria of Ascoli's theorem (Proposition 15.5):

— for all $x \in \omega_B$, every sequence in $\{M_r f(x) : f \in \mathcal{F}\}$ has a convergent subsequence in E; this is property (c) with $\omega = \overset{\circ}{B}(x,r)$ since $M_r f(x) = c \int_{\overset{\circ}{B}(x,r)} f$;

— the family $\{M_r f : f \in \mathcal{F}\}$ is uniformly equicontinuous on ω_B, i.e. for every $\nu \in \mathcal{N}_E$, when $\eta \to 0$,

$$\sup_{f\in\mathcal{F}} \sup_{x_1\in\omega_B,\, x_2\in\omega_B,\, |x_2-x_1|\le\eta} \|M_r f(x_2) - M_r f(x_1)\|_{E;\nu} \to 0;$$

this follows from (a) together with Lemma 15.7 (b).

Hence, $\{M_r f : f \in \mathcal{F}\}$ is precompact in $\mathcal{C}_b(\omega_B; E)$. That is, for every $\nu \in \mathcal{N}_E$ and $\epsilon > 0$, there exists a finite subset $\mathcal{G}$ of $\mathcal{C}_b(\omega_B; E)$ such that, for every $f \in \mathcal{F}$, there exists $g \in \mathcal{G}$ such that

$$\|g - M_r f\|_{\mathcal{C}_b(\omega_B;E);\nu} \le \frac{\epsilon}{|\omega_B|^{1/p}}.$$

2.c. *Conclusion.* It then follows, using the bound from Theorem 11.1 (b),

$$\|g - M_r f\|_{L^p(\omega_B;E);\nu} \le |\omega_B|^{1/p} \|g - M_r f\|_{L^\infty(\omega_B;E);\nu} \le \epsilon. \tag{15.4}$$

Since $\mathcal{K}(\omega_B; E)$ is dense in $L^p(\omega_B; E)$ (Theorem 10.14), for every $g \in \mathcal{G}$, there exists $g' \in \mathcal{K}(\omega_B; E)$ such that $\|g' - g\|_{L^p(\omega_B;E);\nu} \le \epsilon$. With (15.3) and (15.4),

$$\|g' - f\|_{L^p(\omega_B;E);\nu} = \|(g' - g) + (g - M_r f) + (M_r f - f)\|_{L^p(\omega_B;E);\nu} \le 3\epsilon. \tag{15.5}$$

Let $K' = K \cup \bigcup_{g\in\mathcal{G}} \operatorname{supp} g'$, which is a compact set (since it is the finite union of compact sets), and

$$K \subset K' \subset \omega_B.$$

Extend each g' by 0 and denote the result by g' again. Since $g' = 0$ on $\Omega \setminus K'$, which is contained in $\Omega \setminus K$, property (b) gives, with the bound on the restriction from Theorem 11.5 (b),

$$\|g' - f\|_{L^p(\Omega\setminus K';E);\nu} = \|f\|_{L^p(\Omega\setminus K';E);\nu} \le \|f\|_{L^p(\Omega\setminus K;E);\nu} \le \epsilon. \tag{15.6}$$

Since $(\Omega \setminus K') \cup \omega_B = \Omega$, then (15.5) and (15.6) give (Theorem 11.16 (b))

$$\|g' - f\|_{L^p(\Omega;E);\nu} \le \left(\|g' - f\|^p_{L^p(\Omega\setminus K';E);\nu} + \|g' - f\|^p_{L^p(\omega_B;E);\nu}\right)^{1/p} \le 4\epsilon.$$

As the set $\{g' : g \in \mathcal{G}\}$ is finite, this proves (15.2), namely

$$\mathcal{F} \text{ is precompact in } L^p(\Omega; E).$$

*3*º*) Backward implication.* If $\mathcal{F}$ is precompact in $L^p(\Omega; E)$, for every $\nu \in \mathcal{N}_E$ and $\epsilon > 0$, there exists a finite subset $\mathcal{G}$ of $L^p(\Omega; E)$ such that, for every $f \in \mathcal{F}$, there exists $g \in \mathcal{G}$ such that

$$\|g - f\|_{L^p(\Omega;E);\nu} \leq \epsilon.$$

For every g in $\mathcal{G}$:

— property (a) is satisfied by Theorem 15.4 (b);

— since $\mathcal{K}(\Omega; E)$ is sequentially dense in $L^p(\Omega; E)$ (Theorem 10.14), there exists $h \in \mathcal{K}(\Omega; E)$ such that $\|h - g\|_{L^p(\Omega;E);\nu} \leq \epsilon$; property (b) is therefore satisfied for $K = \operatorname{supp} g$.

As $\mathcal{G}$ is finite, properties (a) and (b) hold true for every $f \in \mathcal{F}$.

Finally, since the integral is uniformly continuous (Theorem 5.5), $\{\int_\omega f : f \in \mathcal{F}\}$ is precompact (Proposition A.45 (e)); this is equivalent, because E is a Hilbert space, to relative sequential compactness (Proposition 15.3), namely property (c). □

15.3. Special cases of compactness

For real-valued measures, let us show that criterion (c) of Theorem 15.6 is satisfied when $\mathcal{F}$ is bounded.

Theorem 15.8. — *In $L^p(\Omega; \mathbb{R}^\ell)$, criterion (c) of Theorem 15.6 is satisfied when $\mathcal{F}$ is bounded, whether it be in $L^p(\Omega; \mathbb{R}^\ell)$ or only in $L^p_{\mathrm{loc}}(\Omega; \mathbb{R}^\ell)$.* ▮

Proof of Theorem 15.8. When $E = \mathbb{R}^\ell$, if $\mathcal{F}$ is bounded in $L^p_{\mathrm{loc}}(\Omega; \mathbb{R}^\ell)$ then, for every open set $\omega \Subset \Omega$, $\mathcal{F}$ is bounded in $L^p(\omega; \mathbb{R}^\ell)$, which is contained in $L^1(\omega; \mathbb{R}^\ell)$ (Theorem 11.1).

Hence, $\{\int_\omega f : f \in \mathcal{F}\}$ is bounded in $\mathbb{R}^\ell$ so, by the Bolzano–Weierstrass theorem (Proposition A.37), each of its sequences has a convergent subsequence. □

Let us show that criterion (b) of Theorem 15.6 on equi-integrability is satisfied for any dominated family.

Theorem 15.9. — *Criterion (b) of Theorem 15.6 is satisfied if, for every $\nu \in \mathcal{N}_E$, there exists $g_\nu \in L^p(\Omega)$ such that*

$$\sup_{f \in \mathcal{F}} \|f\|_{E;\nu} \leq g_\nu. \ ▮$$

Proof. The subset $\mathcal{K}(\Omega)$ is dense in $L^p(\Omega)$ (Theorem 10.14). More precisely, given $\epsilon > 0$, there exists $h_\nu \in \mathcal{K}(\Omega)$ such that

$$\|h_\nu - g_\nu\|_{L^p(\Omega)} \leq \epsilon.$$

Thus, for $K = \operatorname{supp} h$, we have

$$\sup_{f \in \mathcal{F}} \|f\|_{L^p(\Omega \setminus K; E); \nu} \leq \|g_\nu\|_{L^p(\Omega \setminus K)} = \|g_\nu - h_\nu\|_{L^p(\Omega \setminus K)} \leq \epsilon. \ \square$$

Let us show that the equi-integrability criterion is also satisfied by every bounded family in $L^q(\Omega; E)$ with $q > p$, if Ω is bounded or, more generally, is of finite measure.

Theorem 15.10. — *Criterion (b) of Theorem 15.6 is satisfied if:*

$$\begin{cases} \Omega \textit{ is bounded or } |\Omega| < \infty, \textit{ and} \\ \textit{there exists } q > p \textit{ such that } \mathcal{F} \textit{ is bounded in } L^q(\Omega; E). \end{cases} \blacksquare$$

Proof. For every $\nu \in \mathcal{N}_E$, suppose that there exists a finite c_ν such that

$$\sup_{f \in \mathcal{F}} \|f\|_{L^q(\Omega; E); \nu} \leq c_\nu.$$

Then, for every compact subset $K \subset \Omega$ (Theorems 11.1 (b) and 11.5),

$$\|f\|_{L^p(\Omega \setminus K; E); \nu} \leq |\Omega \setminus K|^{1/p - 1/q} \|f\|_{L^q(\Omega \setminus K; E); \nu} \leq |\Omega \setminus K|^{1/p - 1/q} c_\nu.$$

This is bounded by ϵ if K is chosen to be large enough when $|\Omega|$ is finite (Theorem 2.12) or, *a fortiori*, when Ω is bounded. $\square$

Let us state some other conditions under which the equi-integrability criterion is satisfied (the definition of a Lipschitz boundary is given on the next page, Definition 15.13).

Theorem 15.11. — *Criteria (a) and (b) of Theorem 15.6 can be replaced by:*

Ω is a bounded open subset of $\mathbb{R}^d$ with a locally Lipschitz boundary

and, for every $\nu \in \mathcal{N}_E$, when $x \to 0$,

$$\sup_{f \in \mathcal{F}} \|\tau_x f - f\|_{L^p(\Omega \cup (\Omega + x); E); \nu} \to 0. \qquad (15.7) \ \blacksquare$$

Before proving Theorem 15.11, let us observe that it provides the following result in one dimension[3].

Theorem 15.12. — *If $\mathcal{F} \in L^p((a,b);E)$, where $-\infty < a < b < +\infty$, then criteria (a) and (b) of Theorem 15.6 can be replaced with: for every $\nu \in \mathcal{N}_E$, when $h \geq 0$ and $h \to 0$,*

$$\sup_{f \in \mathcal{F}} \int_a^{b-h} \|f(y+h) - f(y)\|_{E;\nu}^p \, dy \to 0. \blacksquare$$

Proof. The open set (a,b) is locally Lipschitz, bounded and, according to the integral expression for the $L^p(\Omega;E)$ norm (Theorem 13.4), for $x = h$ or $x = -h$, we have

$$\|\tau_x f - f\|_{L^p(\Omega \cup (\Omega + x);E);\nu} = \left(\int_a^{b-h} \|f(y+h) - f(y)\|_{E;\nu}^p \, dy \right)^{1/p}.$$

Theorem 15.11 then provides the stated result. $\square$

Let us recall the definition of a locally Lipschitz boundary.

Definition 15.13. — *The boundary $\partial\Omega$ of an open subset Ω of $\mathbb{R}^d$ is said to be* ***locally Lipschitz*** *if, for every $z \in \partial\Omega$, there exist real numbers $r_z > 0$ and $\kappa_z < \infty$, and:*

— *a Cartesian coordinate system $(x_1, \ldots, x_d)$ centered at z;*

— *a real-valued function ψ_z on $U_* = \{x_* \in \mathbb{R}_z^{d-1} : |x_*| < r_z\}$ such that $\psi_z(0) = 0$ and, for all x_* and y_* in U,*

$$|\psi_z(x_*) - \psi(y_*)| \leq \kappa_z |x_* - y_*|,$$

and, for all x and y in $U_z = \{x \in \mathbb{R}_z^d : |x| < r_z\}$,

$$x \in \Omega \quad \Leftrightarrow \quad x_d < \psi_z(x_*),$$

$$x \in \partial\Omega \quad \Leftrightarrow \quad x_d = \psi_z(x_*). \blacksquare$$

Let us now prove our compactness result.

Proof of Theorem 15.11. **1°)** *Cone condition.* The boundary $\partial\Omega$ of Ω is closed and bounded, hence it is compact in $\mathbb{R}^d$ by the Borel–Lebesgue theorem (Proposition A.36 (a)). It is covered by the open sets

$$U'_z = \left\{ x \in \mathbb{R}^d : |x - z| < \frac{r_z}{2} \right\},$$

and therefore (Definition 15.1 of compactness) by a finite number of them indexed by a finite set $\mathcal{Z}$.

3 **Historical background of Theorem 15.11.** Jacques SIMON proved in 1987 that the compact subsets of $L^p((0,T);E)$, where E is a Banach space, are characterized by (15.7) and criterion (c) of Theorem 15.6 [68, Theorem 1].

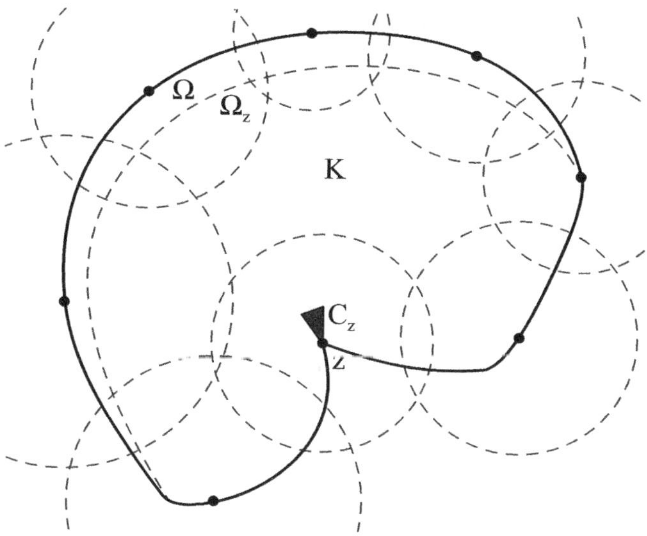

Figure 15.1. *Lipschitz boundary and cone property*

The set $K \stackrel{\text{def}}{=} \Omega \backslash \bigcup_{z\in\mathcal{Z}} U'_z$ is closed and bounded, hence compact, and contained in Ω. By the strong inclusion theorem (Proposition A.19), there exists a ball $B(0,t)$ such that the open set $\omega \stackrel{\text{def}}{=} K + B(0,t)$ is included in Ω. Let

$$\kappa \stackrel{\text{def}}{=} \max_{z\in\mathcal{Z}} \kappa_z, \quad r \stackrel{\text{def}}{=} \min\Big(t, \min_{z\in\mathcal{Z}} \frac{r_z}{2}\Big).$$

Then Ω is covered by a finite number of open sets indexed by $z \in \mathcal{Z}$,

$$\Omega_z \stackrel{\text{def}}{=} \{x \in \mathbb{R}^d : x + C_z \in \Omega\},$$

where the cone C_z is defined, relative to the coordinate system centered at z, by

$$C_z \stackrel{\text{def}}{=} \{x \in \mathbb{R}^d : \kappa|x_*| < x_d,\ |x| < r\}.$$

Indeed, $K \subset \Omega_z$ and $U'_z \cap \Omega \subset \Omega_z$, so

$$\Omega = K \bigcup_{z\in\mathcal{Z}} U'_z \cap \Omega \subset \bigcup_{z\in\mathcal{Z}} \Omega_z \subset \Omega.$$

2°) Preliminary. Let us revisit 2° from the proof of Theorem 15.6, this time replacing M_r with the mapping N_z, defined on each Ω_z by

$$N_z f(x) \stackrel{\text{def}}{=} \frac{1}{|C_z|} \int_{x+C_z} f. \tag{15.8}$$

Observe that, in the place of Lemma 15.7, we obtain $N_z f \in \mathcal{C}(\Omega_z; E)$ and, for all x_1 and x_2 in Ω_z such that $|x_2 - x_1| \leq \eta$,

$$\|N_z f(x_2) - N_z f(x_1)\|_{E;\nu} \leq \frac{1}{|C_z|^{1/p}} \sup_{x \in \Omega_z} \|\tau_x f - f\|_{L^p(\Omega_z;E);\nu} \tag{15.9}$$

and

$$\|N_z f - f\|_{L^p(\Omega_z;E);\nu} \leq \sup_{x \in \Omega_z} \|\tau_x f - f\|_{L^p(\Omega_z;E);\nu}. \tag{15.10}$$

3*º) *A bound. Similarly to (15.2), our aim here is to show that

$$\mathcal{F} \text{ is precompact in } L^p(\Omega; E).$$

So let $\nu \in \mathcal{N}_E$, $\epsilon > 0$, and n be the number of indices in $\mathcal{Z}$. By hypothesis (15.7), there exists $r > 0$ such that, for every $f \in \mathcal{F}$ and $x \in C_z$, $|x| \leq r$, we have

$$\|\tau_x f - f\|_{L^p(\Omega_z;E);\nu} \leq \frac{\epsilon}{2n}.$$

Thus, by (15.10),

$$\|N_z f - f\|_{L^p(\Omega_z;E);\nu} \leq \frac{\epsilon}{2n}. \tag{15.11}$$

4*º) *Precompactness of $N_z f$ in $\mathcal{C}_b(\Omega_z; E)$. Let us show that $\{N_z f|_{\Omega_z} : f \in \mathcal{F}\}$ satisfies the two compactness criteria of Ascoli's theorem (Proposition 15.5):

— for each $x \in \Omega_z$, every sequence in $\{N_z f(x) : f \in \mathcal{F}\}$ has a convergent subsequence in E; this follows from property (c) of Theorem 15.6 with $\omega = x + C_z$, since $N_z f(x) = c \int_{x+C_z} f$ by (15.8);

— the family $\{N_z f : f \in \mathcal{F}\}$ is uniformly equicontinuous on Ω_z, i.e. for every $\nu \in \mathcal{N}_E$, when $\eta \to 0$,

$$\sup_{f \in \mathcal{F}} \sup_{x_1 \in \Omega_z,\, x_2 \in \Omega_z,\, |x_2 - x_1| \leq \eta} \|N_z f(x_2) - N_z f(x_1)\|_{E;\nu} \to 0;$$

this follows from (15.7) using (15.9).

Hence, $\{N_z f : f \in \mathcal{F}\}$ is precompact in $\mathcal{C}_b(\Omega_z; E)$. That is, there exists a finite subset $\mathcal{G}_z$ of $\mathcal{C}_b(\Omega_z; E)$ such that, for every $f \in \mathcal{F}$, there exists $g_z \in \mathcal{G}_z$ such that

$$\|g_z - N_z f\|_{\mathcal{C}_b(\Omega_z;E);\nu} \leq \frac{\epsilon}{2n|\Omega|^{1/p}}.$$

5*º) *Gluing the Ω_z. Using the bounds from Theorems 11.1 (b) and 10.10 (b), it follows that

$$\|g_z - N_z f\|_{L^p(\Omega_z;E);\nu} \leq |\Omega|^{1/p} \|g_z - N_z f\|_{L^\infty(\Omega_z;E);\nu} \leq \frac{\epsilon}{2n}. \tag{15.12}$$

Let $(\alpha_z)_{z\in\mathcal{Z}}$ be a partition of unity subordinate (Proposition A.51) to the covering of Ω by the Ω_z, and let

$$g = \sum_{z\in\mathcal{Z}} \alpha_z g_z.$$

For every $f \in \mathcal{F}$, since $\sum_{z\in\mathcal{Z}} \alpha_z = 1$ and $|\alpha_z| \leq 1$, we have

$$\|g - f\|_{L^p(\Omega;E);\nu} = \Big\|\sum_{z\in\mathcal{Z}} \alpha_z(g_z - f)\Big\|_{L^p(\Omega_z;E);\nu} \leq \sum_{z\in\mathcal{Z}} \|g_z - f\|_{L^p(\Omega_z;E);\nu}.$$

Thus, with (15.11) and (15.12),

$$\|g - f\|_{L^p(\Omega;E);\nu} \leq \sum_{z\in\mathcal{Z}} \|g - N_z f\|_{L^p(\Omega_z;E);\nu} + \|N_z f - f\|_{L^p(\Omega_z;E);\nu} \leq \epsilon.$$

This proves that $\mathcal{F}$ is precompact in $L^p(\Omega; E)$, since the functions g are finite in number, given that there are finitely many g_z for each of the n sets Ω_z.

6º) Converse. If $\mathcal{F}$ is precompact in $L^p(\Omega; E)$ then, for every $\nu \in \mathcal{N}_E$ and $\epsilon > 0$, there exists a finite subset $\mathcal{G}$ of $L^p(\Omega; E)$ such that, for every $f \in \mathcal{F}$, there exists $g \in \mathcal{G}$ such that

$$\|g - f\|_{L^p(\Omega;E);\nu} \leq \epsilon.$$

Property (15.7) is satisfied by Theorem 15.4 (b) for each g in $\mathcal{G}$, which is finite, and hence for every $f \in \mathcal{F}$. □

15.4. Compact subsets of $L^p_{\text{loc}}(\Omega; E)$

Let us characterize the relatively compact subsets of $L^p_{\text{loc}}(\Omega; E)$, where E is a Fréchet space.

Theorem 15.14. — *Let $\mathcal{F} \subset L^p_{\text{loc}}(\Omega; E)$, where*

$$E \textit{ is a Fréchet space}, \quad 1 \leq p < \infty,$$

Ω is an open subset of $\mathbb{R}^d$ and $\{\| \ \|_{E;\nu} : \nu \in \mathcal{N}_E\}$ is the family of semi-norms of E.

Then, the two properties:

(a) *$\forall \nu \in \mathcal{N}_E$, $\forall \omega \Subset \Omega$, when $x \in \mathbb{R}^d$, $\omega + x \subset \Omega$ and $x \to 0$,*

$$\sup_{f\in\mathcal{F}} \|\tau_x f - f\|_{L^p(\omega;E);\nu} \to 0;$$

(b) $\forall \omega \Subset \Omega$,

$$\textit{every sequence in } \left\{ \int_\omega f : f \in \mathcal{F} \right\} \textit{ has a convergent subsequence in } E;$$

(a) and (b) together, are equivalent to:

$$\textit{every sequence in } \mathcal{F} \textit{ has a convergent subsequence in } L^p_{\mathrm{loc}}(\Omega; E),$$

which is itself equivalent to:

$$\mathcal{F} \textit{ is relatively compact in } L^p_{\mathrm{loc}}(\Omega; E). \blacksquare$$

Observe that the second criterion of Theorem 15.6 disappears in this case.

Proof of Theorem 15.14. **1°)** *Preliminary.* Since $L^p_{\mathrm{loc}}(\Omega; E)$ is a Fréchet space (Theorem 11.7), relative sequential compactness is equivalent to relative compactness, which in turn is equivalent to precompactness (Proposition 15.3). It therefore suffices to show that (a) and (b) are equivalent to

$$\mathcal{F} \text{ is precompact in } L^p_{\mathrm{loc}}(\Omega; E).$$

2°) *Forward implication.* Let $\epsilon > 0$ and consider an open set $\omega' \Subset \Omega$. By adapting part 2° from the proof of Theorem 15.6 with $K = \overline{\omega'}$, we observe that the current criteria (a) and (b) (that is, criteria (a) and (c) of Theorem 15.6) imply (15.5).

That is, for every $\nu \in \mathcal{N}_E$, there exists $r > 0$ such that $\omega' \subset \Omega_B$, where B is $B(0, r)$, and also a finite subset $\mathcal{G}$ in $\mathcal{C}_{\mathrm{b}}(\omega_B; E)$ such that, for every $f \in \mathcal{F}$, there exists $g \in \mathcal{G}$ such that

$$\|g - f\|_{L^p(\omega_B; E);\nu} \leq 2\epsilon.$$

In other words,

$$\|g - f\|_{L^p_{\mathrm{loc}}(\Omega; E);\omega';\nu} \leq 2\epsilon,$$

which proves that $\mathcal{F}$ is precompact in $L^p_{\mathrm{loc}}(\Omega; E)$.

3°) *Backward implication.* If $\mathcal{F}$ is precompact in $L^p_{\mathrm{loc}}(\Omega; E)$ then, for every $\nu \in \mathcal{N}_E$, $\omega \Subset \Omega$, and $\eta > 0$, there exists a finite subset $\mathcal{G}$ of $L^p_{\mathrm{loc}}(\Omega; E)$ such that, for every $f \in \mathcal{F}$, there exists $g \in \mathcal{G}$ such that

$$\|g - f\|_{L^p_{\mathrm{loc}}(\Omega; E);\omega,\nu} \leq \eta.$$

For every g in $\mathcal{G}$, property (a) is satisfied by Theorem 15.4 (b), and property (b) is clearly satisfied. Since $\mathcal{G}$ is finite, these properties hold for every $f \in \mathcal{F}$. $\square$

For real-valued measures, condition (b) of Theorem 15.14 is satisfied if $\mathcal{F}$ is bounded.

Theorem 15.15. — *Let Ω be an open subset of $\mathbb{R}^d$,*

$$\mathcal{F} \textit{ be a bounded subset of } L^p_{\mathrm{loc}}(\Omega), \quad 1 \leq p < \infty,$$

such that, for every $\omega \Subset \Omega$, when $x \in \mathbb{R}^d$, $x + \omega \subset \Omega$, and $x \to 0$:

$$\sup_{f \in \mathcal{F}} \|\tau_x f - f\|_{L^p(\omega)} \to 0.$$

Then, every sequence in $\mathcal{F}$ has a convergent subsequence in $L^p_{\mathrm{loc}}(\Omega)$. ∎

Proof. It suffices to verify criterion (b) of Theorem 15.14.

Now, given $\omega \Subset \Omega$, then $\mathcal{F}$ is bounded in $L^p(\omega)$ and therefore (Theorem 11.1 (a)) in $L^1(\omega)$. Hence, $\{\int_\omega f : f \in \mathcal{F}\}$ is bounded (Theorem 15.4) and so, by the Borel–Lebesgue theorem (Proposition A.36), it is included in a compact subset of $\mathbb{R}$. □

15.5. Compactness in intermediate spaces

Let us show that a bounded sequence in a space E_0, which converges to a limit in a space E_1, also converges in some space E, which is said to be **intermediate** between E_0 and E_1, because, typically,

$$E_0 \subsetneq E \subsetneq E_1.$$

Theorem 15.16. — *Let E_0, E_1, and E be three semi-normed spaces whose families of semi-norms are indexed by $\mathcal{N}_{E_0}$, $\mathcal{N}_{E_1}$, and $\mathcal{N}_E$, respectively, such that:*

(a) *E is a Neumann space and $E_0 \subset E \cap E_1$;*

(b) *for every $\nu \in \mathcal{N}_E$, there exist finite subsets N_0 and N_1 of $\mathcal{N}_{E_0}$ and $\mathcal{N}_{E_1}$, respectively, $0 \leq \theta \leq 1$, and $c \in \mathbb{R}$ such that, for every $u \in E_0$,*

$$\|u\|_{E;\nu} \leq c \sup_{\mu \in N_0} \|u\|^{1-\theta}_{E_0;\mu} \sup_{\lambda \in N_1} \|u\|^{\theta}_{E_1;\lambda}.$$

Then, every bounded sequence in E_0 that converges in E_1 also converges in E. ∎

Proof. Let $(u_n)_{n\in\mathbb{N}}$ be a bounded sequence in E_0 and $\nu \in \mathcal{N}_E$. We have

$$\sup_{m\geq n} \|u_n - u_m\|_{E;\nu} \leq c \sup_{m\geq n} \sup_{\mu\in N_0} \|u_n - u_m\|_{E_0;\mu}^{1-\theta} \sup_{\lambda\in N_1} \|u_n - u_m\|_{E_1;\lambda}^{\theta} \leq$$

$$\leq cb^{1-\theta} \sup_{m\geq n} \sup_{\lambda\in N_1} \|u_n - u_m\|_{E_1;\lambda}^{\theta},$$

where $b = 2\sup_{\mu\in N_0} \sup_{n\in\mathbb{N}} \|u_n\|_{E_0;\mu} < \infty$.

The sequence $(u_n)_{n\in\mathbb{N}}$ converges in E_1, and therefore it is Cauchy there (Proposition A.12). It is therefore Cauchy in E, and hence converges there (Definition 1.4) since E is a Neumann space. □

Equivalent formulation. Using Definition 15.2 of relative sequential compactness, Theorem 15.16 can be reformulated as follows. Under the hypotheses (a) and (b) of Theorem 15.16:

$$\begin{cases} \textit{every bounded subset of } E_0 \textit{ that is relatively sequentially compact in } E_1 \\ \textit{is relatively sequentially compact in } E. \end{cases} \quad (15.13)$$

If we were to say that a mapping is *sequentially compacting* if it maps bounded subsets to relatively sequentially compact subsets, then (15.13) can be reformulated as follows:

$$\begin{cases} \textit{if the identity mapping from } E_0 \textit{ into } E_1 \textit{ is sequentially compacting,} \\ \textit{then the identity mapping from } E_0 \textit{ into } E \textit{ is sequentially compacting.} \ \square \end{cases} \quad (15.14)$$

Usefulness of Theorem 15.16. In particular, Hypothesis (b) is satisfied by $L^p(\Omega; E)$ spaces: this is the Riesz convexity inequality (Theorem 11.2). It is also satisfied by $L^p_{\mathrm{loc}}(\Omega; E)$, as well as by $W^{m,p}(\Omega; E)$ and $W^{m,p}_{\mathrm{loc}}(\Omega; E)$.

In these spaces, a bound for p_0 and compactness for p_1 imply compactness for all $p_0 < p < p_1$. This property is used, for example, in the proof of Kondrachov's theorem to deduce compactness up to (but excluding) the exponent p_* of the Sobolev embedding, by using compactness in $L^p(\Omega; E)$. □

Let us state a property that is more general than Theorem 15.16.

Theorem 15.17. — *Let E_0, E_1, and E be three semi-normed spaces whose semi-norms are indexed by $\mathcal{N}_{E_0}$, $\mathcal{N}_{E_1}$, and $\mathcal{N}_E$, respectively, such that:*

(a) *E is a Neumann space and $E_0 \subset E \cap E_1$;*

(b) *for every $\nu \in \mathcal{N}_E$, there exist finite subsets N_0 and N_1 of $\mathcal{N}_{E_0}$ and $\mathcal{N}_{E_1}$, respectively, such that, for every $\epsilon > 0$, there exists $c \in \mathbb{R}$ such that, for every $u \in E_0$,*

$$\|u\|_{E;\nu} \leq \epsilon \sup_{\mu\in N_0} \|u\|_{E_0;\mu} + c \sup_{\lambda\in N_1} \|u\|_{E_1;\lambda};$$

(c) *U is bounded in E_0;*

(d) *every sequence in U has a convergent subsequence in E_1.*

Then, every sequence in U has a convergent subsequence in E. ▌

Proof. Let $(u_n)_{n\in\mathbb{N}}$ be a sequence in U. By (d), one of its subsequences converges in E_1, and hence is Cauchy there (Proposition A.12).

Let $\eta > 0$ and $b = \sup_{v\in U} \sup_{\mu\in N_0} \|v\|_{E_0;\mu}$, which is finite by (c). For the subsequence, (b) together with $\epsilon = \eta/b$, we have

$$\sup_{m\geq n} \|u_n - u_m\|_{E;\nu} \leq \epsilon \sup_{m\geq n} \sup_{\mu\in N_0} \|u_n - u_m\|_{E_0;\mu} + c \sup_{m\geq n} \sup_{\lambda\in N_1} \|u_n - u_m\|_{E_1;\lambda} \leq$$
$$\leq 2\eta + c \sup_{m\geq n} \sup_{\lambda\in N_1} \|u_n - u_m\|_{E_1;\lambda}.$$

This is bounded by 3η as soon as n is large enough. Hence, the subsequence is Cauchy in E and, as E is a Neumann space by (a), it converges there (Definition 1.4). □

Conversely, let us show that hypothesis (b) of Theorem 15.18 is satisfied if the identity mapping from E_0 into E is sequentially compacting. This is known as **Lions'**, or **Ehrling's, compactness lemma**[4].

Theorem 15.18. — *Suppose that we have three spaces such that:*

$$E \text{ is semi-normed,} \quad E_0 \text{ and } E_1 \text{ are normed,} \quad E_0 \subsetneq E \subsetneq E_1,$$

every bounded sequence in E_0 has a convergent subsequence in E.

Then, for every semi-norm $\|\ \|_{E;\nu}$ of E and $\epsilon > 0$, there exists $c \in \mathbb{R}$ such that, for all $u \in E_1$,

$$\|u\|_{E;\nu} \leq \epsilon\|u\|_{E_0} + c\|u\|_{E_1}. \ ▌$$

Proof. If no such c existed then, for every $n \in \mathbb{N}$, there would exist $u_n \in E_1$ such that

$$\|u_n\|_{E;\nu} > \epsilon\|u_n\|_{E_0} + n\|u_n\|_{E_1}.$$

Then $v_n = u_n/\|u_n\|_{E;\nu}$ would satisfy

$$1 = \|v_n\|_{E;\nu} > \epsilon\|v_n\|_{E_0} + n\|v_n\|_{E_1};$$

4 **Historical background of Theorem 15.18.** Jacques-Louis LIONS stated the inequality in 1961 [48, p. 59] when E, E_1, and E_2 are Banach spaces. Still within the framework of Banach spaces, this result was attributed to Gunnar EHRLING by Michael RENARDY and Robert ROGERS [60, Theorem 6.99 p. 217], who do not provide a reference.
As we observe here, completeness of E, E_0, and E_1 is not necessary.

so $\|v_n\|_{E_1} \leq 1/n$, whence

$$v_n \to 0 \text{ in } E_1.$$

Moreover, $\|v_n\|_{E_0} \leq 1/\epsilon$ so, by hypothesis, a subsequence would satisfy, for a certain v,

$$v_n \to v \text{ in } E.$$

Since $E \subsetneq E_1$, this convergence would also take place in E_1, whence $v = 0$ and

$$\|v_n\|_{E;\nu} \to 0.$$

This would contradict $\|v_n\|_{E;\nu} = 1$, therefore such a c must exist. □

Chapter 16

Duals

Here, we present some properties of the duals of $L^p(\Omega; E)$ when E is a Banach space.

We show that $L^p(\Omega; E)$ is uniformly convex if E is a uniformly convex Banach space and, in particular, if it is a Hilbert space when $1 < p < \infty$ (Theorem 16.1). It follows then that $L^p(\Omega; E)$ is reflexive and $L^p(\Omega; E)$-weak is a Neumann space (Theorem 16.2).

We define the canonical injection J from $L^{p'}(\Omega; E')$ into the dual of $L^p(\Omega; E)$ for any Banach space E, and show that it is isometric if p is finite (Theorem 16.7). If E is a uniformly convex Banach space and p is finite, then J is an isomorphism (Theorem 16.8). If E is a Hilbert space and p is finite, then every element of $(L^p(\Omega; E))'$ can be represented by an element of $L^{p'}(\Omega; E)$ (Theorem 16.10), which is the Riesz–Fréchet theorem.

Then we define various families of semi-norms which generate the topology of $L^p(\Omega; E)$-weak (Theorem 16.12) or of $L^\infty(\Omega; E)$-$*$weak (Theorem 16.16). We compare $L^p(\Omega)$-weak to $\mathcal{M}(\Omega)$ (Theorems 16.13 and 16.14), and show that $L^\infty(\Omega; E)$-$*$weak is a Neumann space if E' is a separable Banach space (Theorem 16.17).

These results will be extended upon by Theorems 18.12 and 18.13.

16.1. Uniform convexity of $L^p(\Omega; E)$

We say that a normed space E is **uniformly convex** if, for every $\epsilon > 0$, there exists $\delta > 0$ such that: for all u and v in E,

$$\|u\|_E = \|v\|_E = 1, \ \|u - v\|_E \geq \epsilon \quad \Rightarrow \quad \left\|\frac{u+v}{2}\right\|_E \leq 1 - \delta.$$

Caution. Uniform convexity is not preserved when the original norm is replaced **by an equivalent norm.** For example, $\mathbb{R}^2$ is uniformly convex with its Euclidean norm $\|x\| = (x_1^2 + x_2^2)^{1/2}$, but is no longer so when endowed with either the $\|x\|_1 = |x_1| + |x_2|$ norm or the $\|x\|_\infty = \sup\{|x_1|, |x_2|\}$ norm. □

Let us observe that any Hilbert space is uniformly convex because of the parallelogram equality $\|u+v\|_E^2 + \|u-v\|_E^2 = 2(\|u\|_E^2 + \|v\|_E^2)$ (Proposition A.31), which gives

$$\|u\|_E = \|v\|_E = 1,\ \|u-v\|_E \geq \epsilon \quad \Rightarrow \quad \left\|\frac{u+v}{2}\right\|_E \leq \sqrt{1-\frac{\epsilon^2}{4}} \leq 1-\frac{\epsilon^2}{8}.$$

Let us show that $L^p(\Omega)$, and more generally $L^p(\Omega; E)$ for a uniformly convex Banach space E, is uniformly convex when $1 < p < \infty$, which is **Clarkson's theorem** [1].

Theorem 16.1. — *If E is a uniformly convex Banach space (in particular, if it is a Hilbert space), Ω is an open subset of $\mathbb{R}^d$, and $1 < p < \infty$, then*

$$L^p(\Omega; E) \textit{ is uniformly convex.}$$

In particular, $L^p(\Omega)$ and $L^p(\Omega; L^q(\Lambda))$, when $1 < q < \infty$ and Λ is an open subset of $\mathbb{R}^\ell$, are uniformly convex. ▌

Before going through the proof, observe from this that it implies that $L^p(\Omega; E)$ is reflexive and $L^p(\Omega; E)$-weak is a Neumann space. (This is also true if E is a separable reflexive Banach space, see Theorem 18.13.)

Theorem 16.2. — *If E is a uniformly convex Banach space, Ω is an open subset of $\mathbb{R}^d$, and $1 < p < \infty$, then*

$$\begin{cases} L^p(\Omega; E) \textit{ is a reflexive Banach space, and} \\ L^p(\Omega; E)\text{-weak } \textit{is a Neumann space.} \end{cases}$$

In particular, $L^p(\Omega)$ is reflexive and $L^p(\Omega)$-weak is a Neumann space. ▌

Proof. ***1o) Reflexivity.*** Under these hypotheses, $L^p(\Omega; E)$ is uniformly convex (Theorem 16.1) and therefore reflexive (Definition A.72) by Milman's theorem (Proposition A.74).

2o) Sequential completion. If the space F is reflexive, then F-weak is a Neumann space (Proposition A.75). Thus, $L^p(\Omega; E)$-weak is a Neumann space. □

1 **Historical background of Clarkson's theorem.** James Andrew CLARKSON proved that $L^p(\Omega)$ is uniformly convex when $1 < p < \infty$ in 1936 [17, corollary, p. 403].
Mahlon Marsh DAY showed in 1940, more generally, that $L^p(\Omega; E)$ is uniformly convex if E is a uniformly convex Banach space and $1 < p < \infty$ [18, theorem 2].

We shall now prove Clarkson's theorem in a series of steps.

Proof of Theorem 16.1 for E a Hilbert space. Let E be a Hilbert space.

1º) Case $p \geq 2$. **1.a.** *An inequality in* E. Let us show, for every a and b in E, that

$$\|a+b\|_E^p + \|a-b\|_E^p \leq 2^{p-1}(\|a\|_E^p + \|b\|_E^p). \tag{16.1}$$

Indeed, the first inequality of Proposition 9.3 (comparison of the sums with respect to distinct powers), the parallelogram equality (Proposition A.31 (a)), and the second inequality of Proposition 9.3 give

$$(\|a+b\|_E^p + \|a-b\|_E^p)^{1/p} \leq (\|a+b\|_E^2 + \|a-b\|_E^2)^{1/2} =$$
$$= (2\|a\|_E^2 + 2\|b\|_E^2)^{1/2} \leq 2^{1-1/p}(\|a\|_E^p + \|b\|_E^p)^{1/p}.$$

1.b. *An inequality in* $L^p(\Omega; E)$. Let us show that, for every f and g in $L^p(\Omega; E)$, we obtain **Clarkson's (first) inequality**

$$\|f+g\|_{L^p(\Omega;E)}^p + \|f-g\|_{L^p(\Omega;E)}^p \leq 2^{p-1}(\|f\|_{L^p(\Omega;E)}^p + \|g\|_{L^p(\Omega;E)}^p). \tag{16.2}$$

Indeed, denoting by $(f_n)_{n\in\mathbb{N}}$ and $(g_n)_{n\in\mathbb{N}}$ the p-integrabilizing sequences for f and g (Definition 10.1), from (16.1) we get

$$\int_\Omega \|f_n + g_n\|_E^p + \|f_n - g_n\|_E^p \leq 2^{p-1}\Big(\int_\Omega \|f_n\|_E^p + \int_\Omega \|g_n\|_E^p\Big).$$

Whence, in the limit (Definition 10.2), we obtain Clarkson's inequality.

1.c. *Uniform convexity.* Let $\epsilon > 0$ and

$$\|f\|_{L^p(\Omega;E)} = \|g\|_{L^p(\Omega;E)} = 1, \quad \|f-g\|_{L^p(\Omega;E)} \geq \epsilon.$$

Clarkson's inequality (16.2) gives $\|f+g\|_{L^p(\Omega;E)}^p \leq 2^p - \epsilon^p$, and therefore

$$\Big\|\frac{f+g}{2}\Big\|_{L^p(\Omega;E)} \leq 1-\delta$$

where $\delta = 1 - (1-(\epsilon/2)^p)^{1/p} > 0$, which proves the uniform convexity.

2º) Case $p \leq 2$. **2.a.** *An inequality in* E. Let us show, for every $a \in E$ and $b \in E$, that

$$k\|a-b\|_E^2 \leq (\|a\|_E^p + \|b\|_E^p)^{2/p-1}(\|a\|_E^p + \|b\|_E^p - 2^{1-p}\|a+b\|_E^p), \tag{16.3}$$

where $k = p(p-1)/4$.

2.b. *Case $E = \mathbb{R}$.* Let us begin with this case, from which we shall deduce the general one. For all $s \in \mathbb{R}$ and $t \in \mathbb{R}$, we aim to verify that

$$k|s-t|^2 \leq (|s|^p + |t|^p)^{2/p-1}(|s|^p + |t|^p - 2^{1-p}|s+t|^p). \tag{16.4}$$

It suffices to verify it for $|t| \leq |s|$ (possibly exchanging s and t) and $s = 1$ (by dividing by $|s|$ if $s \neq 0$). Now, Lemma 16.3 (a) below gives, for $|t| \leq 1$,

$$k(1-t)^2 \leq 1 + |t|^p - 2^{1-p}(1+t)^p.$$

We then conclude that, *a fortiori*,

$$k(1-t)^2 \leq (1+|t|^p)^{2/p-1}(1+|t|^p - 2^{1-p}(1+t)^p).$$

2.c. *Case when E is a Hilbert space.* Denote $c = (a+b)/2$ and $d = (a-b)/2$. Inequality (16.4) with $s = \|c\|_E + \|d\|_E$ and $t = \|c\|_E - \|d\|_E$ can be written as

$$\begin{aligned} k\|a-b\|_E^2 \leq ((\|c\|_E + \|d\|_E)^p + |\|c\|_E - \|d\|_E|^p)^{2/p-1} \times \\ \times ((\|c\|_E + \|d\|_E)^p + |\|c\|_E - \|d\|_E|^p - 2^{1-p}(\|a+b\|_E)^p). \end{aligned}$$

We deduce (16.3) from the following bound

$$\begin{aligned} &(\|c\|_E + \|d\|_E)^p + |\|c\|_E - \|d\|_E|^p = \\ &\quad = (\|c\|_E^2 + \|d\|_E^2 + 2\|c\|_E\|d\|_E)^{p/2} + (\|c\|_E^2 + \|d\|_E^2 - 2\|c\|_E\|d\|_E)^{p/2} \leq \\ &\quad \leq (\|c\|_E^2 + \|d\|_E^2 + 2c \cdot d)^{p/2} + (\|c\|_E^2 + \|d\|_E^2 - 2c \cdot d)^{p/2} = \\ &\quad = \|c+d\|_E^p + \|c-d\|_E^p = \\ &\quad = \|a\|_E^p + \|b\|_E^p, \end{aligned}$$

where $\cdot$ denotes the inner product on E, and where the inequality is provided by Lemma 16.3 (b) and $|2c \cdot d| \leq 2\|c\|_E\|d\|_E \leq \|c\|_E^2 + \|d\|_E^2$ (Proposition A.31 (b)).

2.d. *An inequality in $L^p(\Omega; E)$.* Suppose that f and g belong to $L^p(\Omega; E)$, and that $(f_n)_{n\in\mathbb{N}}$ and $(g_n)_{n\in\mathbb{N}}$ are p-integrabilizing sequences (Definition 10.1). Inequality (16.3) then gives, together with Hölder's generalized inequality (Proposition 9.6 with $p_1 = p/(2-p)$, $p_2 = 1$, and the p in the proposition equal to $p/2$),

$$\begin{aligned} &k\Big(\int_\Omega \|f_n - g_n\|_E^p\Big)^{2/p} \leq \\ &\quad \leq \Big(\int_\Omega \|f_n\|_E^p + \|g_n\|_E^p\Big)^{2/p-1}\Big(\int_\Omega \|f_n\|_E^p + \|g_n\|_E^p - 2^{1-p}\|f_n + g_n\|_E^p\Big). \end{aligned}$$

In the limit (Definition 10.2), we obtain the **Morawetz inequality**[2]

$$
\begin{aligned}
k\|f-g\|_{L^p(\Omega;E)}^2 \le (\|f\|_{L^p(\Omega;E)}^p + \|g\|_{L^p(\Omega;E)}^p)^{2/p-1} \times \\
\times (\|f\|_{L^p(\Omega;E)}^p + \|g\|_{L^p(\Omega;E)}^p - 2^{1-p}\|f+g\|_{L^p(\Omega;E)}^p).
\end{aligned} \tag{16.5}
$$

2.e. *Uniform convexity.* Let $\epsilon > 0$ and

$$\|f\|_{L^p(\Omega;E)} = \|g\|_{L^p(\Omega;E)} = 1, \quad \|f-g\|_{L^p(\Omega;E)} \ge \epsilon.$$

The Morawetz inequality then gives $k\epsilon^2 \le 2^{2/p}(1 - 2^{-p}\|f+g\|_{L^p(\Omega;E)}^p)$, and thus

$$\Big\|\frac{f+g}{2}\Big\|_{L^p(\Omega;E)} \le 1-\delta$$

where $\delta = 1-(1-k\epsilon^2 2^{-2/p})^{1/p} > 0$. This once again proves uniform convexity. □

It remains to establish the following inequalities.

Lemma 16.3. — *Let* $1 \le p \le 2$. *For all real* t, s, *and* r, *we have:*

(a) $\quad |t| \le 1 \quad \Rightarrow \quad 1 + |t|^p \ge 2^{1-p}(1+t)^p + p(p-1)(1-t)^2/4;$

(b) $\quad |t| \le |s| \le r \quad \Rightarrow \quad |r+t|^{p/2} + |r-t|^{p/2} \ge |r+s|^{p/2} + |r-s|^{p/2}.$ ▮

Proof. **1°)** *Property* (a). We aim to bound

$$\psi(t) \overset{\text{def}}{=} 1 + |t|^p - 2^{1-p}(1+t)^p - \frac{p(p-1)}{4}(1-t)^2.$$

We have $\psi(1) = \psi'(1) = \psi''(1) = 0$ and, for $t \ne 0$, $t \ne -1$, and $\sigma \overset{\text{def}}{=}$ sign of t,

$$
\begin{aligned}
\psi'(t) &= p\Big(\sigma|t|^{p-1} - 2^{1-p}(1+t)^{p-1} + \frac{p-1}{2}(1-t)\Big),\\
\psi''(t) &= p(p-1)\Big(\sigma|t|^{p-2} - 2^{1-p}(1+t)^{p-2} - \frac{1}{2}\Big),\\
\psi'''(t) &= p(p-1)(p-2)(\sigma|t|^{p-3} - 2^{1-p}(1+t)^{p-3}).
\end{aligned}
$$

---

2 **Historical background of the Morawetz inequality.** Cathleen Synge MORAWETZ showed in 1969 [54, theorem, p. 1300] that there exists $k > 0$ such that (16.5) is satisfied for $E = \mathbb{R}^d$.
In 1984, A. MEIR established, for the same $k = p(p-1)/4$, the inequality [51, inequality (2.1), p. 421]

$$k\|f-g\|_{L^p(\Omega;E)}^2 \le \|\|f\|_E + \|g\|_E\|_{L^p(\Omega)}^{2-p}(\|f\|_{L^p(\Omega;E)}^p + \|g\|_{L^p(\Omega;E)}^p - 2^{1-p}\|f+g\|_{L^p(\Omega;E)}^p).$$

On $(0, 1]$, $\psi''' \leq 0$ (since $t^{p-3} > (1+t)^{p-3}$ as $p - 3 < 0$). Therefore, we have (Proposition A.80 (b)) $\psi'' \geq \psi''(1) = 0$. Whence $\psi' \leq \psi'(1) = 0$, and

$$\psi \geq \psi(1) = 0.$$

It follows by continuity that $\psi(0) \geq 0$.

On $(-1, 0)$, $\psi''' \geq 0$ because $\sigma = -1$. Therefore, ψ'' is increasing and vanishes at most at one point, m.

On $[m, 0)$, $\psi'' \geq 0$, hence $\psi' \leq \psi'(0) \leq 0$, and

$$\psi \geq \psi(0) \geq 0.$$

On $(-1, m]$ (which may be empty), $\psi'' \leq 0$, and $\psi(-1) = 2 - p(p-1) \geq 0$, whence

$$\psi \geq \inf(\psi(-1), \psi(m)) \geq 0.$$

2⁰) Property (b). Denote

$$\phi(t) \stackrel{\text{def}}{=} |r + t|^{p/2} + |r - t|^{p/2}.$$

If $0 \leq t < r$, we have $\phi'(t) = (p/2)(|r+t|^{p/2-1} - |r-t|^{p/2-1})$, which is negative. Thus (Proposition A.80 (b)), if $0 \leq t \leq s \leq r$,

$$\phi(t) \geq \phi(s).$$

This also holds if $|t| \leq |s| \leq r$ since $\phi(-t) = \phi(t)$. □

The proof of Clarkson's theorem (now for a uniformly convex E) will make use of the following properties.

Lemma 16.4. *— If $1 < p < \infty$, then the space ℓ^p of real sequences $u = (u_i)_{i\in\mathbb{N}}$ such that $\sum_{i\in\mathbb{N}} |u_i|^p$ is finite, endowed with the norm $\|u\|_{\ell^p} \stackrel{\text{def}}{=} (\sum_{i\in\mathbb{N}} |u_i|^p)^{1/p}$, is uniformly convex.* ▮

Proof. When $p \geq 2$, then inequality (16.1) together with Hölder's inequality for finite sums (Proposition 9.2) implies that, for all finite sequences u and v (padded by 0),

$$\|u + v\|^p_{\ell^p} + \|u - v\|^p_{\ell^p} \leq 2^{p-1}(\|u\|^p_{\ell^p} + \|v\|^p_{\ell^p}).$$

Since finite real sequences are sequentially dense in ℓ^p, this inequality also holds in this space and implies uniform convexity (as in part 1.c of the proof of Theorem 16.1 when E is a Hilbert space).

When $p \leq 2$, we proceed similarly by using inequality (16.3), which gives

$$k(\|u - v\|_{\ell^p})^2 \leq (\|u\|^p_{\ell^p} + \|v\|^p_{\ell^p})^{2/p-1}(\|u\|^p_{\ell^p} + \|v\|^p_{\ell^p} - 2^{1-p}\|u + v\|^p_{\ell^p}).\ \square$$

Lemma 16.5. — *If E is a uniformly convex Banach space and $1 < p < \infty$, then the space $\ell^p_{\mathrm{f}}(E)$ of finite sequences $b = (b_0, b_1, \ldots, b_I, 0, \ldots)$ in E, endowed with the norm $\|b\|_{\ell^p_{\mathrm{f}}(E)} \stackrel{\text{def}}{=} (\sum_{i\in\mathbb{N}} \|b_i\|^p_E)^{1/p}$, is uniformly convex.* ▌

Proof. To simplify notation, let us write $|||\ |||$ in the place of $\|\ \|_{\ell^p_{\mathrm{f}}(E)}$ and $\|\ \|$ in the place of $\|\ \|_E$.

1º) A particular case. Here, let us suppose that, for $\epsilon > 0$,

$$|||b||| = |||b'||| = 1, \quad |||b - b'||| \geq \epsilon \quad \text{and} \quad \|b_i\| = \|b'_i\| \text{ for all } i. \tag{16.6}$$

***1.a.** Preliminary.* Since the norm $\|\ \|$ of E is uniformly continuous (see Proposition A.10), there exists a strictly increasing positive function δ such that

$$\left\|\frac{b_i + b'_i}{2}\right\| \leq \left(1 - \delta\left(\frac{\|b_i - b'_i\|}{\|b_i\|}\right)\right)\|b_i\|.$$

Therefore,

$$\frac{|||b + b'|||}{2} \leq \left(\sum_{i\in\mathbb{N}}\left(1 - \delta\left(\frac{\|b_i - b'_i\|}{\|b_i\|}\right)\right)^p \|b_i\|^p\right)^{1/p}.$$

Let

$$X \stackrel{\text{def}}{=} \{i \in \mathbb{N} : \|b_i - b'_i\| > \frac{\epsilon}{4}\|b_i\|\}$$

and let Y be the complement of X. If $i \in X$, we have

$$\frac{|||b + b'|||}{2} \leq \left(\left(1 - \delta\left(\frac{\epsilon}{4}\right)\right)^p \sum_{i\in X} \|b_i\|^p + \sum_{i\in Y} \|b_i\|^p\right)^{1/p}.$$

Since $\sum_{i\in X} \|b_i\|^p + \sum_{i\in Y} \|b_i\|^p = |||b|||^p = 1$,

$$\frac{|||b + b'|||}{2} \leq \left(1 - \left(1 - \left(1 - \delta\left(\frac{\epsilon}{4}\right)\right)^p\right) \sum_{i\in X} \|b_i\|^p\right)^{1/p}. \tag{16.7}$$

***1.b.** Conclusion.* If $i \in Y$, we have $\|b_i\| \geq (4/\epsilon)\|b_i - b'_i\|$, so

$$1 = \left(\sum_{i\in\mathbb{N}} \|b_i\|^p\right)^{1/p} \geq \left(\sum_{i\in Y} \|b_i\|^p\right)^{1/p} \geq \frac{4}{\epsilon}\left(\sum_{i\in Y} \|b_i - b'_i\|^p\right)^{1/p}.$$

That is,

$$\left(\sum_{i\in Y} \|b_i - b'_i\|^p\right)^{1/p} \leq \frac{\epsilon}{4}. \tag{16.8}$$

Thus, by the inequalities in (16.6) and (16.8),

$$\Big(\sum_{i\in X}\|b_i-b_i'\|^p\Big)^{1/p}=\Big(\sum_{i\in\mathbb{N}}\|b_i-b_i'\|^p-\sum_{i\in Y}\|b_i-b_i'\|^p\Big)^{1/p}\ge\Big(\epsilon^p-\Big(\frac{\epsilon}{4}\Big)^p\Big)^{1/p}\ge\frac{3\epsilon}{4}.$$

Since $\|b_i-b_i'\|\le\|b_i\|+\|b_i'\|=2\|b_i\|$, we have

$$\Big(\sum_{i\in X}\|b_i\|^p\Big)^{1/p}\ge\frac{3\epsilon}{8}.$$

Then, (16.7) gives

$$\frac{|||b+b'|||}{2}\le\Big(1-\Big(1-\Big(1-\delta\Big(\frac{\epsilon}{4}\Big)\Big)^p\Big)\Big(\frac{3\epsilon}{8}\Big)^p\Big)^{1/p}.$$

This indeed yields uniform convexity, with δ_1 strictly positive and increasing:

$$\frac{|||b+b'|||}{2}\le 1-\delta_1(\epsilon). \tag{16.9}$$

2º) General case. Here, let us suppose that

$$|||b|||=|||b'|||=1,\qquad\frac{|||b+b'|||}{2}>1-\delta_2(\alpha), \tag{16.10}$$

where $0<\alpha\le 2$ and, thanks to Lemma 16.4,

δ_2 is a modulus of uniform convexity in the space ℓ^p.

Since the sequences $(\|b_i\|)_{i\in\mathbb{N}}$ and $(\|b_i'\|)_{i\in\mathbb{N}}$ belong to ℓ^p, (16.10) implies

$$\Big(\sum_{i\in\mathbb{N}}\big|\|b_i\|-\|b_i'\|\big|^p\Big)^{1/p}\le\alpha.$$

Let b'' be the finite sequence in E such that $b''=b_i'\|b_i\|/\|b_i'\|$, $\|b_i''\|=\|b_i\|$, and

$$|||b'-b''|||=\Big(\sum_{i\in\mathbb{N}}\Big|\|b_i'\|-\frac{b_i'}{\|b_i'\|}\|b_i\|\Big|^p\Big)^{1/p}=\Big(\sum_{i\in\mathbb{N}}\big|\|b_i'\|-\|b_i\|\big|^p\Big)^{1/p}\le\alpha. \tag{16.11}$$

Choose α such that

$$0<\alpha<\frac{\epsilon}{2},\quad\delta_2(\alpha)+\frac{\alpha}{2}<\delta_1\Big(\frac{\epsilon}{2}\Big). \tag{16.12}$$

Then, (16.10) implies

$$|||b+b''|||\ge|||b+b'|||-|||b''-b'|||\ge 2\Big(1-\delta_2(\alpha)-\frac{\alpha}{2}\Big)>2\Big(1-\delta_1\Big(\frac{\epsilon}{2}\Big)\Big).$$

From 1°, if (16.9) does not hold, then (16.6) does not hold either, hence

$$|||b - b''||| < \frac{\epsilon}{2}.$$

Finally, using (16.11) and the first condition in (16.12), this yields

$$|||b - b'||| \le |||b - b''||| + |||b'' - b'||| < \frac{\epsilon}{2} + \frac{\epsilon}{2}. \tag{16.13}$$

Define $\delta_3(\epsilon) \stackrel{\text{def}}{=} \delta_2(\alpha)$. Then δ_3 is positive and, if (16.13) is false, then (16.10) is false too, so

$$|||b||| = |||b'||| = 1, \quad |||b - b'||| > \epsilon \quad \Rightarrow \quad \frac{|||b + b'|||}{2} < 1 - \delta_3(\epsilon). \tag{16.14}$$ □

Let us now prove Clarkson's theorem.

Proof of Theorem 16.1. Let $\epsilon > 0$, and consider f and f' in $L^p(\Omega; E)$ such that

$$\|f\|_{L^p(\Omega;E)} = \|f'\|_{L^p(\Omega;E)} = 1, \quad \|f - f'\|_{L^p(\Omega;E)} \ge \epsilon.$$

Let $f_n = \sum_{s \in S_n} 1_{s,n} f_{s,n}$ be the piecewise constant approximation given by Theorem 13.9, and

$$\widehat{f_n} = (2^{-nd} f_{s,n})_{s \in S_n}.$$

Recall that S_n is a finite subset of $\mathbb{Z}^d$ and hence is countable (Proposition A.2 (c), so it can be reindexed as a finite subset of $\mathbb{N}$. Thus, $\widehat{f_n} \in \ell^p_{\mathrm{f}}(E)$ and (Theorem 13.9)

$$\|f_n\|_{L^p(\Omega;E)} = \Big(\sum_{s \in S_n} \|2^{-nd} f_{s,n}\|^p\Big)^{1/p} = \|\widehat{f_n}\|_{\ell^p_{\mathrm{f}}(E)}.$$

Therefore,

$$\|\widehat{f_n}\|_{\ell^p_{\mathrm{f}}(E)} = \|\widehat{f'_n}\|_{\ell^p_{\mathrm{f}}(E)} = 1, \quad \|\widehat{f_n} - \widehat{f'_n}\|_{\ell^p_{\mathrm{f}}(E)} \ge \epsilon.$$

Lemma 16.5 then gives, with $\delta > 0$ independent of n,

$$\left\|\frac{f_n + f'_n}{2}\right\|_{L^p(\Omega;E)} = \left\|\frac{\widehat{f_n} + \widehat{f'_n}}{2}\right\|_{\ell^p_{\mathrm{f}}(E)} \ge 1 - \delta.$$

Taking the limit in n (Theorem 13.9), we obtain

$$\left\|\frac{f + f'}{2}\right\|_{L^p(\Omega;E)} \ge 1 - \delta.$$ □

Other inequalities for $1 < p \leq 2$. Clarkson's second inequality can be written [17] as follows

$$(\|f+g\|_{L^p(\Omega)}^{p'} + \|f-g\|_{L^p(\Omega)}^{p'})^{p/p'} \leq 2^{1-p}(\|f\|_{L^p(\Omega)}^p + \|g\|_{L^p(\Omega)}^p).$$

Using **Beurling's inequality**, also known as **Hanner's inequality** (it was presented by BEURLING in an unpublished seminar; its proof can be found in [HANNER, 34, Theorem 1, p. 239]),

$$(\|f+g\|_{L^p(\Omega)}+\|f-g\|_{L^p(\Omega)})^p+(\|f+g\|_{L^p(\Omega)}-\|f-g\|_{L^p(\Omega)})^p \leq 2^{-p}(\|f\|_{L^p(\Omega)}^p+\|g\|_{L^p(\Omega)}^p).$$

These inequalities, especially the first one, are more difficult to establish than the Morawetz one, because they cannot be deduced from the corresponding real-valued inequalities simply by using Hölder's inequality. □

16.2. Canonical injection from $L^{p'}(\Omega; E')$ into the dual of $L^p(\Omega; E)$

Let us recall some properties of the dual norm [Vol. 1, Theorem 13.9].

Proposition 16.6. — *Let E be a normed space and $B_E = \{e \in E : \|e\|_E \leq 1\}$.*

(a) *The topology on E' is also generated by the* ***dual norm***

$$|||e'|||_{E'} \overset{\text{def}}{=} \sup_{e\in E,\, e\neq 0} \frac{|\langle e', e\rangle|}{\|e\|_E}.$$

(b) $$|||e'|||_{E'} = \sup\nolimits_{e\in B_E} |\langle e', e\rangle|.$$

(c) $$|\langle e', e\rangle| \leq |||e'|||_{E'} \|e\|_E.$$

(d) *The dual norm turns E' into a Banach space.* ▮

Let us state some properties of the **canonical injection** from $L^{p'}(\Omega; E')$ into the dual of $L^p(\Omega; E)$ when E is a Banach space.

Theorem 16.7. — *Let $f \in L^{p'}(\Omega; E')$, where Ω is an open subset of $\mathbb{R}^d$, E is a Banach space, $1 \leq p \leq \infty$, and $1/p + 1/p' = 1$. By endowing the spaces E' and $(L^p(\Omega; E))'$ with their respective dual norms (Proposition 16.6), we have the following results.*

(a) *We define $Jf \in (L^p(\Omega; E))'$ by: for every $g \in L^p(\Omega; E)$,*

$$\langle Jf, g\rangle_{(L^p(\Omega;E))'\times L^p(\Omega;E)} \overset{\text{def}}{=} \int_\Omega \langle f, g\rangle_{E'\times E},$$

where $\langle\ ,\ \rangle_{E'\times E}$ here is the mapping from $L^{p'}(\Omega; E') \times L^p(\Omega; E)$ into $L^1(\Omega)$ given by Definition 12.10.

(b) *The mapping J is linear, continuous, sequentially continuous, and injective from the space $L^{p'}(\Omega; E')$ into $(L^p(\Omega; E))'$.*

(c) *If $1 \leq p < \infty$, then $J(L^{p'}(\Omega; E'))$ is sequentially closed, and hence closed, in $(L^p(\Omega; E))'$, and*

$$|||Jf|||_{(L^p(\Omega;E))'} = \|f\|_{L^{p'}(\Omega;E')}.$$

(d) *If $p = \infty$,*

$$|||Jf|||_{(L^\infty(\Omega;E))'} \leq \|f\|_{L^1(\Omega;E')}. \blacksquare$$

Proof. ***1º)*** *Definition* (a) *of Jf.* As E is a Banach space, we can (Proposition 16.6) endow E' with its dual norm to make it into a Banach space.

The space $L^p(\Omega; E)$ is also a Banach space (Theorem 10.17), so we can also endow its dual with its dual norm.

The mapping $\langle\ ,\ \rangle_{E' \times E}$ is continuous bilinear from $E' \times E$ into $\mathbb{R}$ according to Theorem 16.7 (c), hence (Definition 12.10) $\langle f, g\rangle_{E'\times E}$ belongs to $L^1(\Omega; \mathbb{R}) = L^1(\Omega)$, and its integral is therefore well-defined. By Hölder's inequality (Theorem 12.14 (b)),

$$\left| \int_\Omega \langle f, g\rangle_{E'\times E} \right| \leq \|f\|_{L^{p'}(\Omega;E')} \|g\|_{L^p(\Omega;E)}. \tag{16.15}$$

The mapping $g \mapsto \int_\Omega \langle f, g\rangle_{E'\times E}$ is therefore a continuous linear form on $L^p(\Omega; E)$, i.e.

$$Jf \in (L^p(\Omega; E))'.$$

2º) *Linearity and continuity* (b) *of J.* Linearity is straightforward. By definition of the dual norm (Proposition 16.6 (a)), (16.15) implies

$$\begin{aligned} |||Jf|||_{(L^p(\Omega;E))'} = \sup_{g\in L^p(\Omega;E),\ g\neq 0_E} \frac{|\langle Jf, g\rangle_{(L^p(\Omega;E))'\times L^p(\Omega;E)}|}{\|g\|_{L^p(\Omega;E)}} \leq \\ \leq \|f\|_{L^{p'}(\Omega;E')}. \end{aligned} \tag{16.16}$$

Continuity then follows from the characterization from Proposition 1.9. Sequential continuity follows from continuity by Proposition 1.8.

3º) *Injectivity* (b) *of J.* Suppose $f \neq 0$. Then, there exists $\varphi \in \mathcal{K}(\Omega)$ such that

$$\langle f, \varphi\rangle_{\mathcal{M}(\Omega;E')\times\mathcal{K}(\Omega)} \neq 0_{E'}.$$

Let $a \stackrel{\text{def}}{=} \langle f, \varphi\rangle_{\mathcal{M}(\Omega;E')\times\mathcal{K}(\Omega)}$ and $b \in E$ such that $\langle a, b\rangle_{E'\times E} \neq 0$. Denote $g = \varphi b$, i.e. for every $x \in \Omega$,

$$g(x) = \varphi(x) b.$$

Then, $g \in \mathcal{K}(\Omega; E) \subset L^p(\Omega; E)$ (Theorem 10.8 (a)) and

$$\langle f, \varphi b\rangle_{E' \times E} = \langle f\varphi, b\rangle_{E' \times E} \text{ in } L^1(\Omega)$$

(by Lemma 12.11 on the uniqueness of the image under a bilinear mapping, since this equality holds for every $f \in E'$ and $\varphi \in \mathbb{R}$, and thus for $f \in L^{p'}(\Omega; E')$ and $\varphi \in L^p(\Omega)$). Hence,

$$\int_\Omega \langle f, \varphi b\rangle_{E' \times E} = \int_\Omega \langle f\varphi, b\rangle_{E' \times E} = \Big\langle \int_\Omega f\varphi, b\Big\rangle_{E' \times E}$$

(the second equality follows from the commutation of the integral with a linear mapping, namely Theorem 6.13 with $L = \langle\ , b\rangle_{E' \times E}$).

Since $\int_\Omega f\varphi = \langle f, \varphi\rangle_{\mathcal{M}(\Omega;E') \times \mathcal{K}(\Omega)}$ (Theorem 12.19), which equals a, it follows that

$$\langle Jf, g\rangle_{(L^p(\Omega;E))' \times L^p(\Omega;E)} = \int_\Omega \langle f, \varphi b\rangle_{E' \times E} = \langle a, b\rangle_{E' \times E} \neq 0.$$

Thus, $Jf \neq 0$. Since J is linear, it follows that it is injective.

4°) *Equality* (c) *of norms when* $1 < p < \infty$. **4.a.** *Preliminary.* We must first establish the reverse inequality to (16.16). It suffices to do so when f is a piecewise constant approximation, that is,

$$f = \sum_{s \in S_n} 1_{s,n} f_s, \quad f_s \in E', \tag{16.17}$$

where $n \in \mathbb{N}$, $1_{s,n}$ is the characteristic function on the open cube $\mathring{\Delta}_{s,n}$ of side length 2^{-n} centered at $2^{-n}s$, and $S_n = \{s \in \mathbb{Z}^d : \Delta_{s,n} \subset \Omega \cap B(0,n)\}$.

Indeed, any $f \in L^{p'}(\Omega; E')$ is the limit of such measures (Theorem 13.9), so the inequality for these implies the general inequality, since J is continuous by part (a).

4.b. *Equality.* Let $f \neq 0$ (otherwise the equality is obvious) satisfy (16.17), and $\epsilon > 0$. For every $s \in S_n$, let $e_s \in E$ such that, by definition of the dual norm (Proposition 16.6 (b)),

$$\|e_s\|_E = 1, \quad \langle f_s, e_s\rangle_{E' \times E} \geq (1 - \epsilon) |||f_s|||_{E'}. \tag{16.18}$$

Let

$$g = \sum_{s \in S_n} 1_{s,n} |||f_s|||_{E'}^{p'-1} e_s.$$

On the one hand (Theorem 13.8 on the calculation of piecewise constant measures, applied twice), since $(p' - 1)p = p'$,

$$\|g\|_{L^p(\Omega;E)} = \Big(2^{-nd} \sum_{s \in S_n} |||f_s|||_{E'}^{(p'-1)p}\Big)^{1/p} = \\ = \Big(2^{-nd} \sum_{s \in S_n} |||f_s|||_{E'}^{p'}\Big)^{(p'-1)/p'} = \|f\|_{L^{p'}(\Omega;E')}^{p'-1}.$$

On the other hand,

$$\langle f, g\rangle_{E'\times E} = \sum_{s\in S_n} 1_{s,n} |||f_s|||_{E'}^{p'-1} \langle f_s, e_s\rangle_{E'\times E}.$$

Whence (Theorem 13.8, again),

$$\int_\Omega \langle f, g\rangle_{E'\times E} = 2^{-nd} \sum_{s\in S_n} |||f_s|||_{E'}^{p'-1} \langle f_s, e_s\rangle_{E'\times E} \geq$$

$$\geq (1-\epsilon)2^{-nd} \sum_{s\in S_n} |||f_s|||_{E'}^{p'} = (1-\epsilon)\|f\|_{L^{p'}(\Omega;E')}^{p'}.$$

Thus,

$$|||Jf|||_{(L^p(\Omega;E))'} \geq \frac{\int_\Omega \langle f, g\rangle_{E'\times E}}{\|g\|_{L^p(\Omega;E)}} \geq (1-\epsilon)\|f\|_{L^{p'}(\Omega;E')}.$$

Since this inequality holds for every $\epsilon > 0$, it also holds for $\epsilon = 0$. Hence, equality (c) follows for $1 < p < \infty$.

5º) *Equality* (c) *of norms when* $p = 1$. **5.a.** *Preliminary.* Again, we aim to show the reverse of (16.16), namely for $f \in L^\infty(\Omega; E')$ that

$$\sup_{g\in L^1(\Omega;E),\ g\neq 0_E} \frac{|\int_\Omega \langle f, g\rangle_{E'\times E}|}{\|g\|_{L^1(\Omega;E)}} \geq \|f\|_{L^\infty(\Omega;E)}. \tag{16.19}$$

Note that the above proof is no longer valid because $p' = \infty$.

It suffices to show that, for $f \dot{\in} \mathcal{K}(\Omega; E')$,

$$\sup_{g\in \mathcal{K}(\Omega;E),\ g\neq 0_E} \frac{|\int_\Omega \langle f, g\rangle_{E'\times E}|}{\|g\|_{L^1(\Omega;E)}} \geq \|f\|_{L^\infty(\Omega;E)}. \tag{16.20}$$

Indeed, the regularizations $R_n f$ (Definition 3.17) belong to $\mathcal{K}(\Omega; E')$ and, on the one hand, $\|R_n f\|_{L^\infty(\Omega;E')} \to \|f\|_{L^\infty(\Omega;E')}$ as $n \to \infty$ (Theorem 10.13 (b)); on the other hand, $f \in L^1_{\text{loc}}(\Omega; E')$ (Theorems 11.6 and 11.8), so $R_n f \to f$ in $L^1_{\text{loc}}(\Omega; E')$ (Theorem 10.13 (c)), and thus, if ω is a bounded open subset which contains the support of g and is included in Ω, we have (Theorem 6.19)

$$\int_\Omega \langle R_n f, g\rangle_{E'\times E} = \int_\omega \langle R_n f, g\rangle_{E'\times E} \to \int_\omega \langle f, g\rangle_{E'\times E} = \int_\Omega \langle f, g\rangle_{E'\times E}.$$

5.b. *Equality.* Let $f \in \mathcal{K}(\Omega; E')$, $\epsilon > 0$, and $x \in \Omega$ such that

$$\|f(x)\|_{E'} \geq (1-\epsilon)\|f\|_{L^\infty(\Omega;E')}.$$

Then, by definition of the dual norm (Proposition 16.6 (b)), there exists $e \in E$ such that

$$\|e\|_E = 1, \quad \langle f(x), e\rangle_{E'\times E} \geq (1-\epsilon)\|\|f(x)\|\|_{E'}. \tag{16.21}$$

Since f is continuous, there exists a ball $B(x,\epsilon)$ such that

$$\|f - f(x)\|_{E'} \leq \epsilon\|f\|_{L^\infty(\Omega;E')}. \tag{16.22}$$

Choose

$$\varphi \in \mathcal{K}(\Omega) \text{ with support in } B(x,r), \quad \varphi \geq 0, \quad \int_\Omega \varphi = 1, \tag{16.23}$$

and

$$g = \varphi e.$$

Decompose

$$\frac{\int_\Omega \langle f, g\rangle_{E'\times E}}{\|g\|_{L^1(\Omega;E)}} = \frac{\int_\Omega \langle f(x), g\rangle_{E'\times E}}{\|g\|_{L^1(\Omega;E)}} + \frac{\int_\Omega \langle f - f(x), g\rangle_{E'\times E}}{\|g\|_{L^1(\Omega;E)}}.$$

We have $\|g\|_{L^1(\Omega;E)} = 1$ and, together with (16.21), (16.22), and (16.23), along with Theorems 5.3 and 5.10,

$$\left|\frac{\int_\Omega \langle f - f(x), g\rangle_{E'\times E}}{\|g\|_{L^1(\Omega;E)}}\right| \leq \sup_{B(x,r)} \|f - f(x)\|_{E'} \int_\Omega \varphi\, \|e\|_E \leq \epsilon\|f\|_{L^\infty(\Omega;E')}.$$

Then, again using (16.21) and (16.23) together with Theorems 5.3 and 5.10,

$$\frac{\int_\Omega \langle f, g\rangle_{E'\times E}}{\|g\|_{L^1(\Omega;E)}} \geq \langle f(x), e\rangle_{E'\times E} \int_\Omega \varphi - \epsilon\|f\|_{L^\infty(\Omega;E')} \geq (1-2\epsilon)\|f\|_{L^\infty(\Omega;E')}.$$

Since this inequality holds for every $\epsilon > 0$, it also holds for $\epsilon = 0$. This gives inequality (16.20), hence (16.19); and thus equality (c) for $p = 1$.

6º) Closure (c) *of* $J(L^{p'}(\Omega; E'))$. Denote by V the vector space $J(L^{p'}(\Omega; E'))$ endowed with the topology of $(L^p(\Omega; E))'$.

When p is finite, J is an isomorphism from $L^{p'}(\Omega; E')$ onto V by 3º and the equality of the norms. Since $L^{p'}(\Omega; E')$ is sequentially complete (Theorem 10.16), its image V under J is also sequentially complete (Proposition A.60), and therefore sequentially closed (Proposition A.33). In this case, sequentially closed coincides with closed (Proposition A.25 (a)).

7º) Inequality (d) *when* $p = \infty$. This is a special case of inequality (16.16). □

Case where E is not a Banach space. We do not know how to define the canonical injection from $L^{p'}(\Omega; E')$ into $(L^p(\Omega; E))'$ when E is not normable, as we do not know how to define $\int_\Omega \langle f, g\rangle_{E'\times E}$. Indeed, Definition 12.10 of the image under a bilinear mapping no longer applies, since $\langle\ ,\ \rangle_{E'\times E}$ is no longer continuous [Vol. 1, Theorem 13.22]. □

16.3. Riesz representation theorems

Let us show that the dual of $L^p(\Omega; E)$ is isomorphic to $L^{p'}(\Omega; E')$ if E is a uniformly convex Banach space. (We shall see in Theorem 18.12 that this is also true if E is a Banach space with a separable dual space.)

Theorem 16.8. — *If $1 \le p < \infty$ and E is a uniformly convex Banach space, then the canonical injection J (Theorem 16.7) is an isometric isomorphism from $L^{p'}(\Omega; E')$ onto $(L^p(\Omega; E))'$.*

In particular, if $1 \le p < \infty$, then J is an isometric isomorphism from $L^{p'}(\Omega)$ onto $(L^p(\Omega))'$. ▮

Proof. It suffices to verify that

$$J(L^{p'}(\Omega; E')) \supset (L^p(\Omega; E))'. \tag{16.24}$$

Indeed, the reverse inclusion is satisfied by the definition of J (Theorem 16.7 (b)), and J is an isomorphism from $L^{p'}(\Omega; E')$ onto its image (Theorem 16.7 (c), since p is finite).

1°) *Case of uniformly convex E and $1 < p < \infty$.* It suffices to verify that

$$V = J(L^{p'}(\Omega; E')) \text{ is dense in } (L^p(\Omega; E))'. \tag{16.25}$$

Indeed, since V is closed (Theorem 16.7 (c)), it follows that $V = \overline{V} = (L^p(\Omega; E))'$.

Therefore, it is enough (Proposition A.65) to verify that every continuous linear form S on $(L^p(\Omega; E))'$ which vanishes on V is zero. Such a form S belongs to $(L^p(\Omega; E))''$ and, by Theorem 16.2,

$$L^p(\Omega; E) \text{ is reflexive.}$$

Thus, by Definition A.72 (a) of semi-reflexivity, there exists $g \in L^p(\Omega; E)$ such that, for every $T \in (L^p(\Omega; E))'$,

$$\langle S, T\rangle_{(L^p(\Omega;E))'' \times (L^p(\Omega;E))'} = \langle T, g\rangle_{(L^p(\Omega;E))' \times L^p(\Omega;E)}.$$

For $T = Jf$, by the definition of J (Theorem 16.7 (a)), for every $f \in L^{p'}(\Omega; E')$, we obtain

$$\langle S, Jf\rangle_{(L^p(\Omega;E))'' \times (L^p(\Omega;E))'} = \langle Jf, g\rangle_{(L^p(\Omega;E))' \times L^p(\Omega;E)} =$$
$$= \int_\Omega \langle f, g\rangle_{E' \times E} = 0. \tag{16.26}$$

By the hypothesis on S, we must verify that

$$g = 0.$$

If this were not the case, then there would exist (Theorem 13.9) a piecewise approximation h of g, namely,

$$h = \sum_{s \in S_n} 1_{s,n} h_s, \quad h_s \in E,$$

where $n \in \mathbb{N}$, $1_{s,n}$ is the characteristic measure on the open cube $\mathring{\Delta}_{s,n}$ with side length 2^{-n} centered at $2^{-n}s$, and $S_n = \{s \in \mathbb{Z}^d : \Delta_{s,n} \subset \Omega \cap B(0,n)\}$ such that, for $\epsilon = 1/4$,

$$\|h - g\|_{L^p(0,T;E)} \leq \epsilon \|g\|_{L^p(0,T;E)}.$$

Then, $h \neq 0$ and $\|h - g\|_{L^p(0,T;E)} \leq \epsilon(\|h - g\|_{L^p(0,T;E)} + \|h\|_{L^p(0,T;E)})$, so

$$\|h - g\|_{L^p(0,T;E)} \leq \frac{\epsilon}{1-\epsilon} \|h\|_{L^p(0,T;E)}. \tag{16.27}$$

For all $s \in S_n$, let $f_s \in E'$ such that

$$|||f_s|||_{E'} = 1, \quad \langle f_s, h_s \rangle_{E' \times E} \geq (1-\epsilon)\|h_s\|_E, \tag{16.28}$$

where $||| \ |||_{E'}$ is the dual norm of E' (such an f_s exists by Proposition 16.6 (a)). Let

$$f = \sum_{s \in S_n} 1_{s,n} \|h_s\|_E^{p-1} f_s. \tag{16.29}$$

Then, calculating the norm of a piecewise constant measure (Theorem 13.8, used twice) would give, since $(p-1)p' = p$,

$$\begin{aligned} \|f\|_{L^{p'}(\Omega;E')} &= \Big(2^{-nd} \sum_{s \in S_n} \|h_s\|_E^{(p-1)p'}\Big)^{1/p'} = \\ &= \Big(2^{-nd} \sum_{s \in S_n} \|h_s\|_E^p\Big)^{(p-1)/p} = \|h\|_{L^p(\Omega;E)}^{p-1}. \end{aligned} \tag{16.30}$$

On the other hand,

$$\langle f, h \rangle_{E' \times E} = \sum_{s \in S_n} 1_{s,n} \|h_s\|_E^{p-1} \langle f_s, h_s \rangle_{E' \times E}.$$

Hence (Theorem 13.8, again), using (16.28),

$$\begin{aligned} \int_\Omega \langle f, h \rangle_{E' \times E} &= 2^{-nd} \sum_{s \in S_n} \|h_s\|_E^{p-1} \langle f_s, h_s \rangle_{E' \times E} \geq \\ &\geq (1-\epsilon) 2^{-nd} \sum_{s \in S_n} \|h_s\|_E^p = (1-\epsilon)\|h\|_{L^p(\Omega;E)}^p. \end{aligned} \tag{16.31}$$

Finally, by Hölder's inequality (Theorem 12.14 (b) and Proposition 16.6 (c)) with (16.30) and (16.27),

$$\left|\int_\Omega \langle f, g-h\rangle_{E'\times E}\right| \leq \|f\|_{L^{p'}(\Omega;E')}\|g-h\|_{L^p(\Omega;E)} \leq \frac{\epsilon}{1-\epsilon}\|h\|^p_{L^p(0,T;E)}.$$

From which, using (16.31) (since $\epsilon = 1/4$), we would get

$$\int_\Omega \langle f, g\rangle_{E'\times E} \geq \int_\Omega \langle f, h\rangle_{E'\times E} - \left|\int_\Omega \langle f, g-h\rangle_{E'\times E}\right| \geq$$

$$\geq \left(1-\epsilon-\frac{\epsilon}{1-\epsilon}\right)\|h\|^p_{L^p(\Omega;E)} > 0.$$

This would contradict (16.26), therefore

$$g = 0.$$

It follows then that $S = 0$, which proves (16.25).

2o) Case of uniformly convex E and $p = 1$. Here, we must proceed differently since $L^1(\Omega; E)$ is not reflexive. Let

$$T \in (L^1(\Omega; E))'.$$

2.a. *Case where Ω is bounded.* Here, $L^2(\Omega; E) \underset{\to}{\subsetneq} L^1(\Omega; E)$ (Theorem 11.1 (a)), so the restriction of T to $L^2(\Omega; E)$ is continuous and, by the previous case, there exists $f \in L^2(\Omega; E')$ such that, for every $g \in L^2(\Omega; E)$,

$$\langle T, g\rangle_{(L^1(\Omega;E))'\times L^1(\Omega;E)} = \int_\Omega \langle f, g\rangle_{E'\times E}. \tag{16.32}$$

Let us show that $f \in L^\infty(\Omega; E')$. By the characterization given in Theorem 10.11 (a) (since $f \in L^1_{\mathrm{loc}}(\Omega; E')$), it suffices to verify that the regularizations $R_n f$ satisfy

$$\sup_{n\in\mathbb{N}^*,\, x\in\Omega} \|R_n f(x)\|_{E'} < \infty. \tag{16.33}$$

This is indeed the case since (Definition 3.17 and Theorem 12.19), for every $e \in E$,

$$R_n f(x) = \langle f, \alpha_n \tau_x \rho_n\rangle_{\mathcal{M}(\Omega;E')\times\mathcal{K}(\Omega)} = \int_\Omega f\alpha_n\tau_x\rho_n.$$

Hence, as the integral commutes with $L = \langle\,\cdot\,, e\rangle_{E'\times E}$ (Theorem 6.13),

$$\langle R_n f(x), e\rangle_{E'\times E} = \int_\Omega \langle f\alpha_n\tau_x\rho_n, e\rangle_{E'\times E} = \int_\Omega \langle f, \alpha_n\tau_x\rho_n e\rangle_{E'\times E}.$$

Therefore (equality (16.32) and Proposition 16.6 (c)), since $|\alpha_n| \leq 1$ and $\int_{\mathbb{R}^d} |\rho_n| = 1$,

$$|\langle R_n f(x), e\rangle_{E'\times E}| \leq |||T|||_{(L^1(\Omega;E))'}\|\alpha_n\tau_x\rho_n e\|_{L^1(\Omega;E)} \leq |||T|||_{(L^1(\Omega;E))'}\|e\|_E.$$

Hence we get (16.33), and so

$$f \in L^\infty(\Omega; E').$$

Then, (16.32) extends by continuity (Proposition 11.17 (a)) to every $g \in L^1(\Omega; E)$, since its right-hand side depends continuously on such a g (Theorems 12.14 (b) and Proposition 16.6 (c)), as does its left-hand side, and $L^2(\Omega; E)$ is dense in $L^1(\Omega; E)$ (Theorem 10.14 with $p = 1$ and Proposition 10.7).

Therefore, by the definition of the canonical injection (Theorem 16.7 (a)),

$$Jf = T.$$

This proves (16.24), namely in this case, that

$$J(L^\infty(\Omega; E')) \supset (L^1(\Omega; E))'. \tag{16.34}$$

2.b. *Case where Ω is not bounded.* Let $\omega \subset \Omega$ be a bounded open set. Since the map $g \mapsto \langle T, \widetilde{g}\rangle$, where $\widetilde{\ }$ is the extension to Ω by 0, is continuous linear on $L^1(\omega; E)$ (Theorem 13.6 (b)), it follows from the bounded case that there exists a unique $f_\omega \in L^\infty(\omega; E')$ such that, for every $g \in L^1(\omega; E)$,

$$\langle T, \widetilde{g}\rangle = \int_\omega \langle f_\omega, g\rangle_{E' \times E}. \tag{16.35}$$

It satisfies (Theorem 16.7 (c) on the norm of the canonical injection and Proposition 16.6 defining the dual norm)

$$\|f_\omega\|_{L^\infty(\omega;E')} = |||J_\omega f_\omega|||_{(L^1(\omega;E))'} = \sup_{g \in L^1(\omega;E),\, g \neq 0} \frac{|\langle J_\omega f_\omega, g\rangle|}{\|g\|_{L^1(\omega;E)}}.$$

By the definition of J_ω (Theorem 16.7 (a)) and (16.35), $\langle J_\omega f_\omega, g\rangle = \int_\omega \langle f, g\rangle = \langle T, \widetilde{g}\rangle$, so we obtain

$$\|f_\omega\|_{L^\infty(\omega;E')} = \sup_{g \in L^1(\omega;E),\, g \neq 0} \frac{|\langle T, \widetilde{g}\rangle|}{\|g\|_{L^1(\omega;E)}} \leq |||T|||_{(L^1(\Omega;E))'}$$

(because $|\langle T, \widetilde{g}\rangle| \leq |||T|||_{(L^1(\Omega;E))'} \|\widetilde{g}\|_{L^1(\Omega;E)} = |||T|||_{(L^1(\Omega;E))'} \|g\|_{L^1(\omega;E)}$ by Proposition 16.6 (c) and Theorem 13.6 (b)).

Let $(\omega_n)_{n \in \mathbb{N}}$ be an increasing sequence of bounded open sets whose union is Ω. By Definition 6.16 of the restriction,

$$f_{\omega_{n+1}} = f_{\omega_n} \text{ on } \omega_n.$$

The f_{ω_n} admit (Theorem 11.15) a gluing f which belongs (Theorem 11.18) to the space $L^\infty(\Omega; E')$. It satisfies, for every $g \in \mathcal{B}(\Omega; E)$,

$$\langle T, g\rangle = \int_\Omega \langle f, g\rangle_{E' \times E}$$

(by (16.35) for ω_n, which contains the support of g). By continuity, this holds for every $g \in L^1(\Omega; E)$, which once again proves that $Jf = T$, i.e. (16.34).

3°) Isometry. This is given by Theorem 16.7 (c). □

Simplification of the proof when E is a Hilbert space. In this case, instead of (16.29), one can choose $f = \|h\|_E^{p-1} Ih$, where I is the Riesz–Fréchet isomorphism from E onto E'. Indeed, (16.28) is then satisfied with $e_s = I^{-1} f_s$ and $\epsilon = 0$. Moreover, this choice can be made for $h = g$, without needing to reduce to a step function. □

Let us show that the Riesz Theorem 16.8 does not extend to $p = \infty$.

Theorem 16.9. — *The space $J(L^1(\Omega; E'))$ is strictly included in $(L^\infty(\Omega; E))'$ if Ω is a non-empty open subset of $\mathbb{R}^d$, E is a non-trivial Neumann space, and J is the canonical injection from $L^1(\Omega; E')$ into $(L^\infty(\Omega; E))'$ defined in Theorem 16.7.* ▮

Proof. ***1°)*** *Case $E = \mathbb{R}$.* The Dirac mass δ_a, where $a \in \Omega$, satisfies, for every $\varphi \in \mathcal{K}(\Omega)$,

$$|\langle \delta_a, \varphi \rangle| = |\varphi(a)| \leq \|\varphi\|_{L^\infty(\Omega)}.$$

By the Hahn–Banach theorem (Proposition A.64), δ_a extends to a continuous linear form L on $L^\infty(\Omega)$. Thus, $L \in (L^\infty(\Omega))'$ and, for every $\varphi \in \mathcal{K}(\Omega)$,

$$L(\varphi) = \varphi(a).$$

Suppose there exists $f \in L^1(\Omega)$ such that

$$L = Jf. \tag{16.36}$$

Then (by definition of J from Theorem 16.7 (a)), for every $g \in L^\infty(\Omega)$,

$$L(g) = \langle Jf, g \rangle_{(L^\infty(\Omega))' \times L^\infty(\Omega)} = \int_\Omega fg.$$

The extension of any $\varphi \in \mathcal{K}(\Omega \setminus \{a\})$ by 0 would satisfy (Theorem 6.8, since $\{a\}$ is negligible, see Theorem 2.24 (b))

$$\int_\Omega f\varphi = \int_{\Omega \setminus \{a\}} f\varphi = 0.$$

The Du Bois-Reymond theorem (Proposition 3.13) would then give $f = 0$ on $\Omega \setminus \{a\}$. Again, by Theorem 6.8, for every $\varphi \in \mathcal{K}(\Omega)$,

$$\int_\Omega f\varphi = \int_{\Omega \setminus \{a\}} f\varphi = 0.$$

This would contradict $\int_\Omega f\varphi = L(\varphi) = \varphi(a)$. Therefore, (16.36) is not satisfied.

2°) General case. Let $e \in E$, $e \neq 0_E$, and $e' \in E'$ such that $\langle e', e\rangle_{E'\times E} = 1$ (Proposition A.63). Define $M \in (L^\infty(\Omega; E))'$ by: for $g \in L^\infty(\Omega; E)$,

$$M(g) \stackrel{\text{def}}{=} L\big(\langle g, e'\rangle_{L^\infty(\Omega;E)\times E'}\big)$$

(where $\langle g, e'\rangle_{L^\infty(\Omega;E)\times E'}$ is the image of g under the mapping e', see Theorem 12.1).

If $M = Jf$ where $f \in L^1(\Omega; E')$, then (Theorem 16.7), for $g = g_0 e$ with $g_0 \in L^\infty(\Omega)$,

$$M(g) = L(g_0) = \langle Jf, g\rangle_{(L^\infty(\Omega;E))'\times L^\infty(\Omega;E)} = \int_\Omega \langle f, g\rangle_{E'\times E} = \int_\Omega \langle f, e\rangle_{E'\times E}\, g_0.$$

This is impossible for any $g_0 \in L^\infty(\Omega)$ by 1°, since $\langle f, e\rangle_{E'\times E} \in L^1(\Omega)$. □

16.4. Riesz–Fréchet theorem

If p is finite and E is a Hilbert space, then every linear form on $L^p(\Omega; E)$ can be "represented" by an element of $L^{p'}(\Omega; E')$, according to Theorem 16.8.

Let us show that such a linear form can also be represented by an element of $L^{p'}(\Omega; E)$, which is the content of the **Riesz–Fréchet representation theorem**[3].

Theorem 16.10. — *Let Ω be an open subset of $\mathbb{R}^d$, and*

$$f \in L^{p'}(\Omega; E), \quad 1 \leq p < \infty, \quad E \text{ is a Hilbert space.}$$

We define $Gf \in (L^p(\Omega; E))'$ by: for every $g \in L^p(\Omega; E)$, where $1/p + 1/p' = 1$,

$$\langle Gf, g\rangle_{(L^p(\Omega;E))'\times L^p(\Omega;E)} = \int_\Omega (f, g)_E.$$

The mapping G is an isomorphism from $L^{p'}(\Omega; E)$ onto $(L^p(\Omega; E))'$.

The space $(L^p(\Omega; E))'$ being endowed with its dual norm (Proposition 16.6),

$$|||Gf|||_{(L^p(\Omega;E))'} = \|f\|_{L^{p'}(\Omega;E)}. \blacksquare$$

3 **Historical background of the Riesz–Fréchet theorem.** The fact that $L^2(I)$, where I is a bounded real interval, is isomorphic to its dual was discovered independently by Maurice FRÉCHET [29] and Frigyes RIESZ [62], and published in the same issue of the *Comptes rendus de l'Académie des Sciences de Paris* in 1907.

We shall deduce this from the following theorem, also called the **Riesz–Fréchet representation theorem** for the dual of a Hilbert space [Vol. 1, Theorem 13.14].

Proposition 16.11. — *Let E be a Hilbert space, $(\ ,\ )_E$ be its inner product, and $u \in E$. We define $Hu \in E'$ by: for every $v \in E$,*

$$\langle Hu, v\rangle_{E' \times E} \stackrel{\text{def}}{=} (u, v)_E.$$

Then H is an isomorphism from E onto E' and, for the dual norm (Proposition 16.6),

$$|||Hu|||_{E'} = \|u\|_E. \blacksquare$$

Proof of Theorem 16.10. By composing the definitions of the canonical injection from Theorem 16.7 and Proposition 16.11, we obtain, since J is an isomorphism (Theorem 16.8) as well as the image $\widetilde{H}$ of H on the space $L^p(\Omega; E)$ (Theorem 12.1 (a)),

$$\langle J\widetilde{H}f, g\rangle_{(L^p(\Omega;E))' \times L^p(\Omega;E)} = \int_\Omega \langle \widetilde{H}f, g\rangle_{E' \times E} = \int_\Omega (f, g)_E.$$

Therefore, $G = J\widetilde{H}$, being the composition of the isomorphisms J and $\widetilde{H}$, is itself an isomorphism.

Moreover, Theorems 16.7 (c) and 12.1 yield

$$|||J\widetilde{H}f|||_{(L^p(\Omega;E))'} = |||\widetilde{H}f|||_{L^{p'}(\Omega;E')} = \|f\|_{L^{p'}(\Omega;E)}. \square$$

Identifications. If we identify f with Jf in Theorems 16.7 (a) and 16.8, we would obtain

$$1 \le p \le \infty,\ E \text{ a Banach space} \quad \Rightarrow \quad L^{p'}(\Omega; E') \subsetneq (L^p(\Omega; E))', \tag{16.37}$$

$$1 \le p < \infty,\ E \text{ a uniformly convex Banach space} \quad \Rightarrow \quad L^{p'}(\Omega; E') \underset{\leftrightarrows}{=} (L^p(\Omega; E))', \tag{16.38}$$

$$\Omega \ne \emptyset,\ E \ne \{0_E\} \quad \Rightarrow \quad L^1(\Omega; E') \ne (L^\infty(\Omega; E))'. \tag{16.39}$$

If we identify f with Gf in Theorem 16.10, we would obtain

$$1 \le p < \infty,\ E \text{ a Hilbert space} \quad \Rightarrow \quad L^{p'}(\Omega; E) \underset{\leftrightarrows}{=} (L^p(\Omega; E))'. \tag{16.40}$$

In particular, if $1 \le p < \infty$, we would obtain

$$L^{p'}(\Omega) \underset{\leftrightarrows}{=} (L^p(\Omega))', \tag{16.41}$$
$$L^1(\Omega) \subsetneq (L^\infty(\Omega))', \quad L^2(\Omega) \underset{\leftrightarrows}{=} (L^2(\Omega))', \quad L^\infty(\Omega) \underset{\leftrightarrows}{=} (L^1(\Omega))'.$$

16.5. Weak topology of $L^p(\Omega; E)$

Let us define the semi-norms that generate the topology on $L^p(\Omega; E)$-weak.

Theorem 16.12. — *Let Ω be an open subset of $\mathbb{R}^d$ and $1 \leq p < \infty$.*

(a) *If E is a Banach space which is uniformly convex, the topology of* $L^p(\Omega; E)$-weak *is also generated by the family of semi-norms, indexed by $k \in L^{p'}(\Omega; E')$, where $1/p + 1/p' = 1$,*

$$||||f||||_{L^p(\Omega;E)\text{-weak};k} \stackrel{\text{def}}{=} \Big|\int_\Omega \langle k, f\rangle_{E'\times E}\Big|.$$

(b) *If E is a Hilbert space, the topology of* $L^p(\Omega; E)$-weak *is also generated by the family of semi-norms, indexed by $g \in L^{p'}(\Omega; E)$, where $1/p + 1/p' = 1$,*

$$|||f|||_{L^p(\Omega;E)\text{-weak};g} \stackrel{\text{def}}{=} \Big|\int_\Omega (f, g)_E\Big|.$$

(c) *In particular, the topology of* $L^p(\Omega)$-weak, *where $1 \leq p < \infty$, is also generated by the family of semi-norms, indexed by $g \in L^{p'}(\Omega)$,*

$$|||f|||_{L^p(\Omega)\text{-weak};g} \stackrel{\text{def}}{=} \Big|\int_\Omega fg\Big|. \blacksquare$$

Recall that $L^p(\Omega; E)$-weak is (Definition A.67), for E a Neumann space, the set $L^p(\Omega; E)$ endowed with the semi-norms, indexed by $h \in (L^p(\Omega; E))'$,

$$||f||_{L^p(\Omega;E)\text{-weak};h} \stackrel{\text{def}}{=} |\langle h, f\rangle_{(L^p(\Omega;E))'\times L^p(\Omega;E)}|. \tag{16.42}$$

Proof of Theorem 16.12. ***1º)*** *Property* (a). The definition of J in Riesz's theorem (Theorem 16.7 (a)) gives, together with (16.42), for $k \in L^{p'}(\Omega; E')$,

$$||||f||||_{L^p(\Omega;E)\text{-weak};k} = \Big|\int_\Omega \langle k, f\rangle_{E'\times E}\Big| = |\langle Jk, f\rangle_{(L^p(\Omega;E))'\times L^p(\Omega;E)}| =$$
$$= ||f||_{L^p(\Omega;E)\text{-weak};Jk}.$$

Since J is an isomorphism from $L^{p'}(\Omega; E')$ onto $(L^p(\Omega; E))'$ when E is uniformly convex (Theorem 16.8), this defines the topology of $L^p(\Omega; E)$-weak as k ranges over $L^{p'}(\Omega; E')$, since then Jk ranges over $(L^p(\Omega; E))'$.

2º) *Property* (b). When E is a Hilbert space, the definition of G in the Riesz–Fréchet theorem (Theorem 16.10, with the roles of f and g interchanged) gives, together with (16.42), for $g \in L^{p'}(\Omega; E)$,

$$|||f|||_{L^p(\Omega;E)\text{-weak};g} = \Big|\int_\Omega (f, g)_E\Big| = |\langle Gg, f\rangle_{(L^p(\Omega;E))'\times L^p(\Omega;E)}| =$$
$$= ||f||_{L^p(\Omega;E)\text{-weak};Gg}.$$

Since G is an isomorphism from $L^{p'}(\Omega; E)$ onto $(L^p(\Omega; E))'$ (Theorem 16.10), this defines the topology of $L^p(\Omega; E)$-weak as g ranges over $L^{p'}(\Omega; E)$, since then Gg ranges over $(L^p(\Omega; E))'$.

3º) Property (c). This is a special case of (b), since $(u, v)_{\mathbb{R}} = uv$ (Proposition A.35). □

Let us show that $L^p(\Omega)$-weak is included in $\mathcal{M}(\Omega)$.

Theorem 16.13. — *For every open subset Ω of $\mathbb{R}^d$ and $1 \leq p \leq \infty$,*

$$L^p(\Omega)\text{-weak} \subsetneq \mathcal{M}(\Omega). \blacksquare$$

Proof. We have

$$L^p(\Omega)\text{-weak} = L^p(\Omega) \subset \mathcal{M}(\Omega).$$

For every $f \in L^p(\Omega)$ and $\varphi \in \mathcal{K}(\Omega)$, by Definition 3.2 of the semi-norms of measures, the dual expression on $\mathcal{M}(\Omega) \times \mathcal{K}(\Omega)$ for a p-integrable function (Theorem 12.19), the definition of the canonical injection from $(L^{p'}(\Omega))'$ into $L^p(\Omega)$ (Theorem 16.7 (a)), and Definition A.67 of the semi-norms of $L^p(\Omega)$-weak, we have

$$\|f\|_{\mathcal{M}(\Omega);\varphi} = |\langle f, \varphi \rangle| = \left| \int_\Omega f\varphi \right| = |\langle J\varphi, f \rangle_{(L^p(\Omega))' \times L^p(\Omega)}| = \|f\|_{L^p(\Omega)\text{-weak};J\varphi}. \ \square$$

QUESTION. Does the space $\mathcal{M}(\Omega)$ equal the sequential completion (Definition 5.12) of $L^1(\Omega)$? □

WARNING. The inclusion $L^p(\Omega)$-weak $\subsetneq \mathcal{M}(\Omega)$ does not extend to vector-valued measures in a Neumann space, nor even a Hilbert space, for any p, *if E is infinite dimensional*:

$$L^p(\Omega; E)\text{-weak} \not\subseteq \mathcal{M}(\Omega; E).$$

However, *if E-weak is a Neumann space*, then one can define $\mathcal{M}(\Omega; E\text{-weak})$ (Definition 3.2), and we have

$$L^p(\Omega; E)\text{-weak} \subsetneq \mathcal{M}(\Omega; E\text{-weak}).$$

This is the case if E is semi-reflexive [Vol. 1, Theorem 17.7] and, in particular, if it is a Hilbert space [Vol. 1, Theorem 17.12].

Conversely, if E-weak is not sequentially complete, then this does not make sense as $\mathcal{M}(\Omega; E\text{-weak})$ is not defined in general, see Section 5.5, *Case where E is not a Neumann space.* □

Observe that, for a bounded sequence in $L^p(\Omega)$ with $1 < p < \infty$, weak convergence coincides with convergence in $\mathcal{M}(\Omega)$.

Theorem 16.14. — *Let $(f_n)_{n \in \mathbb{N}}$ be a bounded sequence in $L^p(\Omega)$ and $f \in L^p(\Omega)$, where Ω is an open subset of $\mathbb{R}^d$ and $1 < p < \infty$. Then:*

$$f_n \to f \text{ in } L^p(\Omega)\text{-weak} \quad \Leftrightarrow \quad f_n \to f \text{ in } \mathcal{M}(\Omega). \blacksquare$$

Proof. **1°)** *Forward implication.* By Theorem 16.13, convergence in $L^p(\Omega)$-weak implies convergence in $\mathcal{M}(\Omega)$.

2°) *Backward implication.* Let $g \in L^{p'}(\Omega)$. Then (Theorem 16.12, since $p < \infty$),

$$|||f_n - f|||_{L^p(\Omega)\text{-weak};g} = \left|\int_\Omega (f_n - f)g\right|.$$

Let $\epsilon > 0$ and $a = \sup_{n\in\mathbb{N}} \|f_n\|_{L^p(\Omega)} + \|f\|_{L^p(\Omega)}$. Since $p' < \infty$ (because $p > 1$), $\mathcal{K}(\Omega)$ is dense in $L^{p'}(\Omega)$ (Theorem 10.14), so there exists $\varphi \in \mathcal{K}(\Omega)$ such that

$$\|g - \varphi\|_{L^{p'}(\Omega)} \leq \frac{\epsilon}{2a}.$$

Using Hölder's inequality (Theorem 12.15 (b)), it follows, for all $n \in \mathbb{N}$, that

$$\left|\int_\Omega (f_n - f)(g - \varphi)\right| \leq \|f_n - f\|_{L^p(\Omega)}\|g - \varphi\|_{L^{p'}(\Omega)} \leq \frac{\epsilon}{2}.$$

Suppose now that

$$f_n \to f \ \text{ in } \mathcal{M}(\Omega).$$

Then, there is n_0 such that, for all $n \geq n_0$, we have $|\langle f_n - f, \varphi\rangle_{\mathcal{M}(\Omega)\times\mathcal{K}(\Omega)}| \leq \epsilon/2$, that is (Theorem 12.19, again),

$$\left|\int_\Omega (f_n - f)\varphi\right| \leq \frac{\epsilon}{2}.$$

Then,

$$|||f_n - f|||_{L^p(\Omega)\text{-weak};g} = \left|\int_\Omega (f_n - f)(g - \varphi) + \int_\Omega (f_n - f)\varphi\right| \leq \epsilon.$$

Hence,

$$f_n \to f \ \text{ in } L^p(\Omega)\text{-weak}. \ \square$$

Complement to Theorem 16.14. A sequence $(f_n)_{n\in\mathbb{N}}$ converging in $L^p(\Omega)$-weak is bounded in this space, and hence in $L^p(\Omega)$, by the Banach–Mackey theorem [Vol. 1, Theorem 16.1]. Thus, given an arbitrary sequence in $L^p(\Omega)$ for $1 < p < \infty$,

$$f_n \to f \ \text{ in } L^p(\Omega)\text{-weak} \quad \Leftrightarrow \quad \begin{cases} f_n \to f \ \text{ in } \mathcal{M}(\Omega), \\ (f_n)_{n\in\mathbb{N}} \text{ is bounded in } L^p(\Omega). \ \square \end{cases}$$

16.6. $*$Weak topology of $L^\infty(\Omega; E)$

Let us define the space $L^\infty(\Omega; E)$-$*$weak.

Definition 16.15. — *The space* $L^\infty(\Omega; E)$*-$*$weak, where* E *is a uniformly convex Banach space and* Ω *is an open subset of* $\mathbb{R}^d$*, is defined to be the separated semi-normed space* $L^\infty(\Omega; E)$ *endowed with the semi-norms, indexed by* $k \in L^1(\Omega; E')$,

$$|||f|||_{L^\infty(\Omega;E)\text{-}*\text{weak};k} \stackrel{\text{def}}{=} \left| \int_\Omega \langle k, f \rangle_{E' \times E} \right|. \ \blacksquare$$

Justification. ***1º)*** *Semi-norm.* The mapping $f \mapsto |\int_\Omega \langle k, f \rangle_{E' \times E}|$ trivially satisfies Definition A.9 of a semi-norm.

2º) *Separation.* Suppose that, for all $k \in L^1(\Omega; E')$,

$$|||f|||_{L^\infty(\Omega;E)\text{-}*\text{weak};k} = 0.$$

Given $\varphi \in \mathcal{K}(\Omega)$ and $e' \in E'$, let us denote by $\varphi e'$ the function $x \to \varphi(x)e'$. Then $\varphi e' \in \mathcal{K}(\Omega; E')$, which is included in $L^1(\Omega; E')$ (Theorem 10.8 (a)). Moreover,

$$\langle \varphi e', f \rangle_{E' \times E} = \langle e', f\varphi \rangle_{E' \times E} \text{ in } L^1(\Omega),$$

by Lemma 12.11 on the uniqueness of the image under a bilinear mapping, since this equality holds for every $f \in E$ and $\varphi \in \mathbb{R}$, and thus for $f \in L^\infty(\Omega; E)$ and $\varphi \in L^1(\Omega)$.

Hence, as the integral $\int_\Omega f\varphi$ commutes with the linear mapping $u \mapsto \langle e', u \rangle_{E' \times E}$ (Theorem 6.13), we get

$$\|f\|_{L^\infty(\Omega;E)\text{-}*\text{weak};\varphi e'} = \left| \int_\Omega \langle \varphi e', f \rangle_{E' \times E} \right|$$
$$= \left| \int_\Omega \langle e', f\varphi \rangle_{E' \times E} \right| = \left| \left\langle e', \int_\Omega f\varphi \right\rangle_{E' \times E} \right|.$$

This is zero for every $e' \in E'$ so, according to a corollary (Proposition A.63) of the Hahn–Banach theorem,

$$\int_\Omega f\varphi = 0.$$

That is (Theorem 12.19),

$$\langle f, \varphi \rangle_{\mathcal{M}(\Omega;E) \times \mathcal{K}(\Omega)} = 0.$$

This holds for every $\varphi \in \mathcal{K}(\Omega)$, hence

$$f = 0.$$

Thus, $L^\infty(\Omega; E)$-$*$weak is separated (Definition 1.1). □

Abuse of language. The $*$weak topology is defined only on the dual E' of a semi-normed space E: E'-$*$weak is the set E' endowed with the semi-norms, indexed by $e \in E$,

$$\|e'\|_{E'\text{-}*\text{weak};e} \stackrel{\text{def}}{=} |\langle e', e\rangle|. \tag{16.43}$$

Since $L^\infty(\Omega; E)$ is isomorphic to the dual $(L^1(\Omega; E))'$ by Theorem 16.10, one could identify them and thus speak of the topology of $L^\infty(\Omega; E)$-$*$weak, *i.e.* of $(L^1(\Omega; E))'$-$*$weak. But this identification (which we do not recommend) is not necessary for Definition 16.15. □

Extension. Definition 16.15 extends to $L^p(\Omega; E)$ if $1 < p \leq \infty$, but it is of no interest for finite p since $L^p(\Omega; E)$ is reflexive and the topologies of $L^p(\Omega; E)$-$*$weak and $L^p(\Omega; E)$-weak coincide. □

Let us define other semi-norms that generate the topology of $L^\infty(\Omega; E)$-weak.

Theorem 16.16. — *If E is a Hilbert space and Ω is an open subset of $\mathbb{R}^d$, then the topology of $L^\infty(\Omega; E)$-$*$weak is also generated by the semi-norms, indexed by $g \in L^1(\Omega; E)$,*

$$|||f|||_{L^\infty(\Omega;E)\text{-}*\text{weak};g} \stackrel{\text{def}}{=} \left|\int_\Omega (f, g)_E\right|.$$

In particular, the topology of $L^\infty(\Omega)$-$*$weak is also generated by the semi-norms, indexed by $g \in L^1(\Omega)$,

$$|||f|||_{L^\infty(\Omega)\text{-}*\text{weak};g} \stackrel{\text{def}}{=} \left|\int_\Omega fg\right|. \blacksquare$$

Proof. We define the isometry H from E onto E' by (Proposition 16.11): for every $v \in E$,

$$\langle Hu, v\rangle_{E'\times E} = (u, v)_E \text{ in } \mathbb{R}.$$

Definition 6.9 of the image under a linear mapping for $L(u) = \langle Hu, v\rangle_{E'\times E}$ and $L'(u) = (u, v)_E$, which are equal, yield (Theorem 12.1 (a)), for $g \in L^1(\Omega; E)$,

$$\langle Hg, v\rangle_{E'\times E} = (g, v)_E \text{ in } L^1(\Omega).$$

Thus, for $f \in L^\infty(\Omega; E)$, we have $\langle Hg, f\rangle_{E'\times E} = (g, f)_E$ in $\mathbb{R}$. Hence,

$$|||f|||_{L^\infty(\Omega;E)\text{-}*\text{weak};g} = \left|\int_\Omega (f, g)_E\right| = \left|\int_\Omega \langle Hg, f\rangle_{E'\times E}\right| = ||||f||||_{L^\infty(\Omega;E)\text{-}*\text{weak};Hg}.$$

The mapping $g \mapsto Hg$ is (Theorem 12.1 (b)) an isomorphism from $L^1(\Omega; E)$ onto $L^1(\Omega; E')$, so the semi-norms of Theorem 16.16 are equivalent to those in Definition 16.15. □

Let us state a property of sequential completion.

Theorem 16.17. — *If E is a uniformly convex Banach space with a separable dual E' and Ω is an open subset of $\mathbb{R}^d$, then*

$$L^\infty(\Omega; E)\text{-}{*}\text{weak } \textit{is a Neumann space.}$$

In particular, $L^\infty(\Omega)$-$*$weak *is a Neumann space.* ▮

Proof. If F is a separable Fréchet space, then

$$F'\text{-}{*}\text{weak is a Neumann space.} \tag{16.44}$$

Indeed, every Cauchy sequence in F'-$*$weak is bounded (Proposition A.12), and therefore has a convergent subsequence by the Banach–Bourbaki theorem (Proposition A.76). It then follows that the whole sequence converges (Proposition A.12).

Apply (16.44) to $F = L^1(\Omega; E')$, which is a separable (Theorem 13.12) Banach space (Theorem 10.17, since E' is a Banach space, see Proposition 16.6). Then,

$$(L^1(\Omega; E'))'\text{-}{*}\text{weak is a Neumann space.}$$

Being uniformly convex, E is reflexive (Proposition A.74) and thus isomorphic to E'' (Proposition A.73). Thus, $L^\infty(\Omega; E)$ is isomorphic to $L^\infty(\Omega; E'')$ (Theorem 12.1 (b)). Moreover, $L^\infty(\Omega; E'')$ is isomorphic to $(L^1(\Omega; E'))'$ (Theorem 16.8). Whence,

$$L^\infty(\Omega; E) \text{ is isomorphic to } (L^1(\Omega; E'))'.$$

Thus,

$$L^\infty(\Omega; E)\text{-}{*}\text{weak is isomorphic to } (L^1(\Omega; E'))'\text{-}{*}\text{weak},$$

by Definition 16.15 of $L^\infty(\Omega; E)$-$*$weak using equality (16.43). However, any space that is isomorphic to a Neumann space is itself Neumann (Proposition A.60). □

PART 3

Integrable Functions

Since these mysteries are beyond me,
let me pretend to be their orchestrator.

Jean Cocteau
Mariés de la Tour Eiffel

Chapter 17

Measurable Functions

Here, we define the *classes of integrable functions* equal almost everywhere, and relate them to integrable *measures* when the value space E is a Fréchet space.

We define measurable functions (Definition 17.1) and their integral (Theorem 17.10), which is the limit of the integrals of an increasing sequence of simple functions, and we state (Theorems 17.13 and 17.14) the monotone convergence lemma and Fatou's lemma.

We define the space $\dot{\mathsf{L}}^p(\Omega; E)$ of classes $\dot{\mathsf{f}}$ of p-integrable functions equal almost everywhere (Definition 17.17), and (Definition 17.30) the space $\overline{L}^p(\Omega; E)$ of measures $\overline{f}$ defined by $\langle \overline{f}, \varphi \rangle = \int_\Omega \dot{\mathsf{f}} \varphi$ for every test function φ. We show (Theorem 17.33) that $\overline{L}^p(\Omega; E)$ coincides with the space $L^p(\Omega; E)$ (Definition 10.2) of p-integrable measures.

The construction of $\dot{\mathsf{L}}^p(\Omega; E)$ is largely inspired by *Intégration et espaces de Sobolev à valeurs vectorielles* by Jérôme DRONIOU [19], and *Mesure, intégration, probabilités* by Thierry GALLOUËT and Raphaèle HERBIN [32].

17.1. Measurable functions

Let us define both simple and measurable functions (functions are denoted by f, g, $\mathsf{s}, \ldots$, so as not to be confused with measures denoted by f, g, $s \ldots$).

Definition 17.1. — (a) *A function* s *from a measurable subset* Ω *of* $\mathbb{R}^d$ *into a semi-normed space* E *is said to be* ***simple*** *if there exist a finite number of elements* $e_1, \ldots, e_n$ *in* E *and measurable subsets* $\omega_1, \ldots, \omega_n$ *of* Ω*, each of finite measure, such that*

$$\mathsf{s} = \sum_{i=1}^{n} e_i 1_{\omega_i}. \tag{17.1}$$

(b) *A function* f *from a measurable subset* Ω *of* $\mathbb{R}^d$ *into a semi-normed space* E *is said to be* ***measurable*** *if there exists a sequence* $(\mathsf{s}_n)_{n\in\mathbb{N}}$ *of simple functions such that*

$$\mathsf{s}_n \to \mathsf{f} \ \textit{\textbf{almost everywhere}} \ \textit{in } \Omega,$$

that is, for every $x \in \Omega \setminus \sigma$*, where* σ *has measure zero.* ▮

WARNING. If $E = \mathbb{R}$, then this notion of measurability is equivalent to requiring that the preimage of every open subset of E is measurable. In general, it is **stronger** (that is μ-measurability, where μ denotes the Lebesgue measure). □

Let us now prove that the set where a function exceeds a given constant is measurable.

Theorem 17.2. — *Let* f *be a real-valued measurable function on a measurable subset* Ω *of* $\mathbb{R}^d$ *and* $\mathsf{c} \in \mathbb{R}$. *Then the following sets are measurable:*

$$\{x \in \Omega : \mathsf{f}(x) \geq c\}, \quad \{x \in \Omega : \mathsf{f}(x) \leq c\},$$
$$\{x \in \Omega : \mathsf{f}(x) > c\}, \quad \{x \in \Omega : \mathsf{f}(x) < c\}. \blacksquare$$

Proof. Let

$$\omega \stackrel{\text{def}}{=} \{x \in \Omega : \mathsf{f}(x) \geq c\}.$$

1º) The set λ. Let $(\mathsf{s}_n)_{n\in\mathbb{N}}$ be a sequence of simple functions converging almost everywhere to f, σ be the set of measure zero where this convergence does not hold, and

$$\lambda \stackrel{\text{def}}{=} \bigcap_{m\in\mathbb{N}^*} \bigcap_{N\in\mathbb{N}} \bigcup_{n\geq N} \Big\{x \in \Omega : \mathsf{s}_n(x) \geq c - \frac{1}{m}\Big\}.$$

Since s_n is simple, $\{x \in \Omega : \mathsf{s}_n(x) \geq c - 1/m\}$ is measurable. Since countable intersections and unions of measurable sets are measurable (Theorem 2.15),

$$\lambda \text{ is measurable.}$$

2º) $\omega\backslash\sigma \supset \lambda\backslash\sigma$. If $x \notin \sigma$ and $\mathsf{f}(x) \geq c$ then, for each m, $\mathsf{s}_n(x) \geq c - 1/m$ for all sufficiently large n; thus, x belongs to $\bigcup_{n\geq N}\{x \in \Omega\backslash\sigma : \mathsf{s}_n(x) \geq c - 1/m\}$ for some N large enough, and hence to the corresponding intersection over N. Therefore, $x \in \lambda\backslash\sigma$, that is to say,

$$\omega\backslash\sigma \supset \lambda\backslash\sigma.$$

3º) $\omega\backslash\sigma \subset \lambda\backslash\sigma$. If $x \notin \sigma$ and $\mathsf{f}(x) < c$ then, for sufficiently large m, $\mathsf{f}(x) < c - 2/m$; for all n large enough, we have $\mathsf{s}_n(x) < c - 1/m$, so x does not belong to the set $\bigcup_{n\geq N}\{x \in \Omega\backslash\sigma : \mathsf{s}_n(x) \geq c - 1/m\}$. Therefore, it belongs to neither the intersection over N, nor to that over m, and hence not to $\lambda\backslash\sigma$. Thus,

$$\omega\backslash\sigma \subset \lambda\backslash\sigma.$$

4º) Conclusion. From 2º and 3º, $\omega\backslash\sigma$ coincides with $\lambda\backslash\sigma$, so it is measurable. Hence,

$$\omega \text{ is measurable.}$$

The set $\{x \in \Omega : \mathsf{f}(x) \leq c\} = \{x \in \Omega : -\mathsf{f}(x) \geq -c\}$ is measurable, because the function $-\mathsf{f}$ is measurable (this is clear from Definition 17.1).

Its complement $\{x \in \Omega : \mathsf{f}(x) > c\} = \Omega \setminus \{x \in \Omega : \mathsf{f}(x) \leq c\}$ is also measurable (Theorem 2.11).

Thus, $\{x \in \Omega : \mathsf{f}(x) < c\} = \{x \in \Omega : -\mathsf{f}(x) > -c\}$ is measurable. □

Let us present **Egoroff's theorem**[1].

Theorem 17.3. — *Let* $(\mathsf{f}_n)_{n\in\mathbb{N}}$ *be a sequence of measurable functions and* f *a function from a measurable subset* Ω *of* $\mathbb{R}^d$ *of finite measure into a Fréchet space* E*, such that*

$$\mathsf{f}_n \to \mathsf{f} \ \textit{almost everywhere in } \Omega.$$

Then, for every $\epsilon > 0$*, there exists a measurable set* ω_ϵ *such that* $|\omega_\epsilon| \leq \epsilon$*, and*

$$\mathsf{f}_n \to \mathsf{f} \ \textit{uniformly in } \Omega \setminus \omega_\epsilon. \ \blacksquare$$

Proof. Let $(\| \ \|_{E,m})_{m\in\mathbb{N}}$ be an increasing sequence of semi-norms of E which define its topology (Proposition A.26), and

$$E_{n,k} \stackrel{\text{def}}{=} \left\{x \in \Omega : \|\mathsf{f}_n(x) - \mathsf{f}(x)\|_{E,k} \geq \frac{1}{k}\right\}, \quad F_{N,k} \stackrel{\text{def}}{=} \bigcup_{n\geq N} E_{n,k}.$$

Each $E_{n,k}$ is measurable (Theorem 17.2), as is their union $F_{N,k}$ (Theorem 2.15), and so is the intersection over N of the $F_{N,k}$ (Theorem 2.15, again).

For all $k \geq 1$,

$$\bigcap_{N\geq 1} F_{N,k} \subset \{x \in \Omega : \mathsf{f}_m(x) \not\to \mathsf{f}(x),\ m \to \infty\}.$$

Thus, $|\bigcap_{N\geq 1} F_{N,k}| = 0$. Now, $F_{N,k}$ decreases as N increases and $|\Omega| < \infty$, so (Theorem 2.17), when $N \to \infty$,

$$|F_{N,k}| \to 0.$$

For all $k \geq 1$, there exists $N_k \geq 1$ such that $|F_{N_k,k}| < \epsilon/2^k$. Let

$$\omega_\epsilon \stackrel{\text{def}}{=} \bigcup_{k\geq 1} F_{N_k,k}.$$

1 **Historical background of Egoroff's theorem.** Dmitri EGOROFF proved Theorem 17.3 in 1911 [23] for real-valued functions on $[0, 1]$.

Then (Theorem 2.18 (a)), $|\omega_\epsilon| \leq \sum_{k\geq 1} |F_{N_k,k}| < \sum_{k\geq 1} \epsilon/2^k = \epsilon$, and

$$\Omega \setminus \omega_\epsilon = \bigcap_{k\geq 1} \Omega \setminus F_{N_k,k} = \bigcap_{k\geq 1} \bigcap_{n\geq N_k} \Omega \setminus E_{n,k}.$$

For every $x \notin \omega_\epsilon$ and $n \geq N_k$, we therefore have $x \in \Omega \setminus E_{n,k}$, that is,

$$\|\mathsf{f}_n(x) - \mathsf{f}(x)\|_{E,k} \leq \frac{1}{k}. \quad \square$$

Case of a Banach space. It suffices to choose $\| \ \|_{E;m} = \| \ \|_E$ for all m. □

Case of a Neumann space. The condition "E is a Fréchet space" cannot be weakened to "E is a Neumann space" [BOURBAKI, 8, chap. 4, § 5, no. 4, *Limites de fonctions mesurables*]. □

Let us show that an almost everywhere limit of measurable functions is measurable.

Theorem 17.4. — *Let* $(\mathsf{f}_n)_{n\in\mathbb{N}}$ *be measurable functions and* f *be a function from a measurable subset* Ω *of* $\mathbb{R}^d$ *into a Fréchet space* E*, such that*

$$\mathsf{f}_n \to \mathsf{f} \ \textit{almost everywhere in } \Omega.$$

Then f *is measurable.* ▮

Let us present a lemma that will be used in the proof.

Lemma 17.5. — *Let* $(\mathsf{g}_n)_{n\in\mathbb{N}}$ *be a sequence of measurable functions and* g *be a function from a measurable set* Λ *of finite measure into a Fréchet space* E*, endowed with an increasing sequence of semi-norms* $(\| \ \|_{E;m})_{m\in\mathbb{N}}$*, such that*

$$\mathsf{g}_n \to \mathsf{g} \ \textit{almost everywhere in } \Omega.$$

Then, for every $\epsilon > 0$*, there exist a measurable set* λ_ϵ *such that* $|\lambda_\epsilon| \leq \epsilon$*, and simple functions* $(\mathsf{t}_m)_{m\in\mathbb{N}^*}$ *such that, for all* $m \geq 1$ *and* $x \in \Lambda \setminus \lambda_\epsilon$*,*

$$\|\mathsf{g}(x) - \mathsf{t}_m(x)\|_{E,m} \leq \frac{1}{m}. \ ▮$$

Proof. According to Egoroff's theorem (Theorem 17.3), there exists a measurable set λ_1, of measure less than $\epsilon/2$, such that $\sup_{x\in\Lambda\backslash\lambda_1}\|\mathsf{g}(x)-\mathsf{g}_n(x)\|_{E,m}\to 0$. For all $m\geq 1$, there exists n_m such that

$$\sup_{x\in\Lambda\backslash\lambda_1}\|\mathsf{g}(x)-\mathsf{g}_{n_m}(x)\|_{E,m}\leq\frac{1}{2m}.$$

Since g_{n_m} is measurable, it is the almost everywhere limit of simple functions $(\mathsf{t}_p^m)_{p\in\mathbb{N}^*}$. Again, by Egoroff's theorem, we have a measurable set $\lambda_{2,m}$ of measure less than $(\epsilon/2)/2^m$ such that $\sup_{x\in\Lambda\backslash\lambda_{2,m}}\|\mathsf{g}_{n_m}(x)-\mathsf{t}_p^m(x)\|_{E,m}\to 0$ as $p\to\infty$. Therefore, there exists p_m such that

$$\sup_{x\in\Lambda\backslash\lambda_{2,m}}\|\mathsf{g}_{n_m}(x)-\mathsf{t}_{p_m}^m(x)\|_{E,m}\leq\frac{1}{2m}.$$

The lemma is satisfied by $\mathsf{t}_m\overset{\text{def}}{=}\mathsf{t}_{p_m}^m$ and $\lambda_\epsilon\overset{\text{def}}{=}\lambda_1\cup\bigcup_{m\geq 1}\lambda_{2,m}$, since (Theorem 2.18 (a))

$$|\lambda_\epsilon|\leq\lambda_1|+\sum_{m\geq 1}|\lambda_{2,m}|\leq\frac{\epsilon}{2}+\sum_{m\geq 1}\frac{\epsilon}{2^{m+1}}=\epsilon.\ \square$$

Proof of Theorem 17.4. Let $(\|\ \|_{E,m})_{m\in\mathbb{N}}$ be an increasing sequence of semi-norms of E that defines its topology (Proposition A.26).

1°) Preliminary. When $|\Omega|<\infty$, by induction on the number k, we construct a sequence $(\omega_k)_{k\in\mathbb{N}^*}$ of measurable subsets of Ω such that, for all $k\geq 1$,

$$\omega_k\subset\Omega\setminus\bigcup_{r=1}^{k-1}\omega_r,\quad\Big|\Omega\setminus\bigcup_{r=1}^{k}\omega_r\Big|\leq\frac{1}{k},\tag{17.2}$$

and, for each k, a sequence $(\mathsf{s}_m^k)_{m\in\mathbb{N}^*}$ of simple functions such that, for all $k\geq 1$ and $m\geq 1$,

$$\mathsf{s}_m^k=0\text{ outside }\omega_k,\quad\mathsf{s}_m^k\to\mathsf{f}\text{ almost everywhere in }\omega_k\text{ as }m\to\infty.\tag{17.3}$$

For $k=1$, we use Lemma 17.5 with $\Lambda=\Omega$, $\mathsf{g}=\mathsf{f}$, $\mathsf{g}_n=\mathsf{f}_n$, and $\epsilon=1$, then we define $\omega_1\overset{\text{def}}{=}\Lambda\backslash\lambda_\epsilon$ and $\mathsf{s}_m^1\overset{\text{def}}{=}\mathsf{t}_m$ on ω_1 and 0 outside.

We proceed from k to $k+1$ by using Lemma 17.5 with $\Lambda=\Omega\setminus\bigcup_{r=1}^k\omega_k$, $\mathsf{g}=\mathsf{f}$, $\mathsf{g}_n=\mathsf{f}_n$, and $\epsilon=1/(k+1)$, then define $\omega_{k+1}\overset{\text{def}}{=}\Lambda\backslash\lambda_\epsilon$ and $\mathsf{s}_m^{k+1}\overset{\text{def}}{=}\mathsf{t}_m$ on ω_{k+1} and 0 outside.

2°) Case where $|\Omega|<\infty$. By (17.2), the ω_k are pairwise disjoint, thus we have $|\bigcup_{r\in\mathbb{N}^*}\omega_r|=\sum_{r\in\mathbb{N}^*}|\omega_r|$ (Theorem 2.18) and

$$0=\Big|\Omega\setminus\bigcup_{r\geq 1}\omega_r\Big|=|\Omega|-\Big|\bigcup_{r\geq 1}\omega_r\Big|=|\Omega|-\sum_{r\geq 1}|\omega_r|.\tag{17.4}$$

For all $m \geq 1$, let

$$\mathsf{s}_m \stackrel{\text{def}}{=} \mathsf{s}_m^1 + \mathsf{s}_m^2 + \cdots + \mathsf{s}_m^m,$$

which is a simple function. Since the ω_k are pairwise disjoint and s_m^k vanishes outside ω_k by (17.3), we have $\mathsf{s}_m = \mathsf{s}_m^k$ on ω_k. Thus, (17.3) shows that, when $m \to \infty$,

$$\mathsf{s}_m \to \mathsf{f} \text{ almost everywhere in } \omega_k.$$

This convergence holds almost everywhere on $\bigcup_{r\geq 1} \omega_k$, since a countable union of negligible sets is negligible (Theorem 2.24 (c)), and therefore almost everywhere on Ω by (17.4). Thus,

$$\mathsf{f} \text{ is measurable.}$$

3º) The general case. For all $m \in \mathbb{N}$, let

$$\omega_m \stackrel{\text{def}}{=} \{x \in \Omega : m \leq x < m+1\}.$$

According to 2º, for each m, there exists a sequence $(\mathsf{s}_n^m)_{n\in\mathbb{N}}$ of simple functions converging to f almost everywhere in ω_m; we extend them by 0_E on Ω. For each $m \geq 1$, again let

$$\mathsf{s}_m \stackrel{\text{def}}{=} \mathsf{s}_m^1 + \mathsf{s}_m^2 + \cdots \mathsf{s}_m^m,$$

which is a simple function. When $m \to \infty$, it converges to f almost everywhere in ω_r, and hence in their union Ω, since again any countable union of negligible sets is negligible. Therefore, f is again measurable. □

Let us show that the supremum over a collection of bounded measurable functions is again measurable.

Theorem 17.6. — *Every supremum* $\sup_{n\in\mathbb{N}} \mathsf{f}_n$ *of a sequence* $(\mathsf{f}_n)_{n\in\mathbb{N}}$ *of real-valued measurable functions that is bounded almost everywhere by a function* g *is measurable.* ■

Proof. Let $(\mathsf{s}_n^1)_{n\in\mathbb{N}}$ and $(\mathsf{s}_n^2)_{n\in\mathbb{N}}$ be two sequences of simple functions approximating f_1 and f_2 almost everywhere, respectively (Definition 17.1). Each function $\sup(\mathsf{s}_n^1, \mathsf{s}_n^2)$ is simple, and converges almost everywhere to $\sup(\mathsf{f}_1, \mathsf{f}_2)$, which is therefore measurable.

By induction, $\sup(\mathsf{f}_1, \ldots, \mathsf{f}_n)$ is measurable. When $n \to \infty$, it is bounded above by g, so it converges almost everywhere to $\sup_{n\in\mathbb{N}} \mathsf{f}_n$. Therefore (Theorem 17.4), $\sup_{n\in\mathbb{N}} \mathsf{f}_n$ is measurable. □

17.2. Integral of a positive measurable function

We begin by defining the integral of a simple function, whether it be real-valued or not.

Proposition 17.7. — *Let* $\mathsf{s} = \sum_{i=1}^{n} e_i 1_{\lambda_i}$ *be a simple function (Definition 17.1) from a measurable subset* Ω *of* $\mathbb{R}^d$ *into a Fréchet space* E*, and let* ω *be a measurable subset of* Ω*.*

(a) *The* ***integral*** *of* s *over* ω *is the element of* E *defined by*

$$\int_\omega \mathsf{s} \stackrel{\text{def}}{=} \sum_{i=1}^{n} e_i |\lambda_i \cap \omega|.$$

(b) *The integral of* s *over* ω *does not depend on the choice of the* e_i *and* λ_i*; it is linear and, for every semi-norm* $\| \ \|_{E;\nu}$ *of* E*,* $\|\mathsf{s}\|_{E;\nu}$ *is a real-valued simple function, and*

$$\Big\| \int_\omega \mathsf{s} \Big\|_{E;\nu} \le \int_\omega \|\mathsf{s}\|_{E;\nu} \le \int_\Omega \|\mathsf{s}\|_{E;\nu}.$$

(c) *If* ω' *is a measurable set disjoint from* ω *and included in* Ω*, then*

$$\int_{\omega \cup \omega'} \mathsf{s} = \int_\omega \mathsf{s} + \int_{\omega'} \mathsf{s}.$$

(d) *The integral of a positive real-valued simple function* s *over* ω *is positive, increasing, and*

$$\int_\omega \mathsf{s} \le |\omega| \sup_\omega \mathsf{s}. \ \blacksquare$$

Proof. ***1o)*** *Choice of the* e_i *and* λ_i. The function s admits a finite number of values $(a_j)_{j \in J}$ on Ω, and let $\mu_j = \{x \in \Omega : \mathsf{s}(x) = a_j\}$. By eliminating the values $a_j = 0_E$ and the μ_j of measure zero, if they exist, we obtain a unique representation for s, said to be **canonical**, which consists of pairwise disjoint measurable subsets $(\mu_j)_{j \le J}$.

According to the additivity of the measure (Theorem 2.13 (c)), these operations do not change the value of the integral over ω, so it is therefore uniquely defined.

2o) *Linearity.* As the sum of two simple functions is simple, by choosing its canonical representation, we can verify that its integral is additive and that $t\int_\omega \mathsf{s} = \int_\omega t\mathsf{s}$.

3o) *First inequality of (b).* It suffices to note that, since $\| \ \|_{E;\nu}$ is a semi-norm, (b) is written as $\big\|\sum_{i=1}^{n} e_i |\lambda_i|\big\|_{E;\nu} \le \sum_{i=1}^{n} \|e_i\|_{E;\nu} |\lambda_i|$.

4º) Second inequality of (b). This is clear.

5º) Equality (c). This follows from the fact that the decomposition of s over $\omega \cup \omega'$ is the sum of the decompositions of s over ω and ω' extended by 0_E.

6º) Positivity and monotonicity. If $\mathsf{s} \geq 0$, then all the a_j are positive, and so is the integral over ω. If $\mathsf{s} \geq \mathsf{s}'$, then $\mathsf{s} - \mathsf{s}' \geq 0$, hence $\int_\omega \mathsf{s} - \int_\omega \mathsf{s}' = \int_\omega (\mathsf{s} - \mathsf{s}') \geq 0$.

7º) Inequality (d). By the additivity and monotonicity of the measure (Theorems 2.13 (c) and 2.10), the canonical representation yields

$$\int_\omega \mathsf{s} = \sum_{j \leq J} a_i |\mu_j \cap \omega| \leq \sup_{x \in \Omega} \mathsf{s}(x) \sum_{j \leq J} |\mu_j \cap \omega| \leq \sup_{x \in \Omega} \mathsf{s}(x) |\omega|. \ \square$$

Let us proceed with a lemma that applies to positive real-valued functions.

Lemma 17.8. — *Let f be a positive real-valued measurable function on a measurable subset Ω of $\mathbb{R}^d$. There exists an increasing sequence of simple functions $(\mathsf{s}_n)_{n \in \mathbb{N}}$ such that, when $n \to \infty$,*

$$\mathsf{s}_n \to \mathsf{f} \ \textit{almost everywhere in } \Omega. \ \blacksquare$$

Proof. **1º)** *Case of bounded Ω.* **1.a.** *Construction of s_n.* For $n \in \mathbb{N}$, let

$$\mathsf{s}_n(x) \stackrel{\text{def}}{=} \begin{cases} \dfrac{m}{2^n} & \text{if } \mathsf{f}(x) \in \left[\dfrac{m}{2^n}, \dfrac{m+1}{2^n}\right) \text{ for } m = 0, 1, \ldots, n2^n - 1, \\ n & \text{if } \mathsf{f}(x) \geq n. \end{cases}$$

Since $\omega_c = \{x \in \Omega : \mathsf{f}(x) \geq c\}$ is measurable (Theorem 17.2), the function s_n is simple and thus $\{x \in \Omega : c \leq \mathsf{f}(x) < d\} = \omega_c \setminus \omega_d$ is also measurable (Theorem 2.11).

1.b. *Convergence.* If $\mathsf{f}(x)$ is finite, for $n \geq \mathsf{f}(x)$, we have $|\mathsf{f}(x) - \mathsf{s}_n(x)| \leq 1/2^n$, which tends to 0 as $n \to \infty$.

And, if $\mathsf{f}(x)$ is infinite, then $\mathsf{s}_n(x) = n$, which tends to $\mathsf{f}(x)$ as $n \to \infty$.

1.c. *Monotonicity. First case.* If $\mathsf{f}(x) \geq n+1$, then $\mathsf{s}_n(x) = n < n+1 = \mathsf{s}_{n+1}(x)$.

Second case. If $n \leq \mathsf{f}(x) < n+1$, there exists $i \in [\![n2^{n+1}, \ldots, (n+1)2^{n+1} - 1]\!]$ such that $\mathsf{f}(x) \in [i/2^{n+1}, (i+1)/2^{n+1}]$; then, $\mathsf{s}_n(x) = n \leq i/2^{n+1} = \mathsf{s}_{n+1}(x)$.

Third case. If $\mathsf{f}(x) < n$, then there exists $m \in [\![0, \ldots, n2^n - 1]\!]$ such that

$$\mathsf{f}(x) \in \left[\!\!\left[\frac{m}{2^n}, \frac{m+1}{2^n}\right]\!\!\right] = \left[\!\!\left[\frac{2m}{2^{n+1}}, \frac{2m+2}{2^{n+1}}\right]\!\!\right].$$

If $f(x) \in [\![2m/2^{n+1}, (2m+1)/2^{n+1}]\!]$, then $s_n(x) = m/2^n = 2m/2^{n+1} = s_{n+1}(x)$.

If $f(x) \in [\![(2m+1)/2^{n+1}, (2m+2)/2^{n+1}]\!]$, then $s_n(x) = m/2^n = (2m+1)/\,2^{n+1}$, which equals $s_{n+1}(x)$.

2°) The general case. For s_n, it suffices to choose the above approximation with respect to $\omega_n = \{x \in \Omega : |x| < n\}$ and extend it by 0. □

Let us continue with another lemma.

Lemma 17.9. — *Let $(s_n)_{n\in\mathbb{N}}$ be an increasing sequence of positive real-valued simple functions, and t be a positive real-valued function on a measurable subset Ω of $\mathbb{R}^d$ such that, for all $x \in \Omega$,*

$$t(x) \leq \lim_{n\to\infty} s_n(x).$$

Then,

$$\int_\Omega t \leq \lim_{n\to\infty} \int_\Omega s_n. \tag{17.5}$$ ▮

Proof. 1°) Preliminary. Let $0 < \alpha < 1$ and, for $n \in \mathbb{N}$,

$$\omega_n \stackrel{\text{def}}{=} \{x \in \Omega : \alpha t(x) \leq s_n(x)\}.$$

We shall verify that

$$\Omega = \bigcup_{n\in\mathbb{N}} \omega_n. \tag{17.6}$$

Let $x \in \Omega$. If $t(x) > 0$, then

$$\alpha t(x) < t(x) \leq \lim_{n\to\infty} s_n(x);$$

there exists n_x such that $x \in \omega_n$ if $n \geq n_x$; hence $x \in \bigcup_{n\in\mathbb{N}} \omega_n$.

If $t(x) = 0$, then $x \in \omega_n$ for any n, and thus again $x \in \bigcup_{n\in\mathbb{N}} \omega_n$.

2°) Inequality (17.5). Let $t = \sum_{i=1}^m b_i 1_{\lambda_i}$ be a representation of the simple function t. Then, $\alpha t 1_{\omega_n} = \sum_{i=1}^m \alpha b_i 1_{\lambda_i \cap \omega_n}$ is also a simple function. Hence (Proposition 17.7 (a)),

$$\int_\Omega \alpha t 1_{\omega_n} = \sum_{i=1}^m \alpha b_i |\lambda_i \cap \omega_n|.$$

According to (17.6),

$$\bigcup_{n\in\mathbb{N}} \lambda_i \cap \omega_n = \lambda_i \cap \bigcup_{n\in\mathbb{N}} \omega_n = \lambda_i \cap \Omega = \lambda_i.$$

Now, since $\omega_n \subset \omega_{n+1}$ (as $\mathsf{s}_n \leq \mathsf{s}_{n+1}$), the continuity of the measure (Theorem 2.16) gives $|\lambda_i \cap \omega_n| \to |\lambda_i|$ as $n \to \infty$. Therefore,

$$\int_\Omega \alpha \mathsf{t} 1_{\omega_n} = \sum_{i=1}^m \alpha b_i |\lambda_i \cap \omega_n| \to \sum_{i=1}^m \alpha b_i |\lambda_i| = \int_\Omega \alpha \mathsf{t}. \tag{17.7}$$

By the definition of ω_n, we have $\alpha \mathsf{t} 1_{\omega_n} \leq \mathsf{s}_n$, so the monotonicity of the integral of simple functions (Proposition 17.7 (d)) gives

$$\int_\Omega \alpha \mathsf{t} 1_{\omega_n} \leq \int_\Omega \mathsf{s}_n.$$

Taking the limit, (17.7) yields

$$\int_\Omega \alpha \mathsf{t} \leq \int_\Omega \mathsf{s}_n.$$

By the linearity of the integral of simple functions (Proposition 17.7 (b)), we have $\int_\Omega \alpha \mathsf{t} = \alpha \int_\Omega \mathsf{t}$, so it remains only to let α tend to 1 to conclude the lemma. □

Let us now define the integral of a positive measurable function.

Definition 17.10. — *The **integral** of a positive measurable real-valued function* f *over a measurable subset* Ω *of* $\mathbb{R}^d$ *is the number, positive or infinite, defined by*

$$\int_\Omega \mathsf{f} \stackrel{\text{def}}{=} \lim_{n \to \infty} \int_\Omega \mathsf{s}_n,$$

where $(\mathsf{s}_n)_{n \in \mathbb{N}}$ *is any increasing sequence of simple functions converging almost everywhere to* f *(given by Lemma 17.8).* ▌

Justification. Let us verify that the integral is independent of the choice of the sequence s_n. To this end, let $(\mathsf{t}_m)_{m \in \mathbb{N}}$ be another increasing sequence of simple functions converging almost everywhere to f. For each m, we have $\mathsf{t}_m \leq \mathsf{f} = \lim_{n \to \infty} \mathsf{s}_n$ by the monotonicity, whence, from Lemma 17.9,

$$\lim_{m \to \infty} \mathsf{t}_m \leq \lim_{n \to \infty} \mathsf{s}_n.$$

Since the roles of s_n and t_m can be interchanged, this inequality is in fact an equality. □

Let us present some properties of this integral (vector values will be addressed in Theorem 17.25), including **Markov**'s inequality[2]. For any positive real-valued measurable function f, we define its **essential supremum** to be

$$\sup_\Omega \operatorname{ess} \mathsf{f} \stackrel{\text{def}}{=} \inf\{c \in \mathbb{R} : c \geq \mathsf{f} \text{ almost everywhere on } \Omega\}. \tag{17.8}$$

2 **Markov's inequality.** This inequality was named in honor of Andreï Andreïevitch MARKOV.

Theorem 17.11. — *Let* f *and* g *be positive real-valued measurable functions on a measurable subset* Ω *of* $\mathbb{R}^d$*,* ω *be a measurable subset of* Ω*, and* $c \geq 0$ *(for (d),* $c > 0$*).*

(a)
$$\int_\omega (\mathsf{f}+\mathsf{g}) = \int_\omega \mathsf{f} + \int_\omega \mathsf{g}, \quad \int_\omega c\mathsf{f} = c\int_\Omega \mathsf{f}.$$

(b)
$$\mathsf{f} \leq \mathsf{g} \textit{ almost everywhere} \quad \Rightarrow \quad \int_\omega \mathsf{f} \leq \int_\omega \mathsf{g}.$$

(c)
$$\int_\omega \mathsf{f} \leq |\omega| \sup_\Omega \operatorname{ess} \mathsf{f}.$$

(d)
$$|\{x \in \omega : \mathsf{f}(x) \geq c\}| \leq \frac{1}{c}\int_\omega \mathsf{f}.$$

(e) *If* ω' *is a measurable subset of* Ω *disjoint from* ω*,*

$$\int_{\omega\cup\omega'} \mathsf{f} = \int_\omega \mathsf{f} + \int_{\omega'} \mathsf{f}, \quad \sup_{\omega\cup\omega'} \operatorname{ess} \mathsf{f} = \sup\big\{\sup_\omega \operatorname{ess} \mathsf{f}, \sup_{\omega'} \operatorname{ess} \mathsf{f}\big\}.$$

(f) *If* g *is a positive function in* $\mathcal{B}(\Omega)$*,*

$$\int_\omega \mathsf{fg} \leq \sup_\omega \operatorname{ess} \mathsf{f} \int_\omega \mathsf{g}.$$

(g) *If* σ *is a negligible subset of* Ω*,*

$$\int_{\omega\cup\sigma} \mathsf{f} = \int_\omega \mathsf{f}. \quad \blacksquare$$

Proof. ***1º)*** *Integral properties* (a), (b), (c), *and* (e). These follow from the analogous properties of the integral of simple functions given in Proposition 17.7.

2º) *Property* (d). Let $\lambda \stackrel{\text{def}}{=} \{x \in \omega : \mathsf{f}(x) \geq c\}$, which is measurable (Theorem 17.2). The function $c1_\omega$ is positive, real-valued, measurable, and $c1_\lambda \leq \mathsf{f}$. The calculation of the measure of λ (Proposition 17.7) and the monotonicity (b) yield

$$|\lambda| = \int_\omega 1_\lambda = \frac{1}{c}\int_\omega c1_\lambda \leq \int_\omega \mathsf{f}.$$

3º) *Property* (e) *of* sup ess. For a simple function s, the function $\sup \mathsf{s}$ coincides with $\sup \operatorname{ess} \mathsf{s}$, and $\sup_{\omega\cup\omega'} \mathsf{s} = \sup\big\{\sup_\omega \mathsf{s}, \sup_{\omega'} \mathsf{s}\big\}$. By passing to the limit using the approximations s_n of f, we obtain the stated equality.

4º) *Property* (f). By modifying f on the negligible set where $\mathsf{f} > \sup \operatorname{ess} \mathsf{f}$, we obtain $\mathsf{f} \leq \sup \operatorname{ess} \mathsf{f}$ everywhere. The result then follows from the additivity (a) and monotonicity (b) of the integral.

5º) Property (g). By definition of the integral of a simple function from Proposition 17.7, we have $\int_{\sigma\backslash\omega} \mathsf{f} = 0$. The conclusion then follows from the additivity (e) with respect to the domain. □

17.3. Dominated convergence of positive functions

Let us begin with a result[3] on the limit of functions with bounded integrals.

Theorem 17.12. — *Let $(\mathsf{f}_n)_{n\in\mathbb{N}}$ be an increasing sequence of non-negative measurable real-valued functions on a measurable subset Ω of $\mathbb{R}^d$ and $c \in \mathbb{R}$, such that*

$$\int_\Omega \mathsf{f}_n \leq c. \tag{17.9}$$

Then, the set $\sigma \stackrel{\text{def}}{=} \{x \in \Omega : \lim_{n\to\infty} \mathsf{f}_n(x) = \infty\}$ is negligible and $\lim_{n\to\infty} \mathsf{f}_n$ is a measurable function on $\Omega\backslash\sigma$. ▮

Proof. 1º) The set σ. We have

$$\sigma = \bigcap_{k\in\mathbb{N}} \bigcup_{n\in\mathbb{N}} \{x \in \Omega : \mathsf{f}_n(x) \geq k\}.$$

The set $\{x \in \Omega : \mathsf{f}_n(x) \geq k\}$ is measurable (Theorem 17.2), and Markov's inequality (Theorem 17.11 (d)) yields

$$\left|\{x \in \Omega : \mathsf{f}_n(x) \geq k\}\right| \leq \frac{1}{k}\int_\Omega f_n \leq \frac{c}{k}.$$

The union over $n \in \mathbb{N}$ of the sets $\{x \in \Omega : \mathsf{f}_n(x) \geq k\}$, i.e. their limit as $n \to \infty$ since they are increasing with n, is measurable (Theorem 2.16), and

$$\left|\bigcup_{n\in\mathbb{N}} \{x \in \Omega : \mathsf{f}_n(x) \geq k\}\right| \leq \frac{c}{k}.$$

The intersection σ over $k \in \mathbb{N}$ of this union, *i.e.* the limit since it decreases with k, is also measurable (Theorem 2.17), and $|\sigma| = 0$. Thus, σ is negligible.

2º) The function $\lim_{n\to\infty} \mathsf{f}_n$. On $\Omega\backslash\sigma$, this limit is measurable, as is any function that is the limit of measurable functions (Theorem 17.4). □

---

3 **Historical background of Theorem 17.12.** This theorem **avoids** (throughout this book) the use of functions **with values in $\overline{\mathbb{R}}$**. This is particularly the case for the Beppo Levi theorem and Fatou's lemma (see the variants of these results on p. 343 and 344), and also for establishing property (18.39), p. 386.
I warmly thank Jérôme LEMOINE for providing the present proof, which does not use functions with values in $\overline{\mathbb{R}}$.

Let us now present a variant of Beppo Levi's **monotone convergence theorem**[4].

Theorem 17.13. — *Let* $(\mathsf{f}_n)_{n\in\mathbb{N}}$ *be an increasing sequence of measurable positive real-valued functions on a measurable subset* Ω *of* $\mathbb{R}^d$ *and* $c \in \mathbb{R}$, *such that*

$$\int_\Omega \mathsf{f}_n \leq c.$$

Then, $\sigma \overset{\text{def}}{=} \{x \in \Omega : \lim_{n\to\infty} \mathsf{f}_n(x) = \infty\}$ *is negligible, the sequence* $(\mathsf{f}_n)_{n\in\mathbb{N}}$ *converges in* $\Omega\backslash\sigma$ *to a function* f *that is measurable on* Ω, *and*

$$\int_\Omega \mathsf{f}_n = \int_{\Omega\backslash\sigma} \mathsf{f}_n \to \int_{\Omega\backslash\sigma} \lim_{n\to\infty} \mathsf{f}_n = \int_\Omega \mathsf{f}. \qquad (17.10)$$

Proof. ***1º)*** *The set* σ. It is negligible according to the theorem on the limit of functions with bounded integrals (Theorem 17.12).

2º) *Measurability of* f. For every $x \in \Omega$, the sequence of the $\mathsf{f}_n(x)$ is increasing, hence it has a limit everywhere, which may be either finite or infinite (Proposition A.5). Again, by Theorem 17.12, this limit is finite outside σ. Let

$$\mathsf{f}(x) \overset{\text{def}}{=} \begin{cases} \lim_{n\in\mathbb{N}} \mathsf{f}_n(x) & \text{if } x \in \Omega\setminus\sigma, \\ \text{an arbitrary value} & \text{if } x \in \sigma. \end{cases}$$

Then, $\mathsf{f}_n \to \mathsf{f}$ almost everywhere in Ω, so (Theorem 17.4) f is measurable.

3º) *Bound for* $\int_\Omega \mathsf{f}_n$. The monotonicity of the integral (Theorem 17.11 (b)) gives

$$\lim_{n\to\infty} \int_\Omega \mathsf{f}_n \leq \int_\Omega \mathsf{f}. \qquad (17.11)$$

4º) *Convergence of* $\int_\Omega \mathsf{f}_n$. ***4.1.*** *Approximation of* f. Since f_n is positive and measurable, there exists (Lemma 17.8) an increasing sequence of simple functions $(\mathsf{s}_{n,m})_{m\in\mathbb{N}}$ converging almost everywhere to f_n. Let

$$\mathsf{t}_m \overset{\text{def}}{=} \sup_{n\leq m} \mathsf{s}_{n,m}.$$

Each t_m is a simple function (as is any supremum over a finite number of simple functions), bounded almost everywhere by f (since $\mathsf{t}_m \leq \mathsf{f}_m \leq \mathsf{f}$ almost everywhere), and t_m increases with m because (since $\mathsf{s}_{n,m}$ increases with m)

$$\mathsf{t}_m \leq \sup_{n\leq m} \mathsf{s}_{n,m+1} \leq \sup\{\sup_{n\leq m} \mathsf{s}_{n,m+1}; \mathsf{s}_{m+1,m+1}\} = \mathsf{t}_{m+1}.$$

4 **Historical background of Beppo Levi's Theorem 17.13.** Beppo LEVI proved in 1906 [47] that any increasing and bounded sequence in $L^1(\mathbb{R})$ converges in $L^1(\mathbb{R})$.
The variant presented here allows one to restrict to functions **taking values in** $\mathbb{R}$ (instead of in $\overline{\mathbb{R}}$, as is usually the case).

This defines an almost everywhere limit g for the sequence t_m. We shall now show that, in fact, when $m \to \infty$,

$$\mathsf{t}_m \to \mathsf{f} \text{ almost everywhere in } \Omega. \tag{17.12}$$

First, for $n \le m$, we have $\mathsf{s}_{n,m} \le \mathsf{t}_m$; as $m \to \infty$, we get $\mathsf{f}_n \le \mathsf{g}$ almost everywhere; then, by letting $n \to \infty$, we get $\mathsf{f} \le \mathsf{g}$ almost everywhere. Conversely, $\mathsf{s}_{n,m} \le \mathsf{f}_n \le \mathsf{f}$ almost everywhere; hence $\mathsf{t}_m \le \mathsf{f}$ almost everywhere; then, by letting m tend to infinity, we get $\mathsf{g} \le \mathsf{f}$ almost everywhere. Thus, $\mathsf{g} = \mathsf{f}$ almost everywhere, which proves (17.12).

4.2. *Lower bound for* $\lim_{n\to\infty} \int_\Omega \mathsf{f}_n$. By Definition 17.10 of the integral, (17.12) yields

$$\int_\Omega \mathsf{f} = \lim_{m\to\infty} \int_\Omega \mathsf{t}_m.$$

For all $n \le m$, we have $\mathsf{s}_{n,m} \le \mathsf{f}_n \le \mathsf{f}_m$. Therefore, $\mathsf{t}_m = \sup_{n\le m} \mathsf{s}_{n,m} \le \mathsf{f}_m$, and the monotonicity of the integral (Theorem 17.11 (b)) gives $\int_\Omega \mathsf{t}_m \le \int_\Omega \mathsf{f}_m$. Thus,

$$\int_\Omega \mathsf{f} = \lim_{m\to\infty} \int_\Omega \mathsf{t}_m \le \lim_{m\to\infty} \int_\Omega \mathsf{f}_m.$$

4.3. *Conclusion.* This inequality is the converse to (17.11), which therefore must be an equality. From which (17.10) follows. □

We now present a variant of **Fatou's lemma**[5]. For a real-valued sequence $(c_n)_{n\in\mathbb{N}}$, denote

$$\liminf_{n\to\infty} c_n \stackrel{\text{def}}{=} \lim_{n\to\infty} \inf_{m\ge n} c_m, \quad \limsup_{n\to\infty} c_n \stackrel{\text{def}}{=} \lim_{n\to\infty} \sup_{m\ge n} c_m. \tag{17.13}$$

Theorem 17.14. — *Let* $(\mathsf{f}_n)_{n\in\mathbb{N}}$ *be a sequence of positive real-valued measurable functions on a measurable subset* Ω *of* $\mathbb{R}^d$ *and* $c \in \mathbb{R}$*, such that*

$$\int_\Omega \mathsf{f}_n \le c. \tag{17.14}$$

Then, $\sigma \stackrel{\text{def}}{=} \{x \in \Omega : \liminf_{n\to\infty} \mathsf{f}_n(x) = \infty\}$ *is negligible,* $\liminf_{n\to\infty} \mathsf{f}_n$ *is equal in* $\Omega\backslash\sigma$ *to a measurable function* f *on* Ω*, and*

$$\int_\Omega \mathsf{f} = \int_{\Omega\backslash\sigma} \liminf_{n\to\infty} \mathsf{f}_n \le \liminf_{n\to\infty} \int_\Omega \mathsf{f}_n. \tag{17.15}$$ ▌

5 **Fatou's lemma.** Theorem 17.14 is attributed to Pierre Joseph Louis FATOU (1878–1929), though I have not been able to find a reference.

The version presented here makes it possible to restrict ourselves to functions **taking values in** $\mathbb{R}$ (instead of $\overline{\mathbb{R}}$, as is usually the case).

Proof. ***1º) Preliminary.*** If f and g are measurable functions on Ω, then

$$\text{the function } \inf\{\mathsf{f}, \mathsf{g}\} \text{ is measurable.} \tag{17.16}$$

Indeed, let $(\mathsf{s}_n)_{n\in\mathbb{N}}$ and $(\mathsf{t}_n)_{n\in\mathbb{N}}$ be sequences of simple functions such that $\mathsf{s}_n \to \mathsf{f}$ and $\mathsf{t}_n \to \mathsf{g}$ outside some negligible subsets σ and τ; the function $\inf\{\mathsf{s}_n, \mathsf{t}_n\}$ is simple and $\inf\{\mathsf{s}_n, \mathsf{t}_n\} \to \inf\{\mathsf{f}, \mathsf{g}\}$ outside $\sigma \cup \tau$, which is negligible (Theorem 2.24 (c)).

2º) Measurability of σ. By induction on (17.16), $\inf_{N\geq m\geq n} \mathsf{f}_m$ is measurable for each finite N. This is a positive function that decreases as N increases. Therefore (Theorem A.5), when $N \to \infty$, in all of Ω,

$$\inf_{N\geq m\geq n} \mathsf{f}_m \to \inf_{m\geq n} \mathsf{f}_m.$$

Since any function that is the limit of measurable functions is measurable (Theorem 17.4),

$$\inf_{m\geq n} \mathsf{f}_m \text{ is measurable.}$$

The monotonicity of the integral (Proposition 17.7 (d)) gives, together with (17.14),

$$\int_\Omega \inf_{m\geq n} \mathsf{f}_m \leq \int_\Omega \mathsf{f}_n \leq c.$$

The function $\inf_{m\geq n} \mathsf{f}_m$ increases with n so, according to Theorem 17.12 on the limit of functions with bounded integrals, the set $\{x \in \Omega : \lim_{n\to\infty} \inf_{m\geq n} \mathsf{f}_n = \infty\}$ is negligible. Hence, together with (17.13), we get

$$\{x \in \Omega : \liminf_{n\to\infty} \mathsf{f}_n = \infty\} \text{ is negligible.}$$

3º) Measurability of $\liminf_{n\to\infty} \mathsf{f}_n$. From Theorem 17.12, we also have

$$\liminf_{n\to\infty} \mathsf{f}_n \text{ is measurable on } \Omega\backslash\sigma.$$

4º) Convergence. For all $p \geq n$, we have $\inf_{m\geq n} \mathsf{f}_m \leq \mathsf{f}_p$. The monotonicity of the integral (Proposition 17.11 (b)) therefore gives

$$\int_\Omega \inf_{m\geq n} \mathsf{f}_m \leq \int_\Omega \mathsf{f}_p.$$

Hence,

$$\int_\Omega \inf_{m\geq n} \mathsf{f}_m \leq \inf_{p\geq n} \int_\Omega \mathsf{f}_p.$$

And,

$$\lim_{n\to\infty} \int_\Omega \inf_{m\geq n} \mathsf{f}_m \leq \lim_{n\to\infty} \inf_{p\geq n} \int_\Omega \mathsf{f}_p = \liminf_{n\to\infty} \int_\Omega \mathsf{f}_n.$$

Since the function $\inf_{m\geq n} \mathsf{f}_m$ is measurable, increasing, and has a bounded integral, Beppo Levi's theorem (Theorem 17.13) gives (the second equality below)

$$\int_{\Omega\setminus\sigma} \liminf_{n\to\infty} \mathsf{f}_n = \int_{\Omega\setminus\sigma} \lim_{n\to\infty} \inf_{m\geq n} \mathsf{f}_m = \lim_{n\to\infty} \int_{\Omega} \inf_{m\geq n} \mathsf{f}_m.$$

Together with the previous inequality, this yields (17.15).

5º) The function f. Any function f on Ω equal to $\liminf_{n\to\infty} \mathsf{f}_n$ on $\Omega\setminus\sigma$ satisfies (Proposition 17.11 (g)) $\int_\Omega \mathsf{f} = \int_{\Omega\setminus\sigma} \liminf_{n\to\infty} \mathsf{f}_n$, since σ is negligible. □

Let us turn to a particular case of the dominated convergence theorem (the general result will be given in Theorem 18.5).

Proposition 17.15. — *Let $(\mathsf{f}_n)_{n\in\mathbb{N}}$ be a sequence of real-valued measurable functions, and let f and g be measurable functions on a measurable subset Ω of $\mathbb{R}^d$, such that*

$$\mathsf{f}_n \to \mathsf{f} \text{ almost everywhere}, \quad 0 \leq \mathsf{f}_n \leq \mathsf{g} \text{ almost everywhere}, \quad \int_\Omega \mathsf{g} < \infty.$$

Then, when $n \to \infty$,

$$\int_\Omega |\mathsf{f}_n - \mathsf{f}| \to 0,$$

and therefore,

$$\int_\Omega \mathsf{f}_n \to \int_\Omega \mathsf{f}. \blacksquare$$

Proof. We have $0 \leq \mathsf{f} \leq \mathsf{g}$ almost everywhere, hence $|\mathsf{f}_n - \mathsf{f}| \leq 2\mathsf{g}$. Fatou's lemma (Theorem 17.14) then gives, for a negligible subset σ (Proposition 17.11 (g)),

$$\int_\Omega 2\mathsf{g} = \int_{\Omega\setminus\sigma} 2\mathsf{g} = \int_{\Omega\setminus\sigma} \liminf_{n\to\infty}(2\mathsf{g} - |\mathsf{f}_n - \mathsf{f}|) \leq \liminf_{n\to\infty} \int_\Omega (2\mathsf{g} - |\mathsf{f}_n - \mathsf{f}|).$$

By the additivity of the integral of positive functions (Theorem 17.11 (a)),

$$\int_\Omega (2\mathsf{g} - |\mathsf{f}_n - \mathsf{f}|) + \int_\Omega |\mathsf{f}_n - \mathsf{f}| = \int_\Omega 2\mathsf{g}.$$

The preceding inequality then gives

$$\int_\Omega 2\mathsf{g} \leq \liminf_{n\to\infty}\Big(\int_\Omega 2\mathsf{g} - \int_\Omega |\mathsf{f}_n - \mathsf{f}|\Big) =$$

$$= \int_\Omega 2\mathsf{g} + \liminf_{n\to\infty}\Big(-\int_\Omega |\mathsf{f}_n - \mathsf{f}|\Big) = \int_\Omega 2\mathsf{g} - \limsup_{n\to\infty} \int_\Omega |\mathsf{f}_n - \mathsf{f}|.$$

Whence $\limsup_{n\to\infty} \int_\Omega |\mathsf{f}_n - \mathsf{f}| \leq 0$, and thus

$$\lim_{n\to\infty} \int_\Omega |\mathsf{f}_n - \mathsf{f}| = 0. \ \square$$

17.4. Spaces of classes of integrable functions

Let us define integrable functions and the integral ($\sup \operatorname{ess} \mathsf{f}$ is defined by (17.8)).

Definition 17.16. — *Let* f *be a function from a measurable subset* Ω *of* $\mathbb{R}^d$ *into a Fréchet space* E*, endowed with an increasing sequence of semi-norms* $(\| \ \|_{E;m})_{m\in\mathbb{N}}$ *(given by Proposition A.26), and* $1 \leq p \leq \infty$.

(a) *The function* f *is said to be* ***p-integrable*** *if it is measurable (in which case both* $\|\mathsf{f}\|_{E;m}^p$ *and* $\|\mathsf{f}\|_{E;m}$ *are measurable) and if, for every* $m \in \mathbb{N}$,

$$\int_\Omega \|\mathsf{f}\|_{E;m}^p < \infty \quad \text{if } p < \infty, \tag{17.17}$$

$$\sup_\Omega \operatorname{ess} \|\mathsf{f}\|_{E;m} < \infty \quad \text{if } p = \infty.$$

(b) *If* f *is* ***p-integrable*** *for finite p, there exists a sequence, called a* ***p-approximating sequence****, of simple functions* $(\mathsf{s}_n)_{n\in\mathbb{N}}$ *such that, for every* $m \in \mathbb{N}$*, when* $n \to \infty$,

$$\int_\Omega \|\mathsf{f} - \mathsf{s}_n\|_{E;m}^p \to 0. \tag{17.18}$$

(c) *If* f *is* ***integrable*** *(that is, 1-integrable) and* $(\mathsf{s}_n)_{n\in\mathbb{N}}$ *is a 1-approximating sequence for f, then the* ***integral*** *of* f *is the element of* E*, independent of the 1-approximating sequence,*

$$\int_\Omega \mathsf{f} \stackrel{\text{def}}{=} \lim_{n\to\infty} \int_\Omega \mathsf{s}_n. \ ∎$$

Justification. ***1°)*** *Part (a).* Observe that $\int_\Omega \|\mathsf{f}\|_{E;m}^p$ and $\sup \operatorname{ess}_\Omega \|\mathsf{f}\|_{E;m}$ are well-defined, since $\|\mathsf{f}\|_{E;m}^p$ is measurable. Indeed, f is (Definition 17.1) the limit almost everywhere of a sequence $(\mathsf{s}_n)_{n\in\mathbb{N}}$ of simple functions, and (Proposition A.10) $\| \ \|_{E;m}$ is continuous, so $\|\mathsf{f}\|_{E;m}^p$ is the limit almost everywhere of the sequence of simple functions $(\|\mathsf{s}_n\|_{E;m}^p)_{n\in\mathbb{N}}$.

Hence, there exists (Lemma 17.8) an increasing sequence of simple functions converging almost everywhere to $\|\mathsf{f}\|_{E;m}^p$, which allows us (Definition 17.10) to define $\int_\Omega \|\mathsf{f}\|_{E;m}^p$.

2°) *Part (b).* Let $(\mathsf{t}_n)_{n\in\mathbb{N}}$ be a sequence of simple functions converging almost everywhere to f, given by Definition 17.1 of measurability. For $m \in \mathbb{N}$, consider the simple function

$$\mathsf{s}_{n,m} \stackrel{\text{def}}{=} \mathsf{t}_n 1_{\omega_{n,m}} \quad \text{where } \omega_{n,m} \stackrel{\text{def}}{=} \{x \in \Omega : \|\mathsf{t}_n(x)\|_{E;m} \leq 2\|\mathsf{f}(x)\|_{E;m}\}.$$

If $\mathsf{f}(x) = 0$, then all the $\mathsf{s}_{n,m}(x)$ vanish from some index $m(n)$. Indeed, if $\mathsf{t}_n(x) \neq 0$, there exists $m(n)$ such that for all $m \geq m(n)$, we have $\|\mathsf{t}_n(x)\|_{E;m} \neq 0$ (since the sequence $(\| \ \|_{E;m})$ is increasing), in which case $x \notin \omega_{m,n}$, and therefore $1_{\omega_{m,n}}(x) = 0$.

If $\mathsf{f}(x) \neq 0$ then, from some index m, we have $\|\mathsf{f}(x)\|_{E;m} > 0$. Since, for almost every x, we have $\mathsf{t}_n(x) \to \mathsf{f}(x)$, there exists some $n(x)$ such that, for all $n \geq n(x)$,

$$\|\mathsf{t}_n(x)\|_{E;m} \leq 2\|\mathsf{f}(x)\|_{E;m}. \tag{17.19}$$

Then, $\mathsf{s}_{n,m}(x) = \mathsf{t}_n(x)$, which tends to $\mathsf{f}(x)$ as $n \to \infty$; hence, almost everywhere,

$$\|\mathsf{s}_{n,m} - \mathsf{f}\|_{E;m} \to 0 \ \text{ as } n \to \infty. \tag{17.20}$$

On the other hand, by the construction of $\mathsf{s}_{n,m}$, for all $x \in \Omega$ and $n \in \mathbb{N}$, we have

$$\|\mathsf{s}_{n,m}(x) - \mathsf{f}(x)\|_{E;m} \leq \|\mathsf{s}_{n,m}(x)\|_{E;m} + \|\mathsf{f}(x)\|_{E;m} \leq 3\|\mathsf{f}(x)\|_{E;m}.$$

The dominated convergence theorem for positive real-valued functions (Proposition 17.15), applied to the function $\|\mathsf{s}_{n,m} - \mathsf{f}\|^p_{E;m}$, therefore gives, as $n \to \infty$,

$$\int_\Omega \|\mathsf{s}_{n,m} - \mathsf{f}\|^p_{E;m} \to 0. \tag{17.21}$$

For each $m \in \mathbb{N}$, let $n(m)$ increase with m and be such that

$$\int_\Omega \|\mathsf{s}_{n(m),m} - \mathsf{f}\|^p_{E;m} \leq \frac{1}{m}.$$

Since $\| \ \|_{E;m}$ increases with m, the sequence of the $\mathsf{s}_m = \mathsf{s}_{n(m),m}$ satisfies (17.18).

*3*º*) Part (c).* ***3.a.*** *Convergence of* $\int_\Omega \mathsf{s}_n$. Let $(\mathsf{s}_n)_{n\in\mathbb{N}}$ be a 1-approximating sequence for f. From the bound on the integral of a simple function (Theorem 17.7 (b)) and the monotonicity and the linearity of positive measurable functions (Theorem 17.11 (b)), we have, for all m, n, and n',

$$\left\|\int_\Omega (\mathsf{s}_n - \mathsf{s}_{n'})\right\|_{E;m} \leq \int_\Omega \|\mathsf{s}_n - \mathsf{s}_{n'}\|_{E;m} \leq \int_\Omega \|\mathsf{s}_n - \mathsf{f}\|_{E;m} + \int_\Omega \|\mathsf{f} - \mathsf{s}_{n'}\|_{E;m}.$$

From (17.18), $(\int_\Omega \mathsf{s}_n)_{n\in\mathbb{N}}$ is a Cauchy sequence in E which, being Fréchet, is sequentially complete (Proposition A.32). It therefore has a limit, which we denote ℓ_{s}.

3.b. *Uniqueness of the limit.* If $(\mathsf{s}_n)_{n\in\mathbb{N}}$ and $(\mathsf{t}_n)_{n\in\mathbb{N}}$ are two 1-approximating sequences for f, then the inequality above with t_n in the place of $\mathsf{s}_{n'}$ gives

$$\left\|\int_\Omega (\mathsf{s}_n - \mathsf{t}_n)\right\|_{E;m} \leq \int_\Omega \|\mathsf{s}_n - \mathsf{f}\|_{E;m} + \int_\Omega \|\mathsf{f} - \mathsf{t}_n\|_{E;m} \to 0.$$

Taking the limit, we get $\|\ell_{\mathsf{s}} - \ell_{\mathsf{t}}\|_{E;m} = 0$. This holds for all $m \in \mathbb{N}$, hence $\ell_{\mathsf{s}} = \ell_{\mathsf{t}}$. □

Caution. A p-approximating sequence does not necessarily converge almost everywhere. This is the case for the sequence of *sliding bumps* by Thierry GALLOUËT and Raphaèle HERBIN [32, § 4.7, p. 178]. However, one can extract a subsequence that converges almost everywhere by the converse dominated convergence theorem (Theorem 18.4). The existence of a p-approximating sequence is therefore equivalent to p-integrability. □

Let us define the space of classes of p-integrable functions.

Definition 17.17. — *Given a p-integrable function* f *from a measurable subset* Ω *of* $\mathbb{R}^d$ *into a Fréchet space* E*, where* $1 \le p \le \infty$*, let* $\dot{\mathsf{f}}$ *denote the class of functions equal almost everywhere to* f*. We define the vector space*

$$\dot{\mathsf{L}}^p(\Omega; E) \stackrel{\text{def}}{=} \{\dot{\mathsf{f}} : \mathsf{f} \textit{ is p-integrable }\}$$

endowed with the semi-norms indexed by m*, where* $(\| \ \|_{E;m})_{m\in\mathbb{N}}$ *defines the topology on* E*,*

$$\|\dot{\mathsf{f}}\|_{\dot{\mathsf{L}}^p(\Omega;E);m} \stackrel{\text{def}}{=} \begin{cases} \left(\displaystyle\int_\Omega \|\mathsf{f}\|^p_{E;m}\right)^{1/p} & \textit{if } p < \infty, \\ \displaystyle\sup_\Omega \text{ess} \|\mathsf{f}\|_{E;m} & \textit{if } p = \infty, \end{cases}$$

and which are independent of the choice of the representative f *for* $\dot{\mathsf{f}}$. ▮

Justification. *1°) Vector space.* This follows from the fact that, if f' and g' are almost everywhere equal to f and g, respectively, and if $t \in \mathbb{R}$, then $\mathsf{f}' + \mathsf{g}'$ and $t\mathsf{f}'$ are almost everywhere equal to $\mathsf{f} + \mathsf{g}$ and $t\mathsf{f}$, respectively.

2°) Semi-norms. **2.a.** *Independence of the representative.* If f' is equal almost everywhere to f, and if $(\mathsf{s}_n)_{n\in\mathbb{N}}$ is a p-approximating sequence (Definition 17.16 (b)) for $\|\mathsf{f}\|_{E;m}$, then $(\mathsf{s}_n)_{n\in\mathbb{N}}$ is also a p-approximating sequence for $\|\mathsf{f}'\|_{E;m}$. Therefore, the functions $\|\mathsf{f}\|^p_{E;m}$ and $\|\mathsf{f}'\|^p_{E;m}$ have the same integral given by Definition 17.16 (c).

Moreover, $\|\mathsf{f}\|_{E;m}$ and $\|\mathsf{f}'\|_{E;m}$ are equal almost everywhere, and therefore have the same sup ess (defined by (17.8), p. 340).

2.b. *Semi-norms when* $p < \infty$. Let $(\mathsf{s}_n)_{n\in\mathbb{N}}$ and $(\mathsf{t}_n)_{n\in\mathbb{N}}$ be p-approximating sequences for f and g. Then, $(\|\mathsf{s}_n\|_{E;m})_{n\in\mathbb{N}}$, $(\|\mathsf{t}_n\|_{E;m})_{n\in\mathbb{N}}$, and $(\|\mathsf{s}_n + \mathsf{t}_n\|_{E;m})_{n\in\mathbb{N}}$ converge almost everywhere to $\|\mathsf{f}\|_{E;m}$, $\|\mathsf{g}\|_{E;m}$, and $\|\mathsf{f} + \mathsf{g}\|_{E;m}$, respectively, for each $m \in \mathbb{N}$. By using a common decomposition for s_n and t_n, we obtain sets λ_i such that

$$\mathsf{s}_n = \sum_{i=1}^{j} e_i 1_{\lambda_i}, \quad \mathsf{t}_n = \sum_{i=1}^{j} d_i 1_{\lambda_i}.$$

The variant of Minkowski's inequality for finite sums (Proposition 9.4) yields, by the definition of the integral of a simple function (Proposition 17.7), when $p < \infty$,

$$\Big(\int_\Omega \|\mathsf{s}_n + \mathsf{t}_n\|^p_{E;m}\Big)^{\frac{1}{p}} = \Big(\sum_{i=1}^{j} \|e_i + d_i\|^p_{E;m}|\lambda_i|\Big)^{\frac{1}{p}} \leq$$

$$\leq \Big(\sum_{i=1}^{j} \|e_i\|^p_{E;m}|\lambda_i|\Big)^{\frac{1}{p}} + \Big(\sum_{i=1}^{j} \|d_i\|^p_{E;m}|\lambda_i|\Big)^{\frac{1}{p}} \leq \Big(\int_\Omega \|\mathsf{s}_n\|^p_{E;m}\Big)^{\frac{1}{p}} + \Big(\int_\Omega \|\mathsf{t}_n\|^p_{E;m}\Big)^{\frac{1}{p}}.$$

By the definition of the integral of an integrable function (Definition 17.16 (c)), in the limit, we get

$$\Big(\int_\Omega \|\mathsf{f} + \mathsf{g}\|^p_{E;m}\Big)^{1/p} \leq \Big(\int_\Omega \|\mathsf{f}\|^p_{E;m}\Big)^{1/p} + \Big(\int_\Omega \|\mathsf{g}\|^p_{E;m}\Big)^{1/p}.$$

That is,

$$\|\dot{\overline{\mathsf{f} + \mathsf{g}}}\|_{\dot{\mathsf{L}}^p(\Omega;E);m} \leq \|\dot{\mathsf{f}}\|_{\dot{\mathsf{L}}^p(\Omega;E);m} + \|\dot{\mathsf{g}}\|_{\dot{\mathsf{L}}^p(\Omega;E);m}.$$

Similarly, $\|\dot{\overline{t\mathsf{f}}}\|_{\dot{\mathsf{L}}^p(\Omega;E);m} \leq t\|\dot{\mathsf{f}}\|_{\dot{\mathsf{L}}^p(\Omega;E);m}$.

2.c. *Semi-norms for $p = \infty$.* The inequality $\|u + v\|_{E;m} \leq \|u\|_{E;m} + \|v\|_{E;m}$, for u and v in E, similarly yields, for $p = \infty$, the two inequalities characterizing the semi-norms. □

Let us finally define locally p-integrable functions on an open set.

Definition 17.18. — *We say that a function* f *from an open subset* Ω *of* $\mathbb{R}^d$ *into a Fréchet space* E *is* **locally p-integrable**, *where* $1 \leq p \leq \infty$, *if, for every open subset* $\omega \Subset \Omega$, *the restriction* $\mathsf{f}|_\omega$ *is p-integrable (Definition 17.16).*

We denote by $\dot{\mathsf{f}}$ *the class of functions equal almost everywhere to* f *and*

$$\dot{\mathsf{L}}^p_{\text{loc}}(\Omega; E) \stackrel{\text{def}}{=} \{\dot{\mathsf{f}} : \mathsf{f} \textit{ is locally p-integrable}\}$$

endowed with the semi-norms, indexed by $m \in \mathbb{N}$ *and* $\omega \Subset \Omega$, *where* $(\|\ \|_{E;m})_{m\in\mathbb{N}}$ *defines the topology of* E,

$$\|\dot{\mathsf{f}}\|_{\dot{\mathsf{L}}^p_{\text{loc}}(\Omega;E);m,\omega} \stackrel{\text{def}}{=} \|\dot{\mathsf{f}}|_\omega\|_{\dot{\mathsf{L}}^p(\omega;E);m}. \blacksquare$$

Let us introduce some notions regarding spaces of classes of functions.

Definition 17.19. — *Let Ω be a measurable subset of $\mathbb{R}^d$, ω be a measurable subset included in Ω, E be a Fréchet space, $1 \le p \le \infty$, $\mathsf{f} \in \dot{\mathsf{f}}$, $\mathsf{f}_n \in \dot{\mathsf{f}}_n$, and $\mathsf{g} \in \dot{\mathsf{g}}$.*

(a) *The integral of $\dot{\mathsf{f}} \in \dot{\mathsf{L}}^1(\Omega; E)$ over ω is:* $\int_\omega \dot{\mathsf{f}} \stackrel{\text{def}}{=} \int_\omega \mathsf{f}$.

(b) *The essential supremum of $\dot{\mathsf{f}} \in \dot{\mathsf{L}}^\infty(\Omega; E)$ over ω is:* $\sup \mathrm{ess}_\omega \dot{\mathsf{f}} \stackrel{\text{def}}{=} \sup \mathrm{ess}_\omega \mathsf{f}$.

(c) *The order relation on $\dot{\mathsf{L}}^p_{\mathrm{loc}}(\Omega; E)$ is: $\dot{\mathsf{f}} \le \dot{\mathsf{g}}$ is equivalent to $\mathsf{f} \le \mathsf{g}$ almost everywhere. We also denote $\dot{\mathsf{f}} \ge \dot{\mathsf{g}}$ if $\dot{\mathsf{g}} \le \dot{\mathsf{f}}$.*

(d) *$\dot{\mathsf{f}}_n \to \dot{\mathsf{f}}$ almost everywhere is equivalent to $\mathsf{f}_n \to \mathsf{f}$ almost everywhere.*

(e) *$\dot{\mathsf{f}}_n \to \dot{\mathsf{f}}$ uniformly almost everywhere is equivalent to $\mathsf{f}_n \to \mathsf{f}$ uniformly almost everywhere.* ▌

Justification. These definitions do not depend on the choice of representative for the class: for the integral, this follows from Definition 17.10; for the essential supremum, from Definition (17.8), p. 340; for the order relation and convergence properties, from Definition 17.18 of a class of functions equal almost everywhere. □

17.5. Completion and approximation in spaces of classes of functions

Let us show that $\dot{\mathsf{L}}^p(\Omega; E)$ and $\dot{\mathsf{L}}^p_{\mathrm{loc}}(\Omega; E)$ are Fréchet spaces.

Theorem 17.20. — *The spaces $\dot{\mathsf{L}}^p(\Omega; E)$ and $\dot{\mathsf{L}}^p_{\mathrm{loc}}(\Omega; E)$ are Fréchet spaces if Ω is a measurable subset of $\mathbb{R}^d$, E is a Fréchet space, and $1 \le p \le \infty$.* ▌

Proof. This amounts to verifying (Definitions 1.5 and A.24) that the spaces $\dot{\mathsf{L}}^p(\Omega; E)$ and $\dot{\mathsf{L}}^p_{\mathrm{loc}}(\Omega; E)$ are separated, that their families of semi-norms are equivalent to countable families, and that they are sequentially complete. Let $(\| \ \|_{E;m})_{m \in \mathbb{N}}$ denote the semi-norms that define the topology of E.

1°) Separation of $\dot{\mathsf{L}}^p(\Omega; E)$, $p < \infty$. Let $\dot{\mathsf{f}} \in \dot{\mathsf{L}}^p(\Omega; E)$ and $\mathsf{f} \in \dot{\mathsf{f}}$ such that, for all $m \in \mathbb{N}$,

$$\|\dot{\mathsf{f}}\|_{\dot{\mathsf{L}}^p(\Omega;E);m} = \Big(\int_\Omega \|\mathsf{f}\|^p_{E;m} \Big)^{1/p} = 0.$$

Markov's inequality (Theorem 17.11 (d)) with respect to the function $x \mapsto \|\mathsf{f}(x)\|^p_{E;m}$ shows that, for n and m in $\mathbb{N}$,

$$0 = \int_\Omega \|\mathsf{f}\|^p_{E;m} \ge \frac{1}{n^p} \Big| \Big\{ x \in \Omega : \|\mathsf{f}(x)\|^p_{E;m} \ge \frac{1}{n^p} \Big\} \Big| = \frac{1}{n^p} \Big| \Big\{ x \in \Omega : \|\mathsf{f}(x)\|_{E;m} \ge \frac{1}{n} \Big\} \Big|.$$

Since $\{x \in \Omega : \|\mathsf{f}(x)\|_{E;m} > 0\} \subset \bigcup_{n\in\mathbb{N}^*}\{x \in \Omega : \|\mathsf{f}(x)\|_{E;m} \geq 1/n\}$, the subadditivity of the measure (Theorem 2.18 (a)) gives

$$|\{x \in \Omega : \|\mathsf{f}(x)\|_{E;m} > 0\}| \leq \sum_{n\in\mathbb{N}^*}\left|\left\{x \in \Omega : \|\mathsf{f}(x)\|_{E;m} \geq \frac{1}{n}\right\}\right| = 0.$$

Denote

$$\omega \stackrel{\text{def}}{=} \{x \in \Omega : \mathsf{f}(x) \neq 0\}, \quad \omega_m \stackrel{\text{def}}{=} \{x \in \Omega : \|\mathsf{f}(x)\|_{E;m} > 0\}.$$

We have $|\omega_m| = 0$ and $\omega = \bigcup_{m\in\mathbb{N}}\omega_m$, from which it follows, again by the subadditivity of the measure (Theorem 2.18 (a)), that $|\omega| \leq \sum_{m\in\mathbb{N}}|\omega_n| = 0$. Therefore, f vanishes almost everywhere, and

$$\dot{\mathsf{f}} = 0.$$

2⁰) Separation of $\dot{\mathsf{L}}^\infty(\Omega; E)$. Suppose that, for all $m \in \mathbb{N}$,

$$\|\dot{\mathsf{f}}\|_{\dot{\mathsf{L}}^\infty(\Omega;E);m} = \inf\{c \in \mathbb{R} : \|\mathsf{f}\|_{E;m} \leq c \text{ almost everywhere}\} = 0.$$

Denote

$$\omega_m \stackrel{\text{def}}{=} \{x \in \Omega : \|\mathsf{f}(x)\|_{E;m} > 0\}, \quad \omega_{n,m} \stackrel{\text{def}}{=} \{x \in \Omega : \|\mathsf{f}(x)\|_{E;m} \geq 1/n\}.$$

For $n \in \mathbb{N}^*$, we have $|\omega_{n,m}| = 0$ and $\omega_m = \bigcup_{n\in\mathbb{N}^*}\omega_{n,m}$, so, by the subadditivity of the measure, $|\omega_m| \leq \sum_{n\in\mathbb{N}^*}|\omega_{n,m}| = 0$; since $\{x \in \Omega : f(x) \neq 0\} = \bigcup_{m\in\mathbb{N}^*}\omega_m$, we have $|\{x \in \Omega : \mathsf{f}(x) \neq 0\}| \leq \bigcup_{n\in\mathbb{N}^*}|\omega_m| = 0$. Thus, f vanishes almost everywhere, and

$$\dot{\mathsf{f}} = 0.$$

3⁰) Countability of the semi-norms of $\dot{\mathsf{L}}^p(\Omega; E)$. The set of semi-norms of $\dot{\mathsf{L}}^p(\Omega; E)$ is countable, since it is indexed by $m \in \mathbb{N}$.

4⁰) Completion of $\dot{\mathsf{L}}^p(\Omega; E), p < \infty$. Since the space $\dot{\mathsf{L}}^p(\Omega; E)$ is metrizable, it suffices, by Proposition A.34, to verify that every absolutely convergent series converges. So let $\sum_{n\in\mathbb{N}}\dot{\mathsf{f}}_n$ be an absolutely convergent series, *i.e.* such that for every $m \in \mathbb{N}$, there exists $c_m \in \mathbb{R}$ such that

$$\sum_{n\in\mathbb{N}}\|\dot{\mathsf{f}}_n\|_{\dot{\mathsf{L}}^p(\Omega;E);m} \leq c_m. \tag{17.22}$$

4.a. *Series of semi-norms in E.* For all $n \in \mathbb{N}$, let $\mathsf{f}_n \in \dot{\mathsf{f}}_n$. By (17.22) and Definition 17.17 of the semi-norms of $\dot{\mathsf{L}}^p(\Omega; E)$,

$$\left\|\dot{\overline{\sum_{n\leq N}\|\mathsf{f}_n\|_{E;m}}}\right\|_{\dot{\mathsf{L}}^p(\Omega)} \leq \sum_{n\leq N}\left\|\dot{\overline{\|\mathsf{f}_n\|_{E;m}}}\right\|_{\dot{\mathsf{L}}^p(\Omega)} = \sum_{n\leq N}\|\dot{\mathsf{f}}_n\|_{\dot{\mathsf{L}}^p(\Omega;E);m} \leq c_m.$$

Thus (Definition (17.17) of $L^p(\Omega; E)$),

$$\int_\Omega \Big(\sum_{n\le N} \|\mathsf{f}_n\|_{E;m}\Big)^p \le c_m^p. \tag{17.23}$$

Let

$$\mathsf{g}_m \overset{\text{def}}{=} \sum_{n\in\mathbb{N}} \|\mathsf{f}_n\|_{E;m}.$$

Since $\big(\sum_{n\le N} \|\mathsf{f}_n\|_{E;m}\big)^p$ increases with N, by the monotone convergence theorem (Theorem 17.13), (17.23) implies that its limit g_m^p is a measurable function on $\Omega\backslash\sigma_m$, where σ_m is negligible, and

$$\int_{\Omega\backslash\sigma_m} \mathsf{g}_m^p = \int_{\Omega\backslash\sigma_m} \lim_{N\to\infty} \Big(\sum_{n\le N} \|\mathsf{f}_n\|_{E;m}\Big)^p = \lim_{N\to\infty} \int_\Omega \Big(\sum_{n\le N} \|\mathsf{f}_n\|_{E;m}\Big)^p \le c_m^p.$$

The union $\sigma = \bigcup_{m\in\mathbb{N}} \sigma_m$ is negligible (Theorem 2.24 (c)) and, for all $m \in \mathbb{N}$,

$$\sum_{n\in\mathbb{N}} \|\mathsf{f}_n(x)\|_{E;m} < \infty \ \text{ for every } x \in \Omega\backslash\sigma.$$

4.b. *Convergence.* Since E is a Fréchet space, $\sum_{n\in\mathbb{N}} \mathsf{f}_n(x)$ converges (Proposition A.34, again) to a limit $\mathsf{F}(x)$ for every $x \in \Omega\backslash\sigma$. Define $\mathsf{F}(x)$ by 0_E on σ. Then F is measurable on all of Ω, as the limit almost everywhere of measurable functions (Theorem 17.4) and, when $N \to \infty$,

$$\sum_{n\le N} \mathsf{f}_n(x) \to \mathsf{F}(x) \ \text{ for every } x \in \Omega\backslash\sigma.$$

Inequality $\|\sum_{n\le N} \mathsf{f}_n\|_{E;m} \le \sum_{n\le N} \|\mathsf{f}_n\|_{E;m} \le \|\mathsf{g}_m\|_{E;m}$ implies, when $N \to \infty$, that $\|F\|_{E;m} \le g_m$ outside σ. Hence, with Definitions 17.16 and 17.17 of $\dot{\mathsf{L}}^p(\Omega; E)$,

$$\dot{\mathsf{F}} \in \dot{\mathsf{L}}^p(\Omega; E).$$

Since $\mathsf{F} - \sum_{n=0}^N \mathsf{f}_n = \sum_{n=N+1}^\infty \mathsf{f}_n = \lim_{k\to\infty} \sum_{n=N+1}^k \mathsf{f}_n$, which is measurable, by Fatou's lemma (Theorem 17.14) we get, as σ is negligible,

$$\int_{\Omega\backslash\sigma} \Big\|\mathsf{F} - \sum_{n=0}^N \mathsf{f}_n\Big\|_{E;m}^p = \int_{\Omega\backslash\sigma} \liminf_{k\to\infty} \Big\|\sum_{n=N+1}^k \mathsf{f}_n\Big\|_{E;m}^p \le \liminf_{k\to\infty} \int_\Omega \Big\|\sum_{n=N+1}^k \mathsf{f}_n\Big\|_{E;m}^p =$$

$$= \liminf_{k\to\infty} \Big\|\dot{\overline{\sum_{n=N+1}^k \mathsf{f}_n}}\Big\|_{\dot{\mathsf{L}}^p(\Omega;E);m}^p \le \Big(\sum_{n\ge N+1} \|\dot{\mathsf{f}}_n\|_{\dot{\mathsf{L}}^p(\Omega;E);m}\Big)^p.$$

When $N \to \infty$, this tends to 0 by hypothesis (17.22), so the series $\sum_{n\in\mathbb{N}} \dot{\mathsf{f}}_n$ converges to $\dot{\mathsf{F}}$ in $\dot{\mathsf{L}}^p(\Omega; E)$. Therefore,

$$\dot{\mathsf{L}}^p(\Omega; E) \text{ is a Fréchet space.}$$

5º) Completion of $\dot{\mathsf{L}}^\infty(\Omega; E)$. Let

$$(\dot{\mathsf{f}}_n)_{n\in\mathbb{N}} \text{ be a Cauchy sequence in } \dot{\mathsf{L}}^\infty(\Omega; E).$$

For n, n', and m in $\mathbb{N}$, denote $\mathsf{f}_n \in \dot{\mathsf{f}}_n$, $\mathsf{f}_{n'} \in \dot{\mathsf{f}}_{n'}$, and

$$\sigma_{n,n',m} = \{x \in \Omega : \|\mathsf{f}_n(x) - \mathsf{f}_{n'}(x)\|_{E;m} > \|\dot{\mathsf{f}}_n - \dot{\mathsf{f}}_{n'}\|_{\dot{\mathsf{L}}^\infty(\Omega;E);m}\}.$$

These sets are of measure zero, as well as their union $\sigma = \bigcup_{n,n',m} \sigma_{n,n',m}$ (Theorem 2.24 (c)). Thus, $\mathsf{f}_n(x)$ has a limit $\mathsf{f}(x)$ outside σ, as $n \to \infty$. Defining f to be 0_E in σ, we obtain a function on all of Ω that is the limit of f_n almost everywhere. Therefore (Theorem 17.4),

$$\mathsf{f} \text{ is measurable.}$$

For $x \notin \sigma$, we have $\|\mathsf{f}_n(x) - \mathsf{f}(x)\|_{E;m} = \lim_{n'\to\infty} \|\mathsf{f}_n(x) - \mathsf{f}_{n'}(x)\|_{E;m}$. For every $\epsilon > 0$ and $m \in \mathbb{N}$, there exists $N \in \mathbb{N}$ such that, if $n \geq N$, we have $\|\mathsf{f}_n - \mathsf{f}\|_{E;m} \leq \epsilon$ if $x \notin \sigma$. Whence,

$$\|\dot{\mathsf{f}}_n - \dot{\mathsf{f}}\|_{\dot{\mathsf{L}}^\infty(\Omega;E);m} \leq \epsilon.$$

So $\dot{\mathsf{f}}_n \to \dot{\mathsf{f}}$ in $\dot{\mathsf{L}}^\infty(\Omega; E)$. Therefore,

$$\dot{\mathsf{L}}^\infty(\Omega; E) \text{ is a Fréchet space.}$$

6º) Properties of $\dot{\mathsf{L}}^p_{\mathrm{loc}}(\Omega; E)$. ***6.a.*** *Preliminary.* For $n \in \mathbb{N}^*$, let

$$\omega_n \overset{\text{def}}{=} \{x \in \Omega : |x| < n,\ B(x, 1/n) \subset \Omega\}$$

where $B(x, 1/n)$ denotes the closed ball centered at x of radius $1/n$. Since each open set $\omega \Subset \Omega$ is included in ω_n for n large enough, the topology of $\dot{\mathsf{L}}^p_{\mathrm{loc}}(\Omega; E)$ is generated by the semi-norms $(\|\ \|_{\dot{\mathsf{L}}^p_{\mathrm{loc}}(\Omega;E);m,\omega_n})_{m\in\mathbb{N},\, n\in\mathbb{N}^*}$.

6.b. *Separation.* Suppose that $\dot{\mathsf{f}} \in \dot{\mathsf{L}}^p_{\mathrm{loc}}(\Omega; E)$ satisfies $\|\dot{\mathsf{f}}\|_{\dot{\mathsf{L}}^p_{\mathrm{loc}}(\Omega;E);m,\omega_n} = 0$, for all $m \in \mathbb{N}$ and $n \in \mathbb{N}^*$. By 1º, f vanishes on $\omega_n \backslash \sigma_n$, where σ_n is negligible. Since $\Omega = \bigcup_{n\in\mathbb{N}} \omega_n$, we have $\mathsf{f} = 0$ on $\Omega \backslash \sigma$, where σ is the union of the σ_n and is negligible (Theorem 2.24 (c)). Thus, $\dot{\mathsf{f}} = 0$.

6.c. *Countability of the semi-norms.* The family $(\|\ \|_{\dot{\mathsf{L}}^p_{\mathrm{loc}}(\Omega;E);m,\omega_n})_{m\in\mathbb{N},\, n\in\mathbb{N}^*}$ is indexed by $\mathbb{N} \times \mathbb{N}^*$, which is countable (Proposition A.2 (c)).

6.d. *Completion.* Let $(\dot{\mathsf{f}}_n)_{n\in\mathbb{N}}$ be a Cauchy sequence in $\dot{\mathsf{L}}^p_{\mathrm{loc}}(\Omega; E)$. For all $i \in \mathbb{N}^*$, the sequence of restrictions $(\dot{\mathsf{f}}_n|_{\omega_i})_{n\in\mathbb{N}}$ is Cauchy in $\dot{\mathsf{L}}^p(\omega_i; E)$. By either 4º or 5º, $\dot{\mathsf{f}}_n|_{\omega_i} \to \dot{\mathsf{g}}_i$ in $L^p(\omega_i; E)$. If $\omega_j \subset \omega_i$, we have $\dot{\mathsf{g}}_j = \dot{\mathsf{g}}_i|_{\omega_j}$, so there exists a unique $\dot{\mathsf{f}} \in \dot{\mathsf{L}}^p_{\mathrm{loc}}(\Omega; E)$ such that $\dot{\mathsf{f}}|_{\omega_i} = \dot{\mathsf{g}}_i$. Whence, $\dot{\mathsf{f}}_n \to \dot{\mathsf{f}}$ in $\dot{\mathsf{L}}^p_{\mathrm{loc}}(\Omega; E)$, and therefore,

$$\dot{\mathsf{L}}^p_{\mathrm{loc}}(\Omega; E) \text{ is a Fréchet space. } \square$$

Let us show that the space of classes that contain a simple function is dense in $\dot{\mathsf{L}}^p(\Omega; E)$ when p is finite. Let $\mathsf{S}(\Omega; E)$ denote the set of simple functions and let $\dot{s}$ denote the class of functions equal almost everywhere to s, then

$$\dot{\mathsf{S}}(\Omega; E) = \{\dot{\mathsf{s}} : \mathsf{s} \in \mathsf{S}(\Omega; E)\}. \tag{17.24}$$

Theorem 17.21. — *Let Ω be a measurable subset of $\mathbb{R}^d$, E be a Fréchet space, and $1 \leq p < \infty$. Then the set $\dot{\mathsf{S}}(\Omega; E)$ is dense and sequentially dense in $\dot{\mathsf{L}}^p(\Omega; E)$.* ▮

Proof. ***1°) ****Inclusion* $\dot{\mathsf{S}}(\Omega; E) \subset \dot{\mathsf{L}}^p(\Omega; E)$. Since any simple function s is measurable (Definition 17.1), and (Definition 17.7 (a)) $\int_\Omega \|\mathsf{s}\|^p_{E;m} = \sum_{i=1}^n \|e_i\|^p_{E;m} |\lambda_i| < \infty$, it is therefore p-integrable. Hence,

$$\dot{\mathsf{S}}(\Omega; E) \subset \dot{\mathsf{L}}^p(\Omega; E)$$

***2°) ****Sequential density.* Let $(\| \ \|_{E;m})_{m \in \mathbb{N}}$ be an increasing sequence of semi-norms that defines the topology of E (Proposition A.26), and let $\dot{\mathsf{f}} \in \dot{\mathsf{L}}^p(\Omega; E)$. If $(\mathsf{s}_n)_{n \in \mathbb{N}}$ is a p-approximating sequence for f (Definition 17.16 (b)), we have, for all $m \in \mathbb{N}$, when $n \to \infty$,

$$\|\dot{\mathsf{s}}_n - \dot{\mathsf{f}}\|_{\dot{\mathsf{L}}^p(\Omega;E);m} = \Big(\int_\Omega \|\mathsf{s}_n - \mathsf{f}\|^p_{E;m} \Big)^{1/p} \to 0.$$

Since $\dot{\mathsf{S}}(\Omega; E)$ is included in $\dot{\mathsf{L}}^p(\Omega; E)$ (by 1°), it therefore is sequentially dense there (Definition A.15).

***3°) ****Density.* Since $\dot{\mathsf{L}}^p(\Omega; E)$ is metrizable (Theorem 17.20), density coincides with sequential density (Proposition A.25 (b). □

Let us now show that the classes of functions that contain a uniformly continuous function with bounded support is dense in $\dot{\mathsf{L}}^p(\Omega; E)$, provided that Ω is open. Let us denote

$$\dot{\mathsf{B}}(\Omega; E) = \{\dot{\mathsf{f}} : \mathsf{f} \in \mathcal{B}(\Omega; E)\}. \tag{17.25}$$

Theorem 17.22. — *Let Ω be an open subset of $\mathbb{R}^d$, E be a Fréchet space, and $1 \leq p < \infty$. Then the set $\dot{\mathsf{B}}(\Omega; E)$ is dense and sequentially dense in $\dot{\mathsf{L}}^p(\Omega; E)$.* ▮

Proof. **1º)** *Inclusion* $\dot{\mathsf{B}}(\Omega; E) \subset \dot{\mathsf{L}}^p(\Omega; E)$. Let $\mathsf{f} \in \mathcal{B}(\Omega; E)$. It is the limit almost everywhere of simple functions (the approximations $\sum_{s \in S_n} 1_{s,n} \mathsf{f}_{s,n}$ from Theorem 13.9), and therefore it is measurable (Theorem 17.4).

Let $(\| \ \|_{E;m})_{m \in \mathbb{N}}$ be a sequence of semi-norms that define the topology of E. We have $\|\mathsf{f}\|_{E;m} \le 1_{\operatorname{supp} \mathsf{f}} \sup_\Omega \|\mathsf{f}\|_{E;m}$. By the monotonicity and additivity of the integral of positive functions (Theorem 17.11 (b) and (a)), and by the expression for the integral of a simple function (Theorem 17.7 (a)), we have

$$\int_\Omega \|\mathsf{f}\|^p_{E;m} \le \int_\Omega 1_{\operatorname{supp} \mathsf{f}} \sup_\Omega \|\mathsf{f}\|^p_{E;m} = |\operatorname{supp} \mathsf{f}| \sup_\Omega \|\mathsf{f}\|^p_{E;m}.$$

Thus, f is p-integrable (Definition 17.16 (a)). Whence (Definition 17.17),

$$\dot{\mathsf{B}}(\Omega; E) \subset \dot{\mathsf{L}}^p(\Omega; E). \tag{17.26}$$

2º) *Sequential density.* **2.a.** *Approximation of* $1_\omega e$. Let ω be a measurable subset of Ω of finite measure, and $\epsilon > 0$. There exist (Lemma 2.19) an open set $\mathcal{O}$ and a closed set $\mathcal{F}$ such that $\mathcal{F} \subset \omega \subset \mathcal{O}$ and $|\mathcal{O} \setminus \mathcal{F}| \le \epsilon$. Denote $B_n \stackrel{\text{def}}{=} \{x \in \mathbb{R}^d : |x| \le n\}$. The $B_n \cap \mathcal{F}$ increase to $\mathcal{F}$, so there exists (Theorem 2.16) n such that $|B_n \cap \mathcal{F}| \ge |\mathcal{F}| - \epsilon$. Then the set $K \stackrel{\text{def}}{=} B_n \cap \mathcal{F}$ is compact and

$$K \subset \omega \subset \mathcal{O}, \quad |\mathcal{O} \setminus K| \le 2\epsilon.$$

Urysohn's theorem (Proposition A.50) provides a function $\mathsf{g} \in \mathcal{K}(\mathcal{O})$ such that $\mathsf{g} = 1$ on K and $0 \le \mathsf{g} \le 1$. Let $\widetilde{\mathsf{g}}$ denote its extension to Ω by 0, and $e \in E$. Then,

$$\|e\widetilde{\mathsf{g}} - e 1_\omega\|_{E;m} \le 1_{\mathcal{O} \setminus K} \|e\|_{E;m}.$$

Using the monotonicity of the integral (Theorem 17.11 (b)) and the integral of a simple function (Proposition 17.7 (a)), we obtain

$$\int_\Omega \|e\widetilde{\mathsf{g}} - e 1_\omega\|^p_{E;m} \le \int_\Omega 1_{\mathcal{O} \setminus K} \|e\|^p_{E;m} = |\mathcal{O} \setminus K| \|e\|^p_{E;m} \le 2\epsilon \|e\|^p_{E;m}.$$

Whence,

$$\|\overline{e\widetilde{\mathsf{g}} - e 1_\omega}\|_{\dot{\mathsf{L}}^p(\Omega;E);m} \le (2\epsilon)^{1/p} \|e\|_{E;m}.$$

2.b. *Approximation of* $\dot{\mathsf{L}}^p(\Omega; E)$. A simple function s is a finite sum of functions $e 1_\omega$, so the set $\dot{\mathsf{S}}(\Omega; E)$ of $\dot{\mathsf{s}}$ is the set of limits in $\dot{\mathsf{L}}^p(\Omega; E)$ of functions from $\dot{\mathsf{B}}(\Omega; E)$. Since $\dot{\mathsf{S}}(\Omega; E)$ is sequentially dense in $\dot{\mathsf{L}}^p(\Omega; E)$ (Theorem 17.21), and since every limit of limits of sequences is itself a limit of sequences (Proposition A.27), it follows that

$$\dot{\mathsf{B}}(\Omega; E) \text{ is sequentially dense in } \dot{\mathsf{L}}^p(\Omega; E).$$

3º) *Density.* Since $\dot{\mathsf{L}}^p(\Omega; E)$ is metrizable (Theorem 17.20), density coincides with sequential density (Proposition A.25 (b). □

17.6. Some properties of spaces of classes of functions

Let us begin with a particular case of Hölder's inequality.

Proposition 17.23. — *Let* f_1 *be a* p_1*-integrable function,* f_2 *be a* p_2*-integrable function, both from a measurable subset* Ω *of* $\mathbb{R}^d$ *into* $\mathbb{R}$*,* $1 \le p_1 < \infty$*,* $1 \le p_2 < \infty$*, and* $1/p_1 + 1/p_2 = 1/p$*, where* $1 \le p < \infty$*. Then* $\mathsf{f}_1\mathsf{f}_2$ *is p-integrable, and*

$$\Big(\int_\Omega |\mathsf{f}_1\mathsf{f}_2|^p\Big)^{1/p} \le \Big(\int_\Omega |\mathsf{f}_1|^{p_1}\Big)^{1/p_1}\Big(\int_\Omega |\mathsf{f}_2|^{p_2}\Big)^{1/p_2}. \blacksquare$$

Proof. Note that if f_1 and f_2 are measurable (by Definition 17.16 and 17.17), then the product $\mathsf{f}_1\mathsf{f}_2$ is measurable (Definition 17.1) since it is the limit almost everywhere of the product of simple functions $\mathsf{s}_{1,n}$ and $\mathsf{s}_{2,n}$ that converge to f_1 and f_2, respectively.

For the inequality, we proceed as in Proposition 9.5 for continuous functions: Young's inequality $|\mathsf{fg}| \le |\mathsf{f}|^p/p + |\mathsf{g}|^{p'}/p'$ (Proposition 9.1) implies, by the additivity and monotonicity of the integral of positive functions (Theorem 17.11 (a) and (b)),

$$\int_\Omega |\mathsf{fg}| \le \frac{1}{p}\int_\Omega |\mathsf{f}|^p + \frac{1}{p'}\int_\Omega |\mathsf{g}|^{p'}.$$

Using this inequality for f/a and g/b, where $a = (\int_\Omega |\mathsf{f}|^p)^{1/p}$ and $b = (\int_\Omega |\mathsf{g}|^{p'})^{1/p'}$, we get

$$\frac{1}{ab}\int_\Omega |\mathsf{fg}| \le \frac{1}{pa^p}\int_\Omega |\mathsf{f}|^p + \frac{1}{p'b^{p'}}\int_\Omega |\mathsf{g}|^{p'} = \frac{1}{p} + \frac{1}{p'} = 1.$$

Hence, we have the stated inequality, provided a and b are non-zero. Otherwise, f or g is zero almost everywhere by Markov's inequality (Theorem 17.11 (d)), and the stated inequality is clear. □

Let us compare the $\dot{\mathsf{L}}^p(\Omega; E)$ with respect to p.

Theorem 17.24. — *Given* Ω *a measurable subset of* $\mathbb{R}^d$*,* E *a Fréchet space, and* $1 \le p \le q \le \infty$*, we have*

$$\dot{\mathsf{L}}^p(\Omega; E) \subsetneq \dot{\mathsf{L}}^p_{\mathrm{loc}}(\Omega; E),$$

$$\dot{\mathsf{L}}^\infty_{\mathrm{loc}}(\Omega; E) \subsetneq \dot{\mathsf{L}}^q_{\mathrm{loc}}(\Omega; E) \subsetneq \dot{\mathsf{L}}^p_{\mathrm{loc}}(\Omega; E) \subsetneq \dot{\mathsf{L}}^1_{\mathrm{loc}}(\Omega; E),$$

and, if Ω *has finite measure,*

$$\dot{\mathsf{L}}^\infty(\Omega; E) \subsetneq \dot{\mathsf{L}}^q(\Omega; E) \subsetneq \dot{\mathsf{L}}^p(\Omega; E) \subsetneq \dot{\mathsf{L}}^1(\Omega; E). \blacksquare$$

Proof. **1º)** *Inclusion of* $\dot{\mathsf{L}}^p(\Omega; E)$ *in* $\dot{\mathsf{L}}^p_{\text{loc}}(\Omega; E)$. This follows from their Definitions 17.17 and 17.18.

2º) *Growth of* $\dot{\mathsf{L}}^p(\Omega; E)$ *if* $|\Omega| < \infty$. If f is p-integrable (Definition 17.16), then it is measurable, and hence $\|\mathsf{f}\|^q_{E;\nu}$ is also measurable (1º of the justification of Definition 17.16).

Hölder's inequality (Proposition 17.23) with $\mathsf{f}_1 = \mathsf{f}$, $\mathsf{f}_2 = 1$, and $p_1 = q$ gives, for $1 \leq p \leq q < \infty$ and if f is q-integrable, that

$$\Big(\int_\Omega \|\mathsf{f}\|^p_{E;\nu}\Big)^{1/p} \leq \Big(\int_\Omega \|\mathsf{f}\|^q_{E;\nu}\Big)^{1/q}\Big(\int_\Omega 1^{p_2}\Big)^{1/p-1/q} = \Big(\int_\Omega \|\mathsf{f}\|^q_{E;\nu}\Big)^{1/q}|\Omega|^{1/p-1/q}.$$

Hence,

$$\dot{\mathsf{L}}^q(\Omega; E) \subsetneqq \dot{\mathsf{L}}^p(\Omega; E).$$

The inclusion $\dot{\mathsf{L}}^\infty(\Omega; E) \subsetneqq \dot{\mathsf{L}}^q(\Omega; E)$ then follows (Theorem 17.11 (c)) from the bound

$$\int_\Omega \|\mathsf{f}\|^q_{E;\nu} \leq |\Omega| \sup_\Omega \operatorname{ess} \|\mathsf{f}\|^q_{E;\nu}.$$

3º) *Growth of* $\dot{\mathsf{L}}^p_{\text{loc}}(\Omega; E)$. This follows from 2º since, by Definition 17.18 of the space $\dot{\mathsf{L}}^p_{\text{loc}}(\Omega; E)$, we have $\|\dot{\mathsf{f}}\|_{\dot{\mathsf{L}}^p(\Omega;E);\omega,m} = \|\dot{\mathsf{f}}|_\omega\|_{\dot{\mathsf{L}}^p(\omega;E);m}$ for every $\omega \Subset \Omega$ (in which case ω is bounded and therefore has finite measure). □

Let us present some properties of the vector-valued integral (Definition 17.16 (c)).

Theorem 17.25. — *Let* f *be an integrable function from a measurable subset* Ω *of* $\mathbb{R}^d$ *into a Fréchet space* E, ω *be a measurable subset of* Ω, *and* $c \in \mathbb{R}$.

(a)
$$\int_\omega (\mathsf{f} + \mathsf{g}) = \int_\omega \mathsf{f} + \int_\omega \mathsf{g}, \quad \int_\omega c\mathsf{f} = c\int_\omega \mathsf{f}.$$

(b) *For every semi-norm* $\|\ \|_{E;m}$ *of* E,

$$\Big\|\int_\omega \mathsf{f}\Big\|_{E;m} \leq \int_\omega \|\mathsf{f}\|_{E;m} \leq \int_\Omega \|\mathsf{f}\|_{E;m}.$$

(c) *If* ω' *is a measurable subset of* Ω *disjoint from* ω,

$$\int_{\omega\cup\omega'} \mathsf{f} = \int_\omega \mathsf{f} + \int_{\omega'} \mathsf{f}.$$

(d) *Let* $(\omega_n)_{n\in\mathbb{N}}$ *be a sequence of measurable subsets of* Ω; *if it is increasing and* $\omega \stackrel{\text{def}}{=} \bigcup_{n\in\mathbb{N}} \omega_n$, *or if it is decreasing with* $|\omega_n| < \infty$ *for some* n *and* $\omega \stackrel{\text{def}}{=} \bigcap_{n\in\mathbb{N}} \omega_n$, *then, as* $n \to \infty$,

$$\int_{\omega_n} \mathsf{f} \to \int_\omega \mathsf{f}. \blacksquare$$

Proof. ***1º)*** *Properties* (a), (b), *and* (c). These follow by taking the limits (Definition 17.16 (c) of the integral) of the corresponding properties of the integral of simple functions (Proposition 17.7).

2º) *Property* (d). For every $\epsilon > 0$, there exists (Theorem 17.22) $\mathsf{g} \in \mathcal{B}(\Omega; E)$ such that

$$\int_\Omega \|\mathsf{f} - \mathsf{g}\|_{E;m} \leq \epsilon. \tag{17.27}$$

Using (c), (b), and the triangle inequality, we obtain

$$\left\| \int_{\omega_n} \mathsf{f} - \int_\omega \mathsf{f} \right\|_{E;m} \leq \int_{\omega_n \setminus \omega} \|\mathsf{f}\|_{E;m} \leq \int_{\omega_n \setminus \omega} \|\mathsf{g}\|_{E;m} + \int_{\omega_n \setminus \omega} \|\mathsf{f} - \mathsf{g}\|_{E;m}.$$

Using (17.27), the bound for the real-valued integral (Theorem 17.11 (c)), and the fact that sup ess $\mathsf{h} = \sup \mathsf{h}$ for a positive continuous function h, we obtain

$$\left\| \int_{\omega_n} \mathsf{f} - \int_\omega \mathsf{f} \right\|_{E;m} \leq \epsilon + (|\omega_n| - |\omega|) \sup_\Omega \|\mathsf{g}\|_{E;m}.$$

We have $|\omega_n| \to |\omega|$ (Theorem 2.16), hence $\| \int_{\omega_n} \mathsf{f} - \int_\omega \mathsf{f} \|_{E;m} \leq 2\epsilon$ for n sufficiently large, which proves the stated convergence.

If $(\omega)_{n\in\mathbb{N}}$ is decreasing, we use Theorem 2.17 instead of 2.16. □

17.7. Lebesgue points

Let us show that almost every point is a **Lebesgue point**, that is, a point where the convergence (17.28) holds, provided Ω is open.

Theorem 17.26. — *Let* f *be a locally integrable function from an open subset* Ω *of* $\mathbb{R}^d$ *into a Fréchet space* E*, and let* $B(x, r)$ *be an open ball centered at* x *with radius* r *included in* Ω*. Then, for almost every* $x \in \Omega$*, when* $r \to 0$*,*

$$\frac{1}{|B(x,r)|} \int_{B(x,r)} \mathsf{f} \to \mathsf{f}(x). \tag{17.28}$$ ▮

The proof will use **Vitali's covering lemma**[6], which we now state.

6 **Vitali's covering lemma.** This lemma was established in 1908 by Giuseppe **Vitali** [80].

Lemma 17.27. — *Let $U = \bigcup_{1 \le i \le N} B(x_i, r_i)$ be a finite union of open balls in $\mathbb{R}^d$. Then there exists a subset $S \subset \{1, \dots, N\}$ such that the balls $(B(x_i, r_i))_{i \in S}$ are pairwise disjoint and $U \subset \bigcup_{i \in S} B(x_i, 3r_i)$; then,*

$$|U| \le 3^d \sum_{i \in S} |B(x_i, r_i)|. \blacksquare$$

Proof. Order the balls by decreasing radii, that is, $r_1 \ge r_2 \ge \dots \ge r_N$. Set $i_1 = 1$, discard all balls that intersect B_{i_1}, and let B_{i_2} be the first ball disjoint from it. Then discard all balls that intersect B_{i_2}, and let B_{i_3} be the first ball disjoint from it. Continue this process until exhaustion, and denote by $S = \{i_1, i_2, \dots\}$ the indices consequently obtained from this process.

The balls $(B_i)_{i \in S}$ are pairwise disjoint. Each discarded ball B_i intersects some B_{i_j} and $r_i \le r_{i_j}$, so it is contained in $B(x_{i_j}, 3r_{i_j})$ and hence $U \subset \bigcup_{i \in S} B(x_i, 3r_i)$. Finally, the monotonicity of measure (Theorem 2.10), its additivity (Theorem 2.13 (c)), and its behavior under homothety imply that

$$|U| \le \Big|\sum_{i \in S} B(x_i, 3r_i)\Big| = \sum_{i \in S} |B(x_i, 3r_i)| = 3^d \sum_{i \in S} |B(x_i, r_i)|. \square$$

The proof of Theorem 17.26 will also make use of the **Hardy–Littlewood maximal inequality**[7], stated below. It concerns the **maximal function** $M\mathsf{f}$, where, for a positive integrable function f and an open ball $B(x, r)$,

$$M_r\mathsf{f}(x) \stackrel{\text{def}}{=} \frac{1}{|B(x,r)|} \int_{B(x,r)} \mathsf{f},$$

$$M\mathsf{f}(x) \stackrel{\text{def}}{=} \sup_{r>0} M_r\mathsf{f}(x).$$

Lemma 17.28. — *Let f be a positive integrable function on $\mathbb{R}^d$. Then, for all $c > 0$,*

$$\big|\{x \in \mathbb{R}^d : M\mathsf{f}(x) > c\}\big| \le \frac{3^d}{c} \int_{\mathbb{R}^d} \mathsf{f}. \qquad (17.29) \blacksquare$$

Proof. Let us denote

$$\omega \stackrel{\text{def}}{=} \{x \in \mathbb{R}^d : M\mathsf{f}(x) > c\},$$

and suppose $|\omega| > 0$, since otherwise inequality (17.29) is immediate.

---

7 **Hardy–Littlewood inequality.** It was established in 1930 [35] by these two authors.

1°) Measurability of ω. If $r_n \to r$ is increasing or decreasing, then $B(x, r_n)$ also increases or decreases to $B(x,r)$, so $|B(x,r_n)| \to |B(x,r)|$ (Theorem 2.16 or 2.17) and (Theorem 17.25 (d)) $\int_{B(x,r_n)} \mathsf{f} \to \int_{B(x,r)} \mathsf{f}$. Hence, $M_{r_n}\mathsf{f}(x) \to M_r\mathsf{f}(x)$, and therefore

$$M\mathsf{f}(x) = \sup_{r>0} M_r\mathsf{f}(x) = \sup_{r\in\mathbb{Q},\, r>0} M_r\mathsf{f}(x). \tag{17.30}$$

For fixed r, the function $M_r\mathsf{f}$ is continuous (this follows similarly to the continuity with respect to r above). Therefore, it is measurable (as it is the limit of the simple functions $\mathsf{t}_n = \sum_{s\in S_n} 1_{s,n}(M_r\mathsf{f})_{s,n}$ from Theorem 13.9). Hence (Theorem 17.2),

$$\{x \subset \mathbb{R}^d : M_r\mathsf{f}(x) > c\} \text{ is measurable.}$$

From (17.30),

$$\omega = \{x \in \mathbb{R}^d : \sup_{r\in\mathbb{Q},\, r>0} M_r\mathsf{f}(x) > c\} = \bigcup_{r\in\mathbb{Q},\, r>0} \{x \in \mathbb{R}^d : M_r\mathsf{f}(x) > c\}.$$

Since every countable union of measurable sets is measurable (Theorem 2.15),

$$\omega \text{ is measurable.}$$

2°) Inequality. Let K be a compact subset of ω. For every $x \in K$, there exists $r_x > 0$ such that $M_{r_x}\mathsf{f}(x) > c$; that is,

$$|B(x,r_x)| < \frac{1}{c}\int_{B(x,r_x)} \mathsf{f}.$$

By compactness, there exists a finite number N of x_i and r_i such that the $B(x_i, r_i)$ cover K. Using Lemma 17.27, we can extract a family $(B(x_i,r_i))_{i\in S}$ of pairwise disjoint balls such that

$$|K| \le \Big|\bigcup_{i\le N} B(x_i,r_i)\Big| \le 3^d \sum_{i\in S} |B(x_i,r_i)|.$$

Using Markov's inequality (Theorem 17.11 (d)) and the fact that the $(B(x_i,r_i))_{i\in S}$ are pairwise disjoint, we get

$$|K| \le \frac{3^d}{c}\sum_{i\in S}\int_{B(x_i,r_i)} \mathsf{f} \le \frac{3^d}{c}\int_{\mathbb{R}^d} \mathsf{f}.$$

By considering a sequence $(K_n)_{n\in\mathbb{N}}$ of compact subsets increasing to ω, we obtain (Lemma 2.19) $|K_n| \to |\omega|$, and therefore inequality (17.29). □

Let us now proceed onto the proof of the existence of Lebesgue points.

Proof of Theorem 17.26. **1°)** *Case where* f *is integrable on* $\mathbb{R}^d$. Endow E with an increasing sequence $(\| \ \|_{E;m})_{n\in\mathbb{N}}$ of semi-norms (Proposition A.26), and let f be an integrable function from $\mathbb{R}^d$ into E. Let us denote

$$\fint_{B(x,r)} \mathsf{f} \stackrel{\text{def}}{=} \frac{1}{|B(x,r)|}\int_{B(x,r)} \mathsf{f}.$$

1.a. *Approximation of* f. Let $m \in \mathbb{N}$ and $\epsilon > 0$. As $\dot{\mathsf{B}}(\mathbb{R}^d; E)$ is dense in $\dot{\mathsf{L}}^1(\mathbb{R}^d; E)$ (Theorem 17.22), there exists $\mathsf{g} \in \mathcal{B}(\mathbb{R}^d; E)$ such that

$$\int_{\mathbb{R}^d} \|\mathsf{f} - \mathsf{g}\|_{E;m} \le \epsilon. \tag{17.31}$$

1.b. *Estimate of* $\mathsf{h} \stackrel{\text{def}}{=} \mathsf{f} - \mathsf{g}$. The triangle inequality gives, for $x \in \mathbb{R}^d$ and $r > 0$,

$$\fint_{B(x,r)} \|\mathsf{h} - \mathsf{h}(x)\|_{E;m} \le \|\mathsf{h}(x)\|_{E;m} + \fint_{B(x,r)} \|\mathsf{h}\|_{E;m}.$$

Hence,

$$\limsup_{r\to 0} \fint_{B(x,r)} \|\mathsf{h} - \mathsf{h}(x)\|_{E;m} \le \|\mathsf{h}(x)\|_{E;m} + \sup_{r>0} \fint_{B(x,r)} \|\mathsf{h}\|_{E;m}.$$

If the left-hand side is greater than $2c$, then at least one of the two terms on the right-hand side must be greater than c. Therefore, for $c > 0$,

$$\begin{aligned}\Big|\Big\{x : \limsup_{r\to 0} \fint_{B(x,r)} \|\mathsf{h} - \mathsf{h}(x)\|_{E;m} > 2c\Big\}\Big| &\le \\ \le \big|\{x : \|\mathsf{h}(x)\|_{E;m} > c\}\big| &+ \Big|\Big\{x : \sup_{r>0} \fint_{B(x,r)} \|\mathsf{h}\|_{E;m} > c\Big\}\Big|.\end{aligned}$$

Now, Lemma 17.28 applied to the function $\|\mathsf{h}\|_{E;m}$ gives, together with (17.31),

$$\Big|\Big\{x : \sup_{r>0} \fint_{B(x,r)} \|\mathsf{h}(x)\|_{E;m} > c\Big\}\Big| \le \frac{3^d}{c}\int_{\mathbb{R}^d} \|\mathsf{h}\|_{E;m} \le \frac{3^d}{c}\epsilon$$

and Markov's inequality (Theorem 17.11 (d)) applied to the function $\|\mathsf{h}\|_{E;m}$ gives

$$\big|\{x : \|\mathsf{h}(x)\|_{E;m} > c\}\big| \le \frac{1}{c}\int_{\mathbb{R}^d} \|\mathsf{h}\|_{E;m} \le \frac{1}{c}\epsilon.$$

Therefore,

$$\Big|\Big\{x : \limsup_{r\to 0} \fint_{B(x,r)} \|\mathsf{h} - \mathsf{h}(x)\|_{E;m} > 2c\Big\}\Big| \le \frac{1+3^d}{c}\epsilon. \tag{17.32}$$

1.c. *Estimate of* f. Since g is continuous, we have

$$\lim_{r\to 0} \frac{1}{|B(x,r)|} \int_{B(x,r)} \|\mathsf{g} - \mathsf{g}(x)\|_{E;m} = 0.$$

With (17.32), it follows that

$$\left|\left\{x : \limsup_{r\to 0} \fint_{B(x,r)} \|\mathsf{f} - \mathsf{f}(x)\|_{E;m} > 2c\right\}\right| \leq \frac{1+3^d}{c}\epsilon.$$

Since this holds for every $\epsilon > 0$, the left-hand side is zero and thus the set σ_m where $\limsup_{r\to 0} \fint_{B(x,r)} \|\mathsf{f} - \mathsf{f}(x)\|_{E;m} > 1/m$ is negligible. Since any countable union of negligible sets is negligible (Theorem 2.24 (c)), and since $\|\ \|_{E;m}$ increases with m,

$$\left\{x : \fint_{B(x,r)} \mathsf{f} \not\to \mathsf{f}(x) \text{ as } r \to 0\right\} \text{ is negligible.}$$

2º) *Case where f is integrable on Ω.* The extension $\tilde{\mathsf{f}}$ of f by 0_E is integrable on the entirety of $\mathbb{R}^d$. The set σ of all $x \in \Omega$ where $\fint_{B(x,r)} \mathsf{f}$ does not tend to $\mathsf{f}(x)$ is included in the set of all $x \in \mathbb{R}^d$ where $\fint_{B(x,r)} \tilde{\mathsf{f}}$ does not tend to $\tilde{\mathsf{f}}(x)$, which is negligible. Therefore, σ is negligible.

3º) *General case.* Let f be a locally integrable function from Ω into E and, for $n \in \mathbb{N}^*$,

$$\omega_n = \{x \in \Omega : \overline{B}(x, 1/n) \subset \Omega,\ |x| < n\},$$

where $\overline{B}$ denotes the closed ball. The restriction of f to ω_n is integrable, so the set σ_n of all $x \in \omega_n$ where $\fint_{B(x,r)} \mathsf{f}$ does not tend to $\mathsf{f}(x)$ is negligible. The set σ of all $x \in \omega$ where $\fint_{B(x,r)} \mathsf{f}$ does not tend to $\mathsf{f}(x)$ is included in $\bigcup_{n\in\mathbb{N}^*} \sigma_n$, which is negligible (Theorem 2.24 (c), again). □

17.8. Measures associated to classes of functions

Let us associate a measure to each class of integrable functions.

Theorem 17.29. — *To every element $\dot{\mathsf{f}}$ of $\dot{\mathsf{L}}^p(\Omega; E)$ or $\dot{\mathsf{L}}^p_{\text{loc}}(\Omega; E)$, where Ω is an open subset of $\mathbb{R}^d$, E is a Fréchet space, and $1 \leq p \leq \infty$, we associate a measure $\overline{f} \in \mathcal{M}(\Omega; E)$, independent of the representative* f *for* $\dot{\mathsf{f}}$, *by: for every $\varphi \in \mathcal{K}(\Omega)$,*

$$\langle \overline{f}, \varphi \rangle \stackrel{\text{def}}{=} \int_\Omega \mathsf{f}\varphi. \qquad (17.33)$$ ▮

Proof. It suffices to prove this for $\dot{\mathsf{f}}$ in $\dot{\mathsf{L}}^1_{\text{loc}}(\Omega; E)$, since this space contains both $\dot{\mathsf{L}}^p(\Omega; E)$ and $\dot{\mathsf{L}}^p_{\text{loc}}(\Omega; E)$ (Theorem 17.24).

1º) Meaning of $\langle \overline{f}, \varphi \rangle$. By Definition 17.17 of $\dot{\mathsf{L}}^p(\Omega; E)$, f is an integrable function, that is (Definition 17.16 (a)), there exists a sequence of simple functions such that $\mathsf{s}_n \to \mathsf{f}$ almost everywhere and, for every semi-norm $\| \; \|_{E;m}$ of E, $\int_\Omega \|\mathsf{f}\|_{E;m} < \infty$.

The function φ is measurable, since it is the limit almost everywhere of simple functions (the approximations $\mathsf{t}_n \overset{\text{def}}{=} \sum_{s \in \mathcal{S}_n} 1_{s,n} \varphi_{s,n}$ from Theorem 13.9). Therefore, $\mathsf{f}\varphi$ is measurable, since the $\mathsf{s}_n \mathsf{t}_n$ are simple functions converging almost everywhere to $\mathsf{f}\varphi$ (as the countable union of negligible sets is negligible, see Theorem 2.24 (c)).

One may set the t_n to vanish outside the support ω of φ, which is open and therefore measurable (Theorem 2.7) and bounded. By the monotonicity and homogeneity of the integral of positive functions (Theorem 17.11, (a) and (b)), and by the expression of the semi-norms of $\dot{\mathsf{L}}^1(\Omega; E)$ (Definition 17.18),

$$\int_\Omega \|\mathsf{f}\varphi\|_{E;m} \le \sup_\omega |\varphi| \int_\omega \|\mathsf{f}\|_{E;m} = \sup_\Omega |\varphi| \|\dot{\mathsf{f}}\|_{\dot{\mathsf{L}}^1(\Omega;E);\omega,m} < \infty. \tag{17.34}$$

2º) Measure. The mapping $\varphi \mapsto \int_\Omega \mathsf{f}\varphi$ is linear (this is clear). By (17.34), it satisfies the characterization of measures from Theorem 3.3 with $c = \|\dot{\mathsf{f}}\|_{\dot{\mathsf{L}}^1(\Omega;E);\omega,m}$ and $q = 1$.

3º) Uniqueness. If g is another representative for $\dot{\mathsf{f}}$, then $\mathsf{g} = \mathsf{f}$ almost everywhere, so $\mathsf{g}\varphi = \mathsf{f}\varphi$ almost everywhere. Then (justification of Definition 17.16), $\int_\Omega \mathsf{g}\varphi = \int_\Omega \mathsf{f}\varphi$. That is, $\overline{\mathsf{g}} = \overline{\mathsf{f}}$. □

We may now define the space of *p-integrable measures*, $\overline{L}^p(\Omega; E)$.

Definition 17.30. — *Let Ω be an open subset of $\mathbb{R}^d$, E be a Fréchet space, and $1 \le p \le \infty$. We denote*

$$\overline{L}^p(\Omega; E) = \{\overline{f} \in \mathcal{M}(\Omega; E) : \dot{\mathsf{f}} \in \dot{\mathsf{L}}^p(\Omega; E)\},$$

$$\overline{L}^p_{\text{loc}}(\Omega; E) = \{\overline{f} \in \mathcal{M}(\Omega; E) : \dot{\mathsf{f}} \in \dot{\mathsf{L}}^p_{\text{loc}}(\Omega; E)\},$$

endowed respectively with the semi-norms $\|\overline{f}\|_{\overline{L}^p(\Omega;E);m} = \|\dot{\mathsf{f}}\|_{\dot{\mathsf{L}}^p(\Omega;E);m}$ on the one hand, and $\|\overline{f}\|_{\overline{L}^p_{\text{loc}}(\Omega;E);m,\omega} = \|\dot{\mathsf{f}}\|_{\dot{\mathsf{L}}^p_{\text{loc}}(\Omega;E);m,\omega}$ on the other, where the semi-norms of E are indexed by $m \in \mathbb{N}$ and by open sets $\omega \Subset \Omega$. If $p = 1$, we write

$$\int_\omega \overline{f} \overset{\text{def}}{=} \int_\omega \mathsf{f}. \blacksquare$$

Justification. The mapping $\dot{\mathsf{f}} \mapsto \overline{f}$ is clearly linear, so $\overline{L}^p(\Omega; E)$ and $\overline{L}^p_{\text{loc}}(\Omega; E)$ are both vector spaces and $\| \ \|_{\overline{L}^p(\Omega;E);m}$ and $\| \ \|_{\overline{L}^p_{\text{loc}}(\Omega;E);m,\omega}$ are semi-norms. □

Let us show that the mapping $\dot{\mathsf{f}} \mapsto \overline{f}$ is an isomorphism.

Theorem 17.31. — *The mapping $\dot{\mathsf{f}} \mapsto \overline{f}$ (Theorem 17.29) is an isometric isomorphism from $\dot{\mathsf{L}}^p(\Omega; E)$ onto $\overline{L}^p(\Omega; E)$, and also an isomorphism from $\dot{\mathsf{L}}^p_{\text{loc}}(\Omega; E)$ onto $\overline{L}^p_{\text{loc}}(\Omega; E)$, where Ω is an open subset of $\mathbb{R}^d$, E is a Fréchet space, and $1 \leq p \leq \infty$.* ▮

Proof. ***1°) Injectivity.*** Endow E with an increasing sequence of semi-norms (Proposition A.26), and let $\dot{\mathsf{f}} \in \dot{\mathsf{L}}^1_{\text{loc}}(\Omega; E)$ and $\mathsf{f} \in \dot{\mathsf{f}}$, so f is locally integrable.

For every $m \in \mathbb{N}^*$ and every open ball $B(x, r)$ whose closure is contained in Ω, there exists an annulus $C = \{y : r' < |x - y| < r\}$ and $\varphi \in \mathcal{K}(\Omega)$ such that (Proposition A.50, by Uryshon's) $0 \leq \varphi \leq 1$, $\varphi = 1$ on $B(x, r')$, and $\varphi = 0$ outside $B(x, r)$, and

$$\Big\| \frac{1}{|B(x,r)|} \int_{B(x,r)} \mathsf{f}(1-\varphi) \Big\|_{E;m} \leq \frac{1}{|B(x,r)|} \int_C \|\mathsf{f}\|_{E;m} \leq \frac{1}{m}.$$

Indeed, as r' increases to r, C decreases to the empty set and thus, by the continuity of the integral with respect to the domain of integration (Theorem 17.25 (d)), $\int_C \|\mathsf{f}\|_{E;m}$ converges to 0. Now, assume that

$$\overline{f} = 0.$$

We have $0_E = \int_\Omega \mathsf{f}\varphi = \int_{B(x,r)} \mathsf{f}\varphi$, so

$$\Big\| \frac{1}{|B(x,r)|} \int_{B(x,r)} \mathsf{f} \Big\|_{E;m} \leq \frac{1}{m}.$$

If x is a Lebesgue point, so that $x \in \Omega \backslash \sigma$ where σ is negligible, then this gives (Theorem 17.26), when $r \to 0$,

$$\|\mathsf{f}(x)\|_{E;m} \leq \frac{1}{m}.$$

Since $\| \ \|_{E;m}$ increases with m,

$$\mathsf{f}(x) = 0_E \ \text{ if } x \notin \sigma$$

(otherwise, there would exist $x \notin \sigma$ and $n \in \mathbb{N}$ such that $\|\mathsf{f}(x)\|_{E;n} = a > 0$, hence there would exist $m \geq n$ such that $\|\mathsf{f}(x)\|_{E;m} \geq \|\mathsf{f}(x)\|_{E;n} = a > 1/m$). Therefore, $\dot{\mathsf{f}} = 0$, and so

the mapping $\dot{\mathsf{f}} \mapsto \overline{f}$ is injective.

2º) Isomorphisms. The mapping $\dot{\mathsf{f}} \mapsto \overline{f}$ is an isometric isomorphism from the space $\dot{\mathsf{L}}^p(\Omega; E)$ onto $\overline{L}^p(\Omega; E)$, since it is injective by 1º, and isometric and surjective by Definition 17.30.

It is likewise an isomorphism from $\dot{\mathsf{L}}^p_{\mathrm{loc}}(\Omega; E)$ onto $\overline{L}^p_{\mathrm{loc}}(\Omega; E)$. □

Functions associated to measures. If E is a Fréchet space then, to every measure $f \in L^p(\Omega; E)$ given by Definition 10.2, one can associate a p-integrable function $\underline{\mathsf{f}}$ defined for almost every $x \in \Omega$ by

$$\underline{\mathsf{f}}(x) = \lim_{r \to 0} \frac{1}{|B(x,r)|} \int_{B(x,r)} f.$$

The class $\dot{\mathsf{f}}$ of functions equal to $\underline{\mathsf{f}}$ almost everywhere is then the unique class associated to $\overline{f}$. □

These spaces of measures are sequentially complete.

Theorem 17.32. — *The spaces $\overline{L}^p(\Omega; E)$ and $\overline{L}^p_{\mathrm{loc}}(\Omega; E)$ are Fréchet spaces if Ω is an open subset of $\mathbb{R}^d$, E is a Fréchet space, and $1 \leq p \leq \infty$.* ▮

Proof. These spaces are sequentially complete since (Proposition A.60) they are isomorphic (Theorem 17.31) to sequentially complete spaces. They are moreover metrizable, hence Fréchet. □

17.9. Identity of the spaces of measures

Let us show that the spaces of measures $\overline{L}^p(\Omega; E)$ (Definition 17.30) and $L^p(\Omega; E)$ (Definition 10.2) are identical, when both are defined.

Theorem 17.33. — *Let Ω be an open subset of $\mathbb{R}^d$, E a Fréchet space, and $1 \leq p \leq \infty$.*

(a) $$\overline{L}^p(\Omega; E) \underset{\leftrightarrows}{\equiv} L^p(\Omega; E).$$

(b) $$\overline{L}^p_{\mathrm{loc}}(\Omega; E) \underset{\leftrightarrows}{\equiv} L^p_{\mathrm{loc}}(\Omega; E).$$

(c) *If $\overline{f} \in \overline{L}^1(\Omega; E)$ and ω is a measurable subset of Ω, then Definitions 17.30, 17.16(c) and 5.1 of the integral give*

$$\int_\omega \overline{f} = \int_\omega \dot{\mathsf{f}} = \int_\omega f. \qquad (17.35)$$ ▮

Proof. To avoid ambiguities, let us denote by $\int^*$ the integral in the sense of functions.

1°) Equalities on $\mathcal{B}(\Omega; E)$. Endow E with an increasing sequence $(\| \ \|_{E;m})_{m\in\mathbb{N}}$ of semi-norms (Proposition A.26), and let $\mathsf{f} \in \mathcal{B}(\Omega; E)$. Then, we also have $f \in \mathcal{B}(\Omega; E)$, once it has been identified with a measure via (3.16), p. 65.

1.a. *Equality of integrals.* The integral $\int_\Omega^* \dot{\mathsf{f}}$ is (Definition 17.16 (c)) the limit of $\int_\Omega^* \mathsf{s}_n = \sum_{i=1}^{j} e_i|\omega_i|$ for any approximating sequence $(\mathsf{s}_n)_{n\in\mathbb{N}}$ for f, where we have $\mathsf{s}_n = \sum_{i=1}^{j} e_i 1_{\omega_i}$.

This is equal to $\int_\Omega f$, which is (by (5.4), p. 92) the Cauchy integral, *i.e.* (Definition 1.14) the limit of the approximated integrals $\mathbb{S}_\Omega^n f = \sum_{s\in\mathcal{S}} 2^{-nd} f(2^{-n}s)$, where $e_s = f(2^{-n}s)$ and $\omega_s = \mathring{\Delta}_{s,n}$ (cube of side length 2^{-n} centered at $2^{-n}s$), so that $|\omega_s| = 2^{-nd}$.

1.b. *Equality of semi-norms for p finite.* The semi-norm $\|\overline{f}\|_{\overline{L}^p(\Omega;E);m}$ is equal to $\|\dot{\mathsf{f}}\|_{\dot{\mathsf{L}}^p(\Omega;E);m}$ (Theorem 17.30), which is (Definition 17.17) $(\int_\Omega^* (\|\mathsf{f}\|_{E;m})^p)^{1/p}$; this equals $(\int_\Omega \|f\|_{E;m}^p)^{1/p}$ by 1.a, which is (Proposition 10.7), $\|f\|_{L^p(\Omega;E);m}$.

2°) Equality of spaces for p finite. The spaces $\overline{L}^p(\Omega; E)$ and $L^p(\Omega; E)$ are sequentially complete (Theorems 17.30 and 10.16).

The space $\dot{\mathsf{B}}(\Omega; E)$ is dense in $\dot{\mathsf{L}}^p(\Omega; E)$ (Theorem 17.22), which is isomorphic (Theorem 17.31) to $\overline{L}^p(\Omega; E)$. The image of $\dot{\mathsf{B}}(\Omega; E)$ under this isomorphism, namely (see (3.16), p. 65) $\mathcal{B}(\Omega; E)$, is dense in $\overline{L}^p(\Omega; E)$ (Proposition A.45 (d)).

Finally, $\mathcal{B}(\Omega; E)$ is dense in $L^p(\Omega; E)$ (Theorem 10.14).

Therefore, $\overline{L}^p(\Omega; E)$ and $L^p(\Omega; E)$ are the closures of $\mathcal{B}(\Omega; E)$ with respect to their respective families of semi-norms. From 1.b, these families are equal on $\mathcal{B}(\Omega; E)$, so their closures coincide (Proposition A.28). That is,

$$\overline{L}^p(\Omega; E) = L^p(\Omega; E).$$

3°) Equalities of semi-norms for p finite. The semi-norms of a semi-normed space are continuous (Proposition A.10). Now, since $\| \ \|_{\overline{L}^p(\Omega;E);m}$ and $\| \ \|_{L^p(\Omega;E);m}$ are equal on $\mathcal{B}(\Omega; E)$ by 1.b, they coincide on its closure $L^p(\Omega; E)$ (Proposition A.29).

4°) Equality of integrals. The integrals $\int^*$ and $\int$ coincide on $\mathcal{B}(\Omega; E)$ by 1.a. Now, $\int^*$ is continuous on $\overline{L}^1(\Omega; E)$ since, by Definition 17.30 of this integral, the inequality from Theorem 17.25 (b), and Definitions 17.17 and 17.30 of the semi-norms of $\dot{\mathsf{L}}^1(\Omega; E)$ and $\overline{L}^1(\Omega; E)$, we have

$$\left\| \int_\omega^* \overline{\mathsf{f}} \right\|_{E;m} = \left\| \int_\omega^* \dot{\mathsf{f}} \right\|_{E;m} \leq \int_\Omega^* \|\dot{\mathsf{f}}\|_{E;m} = \|\dot{\mathsf{f}}\|_{\dot{\mathsf{L}}^1(\Omega;E);m} = \|\overline{f}\|_{\overline{L}^1(\Omega;E);m}.$$

Moreover, $\int$ is continuous on $L^1(\Omega; E)$ by Theorem 5.5. Therefore, the two integrals coincide on the closures of $\mathcal{B}(\Omega; E)$ in $\overline{L}^1(\Omega; E)$ and $L^1(\Omega; E)$, which are equal by 2º.

5º) Equality of local spaces for p finite. When $p \leq \infty$ (Definitions 17.30 and 11.3),

$$\overline{L}^p_{\text{loc}}(\Omega; E) = \{f \in \mathcal{M}(\Omega; E) : f|_\omega \in \overline{L}^p(\omega; E)\}, \ \forall \omega \Subset \Omega, \qquad (17.36)$$

$$L^p_{\text{loc}}(\Omega; E) = \{f \in \mathcal{M}(\Omega; E) : f|_\omega \in L^p(\omega; E)\}, \ \forall \omega \Subset \Omega. \qquad (17.37)$$

The topological identity of the spaces $\overline{L}^p_{\text{loc}}(\Omega; E)$ and $L^p_{\text{loc}}(\Omega; E)$ when $p < \infty$ follows therefore from the equality of $\overline{L}^p(\omega; E)$ and $L^p(\omega; E)$, and of their semi-norms, as established in 2º and 3º.

6º) Equality of semi-norms for p infinite. **6.a.** *First inclusion.* Let

$$f \in L^\infty(\Omega; E).$$

We have $f \in L^1_{\text{loc}}(\Omega; E)$ (Theorems 11.6 and 11.8), so 2º shows that $\overline{f} \in \overline{L}^1_{\text{loc}}(\Omega; E)$, and therefore (Definition 17.30)

$$\dot{\mathsf{f}} \in \dot{\mathsf{L}}^1_{\text{loc}}(\Omega; E).$$

Given $m \in \mathbb{N}$ and $\epsilon > 0$, there exists a measurable subset ω of Ω such that $|\omega| > 0$ and, for almost every $x \in \omega$, the function f satisfies

$$\|\mathsf{f}(x)\|_{E;m} \geq \sup_\Omega \text{ess} \|\dot{\mathsf{f}}\|_{E;m} - \epsilon.$$

(Otherwise, we would have $\sup \text{ess}_\Omega \|\dot{\mathsf{f}}\|_{E;m} \leq \sup \text{ess}_\Omega \|\dot{\mathsf{f}}\|_{E;m} - \epsilon$, which is clearly false.) It is always possible to choose such a subset ω included in a compact subset of Ω by defining $\omega_i = \{x \in \Omega : B(x, 1/i) \Subset \Omega, \ |x| < i\}$ and observing that $\bigcup_{i \in \mathbb{N}} \omega_i = \omega$, so (Theorem 2.16) $|\omega_i| \to |\omega|$.

Let $(f_n)_{n \in \mathbb{N}}$ be an ∞-integrabilizing sequence for the distribution f, namely (Theorem 11.12), that $f_n \in \mathcal{B}(\Omega; E)$ and

$$(f_n)_{n \in \mathbb{N}} \text{ is bounded in } \mathcal{C}_\text{b}(\Omega; E), \ \ f_n \to f \text{ in } L^1_{\text{loc}}(\Omega; E). \qquad (17.38)$$

There exists $n_0 \in \mathbb{N}$ such that, for all $n \geq n_0$, there exists $x_n \in \omega$ such that

$$\|f_n(x_n)\|_{E;m} \geq \sup_\Omega \text{ess} \|\dot{\mathsf{f}}\|_{E;m} - 2\epsilon.$$

(Otherwise, for some subsequence, $\|f_{\sigma(n)} - f\|_{E;m} \geq \epsilon$ almost everywhere on ω which, by the monotonicity of the integral (Theorem 17.11 (b)), would then give $\int_\omega \|f_{\sigma(n)} - f\|_{E;m} \geq \epsilon|\omega|$, and would not tend to 0 as $n \to \infty$.) Thus,

$$\sup_{n \in \mathbb{N}, \, x \in \Omega} \|f_n(x)\|_{E;m} \geq \sup_\Omega \text{ess} \|\dot{\mathsf{f}}\|_{E;m} - 2\epsilon.$$

Hence, by Definition 10.2 of the semi-norms of $L^\infty(\Omega; E)$,

$$\|f\|_{L^\infty(\Omega;E);m} = \inf_{(f_n)_{n\in\mathbb{N}}} \sup_{n\in\mathbb{N},\, x\in\Omega} \|f_n(x)\|_{E;m} \geq \sup_\Omega \text{ess} \|\dot{\mathsf{f}}\|_{E;m} - 2\epsilon.$$

Since this inequality holds for every $\epsilon > 0$, it also holds for $\epsilon = 0$. Therefore, for all $m \in \mathbb{N}$, using Definitions 17.17 of the semi-norms of $\dot{\mathsf{f}}$ and 17.30 of the semi-norms of $\overline{f}$,

$$\|f\|_{L^\infty(\Omega;E);m} \geq \sup_\Omega \text{ess} \|\dot{\mathsf{f}}\|_{E;m} = \|\dot{\mathsf{f}}\|_{\dot{\mathsf{L}}^\infty(\Omega;E);m} = \|\overline{f}\|_{\overline{L}^\infty(\Omega;E);m}. \tag{17.39}$$

6.b. *Reverse inclusion.* Now,

$$\overline{f} \in \overline{L}^\infty(\Omega; E). \tag{17.40}$$

We define the regularizations $R_n f$ in $\mathcal{M}(\Omega; E)$ (Definition 3.17), and hence also in $L^1_{\text{loc}}(\Omega; E)$, which coincides with $\overline{L}^1_{\text{loc}}(\Omega; E)$ by 2°, and thus in $L^\infty(\Omega; E)$ (Theorems 11.6 and 11.8), and in $\overline{L}^\infty(\Omega; E)$ (Theorem 17.24 and Definition 17.31).

According to the expression of the semi-norms of $L^\infty(\Omega; E)$ (Theorem 10.11) and Definition 3.17 of $R_n f$, for all $m \in \mathbb{N}$,

$$\|f\|_{L^\infty(\Omega;E);m} = \sup_{n\in\mathbb{N}^*,\, x\in\Omega} \|R_n f(x)\|_{E;m} = \sup_{n\in\mathbb{N}^*,\, x\in\Omega} \|\langle f, \alpha_n \tau_x \rho_n\rangle\|_{E;m}.$$

If $f \in L^1(\Omega; E)$, we have (Theorem 12.19)

$$\langle f, \alpha_n \tau_x \varphi\rangle = \int_\Omega f \alpha_n \tau_x \rho_n; \tag{17.41}$$

if $f \notin L^1(\Omega; E)$, then we obtain this inequality as follows: since (Definition 3.16) α_n vanishes outside the compact set $\overline{\Omega^n_{2/n}}$, we introduce (Proposition A.20) an intermediate open subset ω' such that $\Omega^n_{2/n} \Subset \omega' \Subset \Omega$; we then have $f|_{\omega'} \in L^1(\omega'; E)$ and (Definition 6.16 of the restriction and Theorem 6.19 on the contribution to an integral of the domain where the function vanishes),

$$\langle f, \alpha_n \tau_x \varphi\rangle_\Omega = \langle f|_{\omega'}, \alpha_n \tau_x \varphi\rangle_{\omega'} = \int_{\omega'} f|_{\omega'} \alpha_n \tau_x \rho_n = \int_\Omega f \alpha_n \tau_x \rho_n.$$

By 4°, we have $\int g = \int^* \dot{\mathsf{g}}$ for all $g \in L^1_{\text{loc}}(\Omega; E)$, so the inequality of Theorem 17.25 (b) implies

$$\|f\|_{L^\infty(\Omega;E);m} = \sup_{n\in\mathbb{N}^*,\, x\in\Omega} \left\| \int_\Omega^* \dot{\mathsf{f}} \alpha_n \tau_x \rho_n \right\|_{E;m} \leq \sup_{n\in\mathbb{N}^*,\, x\in\Omega} \int_\Omega^* \|\dot{\mathsf{f}}\|_{E;m} |\alpha_n \tau_x \rho_n|.$$

Since α_n and ρ_n are positive, $\alpha_n \leq 1$ and $\int_\Omega^* \tau_x \rho_n = \int_\Omega \tau_x \rho_n \leq \int_{\mathbb{R}^d} \rho_n = 1$ (Definition 3.16), the inequality from Theorem 17.11 (f) gives

$$\|f\|_{L^\infty(\Omega;E);m} \leq \sup_\Omega \text{ess} \|\dot{\mathsf{f}}\|_{E;m} = \|\dot{\mathsf{f}}\|_{\dot{\mathsf{L}}^\infty(\Omega;E);m} = \|\overline{f}\|_{\overline{L}^\infty(\Omega;E);m}.$$

This is the inverse inequality to (17.39), which therefore is an equality.

7º) *Equality of the spaces for p infinite.* Since their semi-norms are equal by 6º, the subspaces $L^\infty(\Omega; E)$ and $\overline{L}^\infty(\Omega; E)$ of $\mathcal{M}(\Omega; E)$ coincide. By Definition 1.4 (a) of a topological identity,

$$\overline{L}^\infty(\Omega; E) \underset{\leftrightarrows}{\equiv} L^\infty(\Omega; E).$$

It follows that $\overline{L}^\infty_{\text{loc}}(\Omega; E) \underset{\leftrightarrows}{\equiv} L^\infty_{\text{loc}}(\Omega; E)$ by (17.36) and (17.37) when $p = \infty$. □

Chapter 18

Applications

Here, we present some additional properties of $\dot{L}^p(\Omega; E)$, and hence of $L^p(\Omega; E)$, which follow from convergence almost everywhere.

We show that the image under certain continuous functions remains p-integrable (Theorem 18.6), that the spaces $L^p(\Omega; E)$ depend continuously on increasing p (Theorem 18.9), and that $\dot{L}^{p'}(\Omega; E')$ is isomorphic to $(\dot{L}^p(\Omega; E))'$ when E is a Banach space with a separable dual (Theorem 18.12).

These properties follow from Lebesgue's dominated convergence theorem and its converse (Theorems 18.5 and 18.4), which we derive from Vitali's convergence theorem (Theorem 18.3) using equi-integrability (Definition 18.1).

18.1. Equi-integrability

Let us define equi-integrability (the properties of classes of functions were introduced in Definition 17.19).

Definition 18.1. — *A family $\mathcal{F}$ of $\dot{L}^p(\Omega; E)$, where Ω is a measurable subset of $\mathbb{R}^d$, E is a Fréchet space, and $1 \leq p < \infty$, is said to be **p-equi-integrable** if, for every semi-norm $\| \ \|_{E;\nu}$ of E and $\epsilon > 0$:*

(a) *there exists a measurable set K of finite measure such that, for all $\dot{f} \in \mathcal{F}$,*

$$\int_{\Omega \setminus K} \|\dot{f}\|_{E;\nu}^p \leq \epsilon;$$

(b) *there exists $\delta > 0$ such that, for every measurable subset ω of Ω such that $|\omega| \leq \delta$ and $\dot{f} \in \mathcal{F}$,*

$$\int_{\omega} \|\dot{f}\|_{E;\nu}^p \leq \epsilon. \blacksquare$$

Note that any finite or convergent family is p-equi-integrable.

Theorem 18.2. — *Every finite family and every convergent sequence in $\dot{\mathrm{L}}^p(\Omega; E)$, where Ω is a measurable subset of $\mathbb{R}^d$, E is a Fréchet space, and $1 \le p < \infty$, is p-equi-integrable.* ▌

Proof. ***1°)*** *Unique function.* We need to verify conditions (a) and (b) of Definition 18.1. Let

$$\dot{\mathsf{f}} \in \dot{\mathrm{L}}^p(\Omega; E).$$

1.a. *Preliminary.* Let $\| \ \|_{E;m}$ be a semi-norm of E; then $\|\mathsf{f}\|^p_{E;m}$ is 1-integrable (Definition 17.16 (a)). By Lemma 17.8, there exists an increasing sequence $(\mathsf{s}_n)_{n\in\mathbb{N}}$ of simple functions increasing to $\|\mathsf{f}\|^p_{E;m}$ and, by Definitions 17.10 of the integral of functions and 17.19 of the integral of classes, $\int_\Omega \dot{\mathsf{s}}_n \to \int_\Omega \|\dot{\mathsf{f}}\|^p_{E;m}$. For $\epsilon > 0$, there exists $n \in \mathbb{N}$ such that the class $\dot{\mathsf{s}}_n$ of $\mathsf{s}_n = \sum_{i=1}^j t_i 1_{\omega_i}$, where $t_i \in \mathbb{R}$, satisfies

$$0 \le \dot{\mathsf{s}}_n \le \|\dot{\mathsf{f}}\|^p_{E;m}, \quad 0 \le \int_\Omega \big(\|\dot{\mathsf{f}}\|^p_{E;m} - \dot{\mathsf{s}}_n\big) \le \frac{\epsilon}{2}. \tag{18.1}$$

1.b. *Condition (a).* It is satisfied by $K = \bigcup_{i=1}^j \omega_i$, whose measure satisfies (Theorem 2.13 (b)) $|K| \le \sum_{i=1}^j |\omega_i|$, which is finite. Indeed, $\int_{\Omega\setminus K} \dot{\mathsf{s}}_n = 0$, and thus the additivity and monotonicity of the integral of positive functions (Theorem 17.11 (a) and (b)) gives, together with (18.1),

$$\int_{\Omega\setminus K} \|\dot{\mathsf{f}}\|^p_{E;m} \le \int_{\Omega\setminus K} \big(\|\dot{\mathsf{f}}\|^p_{E;m} - \dot{\mathsf{s}}_n\big) + \int_{\Omega\setminus K} \dot{\mathsf{s}}_n \le \int_\Omega \big(\|\dot{\mathsf{f}}\|^p_{E;m} - \dot{\mathsf{s}}_n\big) \le \frac{\epsilon}{2}.$$

1.c. *Condition (b).* For every measurable subset ω of Ω, we similarly have

$$\int_\omega \|\dot{\mathsf{f}}\|^p_{E;m} \le \int_\omega \big(\|\dot{\mathsf{f}}\|^p_{E;m} - \dot{\mathsf{s}}_n\big) + \int_\omega \dot{\mathsf{s}}_n.$$

If $\dot{\mathsf{s}}_n$ is zero, (18.1) provides condition (b). Otherwise, for $|\omega| \le \epsilon/(2 \sup_{x\in\Omega} |\dot{\mathsf{s}}_n(x)|)$, by bounding the integral (Theorem 17.11 (c) and Definition 17.19 of the integral), we get $\int_\omega \mathsf{s}_n \le |\omega| \sup_{x\in\Omega} |\dot{\mathsf{s}}_n(x)| \le \epsilon/2$, and (18.1) again provides condition (b).

2°) *Finite family.* Conditions (a) and (b) are satisfied for all f in a finite family $\mathcal{F}$ by taking $K = \bigcup_{\mathsf{f}\in\mathcal{F}} K_\mathsf{f}$ and $\delta = \inf_{\mathsf{f}\in\mathcal{F}} \delta_\mathsf{f}$, respectively.

3°) *Convergent sequence.* Let

$$\dot{\mathsf{f}}_n \to \dot{\mathsf{f}} \ \text{ in } \dot{\mathrm{L}}^p(\Omega; E).$$

For every measurable subset ω of Ω, Minkowski's inequality (which follows from Definition 17.17 of $\dot{\mathsf{L}}^p(\omega; E)$), the monotonicity of the integral (Theorem 17.11 (b)), and Definitions 17.17 of $\| \; \|_{\dot{\mathsf{L}}^p(\Omega;E);m}$ and 17.19 of the integral of classes of functions, give

$$\Big(\int_\omega \|\dot{\mathsf{f}}_n\|^p_{E;m}\Big)^{1/p} \leq \Big(\int_\omega \|\dot{\mathsf{f}}_n - \dot{\mathsf{f}}\|^p_{E;m}\Big)^{1/p} + \Big(\int_\omega \|\dot{\mathsf{f}}\|^p_{E;m}\Big)^{1/p} \leq$$
$$\leq \|\dot{\mathsf{f}}_n - \dot{\mathsf{f}}\|_{\dot{\mathsf{L}}^p(\Omega;E);m} + \Big(\int_\omega \|\dot{\mathsf{f}}\|^p_{E;m}\Big)^{1/p}.$$

The class $\dot{\mathsf{f}}$ is p-equi-integrable by 1°. Thus, conditions (a) and (b) are satisfied by $\dot{\mathsf{f}}_n$, provided that n is sufficiently large to ensure that $\|\dot{\mathsf{f}}_n - \dot{\mathsf{f}}\|_{\dot{\mathsf{L}}^p(\Omega;E);m} \leq \epsilon/2$. The finitely many $\dot{\mathsf{f}}_n$ that remain are p-equi-integrable by 2°. □

Let us state **Vitali's convergence theorem** [1].

Theorem 18.3. — *Let $(\dot{\mathsf{f}}_n)_{n\in\mathbb{N}}$ be a sequence in $\dot{\mathsf{L}}^p(\Omega; E)$, where Ω is a measurable subset of $\mathbb{R}^d$, E is a Fréchet space, $1 \leq p < \infty$, and $\dot{\mathsf{f}}$ is a class of functions from Ω into E equal almost everywhere, such that*

$$(\dot{\mathsf{f}}_n)_{n\in\mathbb{N}} \textit{ is p-equi-integrable,} \quad \dot{\mathsf{f}}_n \to \dot{\mathsf{f}} \textit{ almost everywhere in } \Omega.$$

Then,

$$\dot{\mathsf{f}}_n \to \dot{\mathsf{f}} \textit{ in } \dot{\mathsf{L}}^p(\Omega; E). \blacksquare$$

Proof. ***1°)*** *p-integrability of* f. Let $\mathsf{f} \in \dot{\mathsf{f}}$. The function f is measurable, since it is the limit almost everywhere of a sequence of measurable functions (Theorem 17.4). Let $\epsilon > 0$, and suppose K and δ satisfy Definition 18.1 of p-equi-integrability.

Since K has finite measure, for all $\delta > 0$, by Egoroff's theorem (Theorem 17.3), there exists a measurable set $\omega_\delta \subset K$ such that $|\omega_\delta| \leq \delta$, and $\mathsf{f}_n \to \mathsf{f}$ uniformly in $K \setminus \omega_\delta$. For any semi-norm $\| \; \|_{E;m}$ of E (Theorem 17.11 (c)), we have

$$\int_{K\setminus\omega_\delta} \|\mathsf{f} - \mathsf{f}_n\|^p_{E;m} \leq |K| \operatorname*{sup\,ess}_{K\setminus\omega_\delta} \|\mathsf{f} - \mathsf{f}_n\|^p_{E;m}. \tag{18.2}$$

This tends to 0 as $n \to \infty$, so there exists $n_0 \in \mathbb{N}$ such that $\int_{K\setminus\omega_\delta} \|\mathsf{f} - \mathsf{f}_{n_0}\|^p_{E;m}$ is finite. Then,

$$\int_{K\setminus\omega_\delta} \|\mathsf{f}\|^p_{E;m} \leq \int_{K\setminus\omega_\delta} \|\mathsf{f} - \mathsf{f}_{n_0}\|^p_{E;m} + \int_{K\setminus\omega_\delta} \|\mathsf{f}_{n_0}\|^p_{E;m} \tag{18.3}$$

is also finite.

---

1 **Vitali's theorem.** The origin of this theorem is attributed to Giuseppe VITALI (1875–1932), but I have not been able to find a precise reference.

Fatou's lemma (Theorem 17.14, since the integral is invariant under the addition of negligible sets) and Definition 18.1 of equi-integrability yield, for $\omega = \Omega \backslash K$ and $\omega = \omega_\delta$,

$$\int_\omega \|\mathsf{f}\|_{E;m}^p = \int_\omega \liminf_{n\to\infty} \|\mathsf{f}_n\|_{E;m}^p \leq \liminf_{n\to\infty} \int_\omega \|\mathsf{f}_n\|_{E;m}^p \leq \epsilon. \tag{18.4}$$

By the additivity of the measure (Theorem 17.11 (e)), we have

$$\int_\Omega \|\mathsf{f}\|_{E;m}^p = \int_{\Omega\backslash K} \|\mathsf{f}\|_{E;m}^p + \int_{K\backslash\omega_\delta} \|\mathsf{f}\|_{E;m}^p + \int_{\omega_\delta} \|\mathsf{f}\|_{E;m}^p.$$

This is finite by inequalities (18.3) and (18.4), and so

f is p-integrable.

2º) Convergence in $L^p(\Omega; E)$. Since $\dot{\mathsf{f}} - \dot{\mathsf{f}}_n \in \dot{\mathsf{L}}^p(\Omega; E)$, we have (Definition 17.17 and Theorem 17.11 (e), again),

$$\int_\Omega \|\mathsf{f}-\mathsf{f}_n\|_{E;m}^p = \int_{\Omega\backslash K} \|\mathsf{f}-\mathsf{f}_n\|_{E;m}^p + \int_{K\backslash\omega_\delta} \|\mathsf{f}-\mathsf{f}_n\|_{E;m}^p + \int_{\omega_\delta} \|\mathsf{f}-\mathsf{f}_n\|_{E;m}^p < \infty. \tag{18.5}$$

From (18.2) and the uniform convergence of f_n to f in $K\backslash\omega_\delta$, as shown in 1º, for n sufficiently large,

$$\int_{K\backslash\omega_\delta} \|\mathsf{f} - \mathsf{f}_n\|_{E;m}^p \leq \epsilon. \tag{18.6}$$

From (18.4) and Definition 18.1 of p-equi-integrability, for all $n \in \mathbb{N}$ with $\omega = \Omega\backslash K$ and $\omega = \omega_\delta$, we have

$$\int_\omega \|\mathsf{f} - \mathsf{f}_n\|_{E;m}^p \leq \int_\omega \|\mathsf{f}\|_{E;m}^p + \int_\omega \|\mathsf{f}_n\|_{E;m}^p \leq 2\epsilon. \tag{18.7}$$

Using (18.6) and (18.7), (18.5) gives us $\int_\Omega \|\mathsf{f} - \mathsf{f}_n\|_{E;m}^p \leq 5\epsilon$ for every semi-norm of E and n large enough. Hence (Definition 17.17 of the norm in $\dot{\mathsf{L}}^p(\Omega; E)$),

$$\dot{\mathsf{f}}_n \to \dot{\mathsf{f}} \text{ in } \dot{\mathsf{L}}^p(\Omega; E). \ \square$$

18.2. Dominated convergence

Let us first state the **converse to the dominated convergence theorem**.

Theorem 18.4. — *Let $(\dot{\mathsf{f}}_n)_{n\in\mathbb{N}}$ be a sequence and $\dot{\mathsf{f}}$ be an element of $\dot{\mathsf{L}}^p(\Omega; E)$, where Ω is a measurable subset of $\mathbb{R}^d$, E is a Fréchet space endowed (Proposition A.26)* *with an increasing sequence of semi-norms* $(\| \|_{E;m})_{m\in\mathbb{N}}$ *that defines its topology, and* $1 \le p < \infty$, *such that*

$$\dot{\mathsf{f}}_n \to \dot{\mathsf{f}} \ \textit{in} \ \dot{\mathsf{L}}^p(\Omega; E).$$

Then, there exists a subsequence $(\dot{\mathsf{f}}_{n_k})_{k\in\mathbb{N}}$ *and, for every* $\epsilon > 0$, *a measurable set* ω_ϵ *such that* $|\omega_\epsilon| \le \epsilon$, *and*

$$\dot{\mathsf{f}}_{n_k} \to \dot{\mathsf{f}} \ \textit{almost everywhere in } \Omega, \textit{ and uniformly almost everywhere in } \Omega \setminus \omega_\epsilon,$$

and, for every $m \in \mathbb{N}$, *there exists* $\dot{\mathsf{g}}_m \in \dot{\mathsf{L}}^p(\Omega)$, *such that*

$$\|\dot{\mathsf{f}}_{n_k}\|_{E;m} \le \dot{\mathsf{g}}_m. \ \blacksquare$$

Proof. ***1o)*** *Construction of* $\dot{\mathsf{f}}_{n_k}$. Since $\| \|_{E;k}$ increases with k, for each k, there exists n_k (increasing with k) such that $\|\dot{\mathsf{f}}_{n_k} - \dot{\mathsf{f}}\|_{\dot{\mathsf{L}}^p(\Omega;E);k} \le 2^{-2k-3}$. Then,

$$\|\dot{\mathsf{f}}_{n_k} - \dot{\mathsf{f}}_{n_{k-1}}\|_{\dot{\mathsf{L}}^p(\Omega;E);k} \le \frac{1}{2^{2k}}. \tag{18.8}$$

2o) *Construction of* $\dot{\mathsf{g}}_m$. For $K \in \mathbb{N}$, in $\dot{\mathsf{L}}^1_{\text{loc}}(\Omega)$, we define

$$\dot{\mathsf{g}}_{m,K} \stackrel{\text{def}}{=} \|\dot{\mathsf{f}}_{n_1}\|_{E;m} + \sum_{2\le k\le K} \|\dot{\mathsf{f}}_{n_k} - \dot{\mathsf{f}}_{n_{k-1}}\|_{E;m}. \tag{18.9}$$

For $L > K$, we have $0 \le \dot{\mathsf{g}}_{m,L} - \dot{\mathsf{g}}_{m,K} \le \sum_{K+1\le k\le L} \|\dot{\mathsf{f}}_{n_k} - \dot{\mathsf{f}}_{n_{k-1}}\|_{E;m}$. By Definition 17.17 of $\dot{\mathsf{L}}^p(\Omega; E)$, we have $\|\dot{\mathsf{f}}\|_{\dot{\mathsf{L}}^p(\Omega;E);m} = \|\|\dot{\mathsf{f}}\|_{E;m}\|_{\dot{\mathsf{L}}^p(\Omega)}$, whence

$$\|\dot{\mathsf{g}}_{m,K} - \dot{\mathsf{g}}_{m,L}\|_{\dot{\mathsf{L}}^p(\Omega)} \le \sum_{K+1\le k\le L} \|\dot{\mathsf{f}}_{n_k} - \dot{\mathsf{f}}_{n_{k-1}}\|_{\dot{\mathsf{L}}^p(\Omega;E);m}.$$

For $k \ge m$, we have $\|\dot{\mathsf{f}}_{n_k} - \dot{\mathsf{f}}_{n_{k-1}}\|_{\dot{\mathsf{L}}^p(\Omega;E);m} \le \|\dot{\mathsf{f}}_{n_k} - \dot{\mathsf{f}}_{n_{k-1}}\|_{\dot{\mathsf{L}}^p(\Omega;E);k}$ since $\| \|_{E;m}$ increases with m. By (18.8), it is bounded by $1/2^{2k}$ so, for $L > K \ge m$, we have

$$\|\dot{\mathsf{g}}_{m,K} - \dot{\mathsf{g}}_{m,L}\|_{\dot{\mathsf{L}}^p(\Omega)} \le \sum_{K+1\le k\le L} \frac{1}{2^{2k}} \le \frac{1}{2^{2K}}.$$

The sequence $(\dot{\mathsf{g}}_{m,K})_{K\in\mathbb{N}}$ is increasing and Cauchy in $\dot{\mathsf{L}}^p(\Omega)$, which is sequentially complete (Theorem 17.20), so it converges to an element $\dot{\mathsf{g}}_m \ge \dot{\mathsf{g}}_{m,K}$ with

$$\dot{\mathsf{g}}_m \in \dot{\mathsf{L}}^p(\Omega).$$

3o) *Bound for* $\dot{\mathsf{f}}_{n_k}$. By (18.9), we have

$$\|\dot{\mathsf{f}}_{n_K}\|_{E;m} = \|\dot{\mathsf{f}}_{n_1} + \sum_{2\le k\le K} (\dot{\mathsf{f}}_{n_k} - \dot{\mathsf{f}}_{n_{k-1}})\|_{E;m} \le$$

$$\le \|\dot{\mathsf{f}}_{n_1}\|_{E;m} + \sum_{2\le k\le K} \|\dot{\mathsf{f}}_{n_k} - \dot{\mathsf{f}}_{n_{k-1}}\|_{E;m} = \dot{\mathsf{g}}_{m,K} \le \dot{\mathsf{g}}_m.$$

4°) Convergence of $\dot{\mathsf{f}}_{n_k}$. **4.a.** *Preliminary.* Every function $\mathsf{f}_n \in \dot{\mathsf{f}}_n$ is measurable (by Definition 17.16), so let

$$X_k \stackrel{\text{def}}{=} \left\{x \in \Omega : \|(\mathsf{f}_{n_k} - \mathsf{f}_{n_{k-1}})(x)\|_{E;k} \geq \frac{1}{2^k}\right\}. \tag{18.10}$$

Applying Markov's inequality (Theorem 17.11 (d) to $x \mapsto \|(\mathsf{f}_{n_k} - \mathsf{f}_{n_{k-1}})(x)\|^p_{E;k}$) and using (18.8) gives

$$|X_k| = \left|\left\{x \in \Omega : \|(\mathsf{f}_{n_k} - \mathsf{f}_{n_{k-1}})(x)\|^p_{E;k} \geq \frac{1}{2^{kp}}\right\}\right| \leq 2^{kp} \int_{X_k} \|\mathsf{f}_{n_k} - \mathsf{f}_{n_{k-1}}\|^p_{E;k} \leq \frac{1}{2^{kp}}.$$

4.b. *The subset* ω_{k_ϵ}. Let

$$\omega_k \stackrel{\text{def}}{=} X_k \cup X_{k+1} \cup \ldots$$

This set is measurable, as it is the countable union of measurable sets (Theorem 2.15). We have (Theorem 2.13 (b) on the measure of a union and Proposition A.6 (b) on the sum of a geometric series)

$$|\omega_k| \leq \frac{1}{2kp} + \frac{1}{2^{(k+1)p}} \cdots = \frac{1}{2^{kp}} \frac{1}{1 - 2^{-p}}.$$

For k_ϵ sufficiently large (*i.e.* when $k_\epsilon \geq \log(1/(\epsilon(1 - 2^{-p})))/(p \log 2)$), we get

$$|\omega_{k_\epsilon}| \leq \epsilon. \tag{18.11}$$

4.c. *Uniform convergence.* We have

$$\dot{\mathsf{f}}_{n_1} + \sum_{1 \leq k \leq \ell} (\dot{\mathsf{f}}_{n_k} - \dot{\mathsf{f}}_{n_{k-1}}) = \dot{\mathsf{f}}_{n_\ell}.$$

From (18.8), $\dot{\mathsf{f}}_{n_\ell} \to \dot{\mathsf{f}}$ in $\dot{\mathrm{L}}^p(\Omega; E)$ as $n_\ell \to \infty$. Furthermore,

$$\|\dot{\mathsf{f}}_{n_\ell} - \dot{\mathsf{f}}_{n_K}\|_{E;m} = \left\| \sum_{K+1 \leq k \leq \ell} (\dot{\mathsf{f}}_{n_k} - \dot{\mathsf{f}}_{n_{k-1}}) \right\|_{E;m} \leq \sum_{K+1 \leq k \leq \ell} \|\dot{\mathsf{f}}_{n_k} - \dot{\mathsf{f}}_{n_{k-1}}\|_{E;m}.$$

If $K \geq m + 1$, then by (18.10), outside X_k for $k \geq k_\epsilon$, and hence outside ω_{k_ϵ},

$$\|\dot{\mathsf{f}}_{n_\ell} - \dot{\mathsf{f}}_{n_K}\|_{E;m} \leq \sum_{K+1 \leq k \leq \ell} \|\dot{\mathsf{f}}_{n_k} - \dot{\mathsf{f}}_{n_{k-1}}\|_{E;k} \leq \sum_{K+1 \leq k \leq \ell} \frac{1}{2^k} \leq \frac{1}{2^K}.$$

This proves that

$$\dot{\mathsf{f}}_{n_k} \to \dot{\mathsf{f}} \text{ uniformly almost everywhere in } \Omega \setminus \omega_{k_\epsilon}.$$

4.d. *Almost everywhere convergence.* We have $\dot{\mathsf{f}}_{n_k} \to \dot{\mathsf{f}}$ almost everywhere in each $\Omega \setminus \omega_{k_\epsilon}$, and thus almost everywhere outside $\bigcap_{\epsilon > 0} \omega_{k_\epsilon}$, since a countable union of negligible sets is negligible (Theorem 2.24 (c)). For every $\epsilon > 0$, this intersection has measure less than ϵ by (18.11), and therefore it has measure zero. It follows then that

$$\dot{\mathsf{f}}_{n_k} \to \dot{\mathsf{f}} \text{ almost everywhere in } \Omega. \ \square$$

Let us now state Lebesgue's **dominated convergence theorem**[2].

Theorem 18.5. — *Let $(\dot{\mathsf{f}}_n)_{n\in\mathbb{N}}$ be a sequence in $\dot{\mathsf{L}}^p(\Omega; E)$, where Ω is a measurable subset of $\mathbb{R}^d$, E is a Fréchet space, and $1 \leq p < \infty$, such that, as $n \to \infty$, and for every semi-norm $\| \ \|_{E;m}$ of E and $\dot{\mathsf{g}}_m \in \dot{\mathsf{L}}^p(\Omega)$ such that, for all $n \in \mathbb{N}$,*

$$\|\dot{\mathsf{f}}_n\|_{E;m} \leq \dot{\mathsf{g}}_m \text{ almost everywhere in } \Omega. \tag{18.12}$$

Let $\dot{\mathsf{f}}$ be a class of functions equal almost everywhere from Ω into E such that, as $n \to \infty$,

$$\dot{\mathsf{f}}_n \to \dot{\mathsf{f}} \text{ almost everywhere in } \Omega.$$

Then, $\dot{\mathsf{f}} \in \dot{\mathsf{L}}^p(\Omega; E)$ and, as $n \to \infty$,

$$\dot{\mathsf{f}}_n \to \dot{\mathsf{f}} \text{ in } \dot{\mathsf{L}}^p(\Omega; E). \ \blacksquare$$

Proof. Since $\dot{\mathsf{g}}_m$ belongs to $\dot{\mathsf{L}}^p(\Omega)$, it satisfies (Theorem 18.2) the conditions for p-equi-integrability (Definition 18.1). For each semi-norm $\| \ \|_{E;m}$ of E, every $\|\dot{\mathsf{f}}_n\|_{E;m}$ is bounded by $\dot{\mathsf{g}}_m$, so they all satisfy the same conditions. Thus,

$$(\dot{\mathsf{f}}_n)_{n\in\mathbb{N}} \text{ is } p\text{-equi-integrable.}$$

Vitali's theorem (Theorem 18.3) then asserts that $\dot{\mathsf{f}}_n \to \dot{\mathsf{f}}$ in $\dot{\mathsf{L}}^p(\Omega; E)$. □

Example of a non-dominated convergent sequence. The class of functions $\dot{\mathsf{f}}_n$ on $(0, 1)$ associated with the function defined by $\mathsf{f}_n(x) = |x - 1/n|^{-1/2}$ converges in $\dot{\mathsf{L}}^1(0, 1)$ to the class associated with the function f defined by $\mathsf{f}(x) = x^{-1/2}$ (since a translation is continuous on $L^1(0, 1)$).

However, the f_n are not dominated by a single function g, otherwise it would be bounded below by n on $[1/(n+1), 1/n]$, which would imply $\int_{(0,1)} \mathsf{g} \geq (1/2) \sum_{n\geq 1} n^{-1}$, which is infinite. □

Comparison of convergences in $L^1(\Omega)$ and $\mathcal{M}(\Omega)$. Convergence in $\mathcal{M}(\Omega)$ is weaker than that in $L^1(\Omega)$, even for a dominated sequence: it does not imply convergence *almost everywhere*.

For example, the functions $\mathsf{f}_n \in \mathbf{C}_b(0, 1)$ defined by $\mathsf{f}_n(x) = \sin(2\pi n x)$ converge to 0 in $\mathcal{M}(0, 1)$, and they are dominated by the function with constant value 1; however, they converge neither in $L^1(0, 1)$ (since it would imply it converges to 0, whereas $\int_0^1 |\mathsf{f}_n| = \int_0^1 |\mathsf{f}_1| \neq 0$), nor almost everywhere. □

18.3. Image under a continuous mapping

Let us prove the integrability and the convergence of the composite function of certain continuous mappings. We denote by

$$T(\dot{\mathsf{f}}) \text{ the class of functions equal almost everywhere to } T(\mathsf{f}), \text{ where } \mathsf{f} \in \dot{\mathsf{f}}. \tag{18.13}$$

2 **Historical background of Lebesgue's theorem.** Henri LEBESGUE proved the dominated convergence theorem in 1910 [45, p. 375].

Theorem 18.6. — *Let E and F be Fréchet spaces endowed with the sequences of semi-norms $(\| \|_{E;m})_{m\in\mathbb{N}}$ and $(\| \|_{F;m})_{m\in\mathbb{N}}$ respectively, and let T be a continuous mapping from E into F such that, for every $i \in \mathbb{N}$, there exists $m \in \mathbb{N}$ and real constants $c_0 \geq 0$, $c_1 \geq 0$, and $s > 0$, such that, for all $e \in E$,*

$$\|T(e)\|_{F;i} \leq c_0 + c_1\|e\|_{E;m}^s.$$

Let Ω be a measurable subset of $\mathbb{R}^d$ and $1 \leq p \leq \infty$, with $p \geq s$. Suppose

$$\Omega \text{ has finite measure, or } c_0 = 0, \text{ or } p = \infty.$$

(a) *If $\dot{f} \in \dot{L}^p(\Omega; E)$, then $T(\dot{f}) \in \dot{L}^{p/s}(\Omega; F)$ and, whether p is finite or infinite,*

$$\|T(\dot{f})\|_{\dot{L}^{p/s}(\Omega;F);i} \leq c_0|\Omega|^{s/p} + c_1\|\dot{f}\|_{\dot{L}^p(\Omega;E);m}^s, \tag{18.14}$$

$$\|T(\dot{f})\|_{\dot{L}^\infty(\Omega;F);i} \leq c_0 + c_1\|\dot{f}\|_{\dot{L}^\infty(\Omega;E);m}^s. \tag{18.15}$$

(b) *Moreover, if $\dot{f}_n \to \dot{f}$ in $\dot{L}^p(\Omega; E)$ when $p < \infty$ then, when $n \to \infty$,*

$$T(\dot{f}_n) \to T(\dot{f}) \text{ in } \dot{L}^{p/s}(\Omega; F). \blacksquare$$

Proof. ***1°)*** *Property* (a). **1.a.** *Measurability.* Every $f \in \dot{f}$ is measurable (Definitions 17.16 and 17.17) and, since T is continuous, $T(f)$ is measurable. Indeed, f is (Definition 17.1) the almost everywhere limit of a sequence $(s_n)_{n\in\mathbb{N}}$ of simple functions, so $T(f)$ is the limit almost everywhere of the sequence of simple functions $(T(s_n))_{n\in\mathbb{N}}$. Hence, the functions in $T(\dot{f})$ are measurable.

Similarly, $\|f\|_{E;m}$ is measurable, hence $\|f\|_{E;m}^s$ is too, since raising to a positive power is continuous (Proposition A.82). And c_0 is clearly measurable.

1.b. *p/s-integrability.* The function $\|f\|_{E;m}^s$ is p/s-integrable (Definition 17.16) as, since f is p-integrable,

$$\int_\Omega (\|f\|_{E;m}^s)^{p/s} = \int_\Omega \|f\|_{E;m}^p < \infty. \tag{18.16}$$

Therefore, $c_0 + c_1\|\dot{f}\|_{E;m}^s$ is also p/s-integrable. Since

$$\|T(f)\|_{F;i} \leq c_0 + c_1\|f\|_{E;m}^s, \tag{18.17}$$

the monotonicity of the integral of measurable functions (Theorem 17.11 (b)) implies that $\int_\Omega \|T(f)\|_{F;i} < \infty$. Namely, $T(f)$ is p/s-integrable. Hence (Definition 17.17),

$$T(\dot{f}) \in L^{p/s}(\Omega; F).$$

***1.c.** Semi-norm when $p < \infty$.* Inequality (18.17) gives, together with (18.16) and Proposition 17.7 (d) (for the norm of c_0),

$$\|T(\dot{\mathsf{f}})\|_{\dot{\mathsf{L}}^{p/s}(\Omega;F);i} \leq \|c_0\|_{\dot{\mathsf{L}}^{p/s}(\Omega)} + c_1 \big\| \|\dot{\mathsf{f}}\|^s_{E;m} \big\|_{\dot{\mathsf{L}}^{p/s}(\Omega)} = c_0|\Omega|^{s/p} + c_1 \|\dot{\mathsf{f}}\|^s_{\dot{\mathsf{L}}^p(\Omega;E);m}.$$

***1.d.** Semi-norm when $p = \infty$.* Inequality (18.15) is (Definition 17.17) simply the reformulation of

$$\sup_{\Omega} \text{ess}\, \|T(\mathsf{f})\|_{F;i} \leq c_0 + c_1 \sup_{\Omega} \text{ess}\, \|\mathsf{f}\|^s_{E;m}.$$

2º) Convergence. ***2.a.** Choice of a subsequence.* Let $(\dot{\mathsf{f}}_n)_{n\in\mathbb{N}}$ be a sequence in $\dot{\mathsf{L}}^p(\Omega; E)$ converging to $\dot{\mathsf{f}}$, and $\mathsf{f}_n \in \dot{\mathsf{f}}_n$. According to the converse of the dominated convergence theorem (Theorem 18.4), there exists a subsequence $(\dot{\mathsf{f}}_{n_k})_{k\in\mathbb{N}}$ such that $\mathsf{f}_{n_k} \to \mathsf{f}$ almost everywhere and, for each $m \in \mathbb{N}$, there exists $\dot{\mathsf{g}}_m \in \dot{\mathsf{L}}^p(\Omega)$ such that

$$\|\mathsf{f}_{n_k}\|_{E;m} \leq \mathsf{g}_m$$

almost everywhere. As T is continuous, $T(\mathsf{f}_{n_k}) \to T(\mathsf{f})$ almost everywhere. Moreover,

$$\|T(\mathsf{f}_{n_k})\|_{F;i} \leq c_0 + c_1 \|\mathsf{f}_{n_k}\|^s_{E;m} \leq c_0 + c_1 \mathsf{g}^s_m.$$

Since $\dot{\mathsf{g}}^s_m \in \dot{\mathsf{L}}^{p/s}(\Omega)$ (Theorem 13.3, as $p \geq s$, which extends easily to a measurable Ω), the dominated convergence theorem (Theorem 18.5) yields

$$T(\dot{\mathsf{f}}_{n_k}) \to T(\dot{\mathsf{f}}) \text{ in } \dot{\mathsf{L}}^{p/s}(\Omega; F).$$

***2.b.** Convergence of the full sequence.* Suppose that there exist $i \in \mathbb{N}$, $\epsilon > 0$, and a subsequence $(\dot{\mathsf{f}}_{n_{k'}})_{k\in\mathbb{N}}$ such that, for all k' beyond a certain index, we have

$$\|T(\dot{\mathsf{f}}_{n_{k'}}) - T(\dot{\mathsf{f}})\|_{\dot{\mathsf{L}}^{p/s}(\Omega;F);i} \geq \epsilon. \tag{18.18}$$

Then, reasoning as above, we could extract a further subsequence $(\dot{\mathsf{f}}_{n_{k''}})_{k'\in\mathbb{N}}$ such that $T(\dot{\mathsf{f}}_{n_{k''}}) \to T(\dot{\mathsf{f}})$ in $\dot{\mathsf{L}}^{p/s}(\Omega; F)$ as $k'' \to \infty$, which would contradict (18.18). Therefore, (18.18) never holds. Thus,

$$T(\dot{\mathsf{f}}_n) \to T(\dot{\mathsf{f}}) \text{ in } \dot{\mathsf{L}}^{p/s}(\Omega; F). \;\square$$

Let us express Theorem 18.6 in terms of the measures from Definition 10.2.

Theorem 18.7. — *Let E, F, T, and Ω be as in Theorem 18.6, with, moreover,*

$$\Omega \textit{ is an open subset of } \mathbb{R}^d.$$

(a) *If* $f \in L^p(\Omega; E)$, *there exists a unique distribution*

$$T(f) \in L^{p/s}(\Omega; F)$$

such that, for every p*-integrabilizing sequence (Definition 10.1)* $(f_k^*)_{k\in\mathbb{N}}$ *for* f,

$$T(f) \stackrel{\text{def}}{=} \lim_{k\to\infty} T(f_k^*) \quad \text{in} \begin{cases} L^{p/s}(\Omega; F) & \text{if } p < \infty, \\ L^1_{\text{loc}}(\Omega; F) & \text{if } p = \infty. \end{cases}$$

(b) *Moreover, if* $f_n \to f$ *in* $L^p(\Omega; E)$ *when* $n \to \infty$,

$$T(f_n) \to T(f) \quad \text{in } L^{p/s}(\Omega; F). \blacksquare$$

Proof. ***1*°*)*** *Preliminary*. For every Fréchet space E and $1 \le p \le \infty$,

$$f \in L^p(\Omega; E) \text{ is equivalent to } \dot{f} \in \dot{L}^p(\Omega; E). \tag{18.19}$$

Indeed, $f \in L^p(\Omega; E)$ is equivalent to $\overline{f} \in \overline{L}^p(\Omega; E)$ (Theorem 17.33), and thus (Theorem 17.31) to $\dot{f} \in \dot{L}^p(\Omega; E)$. Similarly,

$$f_n \to f \text{ in } L^p(\Omega; E) \text{ is equivalent to } \dot{f}_n \to \dot{f} \text{ in } \dot{L}^p(\Omega; E). \tag{18.20}$$

2*°*) *Property* (a) *when* $p < \infty$. Let $f \in L^p(\Omega; E)$. By (18.19), we get $\dot{f} \in \dot{L}^p(\Omega; E)$. Theorem 18.6 (a) then gives $(T(f))^{\cdot} \in \dot{L}^{p/s}(\Omega; F)$. That is, again by (18.19),

$$T(f) \in L^{p/s}(\Omega; F). \tag{18.21}$$

Let $(f_k^*)_{k\in\mathbb{N}}$ be a p-integrabilizing sequence for f; then, we have (Proposition 10.9) $f_k^* \in \mathcal{B}(\Omega; E)$ and $f_k^* \to f$ in $L^p(\Omega; E)$ when $k \to \infty$. Since T is continuous, then $T(f_k^*) \in \mathcal{B}(\Omega; F)$ and, by (18.20), $\dot{f}_k^* \to \dot{f}$ in $\dot{L}^p(\Omega; E)$; so, by Theorem 18.6 (b), $T(\dot{f}_k^*) \to T(\dot{f})$ in $\dot{L}^{p/s}(\Omega; F)$; i.e. by (18.20) again,

$$T(f_k^*) \to T(f) \quad \text{in } L^{p/s}(\Omega; F). \tag{18.22}$$

3*°*) *Property* (a) *when* $p = \infty$. In this case, (18.21) is still satisfied and (18.22) holds in $L^1_{\text{loc}}(\Omega; F)$, which allows for $T(f)$ to be defined.

4*°*) *Property* (b). This is proved in the same way as (18.22), just now with f_n in the place of f_k^*. □

18.4. Continuity with respect to increasing p (again)

Let us state some properties of continuity with respect to increasing p.

Theorem 18.8. — *Let* $\dot{\mathsf{f}} \in \dot{\mathsf{L}}^p(\Omega; E)$ *for every* $p \in (p_0, p_1)$, *where* $1 \le p_0 < p_1 \le \infty$, Ω *is a measurable subset of* $\mathbb{R}^d$, E *is a Banach space, and* $c \in \mathbb{R}$, *such that*

$$\|\dot{\mathsf{f}}\|_{\dot{\mathsf{L}}^p(\Omega;E)} \le c.$$

Then, $\dot{\mathsf{f}} \in \dot{\mathsf{L}}^{p_1}(\Omega; E)$ *and*

$$\|\dot{\mathsf{f}}\|_{\dot{\mathsf{L}}^{p_1}(\Omega;E)} \le c. \blacksquare$$

Proof. **1°)** *Case of* p_1 *finite.* **1.a.** *Truncation.* Let T_r, where $r > 0$, be the truncation from E into itself which is defined, for $b \in E$, by

$$T_r(b) \stackrel{\text{def}}{=} \begin{cases} b & \text{if } \|b\|_E \le r, \\ rb/\|b\|_E & \text{if } \|b\|_E \ge r. \end{cases}$$

For $\dot{\mathsf{f}} \in \dot{\mathsf{L}}^1_{\text{loc}}(\Omega; E)$, we define $T_r(\dot{\mathsf{f}}) \in \dot{\mathsf{L}}^1_{\text{loc}}(\Omega; E)$ (see (18.13), p. 377, and Definition 18.6) by, for $\mathsf{f} \in \dot{\mathsf{f}}$,

$$(T_r(\mathsf{f}))(x) \stackrel{\text{def}}{=} T_r(\mathsf{f}(x)).$$

Inequality (18.14), p. 378, with $s = 1$, $c_0 = 0$, and $c_1 = 1$, together with inequality (18.15) with $c_0 = n$ and $c_1 = 0$, yield

$$\|T_n(\dot{\mathsf{f}})\|_{\dot{\mathsf{L}}^p(\Omega;E)} \le \|\dot{\mathsf{f}}\|_{\dot{\mathsf{L}}^p(\Omega;E)} \le c, \tag{18.23}$$

$$\|T_n(\dot{\mathsf{f}})\|_{\dot{\mathsf{L}}^\infty(\Omega;E)} \le n. \tag{18.24}$$

1.b. *Convexity.* The Riesz convexity theorem (Theorem 11.2, which readily extends to a measurable Ω, with p, p_1, ∞ replacing p_0, p, p_1) and the equivalence (18.19) imply $\dot{\mathsf{L}}^p(\Omega; E) \cap \dot{\mathsf{L}}^\infty(\Omega; E) \subsetneq \dot{\mathsf{L}}^{p_1}(\Omega; E)$ and, with (18.23), (18.24), and $\theta = 1 - p/p_1$,

$$\|T_n(\dot{\mathsf{f}})\|_{\dot{\mathsf{L}}^{p_1}(\Omega;E)} \le \|T_n(\dot{\mathsf{f}})\|^{1-\theta}_{\dot{\mathsf{L}}^p(\Omega;E)} \|T_n(\dot{\mathsf{f}})\|^{\theta}_{\dot{\mathsf{L}}^\infty(\Omega;E)} \le c^{1-\theta} n^\theta.$$

When $p \to p_1$, we have $\theta \to 0$, hence $n^\theta \to 1$. Taking the limit, we get

$$\|T_n(\dot{\mathsf{f}})\|_{\dot{\mathsf{L}}^{p_1}(\Omega;E)} \le c. \tag{18.25}$$

1.c. *Convergence.* For $\mathsf{f} \in \dot{\mathsf{f}}$, we have $\int_\Omega \|\mathsf{f}\|^p_E < \infty$, so $\|\mathsf{f}\|^p_E < \infty$ almost everywhere. Hence, $\|T_n(\mathsf{f})\|^p_E \to \|\mathsf{f}\|^p_E$ almost everywhere as $n \to \infty$; therefore,

$$\|T_n(\mathsf{f})\|^{p_1}_E \to \|\mathsf{f}\|^{p_1}_E \ \text{ almost everywhere.} \tag{18.26}$$

The monotone convergence theorem (Theorem 17.13) then yields, from (18.25), *i.e.* $\int_\Omega \|T_n(\mathsf{f})\|^{p_1}_E \le c^{p_1}$, the growth of $\|T_n(\dot{\mathsf{f}})\|^{p_1}_E$ as n grow, and the convergence (18.26),

$$\int_\Omega \|T_n(\mathsf{f})\|^{p_1}_E \to \int_\Omega \|\mathsf{f}\|^{p_1}_E. \tag{18.27}$$

Thus, $\dot{\mathsf{f}} \in \dot{\mathsf{L}}^{p_1}(\Omega; E)$. Using the convergence (18.27) and inequality (18.25), we get

$$\|\dot{\mathsf{f}}\|_{\dot{\mathsf{L}}^{p_1}(\Omega;E)} = \lim_{n\to\infty} \|T_n(\dot{\mathsf{f}})\|_{\dot{\mathsf{L}}^{p_1}(\Omega;E)} \leq c.$$

2º) Case of $p_1 = \infty$. **2.a.** *Case* $|\Omega| < \infty$. Suppose $|\Omega| < \infty$, $k > c$, and let

$$\sigma \stackrel{\text{def}}{=} \{x \in \Omega : \|\mathsf{f}(x)\|_E > k\}.$$

Since $\sigma = \{x \in \Omega : \|\mathsf{f}(x)\|_E^p > k^p\}$, Markov's inequality (Theorem 17.11 (d)) yields

$$|\sigma| \leq \frac{1}{k^p} \int_\Omega \|f\|_E^p \leq \left(\frac{c}{k}\right)^p.$$

This holds for all finite p, so

$$|\sigma| = 0.$$

So (Definition 17.17 of $\dot{\mathsf{L}}^\infty(\Omega; E)$), $\dot{\mathsf{f}} \in \dot{\mathsf{L}}^\infty(\Omega; E)$, and the norm of $\dot{\mathsf{f}}$ is less than or equal to all $k > c$, hence

$$\|\dot{\mathsf{f}}\|_{\dot{\mathsf{L}}^\infty(\Omega;E)} \leq c. \tag{18.28}$$

2.b. *General case.* Let $n \in \mathbb{N}$ and

$$\sigma_n \stackrel{\text{def}}{=} \{x \in \Omega \cap \mathring{B}(0,n) : \|\mathsf{f}(x)\|_E > k\}.$$

Step 2.a shows that σ is negligible. Since a countable union of negligible sets is negligible (Theorem 2.24 (c)), the union $\{x \in \Omega : \|\mathsf{f}(x)\|_E > k\}$ of the σ_n is also negligible. Thus, once again, $\dot{\mathsf{f}} \in \dot{\mathsf{L}}^\infty(\Omega; E)$ with (18.28). □

Let us extend Theorem 11.20 (b) on the continuity with respect to increasing p to include the case when E is an arbitrary Banach space.

Theorem 18.9. — *Let* $f \in L^p(\Omega; E)$ *for any* $p \in (p_0, p_1)$, *where* $1 \leq p_0 < p_1 \leq \infty$, Ω *is an open subset of* $\mathbb{R}^d$, E *is a Banach space, and* $c \in \mathbb{R}$, *such that*

$$\|f\|_{L^p(\Omega;E)} \leq c.$$

Then, $f \in L^{p_1}(\Omega; E)$ *and*

$$\|f\|_{L^{p_1}(\Omega;E)} \leq c. \blacksquare$$

Proof. ***1º) Preliminary.*** By the equivalence (18.19), for any $p \in (p_0, p_1)$, we have $\dot{\mathsf{f}} \in \dot{\mathsf{L}}^p(\Omega; E)$ and, by Definition 17.30 of $\overline{L}^p(\Omega; E)$ and Theorem 17.33 (a) on the identity of the spaces of measures,

$$\|\dot{\mathsf{f}}\|_{\dot{\mathsf{L}}^p(\Omega;E)} = \|\overline{f}\|_{\overline{L}^p(\Omega;E)} = \|f\|_{L^p(\Omega;E)} \leq c.$$

Theorem 18.8 then gives $\dot{\mathsf{f}} \in \dot{\mathsf{L}}^{p_1}(\Omega; E)$ and $\|\dot{\mathsf{f}}\|_{\dot{\mathsf{L}}^{p_1}(\Omega;E)} \leq c$. By using the equivalence (18.19), we get $f \in L^{p_1}(\Omega; E)$ and, with Theorem 17.33 (a) and Definition 17.30,

$$\|f\|_{L^{p_1}(\Omega;E)} = \|\overline{f}\|_{\overline{L}^{p_1}(\Omega;E)} = \|\dot{\mathsf{f}}\|_{\dot{\mathsf{L}}^{p_1}(\Omega;E)} \leq c. \ \square$$

18.5. Riesz representation theorem (again)

Let us first characterize the open subsets of a separable Banach space.

Lemma 18.10. — *Every open subset of a separable Banach space is a countable union of open balls $\mathring{B}_E(e, 1/k_e)$, where $e \in \mathcal{E}_0$, $k_e \in \mathbb{N}$, $\mathcal{E}_0 \subset \mathcal{E}$, and $\mathcal{E}$ is a countable dense subset of E.* ▮

Proof. Let $x \in \Omega$. There exists $k \in \mathbb{N}$ such that $B(x, 1/k) \subset \Omega$ and $e \in \mathcal{E}$ such that $\|e - x\|_E < 1/(2k)$. Then, $k_e = 2k$, since $x \in B(e, 1/(2k))$ and $B(e, 1/(2k))$ is included in $B(x, 1/k)$, and therefore in Ω. □

Next, let us characterize the measurability for values in such a space.

Lemma 18.11. — *Let f be a function from a measurable subset Ω of $\mathbb{R}^d$ into a separable Banach space E, and let $\mathcal{E}$ be a countable dense subset of E such that, for every $e \in \mathcal{E}$, the real-valued function $\|\mathsf{f} - e\|_E$ is measurable. Then, f is measurable.* ▮

Proof. **1°)** *Preliminary.* Every open subset U of E is a countable union of open balls $\mathring{B}_E(e, 1/k_e)$, where $e \in \mathcal{E}_0$, $k_e \in \mathbb{N}$, and $\mathcal{E}_0 \subset \mathcal{E}$ (Lemma 18.10). Hence,

$$\mathsf{f}^{-1}(U) = \bigcup_{e \in \mathcal{E}_0} \mathsf{f}^{-1}(\mathring{B}_E(e, 1/k_e)) = \bigcup_{e \in \mathcal{E}_0} \{x \in \Omega : \|\mathsf{f}(x) - e\|_E < 1/k_e\}.$$

Since each set $\{x \in \Omega : \|\mathsf{f}(x) - e\|_E < 1/k_e\}$ is measurable (Theorem 17.2, because $\|\mathsf{f} - e\|_E$ is measurable), the countable union $\mathsf{f}^{-1}(U)$ is measurable (Theorem 2.15).

2°) *Case of bounded Ω.* Let $(e_k^n)_{k \in \mathbb{N}^*}$ be a dense sequence in the ball $\mathring{B}_E(0, n)$. Define subsets of E by $V_k^n \stackrel{\text{def}}{=} \mathring{B}_E(e_k^n, 1/n) \cap \mathring{B}_E(0, n)$ and, by induction on $k \geq 1$,

$$U_k^n \stackrel{\text{def}}{=} V_k^n \setminus (U_0^n \cup \ldots \cup U_{k-1}^n), \quad U_0^n = \emptyset.$$

We have

$$\mathring{B}_E(0, n) = \bigcup_{k \geq 1} U_k^n; \tag{18.29}$$

indeed, if $u \in \mathring{B}_E(0, n)$, there exists a subsequence such that $e_{k_p}^n \to u$, and therefore $\|e_{k_p}^n - u\|_E < 1/n$ for p large enough; for the first such p satisfying this property, we have $u \in U_{k_p}^n$. Let

$$C_k^n \stackrel{\text{def}}{=} \mathsf{f}^{-1}(U_k^n) = \mathsf{f}^{-1}(V_k^n) \setminus \big(\mathsf{f}^{-1}(U_1^n) \cup \ldots \cup \mathsf{f}^{-1}(U_{k-1}^n)\big).$$

The sets $f^{-1}(U_k^n)$ are measurable by induction on k; indeed, by 1°, $f^{-1}(V_k^n)$ is measurable since V_k^n is open; and if $f^{-1}(U_k^n)$ is measurable for all $k \leq K$, then $f^{-1}(U_{K+1}^n)$ is the difference of measurable sets, hence is itself measurable (Theorem 2.11).

We have $U_k^n \cap U_\ell^n = \emptyset$ when $k \neq \ell$, so $C_k^n \cap C_\ell^n = \emptyset$. Let

$$\mathsf{s}_N \stackrel{\text{def}}{=} \sum_{k=1}^{N} e_k^n 1_{C_k^n}.$$

Then, s_N is a simple function and, as $N \to \infty$,

$$\mathsf{s}_N \to \mathsf{f}_n, \quad \mathsf{f}_n \stackrel{\text{def}}{=} \sum_{k \geq 1} e_k^n 1_{C_k^n}. \tag{18.30}$$

The function f_n is therefore measurable (Definition 17.1). Let us now show that, in the whole Ω,

$$\mathsf{f}_n \to \mathsf{f}. \tag{18.31}$$

Indeed, if $x \in \Omega$, then $\mathsf{f}(x) \in \mathring{B}_E(0,n)$ for n large enough. According to (18.29), there exists k_n such that $\mathsf{f}(x) \in U_{k_n}^n$. Hence, $x \in C_{k_n}^n$, so $\mathsf{f}_n(x) = e_{k_n}^n$ by (18.30) and $\|e_{k_n}^n - \mathsf{f}(x)\|_E < 1/n$, which proves (18.31). As any limit of measurable functions is measurable (Theorem 17.4),

$$\mathsf{f} \text{ is measurable.}$$

3°) Case of unbounded Ω. Let $\omega_n \stackrel{\text{def}}{=} \Omega \cap B_{\mathbb{R}^d}(0,n)$. By 2°, $\mathsf{f}_n \stackrel{\text{def}}{=} \mathsf{f}|_{\omega_n}$ is measurable, therefore its extension $\tilde{\mathsf{f}}_n$ by 0_E is measurable too. When $n \to \infty$, we get $\tilde{\mathsf{f}}_n \to \mathsf{f}$, and therefore the limit f is measurable (Theorem 17.4). □

Let us now complete the Riesz theorem (Theorem 16.8, which shows that J is an isomorphism when E is a uniformly convex Banach space and $p < \infty$).

Theorem 18.12. — *Let E be a Banach space whose dual E' is separable, Ω be an open subset of $\mathbb{R}^d$, and $1 \leq p < \infty$. Then the canonical injection J defined in Theorem 16.7 is an isometric isomorphism from $L^{p'}(\Omega; E')$ onto $(L^p(\Omega; E))'$.* ∎

Proof. By Theorem 16.7, J is a linear isometry from $L^{p'}(\Omega; E')$ into $(L^p(\Omega; E))'$, so it remains to show that it is surjective. To this end, let

$$F \in (L^p(\Omega; E))'. \tag{18.32}$$

1°) Construction of the measures $\underline{f_n}$. Since E' is separable, E is separable (Proposition A.66), so we can write $E = \overline{\bigcup_{n\in\mathbb{N}} E_n}$ where each E_n is finite-dimensional and increasing with n; endow each E_n with the norm inherited from E by restriction.

We define $F_n \in (L^p(\Omega; E_n))'$ by: for every $g \in L^p(\Omega; E_n)$,

$$F_n(g) \stackrel{\text{def}}{=} F(g). \tag{18.33}$$

It satisfies

$$\|F_n\|_{(L^p(\Omega;E_n))'} \leq \|F\|_{(L^p(\Omega;E))'}. \tag{18.34}$$

Since each E_n is finite-dimensional, every family of semi-norms that makes E_n separated are equivalent (Proposition A.38). Therefore, they are equivalent to the Euclidean norm, which makes E_n uniformly convex. By the Riesz theorem (Theorem 16.8), there exists a unique $f_n \in L^{p'}(\Omega; E_n')$ such that, for all $g \in L^p(\Omega; E_n)$,

$$F_n(g) = \int_\Omega \langle f_n, g\rangle_{E_n' \times E_n}.$$

2°) Construction of the functions f_n. From the equivalence (18.19), p. 380 and the equality of the integrals (17.35), p. 366,

$$\int_\Omega \langle f_n, g\rangle_{E_n' \times E_n} = \int_\Omega \overset{\cdot}{\overline{\langle \mathsf{f}_n, \mathsf{g}\rangle_{E_n' \times E_n}}}.$$

Now (Definition 17.15 (b)), f_n is the limit of the simple functions s_n^m in $\dot{\mathsf{L}}^{p'}(\Omega; E_n')$, and g is the limit of the simple functions t^m in $\dot{\mathsf{L}}^p(\Omega; E_n)$. For these simple functions,

$$\overset{\cdot}{\overline{\langle \mathsf{s}_n^m, \mathsf{t}^m\rangle_{E_n' \times E_n}}} = \langle \dot{\mathsf{s}}_n^m, \dot{\mathsf{t}}^m\rangle_{E_n' \times E_n}.$$

Taking the limit in m, by Definition 12.10 of the product $\langle\ ,\ \rangle_{E_n' \times E_n}$ with the equivalence (12.19), we get

$$\overset{\cdot}{\overline{\langle \mathsf{f}_n, \mathsf{g}\rangle_{E_n' \times E_n}}} = \langle \dot{\mathsf{f}}_n, \dot{\mathsf{g}}\rangle_{E_n' \times E_n}.$$

Whence,

$$F_n(g) = \int_\Omega \langle f_n, g\rangle_{E_n' \times E_n} = \int_\Omega \langle \dot{\mathsf{f}}_n, \dot{\mathsf{g}}\rangle_{E_n' \times E_n} = \int_\Omega \langle \mathsf{f}_n, \mathsf{g}\rangle_{E_n' \times E_n}. \tag{18.35}$$

3°) Construction of f. We have (equivalence (18.19), Theorem 16.8, and (18.34))

$$\left(\int_\Omega \|\mathsf{f}_n\|_{E_n'}^{p'}\right)^{1/p'} = \|\dot{\mathsf{f}}_n\|_{\dot{\mathsf{L}}^{p'}(\Omega;E_n')} = \|F_n\|_{(L^p(\Omega;E_n))'} \leq \|F\|_{(L^p(\Omega;E))'}. \tag{18.36}$$

If $n \leq m$, then $E_n \subset E_m$ and, by the uniqueness part of the Riesz theorem (Theorem 16.8, again), we have $\mathsf{f}_m|_{E_n} = \mathsf{f}_n$ and

$$\|\mathsf{f}_n\|_{E_n'} = \|\mathsf{f}_m|_{E_n}\|_{E_n'} \leq \|\mathsf{f}_m\|_{E_m'}. \tag{18.37}$$

For every $x \in \Omega$, we define $\mathsf{f}(x)$ from $\bigcup_{n\geq 1} E_n$ into $\mathbb{R}$ by: for all $n \in \mathbb{N}$ and $e \in E_n$,

$$(\mathsf{f}(x))(e) \stackrel{\text{def}}{=} \langle \mathsf{f}_n(x), e\rangle_{E'_n \times E_n}. \tag{18.38}$$

4⁰) Norm of $\|\mathsf{f}\|_{E'}$. The sequence $(\|\mathsf{f}_n\|_{E'_n})_{n\in\mathbb{N}}$, and thus also $(\|\mathsf{f}_n\|^{p'}_{E'_n})_{n\in\mathbb{N}}$, is increasing almost everywhere by (18.37); moreover, $\|\mathsf{f}_n\|^{p'}_{E'_n}$ is measurable with bounded integral by (18.36). Thus, by the theorem on the limit of functions with bounded integrals (Theorem 17.12), the set $\sigma \stackrel{\text{def}}{=} \{x \in \Omega : \lim_{n\to\infty} \|\mathsf{f}_n(x)\|^{p'}_{E'_n} = \infty\}$ is negligible. That is,

$$\sigma = \{x \in \Omega : \lim_{n\to\infty} \|\mathsf{f}_n(x)\|_{E'_n} = \infty\} \text{ is negligible.} \tag{18.39}$$

This remains true even when $p' = \infty$ since, by (18.36) in the $p = 1$ case,

$$\operatorname*{sup\,ess}_{x\in\Omega} \lim_{n\to\infty} \|\mathsf{f}_n(x)\|_{E'_n} \leq \|F\|_{(L^1(\Omega;E))'}. \tag{18.40}$$

By Definition (18.38) of f,

$$\mathsf{f}|_{E_n} = \mathsf{f}_n. \tag{18.41}$$

Thus, for almost every x,

$$\lim_{n\to\infty} \|\mathsf{f}(x)|_{E_n}\|_{E'_n} < \infty.$$

For such x, $\mathsf{f}(x)$ is continuous from $\bigcup_{n\geq 1} E_n$ into $\mathbb{R}$ by (18.38); by density, it extends to a continuous linear form on E, still denoted by $\mathsf{f}(x)$, whose norm satisfies $\|\mathsf{f}(x)\|_{E'} \leq \lim_{n\to\infty} \|\mathsf{f}(x)|_{E_n}\|_{E'_n}$ (since the latter increases with n). The reverse inequality holds by (18.38). Therefore,

$$\lim_{n\to\infty} \|\mathsf{f}(x)|_{E_n}\|_{E'_n} = \|\mathsf{f}(x)\|_{E'}. \tag{18.42}$$

5⁰) Integrability of $\|\mathsf{f}\|_{E'}$. When $1 < p' < \infty$, by Fatou's lemma (Theorem 17.14) (since the limit here coincides with the lower limit) and using (18.36), we get

$$\int_\Omega \lim_{n\to\infty} \|\mathsf{f}|_{E_n}\|^{p'}_{E'_n} = \int_\Omega \lim_{n\to\infty} \|\mathsf{f}_n\|^{p'}_{E'_n} \leq \lim_{n\to\infty} \int_\Omega \|\mathsf{f}_n\|^{p'}_{E'_n} \leq \|F\|^{p'}_{(L^{p'}(\Omega;E))'}.$$

This inequality and (18.40), together with (18.41) and (18.42), give

$$\Big(\int_\Omega \|\mathsf{f}\|^{p'}_{E'}\Big)^{1/p'} \leq \|F\|_{(L^{p'}(\Omega;E))'} \quad \text{if } p < \infty, \tag{18.43}$$

$$\operatorname*{sup\,ess}_{\Omega} \|\mathsf{f}\|_{E'} \leq \|F\|_{(L^1(\Omega;E))'} \quad \text{if } p = \infty. \tag{18.44}$$

6⁰) Integrability of f. Let $X \stackrel{\text{def}}{=} \{e \in \bigcup_{n\geq 1} E_n : \|e\|_E \leq 1\}$, and suppose $(e_k)_{k\in\mathbb{N}}$ is a dense sequence in X. This sequence is dense in the unit ball B_E of E, since

$\bigcup_{n\geq 1} E_n$ is dense in E. Each e_k belongs to one of the E_n, say E_{n_k}. For $\ell \in E'$, by the characterization of the norm of E' from Proposition 16.6 (b), we have

$$\|\mathsf{f} - \ell\|_{E'} = \sup_{e\in B_E} |\langle \mathsf{f} - \ell, e\rangle| = \sup_{k\geq 1} |\langle \mathsf{f} - \ell, e_k\rangle_{E\times E'}| = \\ = \sup_{k\geq 1} |\langle \mathsf{f}_{n_k} - \ell|_{E'_{n_k}}, e_k\rangle_{E_{n_k}\times E'_{n_k}}|.$$

Thus, $\|\mathsf{f} - \ell\|_{E'}$ is measurable (Theorem 17.6), since it is the supremum of a countable family of measurable functions bounded almost everywhere by a function (by (18.39)). Hence (Lemma 18.11), f is measurable; then, together with (18.43) and (18.44),

$$\dot{\mathsf{f}} \in \dot{\mathsf{L}}^{p'}(\Omega; E').$$

7º) Surjectivity. For almost every $x \in \Omega$, we have $\mathsf{f}(x) \in E'$, so the left-hand side of (18.38) equals $\langle \mathsf{f}(x), e\rangle_{E'\times E}$. Then, for all $e \in \bigcup_{n\geq 1} E_n$, all measurable sets ω of finite measure, and $k \geq 1$ such that $e \in E_k$, by (18.33), (18.35), and the equality $\langle \mathsf{f}_k, e1_\omega\rangle_{E'_k\times E_k} = \langle \mathsf{f}_k, e\rangle_{E'_k\times E_k} 1_\omega$ (which follows by passing to the limit in the approximations of f_k using simple functions), followed by (18.38), we have

$$F(e1_\omega) = F_k(e1_\omega) = \int_\Omega \langle \mathsf{f}_k, e1_\omega\rangle_{E'_k\times E_k} = \int_\Omega \langle \mathsf{f}_k, e\rangle_{E'_k\times E_k} 1_\omega = \int_\Omega \langle \mathsf{f}, e\rangle_{E'\times E} 1_\omega.$$

Using the equality $\langle \mathsf{f}, e\rangle_{E'\times E} 1_\omega = \langle \mathsf{f}, e1_\omega\rangle_{E'\times E}$, the last two equalities of (18.35) (with f instead of f_n) and the definition of J from Theorem 16.7, it follows that

$$F(e1_\omega) = \int_\Omega \langle \mathsf{f}, e1_\omega\rangle_{E'\times E} = \int_\Omega \langle f, e1_\omega\rangle_{E'\times E} = (Jf)(e1_\omega).$$

By (18.32) and Theorem 16.7, F and Jf are continuous from $L^p(\Omega; E)$ into $\mathbb{R}$. Since $\bigcup_{n\geq 1} E_n$ is dense in E, this equality therefore extends to $e \in E$, and also to every simple function $\mathsf{s} \in \mathsf{S}(\Omega; E)$ via linear combinations.

However, $\dot{\mathsf{S}}(\Omega; E)$ is sequentially dense in $\dot{\mathsf{L}}^p(\Omega; E)$ (Theorem 17.21), so $\mathsf{S}(\Omega; E)$ is sequentially dense in $L^p(\Omega; E)$ (by the equivalence (18.20)). Hence, we finally have $F = Jf$. This proves that

J is surjective. □

Remark[3] on the separability hypothesis of E'. In Theorem 18.12, one cannot remove the separability hypothesis placed on E' (though perhaps it can be weakened?).

For example, J is not an isomorphism from $L^\infty(\Omega; (L^1(\Lambda))')$ onto $(L^1(\Omega; L^1(\Lambda)))'$. This confirms Theorem 13.13, which states that the dual $L^\infty(\Lambda)$ of $L^1(\Lambda)$ is not separable.

Indeed, if $L^\infty(\Lambda) = (L^1(\Lambda))'$ were separable, then $L^\infty(\Omega; L^\infty(\Lambda)) = (L^\infty(\Omega; (L^1(\Lambda))')$ would be $(L^1(\Omega; L^1(\Lambda))'$ after identifying these spaces via the canonical isomorphism J (Theorem 18.12). This

3 **Concerning the remark on the separability of E'.** This was pointed out by Jérôme LEMOINE.

space is $(L^1(\Omega\times\Lambda))'$ by (14.17), p. 273, *i.e.* $L^\infty(\Omega\times\Lambda)$. However, $L^\infty(\Omega\times\Lambda)$ is strictly larger than $L^\infty(\Omega; L^\infty(\Lambda))$ according to (14.18) and (14.19). Therefore, $L^\infty(\Lambda)$ is not separable. □

Let us extend the reflexivity result for $L^p(\Omega; E)$ (Theorem 16.2, which assumes E to be uniformly convex).

Theorem 18.13. — *If E is a separable reflexive Banach space, Ω is an open subset of $\mathbb{R}^d$, and $1<p<\infty$, then $L^p(\Omega;E)$ is a reflexive Banach space and $L^p(\Omega;E)$-weak is a Neumann space.* ▌

Proof. **1º)** *Preliminary.* It suffices to verify that the canonical injection I from the space $L^p(\Omega; E)$ into $(L^p(\Omega; E))''$ is an isomorphism, where: for $g \in (L^p(\Omega; E))'$,

$$\langle If, g\rangle_{(L^p(\Omega;E))''\times(L^p(\Omega;E))'} \overset{\text{def}}{=} \langle g, f\rangle_{(L^p(\Omega;E))'\times L^p(\Omega;E)}. \tag{18.45}$$

Then (Proposition A.73), $L^p(\Omega; E)$ is reflexive, and thus $L^p(\Omega; E)$-weak is (Proposition A.75) a Neumann space.

2º) *Definitions.* Here, we will show that

$$I = (J^{-1})^* \circ J' \circ \hat{\imath}, \tag{18.46}$$

where: J is the injection from $L^{p'}(\Omega; E')$ into $(L^p(\Omega; E))'$ defined, for $k \in L^{p'}(\Omega; E')$ and $\ell \in L^p(\Omega; E)$, by

$$\langle Jk, \ell\rangle_{(L^p(\Omega;E))'\times L^p(\Omega;E)} \overset{\text{def}}{=} \int_\Omega \langle k, \ell\rangle_{E'\times E}; \tag{18.47}$$

here, $(J^{-1})^*$ is the transpose of J^{-1} from $(L^{p'}(\Omega; E'))'$ into $(L^p(\Omega; E))''$ defined, for $q \in (L^{p'}(\Omega; E'))'$ and $r \in (L^p(\Omega; E))'$, by

$$\langle (J^{-1})^* q, r\rangle_{(L^p(\Omega;E))''\times(L^p(\Omega;E))'} \overset{\text{def}}{=} \langle q, J^{-1} r\rangle_{(L^{p'}(\Omega;E'))'\times L^{p'}(\Omega;E')}; \tag{18.48}$$

J' is the injection from $L^p(\Omega; E'')$ into $(L^{p'}(\Omega; E'))'$ defined, for $s \in L^p(\Omega; E'')$ and $t \in L^{p'}(\Omega; E')$, by

$$\langle J's, t\rangle_{(L^{p'}(\Omega;E'))'\times L^{p'}(\Omega;E')} \overset{\text{def}}{=} \int_\Omega \langle s, t\rangle_{E''\times E'}; \tag{18.49}$$

$\hat{\imath}$ is the image (Theorem 12.1 (a)) of $L^p(\Omega; E)$ in $L^p(\Omega; E'')$ under the injection ß from E into E'' defined, for $e' \in E'$, by

$$\langle \hat{\imath} e, e'\rangle_{E''\times E'} \overset{\text{def}}{=} \langle e', e\rangle_{E'\times E}. \tag{18.50}$$

3°) Isomorphisms. Since E is separable, its image E'' under the canonical injection is separable (Proposition A.45 (a)), thus E' is separable (Proposition A.66). Then, Theorem 18.12 shows that the canonical injections J and J' are isomorphisms. Moreover, the transpose $(J^{-1})^*$ of the isomorphism J^{-1} is also an isomorphism (Proposition A.61).

Since E is reflexive, î is an isomorphism from E onto E'' (Proposition A.73), and hence î is an isomorphism from $L^p(\Omega; E)$ onto $L^p(\Omega; E'')$ (Theorem 12.1 (b)).

The composition of isomorphisms $(J^{-1})^* \circ J' \circ \hat{\imath}$ is hence an isomorphism.

4°) Equality (18.46) and conclusion. Since J is an isomorphism from $L^{p'}(\Omega; E')$ onto $(L^p(\Omega; E))'$, there exists a unique $h \in L^{p'}(\Omega; E')$ such that $g = Jh$. Then, from (18.45) and (18.47), we obtain

$$\langle If, g\rangle_{(L^p(\Omega;E))''\times(L^p(\Omega;E))'} = \langle Jh, f\rangle_{(L^p(\Omega;E))'\times L^p(\Omega;E)} = \int_\Omega \langle h, f\rangle_{E'\times E};$$

using Definitions (18.50) and (18.49) of î and J', this is equal to

$$= \int_\Omega \langle \hat{\imath}f, h\rangle_{E''\times E'} = \langle J'\circ\hat{\imath}f, h\rangle_{(L^{p'}(\Omega;E'))'\times L^{p'}(\Omega;E')};$$

and, using the definition of h and that of the transposition (18.48), this is equal to

$$= \langle J'\circ\hat{\imath}f, J^{-1}g\rangle_{(L^{p'}(\Omega;E'))'\times L^{p'}(\Omega;E')} = \langle (J^{-1})^*\circ J'\circ\hat{\imath}f, g\rangle_{(L^p(\Omega;E))''\times(L^p(\Omega;E))'}.$$

This proves (18.46) and, therefore, I is an isomorphism by 3°. □

Extension. The essential results obtained in Chapters 4 to 16 for integrable measures when Ω is *open* extend to classes of functions on a *measurable* set Ω with values in a Fréchet space E. However, this is of no use to us here and would unduly lengthen this book. □

Identification. When Ω is open, one may **identify** $\dot{L}^p(\Omega; E)$ (Definition 17.17) with $\overline{L}^p(\Omega; E)$ (Definition 17.30), *i.e.* with $L^p(\Omega; E)$ (Definition 10.2) according to Theorem 17.33, as is usually done in the study of PDEs. We have not done so here in order to avoid ambiguities in the proofs. □

Appendix

Reminders

Here, we provide the precise statements of some definitions and results used from Volume 1, *Banach, Fréchet, Hilbert and Neumann Spaces* [72], Volume 2, *Continuous Functions* [73], and Volume 3, *Distributions* [74] of the *Analysis for PDEs* set, which are used in this book in order to make it self-contained [1]. The proofs are given in the aforementioned volumes, to which we refer, for example by [Vol. 1, Theorem 12.2]. Some statements are restricted to the cases relevant to us when this simplifies them, whereas others are completed and then subsequently proven.

A.1. Notation and numbering

Set theory notation

We use the standard set theory notation, namely, $=$ for "*is equal to*", $\in$ for "*belongs to*", $\forall$ for "*for all*", $\exists$ for "*there exists*", $\Rightarrow$ for "*implies*", $\Leftarrow$ for "*follows from*", $\Leftrightarrow$ for "*is equivalent to*". A barred symbol denotes the negation of a property; for example, $\neq$ means "*is not equal to*".

We denote by $\{a, b, \dots, z\}$ the set consisting of a, b, …, z, by $\{a : P(a)\}$ or $\{a\}_{P(a)}$ the set of all a satisfying the property $P(a)$, and by $\emptyset$ the empty set. We write $\subset$ for set inclusion, $\cup$ for union, and $\cap$ for intersection. We write $\supset$ for "*contains*", and $U \setminus V = \{u \in U : u \notin V\}$.

We denote by $(a, b, \dots, z)$ the ordered set consisting of a, b, …, z. Thus, $(a, b) \neq (b, a)$ if $a \neq b$. In contrast, we have $\{a, b\} = \{b, a\}$.

We denote by $U_1 \times U_2 \times \cdots \times U_d = \{(u_1, u_2, \dots, u_d) : \forall i,\ u_i \in U_i\}$ the *product* of the sets U_i. We denote by U^d the d-fold Cartesian product of U with itself.

Given a set I and, for each $i \in I$, a set U_i, we denote by $\bigcup_{i \in I} U_i$ the union of the U_i, by $\bigcap_{i \in I} U_i$ their intersection, and by $\prod_{i \in I} U_i$ their product, namely, the set $\{(u_i)_{i \in I} : u_i \in U_i,\ \forall i\}$.

A **mapping** from a set U into a set X is the assignment, to each $u \in U$, of an element $T(u) \in X$. We say that $T(u)$ is the **image** under T of the point u. The **image** of a subset V of U under T is the set $T(V) = \{T(u) : u \in V\}$.

---

1 **Historical note.** In Volumes 1, 2, and 3, the reader will find the historical background of many of the definitions and properties recalled here.

We say that T is **injective** if $T(u) = T(v)$ implies $v = u$, that it is **surjective** if $T(U) = X$, and that it is **bijective** if it is both injective and surjective. We also say that T is an **injection**, **surjection**, or a **bijection**, respectively.

The **inverse image** of a subset Y of X under T is the set $T^{-1}(Y) = \{u \in U : T(u) \in Y\}$. If T is bijective, there exists a unique mapping T^{-1} from X into U, called the **inverse** of T, such that $T^{-1}(T(u)) = u$ for every $u \in U$. Then T is the inverse of T^{-1}, that is, we have $T(T^{-1}(u)) = u$.

Countability

We assume familiarity with the set[2] $\mathbb{N} = \{0, 1, 2, 3, \ldots\}$ of natural integers and also with the set $\mathbb{Z} = \{\ldots, -2, -1, 0, 1, 2, \ldots\}$ of integers, as well as with their addition, subtraction, multiplication, absolute value, and order. We denote $\mathbb{N}^* = \{1, 2, 3, \ldots\}$.

Definition A.1. — (a) *A set U is said to be **countable** if there exists a bijection from U onto a subset of $\mathbb{N}$.*

(b) *A set U is said to be **finite** if there exists $n \in \mathbb{N}$ and a bijection from U onto the integer interval $[\![1, n]\!] = \{i \in \mathbb{N} : 1 \le i \le n\}$.*

(c) *A set is said to be **infinite** if it is not finite.* □

Let us state some properties of countability [Vol. 1, Theorem 1.1].

Proposition A.2. — (a) *Every subset of a countable set is countable.*

(b) *The image of a countable set under any mapping is countable.*

(c) *Any finite product of countable sets is countable.*

(d) *Any countable union of countable sets is countable.* □

A **sequence** in a set U is defined as the assignment, for each $n \in \mathbb{N}$, of an element $u_n \in U$. It is an ordered set denoted by $(u_n)_{n\in\mathbb{N}}$. A **subsequence** of $(u_n)_{n\in\mathbb{N}}$ is any sequence of the form $(u_{\sigma(n)})_{n\in\mathbb{N}}$ such that $\sigma(n) \in \mathbb{N}$ and $\sigma(n+1) > \sigma(n)$ for all $n \in \mathbb{N}$.

Proposition A.3. — *A non-empty open subset of $\mathbb{R}^d$ is not countable.* □

Proof. Suppose that a non-empty open subset Ω of $\mathbb{R}^d$ is countable. The set $\{x_1 : x \in \Omega\}$, up to a translation and dilation, which do not affect countability (Proposition A.2 (b)), contains the interval $(0, 1)$. Then this interval would be countable (Proposition A.2 (a)).

Suppose $(0, 1) = \{x_n : n \ge 1\}$. Write the decimal expansion of each x_n as $x_n = \sum_{k\in\mathbb{N}} x_{k,n} 10^{-k}$ (proper expansion, meaning that not all the $x_{k,n}$ can equal 9 from some index k onward, so that the expansion for x_n is unique). For each n, define $a_n = 2$ if $x_{n,n} = 1$, and $a_n = 1$ if $x_{n,n} \neq 1$. Then the number $a = \sum_{k\in\mathbb{N}} a_k 10^{-k}$ belongs to $(0, 1)$ yet, for all $n \ge 1$, we have $a \neq x_n$ since $a_n \neq x_{n,n}$. This is a contradiction. □

---

2 **CAUTION.** We denote $\mathbb{N} \stackrel{\text{def}}{=} \{0, 1, 2, \ldots\}$ and $\mathbb{N}^* \stackrel{\text{def}}{=} \{1, 2, \ldots\}$, conforming to the ISO 80000-2 standard for mathematical and physics notation (edited in 2009).
Any possible confusion will be of no consequence, apart from surprising the reader used to the opposite notation when seeing a term u_0 of a series indexed by $\mathbb{N}$, or the inverse $1/n$ of a number n in $\mathbb{N}^*$.

Real numbers

We assume familiarity with the set $\mathbb{R}$ of real numbers, as well as its addition, subtraction, multiplication, absolute value, and order. Let us define convergent real sequences.

Definition A.4. — (a) *A sequence $(x_n)_{n\in\mathbb{N}}$ in $\mathbb{R}$ is said to* ***converge*** *to a* ***limit*** *$x \in \mathbb{R}$ if, for every $\epsilon > 0$, there exists $m \in \mathbb{N}$ such that: $n \geq m$ implies $|x_n - x| \leq \epsilon$. We then write $x_n \to x$.*

(b) *A sequence $(x_n)_{n\in\mathbb{N}}$ in $\mathbb{R}$ is said to* ***tend to infinity*** *if, for every $c \in \mathbb{R}$, there exists $m \in \mathbb{N}$ such that: $n \geq m$ implies $x_n \geq c$. We then write $x_n \to \infty$.* □

A real sequence is said to be **increasing** if $u_{n+1} \geq u_n$, and **decreasing** if $u_{n+1} \leq u_n$. Such sequences, if bounded above or below respectively, converge [Vol. 1, Theorem 1.3].

Proposition A.5. — *Every increasing real sequence that is bounded above, or decreasing real sequence that is bounded below, converges.*

Every increasing real sequence converges to a finite or infinite limit. □

Let us give the sum of a **geometric series** [Vol. 1, Property (1.10), p. 7].

Proposition A.6. — *Let $x \in \mathbb{R}$, $x \neq 1$ and consider two integers such that $n \leq m$. Then:*

(a)
$$x^n + x^{n+1} + \cdots + x^m = \frac{x^n - x^{m+1}}{1-x};$$

(b) *if $0 < x < 1$,*
$$0 < x^n + x^{n+1} + \cdots + x^m < \frac{x^n}{1-x}. \quad \square$$

Let us define upper bounds and other various related notions.

Definition A.7. — (a) *An element m is said to be an* ***upper bound*** *of a subset V of $\mathbb{R}$ if every $v \in V$ satisfies $v \leq m$.*

(b) *An element m is said to be the* ***supremum*** *of V if m is an upper bound of V and if every other upper bound v of V satisfies $m \leq v$. If it exists, it is unique and denoted by $\sup V$.*

(c) *An element m is said to be the* ***maximum*** *of V if m belongs to V and is an upper bound of V. If it exists, it is unique and denoted by $\max V$.*

(d) *The terms* ***lower bound, infimum****, and* ***minimum*** *are defined analogously by replacing $\leq$ with $\geq$. If they exist, the infimum is denoted by $\inf V$, and the minimum by $\min V$.* □

A.2. Banach, Hilbert, Fréchet, and Neumann spaces

Norms and semi-norms on a vector space

Let us define vector spaces and their subspaces [Vol. 1, Justification of Definition 3.5].

Definition A.8. — (a) *A* ***vector space*** *is any set E endowed with an* ***addition****, denoted $+$, from $E \times E$ into E, a* ***scalar multiplication*** *from $\mathbb{R} \times E$ into E, and an element $0_E \in E$ such that: for all u, v, and w in E, and s and t in $\mathbb{R}$,*

$$(u+v)+w = u+(v+w), \quad u+v = v+u, \quad u+0_E = u,$$
$$t(u+v) = tu+tv, \quad (s+t)u = su+tu, \quad s(tu) = (st)u, \quad 1u = u,$$

there exists an element $-u \in E$ such that $-u+u = 0_E$.

(b) *A **vector subspace** of E is any subset F of E that is a vector space for the addition and scalar multiplication inherited from E.* □

We sometimes write 0 instead of 0_E when there is no ambiguity. Let us now define norms and semi-norms.

Definition A.9. — (a) *A **semi-norm** is any mapping p from a vector space E into $\mathbb{R}$ such that: for all u and v in E and $t \in \mathbb{R}$,*

$$p(u) \geq 0, \quad p(tu) = |t|\, p(u), \quad p(u+v) \leq p(u) + p(v).$$

(b) *A **norm** is any semi-norm p such that $p(u) > 0$ whenever $u \neq 0_E$.* □

The semi-norms of a semi-normed space are continuous [Vol. 1, Theorem 7.10].

Proposition A.10. — *Every semi-norm $\| \ \|_{E;\nu}$ of a semi-normed space E is uniformly continuous.* □

Recall that every semi-norm is *contracting* [Vol. 1, Theorem 2.5].

Proposition A.11. — *Let p be a semi-norm on a vector space E, $u \in E$ and $v \in E$. Then,*

$$|p(v) - p(u)| \leq p(v-u). \ \square$$

Convergent sequences, bounded, open, and closed sets

We state some properties of sequences [Vol. 1, Theorems 2.7 and 2.8].

Proposition A.12. — *Every convergent sequence in a separated semi-normed space is bounded, a Cauchy sequence, and has a unique limit.*

Every Cauchy sequence is bounded.

A Cauchy sequence that has a convergent subsequence converges. □

Let us define open and closed sets in a semi-normed space.

Definition A.13. — *Let U be a subset of a separated semi-normed space E.*

(a) *U is said to be **open** if, for every $u \in U$, there exists a finite set $N \subset \mathcal{N}_E$ and $\eta > 0$ such that U contains the set $\{v \in E : \|v-u\|_{E;\nu} \leq \eta\}$.*

(b) *U is said to be **closed** if $E \setminus U$ is open.*

(c) *U is said to be **sequentially closed** if every sequence in U that converges in E has its limit in U.* □

Closed implies sequentially closed [Vol. 1, Theorem 2.10]. Open sets are stable under unions and finite intersections [Vol. 1, Theorem 2.11], while closed and bounded sets are stable under finite unions and intersections [Vol. 1, Theorem 2.12].

Let us define the interior and closure of a set of a semi-normed space.

Definition A.14. — *Let U be a subset of a separated semi-normed space.*

(a) *The* ***interior*** *of U is the largest open set contained in U. We denote it by $\mathring{U}$.*

(b) *The* ***closure*** *of U is the smallest closed set containing U. We denote it by $\overline{U}$.* □

There indeed exist a largest open set contained in U, and a smallest closed set containing U, and a smallest sequentially closed set containing U [Vol. 1, Justification of Definition 2.15].

Let us define the dense subsets of a separated semi-normed space.

Definition A.15. — *Let U and V be subsets of a separated semi-normed space.*

(a) *We say that V is* ***dense*** *in U if $V \subset U \subset \overline{V}$.*

(b) *We say that V is* ***sequentially dense*** *in U if $V \subset U$ and every $u \in U$ is the limit of a sequence in V.* □

Let us compare these notions of density [Vol. 1, Theorem 2.20].

Proposition A.16. — *Every subset of a separated semi-normed space, that is sequentially dense in another set, is dense in this set.* □

Connectedness, separation of compact sets

Let us define connected sets [Vol. 1, Definition 2.30 and its justification].

Definition A.17. — *A set U of a separated semi-normed space is said to be* ***connected*** *if it cannot be covered by two open sets whose intersections with U are disjoint and non-empty.* □

Let us state a **separation theorem** of a closed set and a disjoint compact set [Vol. 1, Theorem 3.19].

Proposition A.18. — *Let U be a closed set and K be a compact set in a normed space E such that $U \cap K = \emptyset$. Then, there exists a ball $B = \{u \in E : \|u\|_E \leq r\}$, where $r > 0$, in E such that*

$$(U + B) \cap K = \emptyset. \ \square$$

We now state the **strong inclusion theorem** of a compact set in an open set [Vol. 3, Theorem A.22].

Proposition A.19. — *Let U be an open set and K be a compact set in a normed space E such that $K \subset U$. Then, there exists a ball $B = \{u \in E : \|u\|_E \leq r\}$, where $r > 0$, such that*

$$K + B \subset U. \ \square$$

Let us show that intermediate sets can always be inserted between compact inclusions.

Proposition A.20. — *If $\omega \Subset \Omega \subset \mathbb{R}^d$, then there exists an open set ω' such that*

$$\omega \Subset \omega' \Subset \Omega.$$

In particular, this is achieved if ω is a compact set contained in an open set Ω. □

Proof. According to the strong inclusion theorem (Proposition A.19), there exists a closed ball such that $\omega + B(0, r) \subset U$, so the property is satisfied by $\omega' = \omega + \mathring{B}(0, r/2)$.

If ω is compact and Ω is open, then $\omega \Subset \Omega$ because ω is bounded and $\overline{\omega} = \omega \subset \Omega = \mathring{\Omega}$. □

Topological inclusions and metrizable spaces

Let us define topological subspaces.

Definition A.21. — *Let E and F be two semi-normed spaces.*

We say that E is a ***topological subspace*** *of F if it is a vector subspace of F, and if it is endowed with the restrictions of the semi-norms of F, or with a family equivalent to the family of these restrictions.* □

Let us state some properties of topological inclusions [Vol. 1, Theorem 3.8].

Proposition A.22. — *Let E and F be two separated semi-normed spaces such that $E \subsetneqq F$. Then, every Cauchy sequence in E is a Cauchy sequence in F.* □

Every semi-normed space can be endowed with a filtering family [Vol. 1, Theorem 3.15].

Proposition A.23. — *Every family of semi-norms on a vector space is equivalent to a filtering family.* □

Let us define normability.

Definition A.24. — *A semi-normed space is said to be* ***normable*** *if its family of semi-norms is equivalent to (a family reduced to) a single norm.* □

In a metrizable space, sequential properties coincide with topological properties [Vol. 1, Theorem 4.6], including the following.

Proposition A.25. — *For any subset U of a metrizable semi-normed space:*

(a) *closed* $\Leftrightarrow$ *sequentially closed;*

(b) *dense* $\Leftrightarrow$ *sequentially dense;*

(c) *separable* $\Leftrightarrow$ *sequentially separable;*

(d) $\overline{U}$ = *set of limits of sequences in U.* □

A metrizable space can be endowed with an increasing sequence of semi-norms [Vol. 1, Theorem 4.4].

Proposition A.26. — *The family of semi-norms of any metrizable semi-normed space is equivalent to an increasing sequence of semi-norms.* □

In a metrizable space, every limit of limits of sequences is a limit of sequences.

Proposition A.27. — *In a metrizable semi-normed space, every limit of limits of sequences of a set U is a limit of sequences of U.* □

Proof. From A.25 (d), $\overline{U}$ is the set of limits of sequences in U. Thus, $\overline{U} = \overline{\overline{U}}$ is also the set of limits of limits of sequences in U. □

Sequential closure is preserved under the equivalence of semi-norms [Vol.1, Theorem 3.4 (d)].

Proposition A.28. — *Two equivalent families of semi-norms on the same separated semi-normed space E generate the same closure $\overline{U}$ for any subset U of E.* □

We may extend an equality of mappings by using continuity [Vol. 1, Theorem 1.10].

Proposition A.29. — *Let T_1 and T_2 be two continuous mappings from a separated semi-normed space E into a separated semi-normed space F. If $T_1 = T_2$ on a subset U of E, then $T_1 = T_2$ on the closure $\overline{U}$.* □

Hilbert, Banach, Fréchet, and Neumann spaces

Let us define Hilbert spaces.

Definition A.30. — *A **scalar product** on a vector space E is any bilinear mapping b from $E \times E$ into $\mathbb{R}$ such that: for all u and v in E,*

$$b(v, u) = b(u, v), \quad b(v, v) > 0 \text{ if } v \neq 0_E.$$

*A **pre-Hilbert space** is any vector space E endowed with a scalar product $(\ ,\ )_E$ and with the norm defined by $\|u\|_E = (u, u)_E^{1/2}$.*

*A **Hilbert space** is any pre-Hilbert space that is sequentially complete.* □

Let us state the **parallelogram identity** and the **Cauchy–Schwarz** inequality [Vol. 1, Theorems 5.2 and 5.3].

Proposition A.31. — *Let E be a pre-Hilbert space and $u \in E$, $v \in E$. Then:*

(a) $$\|u + v\|_E^2 + \|u - v\|_E^2 = 2(\|u\|_E^2 + \|v\|_E^2);$$

(b) $$|(u, v)_E| \leq \|u\|_E \|v\|_E. \ \square$$

Let us compare various spaces [Vol. 1, Theorem 5.4].

Proposition A.32. — *Hilbert $\Rightarrow$ Banach $\Rightarrow$ Fréchet $\Rightarrow$ Neumann.* □

Sequentially complete subspaces are sequentially closed [Vol. 1, Theorem 4.12].

Proposition A.33. — *Every sequentially complete topological subspace of a separated semi-normed space is sequentially closed.* □

Let us state a condition for a space to be Fréchet [Vol. 1, Theorem 4.23] (a series $\sum_{n \in \mathbb{N}} u_n$ is said to be **absolutely convergent** if $\sum_{n \in \mathbb{N}} \|u_n\|_{E;\nu} < \infty$ for every semi-norm of E).

Proposition A.34. — *Every metrizable semi-normed space in which every absolutely convergent series converges is a Fréchet space.* □

Properties of $\mathbb{R}$ and $\mathbb{R}^d$

The space $\mathbb{R}$ is sequentially complete and a Hilbert space [Vol. 1, Theorems 4.15 and 5.5].

Proposition A.35. — *The space $\mathbb{R}$ is a Banach space, and it is a Hilbert space whose scalar product is the usual product.* □

Let us state the **Borel–Lebesgue theorem** (a) [Vol. 1, Theorem 5.13].

Proposition A.36. — *For every subset of* $\mathbb{R}^d$*:*

(a) *compact* $\Leftrightarrow$ *sequentially compact* $\Leftrightarrow$ *bounded and closed;*

(b) *relatively compact* $\Leftrightarrow$ *relatively sequentially compact* $\Leftrightarrow$ *precompact* $\Leftrightarrow$ *bounded.* □

Let us state the **Bolzano–Weierstrass theorem** [Vol. 1, Theorem 5.14].

Proposition A.37. — *Every bounded sequence in* $\mathbb{R}^d$ *has a convergent subsequence.* □

Let us state a result on the equivalence of families of semi-norms.

Proposition A.38. — *All families of semi-norms that make* $\mathbb{R}^d$ *separate are equivalent.* □

Proof. Let $\{\| \ \|_\nu\}_{\nu\in\mathcal{N}}$ and $\{||| \ |||_\mu\}_{\nu\in\mathcal{M}}$ be two families of semi-norms that make $\mathbb{R}^d$ separated. The set $B \stackrel{\text{déf}}{=} \{x \in \mathbb{R}^d : \sum_{i=1}^d |x_i| = 1\}$ is compact. Since $\mathbb{R}^d$ is separated by $\{\| \ \|_\nu\}_{\nu\in\mathcal{N}}$, for every $a \in B$, there exists a finite ν_a such that $\|a\|_{\nu_a} = \epsilon_a > 0$. The open sets $\{x \in \mathbb{R}^d : \|x\|_{\nu_a} > \epsilon_a/2\}$ cover the compact set B, so there exists a finite subcover A. Then, $\sup_{a\in A} \|x\|_{\nu_a} \geq \inf_{a\in A} \epsilon_a/2 = \epsilon > 0$ for all x in B.

On the other hand, for every $\mu \in \mathcal{M}$, we have $\sup_{x\in B} |||x|||_\mu \leq c \leq c/\epsilon \sup_{a\in A} \sup_{x\in B} \|x\|_{\nu_a}$. By homogeneity, this remains true for all $x \in \mathbb{R}^d$. Therefore, $\{||| \ |||_\mu\}_{\nu\in\mathcal{M}}$ dominates $\{\| \ \|_\nu\}_{\nu\in\mathcal{N}}$; and the converse also holds. □

A.3. Continuous mappings

Continuity in a semi-normed space

We say that a mapping T is **bounded** if its image $T(U)$ is bounded. We say that "T *is continuous on* U" to mean that "*the restriction of* T *to* U *is continuous*". We denote by $S \circ T$ a **composite mapping**, i.e. $(S \circ T)(u) = S(T(u))$.

The composition of continuous mappings is continuous [Vol. 1, Theorem 7.9].

Proposition A.39. — *Let* E*,* F*, and* G *be separated semi-normed spaces,* $U \subset E$*,* $Y \subset F$*,* T *be a mapping from* U *into* Y*, and* S *be a mapping from* Y *into* G*.*

If S *and* T *are continuous, or uniformly continuous, or sequentially continuous, then* $S \circ T$ *has the same property.* □

In particular, topological embeddings preserve continuity [Vol. 1, Theorems 7.6 and 7.7].

Proposition A.40. — *Let* E *and* F *be separated semi-normed spaces such that* $E \subsetneq F$*.*

(a) *If a mapping is continuous, uniformly continuous, or sequentially continuous from a subset of a separated semi-normed space that takes values in* E*, then it has the same property when considered to take values in* F*.*

(b) *If a mapping is continuous, uniformly continuous, or sequentially continuous from* F *into a separated semi-normed space* G*, then it has the same property from* E *into* G*.* □

Let us state **Heine's theorem** [Vol. 1, Theorem 8.10].

Proposition A.41. — *Every continuous mapping from a compact subset of a separated semi-normed space into a separated semi-normed space is uniformly continuous and bounded.* □

Let us state a corollary to it [Vol. 2, Theorem 2.20 (b)].

Proposition A.42. — *Any continuous function defined on an open subset Ω of $\mathbb{R}^d$, that vanishes outside a compact subset contained in Ω, is uniformly continuous.* □

Let us state a property of the composition with a linear mapping [Vol. 1, Theorem 9.4].

Proposition A.43. — *If f is a uniformly continuous function from a subset of $\mathbb{R}^d$ into a separated semi-normed space E, and if L is a sequentially continuous linear mapping from E into a separated semi-normed space, then $L \circ f$ is uniformly continuous.* □

The semi-norms of a semi-normed space are continuous [Vol. 1, Theorem 7.10].

Proposition A.44. — *Every semi-norm $\| \ \|_{E;\nu}$ of a separated semi-normed space E is uniformly continuous from E into $\mathbb{R}$.* □

Let us state some properties of images [Vol. 1, Theorems 8.5 for (a), 8.4 for (d), and 8.7].

Proposition A.45. — *Let T be a continuous mapping from a subset X of a separated semi-normed space E into a separated semi-normed space F, and $U \subset X$.*

(a) *If U is separable, then $T(U)$ is separable.*

(b) *If U is compact, then $T(U)$ is compact.*

(c) *If U is connected, then $T(U)$ is connected.*

(d) *If V is dense in U, then $T(V)$ is dense in $T(U)$.*

(e) *If T is uniformly continuous and U is precompact, then $T(U)$ is precompact.* □

Let us state a property of the inverse image [Vol. 2, Theorem 8.2 (a)].

Proposition A.46. — *If T is a continuous mapping from an open subset U of a separated semi-normed space E into a separated semi-normed space F, then the inverse image $T^{-1}(W)$ of a subset W of F is open in E if W is open, and closed if W is closed.* □

Continuous functions

Every function (*i.e.* any mapping defined on a subset of $\mathbb{R}^d$) that is sequentially continuous is continuous [Vol. 1, Theorem 9.1].

Proposition A.47. — *Every sequentially continuous function from an open subset of $\mathbb{R}^d$ into a separated semi-normed space is continuous.* □

Let us give a property of the product of continuous functions [Vol. 2, Theorem 3.3, with $T(u, t) = tu$].

Proposition A.48. — *If $f \in \mathcal{C}(\Omega; E)$ and $g \in \mathcal{C}(\Omega)$, where Ω is an open subset of $\mathbb{R}^d$ and E is a separated semi-normed space, then*

$$fg \in \mathcal{C}(\Omega; E). \ \square$$

Let us state a property of sequential completeness [Vol. 2, Theorem 1.16 (e)].

Proposition A.49. — *The space $\mathcal{C}_K(\Omega)$ is a Banach space, for every open subset Ω of $\mathbb{R}^d$ and every compact subset $K \subset \Omega$.* □

Let us present a special case of **Urysohn**'s theorem [Vol. 2, Theorem 3.21].

Proposition A.50. — *For every compact set ω contained in an open subset Ω of $\mathbb{R}^d$, there exists a function α such that*

$$\alpha \in \mathcal{K}(\Omega), \quad 0 \leq \alpha \leq 1, \quad \alpha = 1 \textit{ on } \omega. \ \square$$

Let us construct a **partition of unity** $(\alpha_k)_{k\in\mathcal{K}}$ subordinate to an open cover $(\omega_k)_{k\in\mathcal{K}}$ of an open subset Ω of $\mathbb{R}^d$ [Vol. 2, Theorem 7.18].

Proposition A.51. — *Let Ω and $(\omega_k)_{k\in\mathcal{K}}$ be open subsets of $\mathbb{R}^d$ such that $\Omega = \bigcup_{k\in\mathcal{K}} \omega_k$. Then there exists a family $(\alpha_k)_{k\in\mathcal{K}}$ of functions such that, for each k,*

$$\alpha_k \in \mathcal{C}(\Omega), \quad 0 \leq \alpha_k \leq 1, \quad \operatorname{supp} \alpha_k \subset \omega_k, \quad \sum_{k\in\mathcal{K}} \alpha_k = 1 \textit{ on } \Omega. \ \square$$

Let us define the **support** of a function.

Definition A.52. — *For a function f from $\Omega \subset \mathbb{R}^d$ into a separated semi-normed space, we denote*

$$\operatorname{supp} f \stackrel{\text{def}}{=} \overline{\{x \in \Omega : f(x) \neq 0\}} \cap \Omega. \ \square$$

Let us give a condition under which the support of a function is compact [Vol. 2, Theorem 1.18].

Proposition A.53. — *The support of a function from a subset Ω of $\mathbb{R}^d$ into a separated semi-normed space is compact if and only if it is contained in a compact set included in Ω.* $\square$

Let us state a property of the set where a function is non-zero [Vol. 2, Theorem 1.6 (b)].

Proposition A.54. — *If f is a continuous function from an open subset Ω of $\mathbb{R}^d$ into a separated semi-normed space E, then*

$$\{x \in \Omega : f(x) \neq 0_E\} \textit{ is open. } \square$$

Linear and multilinear mappings

Let us define linearity.

Definition A.55. — *A mapping L from a vector space E into another is said to be* ***linear*** *if, for all $u \in E$, $v \in E$, and $t \in \mathbb{R}$,*

$$L(u+v) = L(u) + L(v), \quad L(tu) = t\,L(u). \ \square$$

When there is no ambiguity, we sometimes write Lu instead of $L(u)$ when L is linear.

A mapping is called an **isomorphism** (from one separated semi-normed space onto another) if it is a continuous linear bijection whose inverse is also continuous.

Let us now define multilinear mappings.

Definition A.56. — *A mapping T on a product of vector spaces is said to be* ***multilinear*** *if its partial mappings $u_i \mapsto T(u_1, \ldots, u_d)$ are linear. In the case of two variables, one says* ***bilinear****.* $\square$

Let us now define the space $\mathcal{L}(E; F)$ of continuous linear mappings.

Definition A.57. — *Let E and F be two separated semi-normed spaces.*

(a) *We denote by $\mathcal{L}(E; F)$ the set of continuous linear mappings from E into F, endowed with the semi-norms indexed by the bounded subsets B of E and $\mu \in \mathcal{N}_F$,*

$$\|L\|_{\mathcal{L}(E;F);B,\mu} \stackrel{\text{def}}{=} \sup_{u\in B} \|Lu\|_{F;\mu}.$$

(b) *We denote by* $\mathcal{L}(E;F)$-simple *the same space, endowed with the semi-norms indexed by* $u \in E$ *and* $\mu \in \mathcal{N}_F$,

$$\|L\|_{\mathcal{L}(E;F)\text{-simple};u,\mu} \stackrel{\text{def}}{=} \|Lu\|_{F;\mu}. \ \square$$

For mappings taking values in a vector space, whether they are linear or not, addition and multiplication (by a real number t) are defined by

$$(L+M)(u) = L(u) + M(u), \quad (tL)(u) = tL(u).$$

Multiplication in a separated semi-normed space is continuous [Vol. 1, Theorem 7.24].

Proposition A.58. — *For every separated semi-normed space* E*, the mapping* $(t,u) \mapsto tu$ *is a continuous bilinear mapping from* $\mathbb{R} \times E$ *into* E. $\square$

A continuous multilinear mapping is uniformly continuous on bounded sets [Vol. 1, Theorem 7.22].

Proposition A.59. — *Every continuous multilinear mapping from a product of separated semi-normed spaces into another is uniformly continuous on every bounded set.* $\square$

Isomorphisms preserve sequential completeness [Vol. 1, Theorem 7.18].

Proposition A.60. — *Every separated semi-normed space that is isomorphic to a Neumann space is itself a Neumann space.* $\square$

The transpose of an isomorphism is an isomorphism.

Proposition A.61. — *If* T *is an isomorphism from a normed space* E *onto a normed space* F*, its transpose, defined from* F' *into* E' *by* $\langle T^*\ell, e\rangle_{E'\times E} \stackrel{\text{def}}{=} \langle \ell, Te\rangle_{F'\times F}$ *for all* $e \in E$*, is an isomorphism.* $\square$

Proof. For every $\ell \in F'$, we have $(T^{-1})^*(T^*(\ell)) = (T^{-1})^*(\ell \circ T) = \ell \circ T \circ T^{-1} = \ell$ and, for every $u \in E'$, we have $T^*((T^{-1})^*(u)) = T^*(u \circ T^{-1}) = u \circ T^{-1} \circ T = u$. Thus, $(T^{-1})^*$ is the inverse mapping of T^*. $\square$

A.4. Duals, weak topology, and reflexivity

Duals and the weak topology

A **linear form** on a separated semi-normed space E is any linear mapping from E into $\mathbb{R}$. We now define the **dual** E' of a separated semi-normed space E.

Definition A.62. — *We denote by* E' *the space of continuous linear forms on a separated semi-normed space* E*, endowed with the family of semi-norms indexed by the bounded subsets* B *of* E*,*

$$\|e'\|_{E';B} \stackrel{\text{def}}{=} \sup_{e\in B} |\langle e', e\rangle|. \ \square$$

Let us state a corollary to the **Hahn–Banach theorem** [Vol. 1, Theorem 14.4].

Proposition A.63. — *Let* E *be a semi-normed space and* $e \in E$*. Then:*

$$[\, \langle e', e\rangle_{E'\times E} = 0, \ \forall e' \in E' \,] \quad \Rightarrow \quad e = 0_E. \ \square$$

Let us state another consequence of the **Hahn–Banach theorem** [Vol. 1, Theorem 14.5].

Proposition A.64. — *Every continuous linear form on a topological vector subspace of* E *has a continuous linear extension to* E $\square$

Let us state another corollary to the **Hahn–Banach theorem** [Vol. 1, Theorem 14.6].

Proposition A.65. — *Let F be a vector subspace of a separated semi-normed space E. Then:*

$$\overline{F} = E \quad \Leftrightarrow \quad [\forall f' \in E',\ f'|_F = 0_F \Rightarrow f' = 0_E].\ \square$$

A consequence of this result is the following.

Proposition A.66. — *Every Banach space with a separable dual is separable.* $\square$

Proof. Let $(e'_n)_{n\in\mathbb{N}}$ be a dense sequence in the dual E' of a Banach space E. By the definition of the dual norm (Theorem 16.7 (b)), $|||e'_n|||_{E'} = \sup_{e\in E,\ \|e\|=1} |\langle e'_n, e\rangle|$, so there exists $e_n \in E$ such that

$$\|e_n\|_E = 1, \quad \langle e'_n, e_n\rangle \geq |||e'_n|||/2.$$

Let X_0 be the set of linear combinations of the e_n with coefficients in $\mathbb{Q}$. It is countable and dense in the vector space X generated by the e_n. We will show that X, and hence also X_0, are dense in E.

Let $e' \in E'$ such that $\langle e', e\rangle = 0$ for all $e \in X$. Then $\langle e', e_n\rangle = 0$ for all $n \in \mathbb{N}$. Let $\epsilon > 0$. For each $n \in \mathbb{N}$, there exists e'_n such that $|||e'_n - e'|||_{E'} \leq \epsilon$. But then,

$$|||e'_n|||_{E'} \leq 2\,|\langle e'_n, e_n\rangle| = 2\,|\langle e'_n - e', e_n\rangle| \leq 2\,|||e'_n - e'|||_{E'}\,\|e_n\|_E \leq 2\epsilon.$$

Hence,

$$|||e'|||_{E'} \leq |||e' - e'_n|||_{E'} + |||e'_n||| \leq 3\epsilon.$$

This holds for all $\epsilon > 0$, thus $e' = 0$. Proposition A.65 then implies that X is dense in E. $\square$

Let us define the weak topology on a separated semi-normed space.

Definition A.67. — *Let E be a separated semi-normed space. We denote by E-weak the space E endowed with the semi-norms, indexed by $e' \in E'$,*

$$\|e\|_{E\text{-weak};e'} \stackrel{\text{def}}{=} |\langle e', e\rangle|.\ \square$$

The weak topology is coarser than the initial topology [Vol. 1, Theorem 15.2].

Proposition A.68. — *We have $E \subsetneqq E$-weak.* $\square$

Let us state the **Banach–Dieudonné theorem** [Vol. 1, Theorem 15.7].

Proposition A.69. — *Every continuous linear mapping from a separated semi-normed space E into a separated semi-normed space F is continuous from E-weak into F-weak.* $\square$

The space E-weak is topologically included in F-weak if E is included in F [Vol. 1, Theorem 15.8].

Proposition A.70. — *Let E and F be two separated semi-normed spaces such that $E \subsetneqq F$. Then,*

$$E\text{-weak} \subsetneqq F\text{-weak}.\ \square$$

Let us bound the semi-norms of a weak limit [Vol. 1, Theorem 15.16].

Proposition A.71. — *Let p be a continuous semi-norm on a separated semi-normed space E, and let $(u_n)_{n\in\mathbb{N}}$ be a sequence such that $u_n \to u$ in E-weak and $p(u_n) \leq c$ for all $n \in \mathbb{N}$. Then,*

$$p(u) \leq c.\ \square$$

Reflexivity and uniform convexity

Let us define reflexivity, along with its variants.

Definition A.72. — *Let E be a separated semi-normed space.*

(a) *We say that E is **semi-reflexive** if, for every $e'' \in E''$, there exists $e \in E$ such that, for every $e' \in E'$,*

$$\langle e'', e'\rangle_{E''\times E'} = \langle e', e\rangle_{E'\times E}.$$

(b) *We say that E is* ***pre-reflexive*** *if, for every bounded subset B' of E', there exist a finite set $N \subset \mathcal{N}_E$ and a constant $c \in \mathbb{R}$ such that, for every $e \in E$,*

$$\sup_{e' \in B'} |\langle e', e \rangle| \leq c \sup_{\nu \in N} \|e\|_{E;\nu}.$$

(c) *We say that E is* ***reflexive*** *if it is both semi-reflexive and pre-reflexive.* □

Let us characterize reflexivity by means of the canonical injection [Vol. 1, Theorem 17.4].

Proposition A.73. — *A separated semi-normed space E is reflexive if and only if the canonical injection i is an isomorphism from E onto E'', where: for every $e' \in E'$,*

$$\langle ie, e' \rangle_{E'' \times E'} \stackrel{\text{def}}{=} \langle e', e \rangle_{E' \times E}. \ \square$$

Let us state **Milman's theorem** [Vol. 1, Theorem 17.15].

Proposition A.74. — *Every uniformly convex Banach space is reflexive.* □

Let us give another result [Vol. 1, Theorem 17.7].

Proposition A.75. — *If E is a separated reflexive semi-normed space, then E-*weak *is a Neumann space.* □

Let us state the **Banach–Bourbaki theorem** [Vol. 1, Theorem 13.16].

Proposition A.76. — *If E is a separable Fréchet space, then every bounded sequence in E'-*$*$weak *has a convergent subsequence.* □

Let us give yet another result [Vol. 1, Theorem 18.7 (or Theorem 18.9)].

Theorem A.77. — *If E is a reflexive Banach space, then every bounded sequence in E has a subsequence converging in E-*weak. □

A.5. Differentiable mappings

Let us define differentiability [Vol. 1, Definition 19.1].

Definition A.78. — *A mapping T from an open subset X of a separated semi-normed space E into a separated semi-normed space F is said to be* ***differentiable*** *at a point u of X if there exists a mapping $\mathrm{d}T(u)$ in $\mathcal{L}(E; F)$, and if, for every $\mu \in \mathcal{N}_F$, there exists a finite set $N \subset \mathcal{N}_E$ such that: for every $\epsilon > 0$, there exists $\eta > 0$ such that $z \in E$ and $\sup_{\nu \in N} \|z\|_{E;\nu} \leq \eta$ implies $u + z \in X$ and*

$$\|T(u+z) - T(u) - \mathrm{d}T(u)(z)\|_{F;\mu} \leq \epsilon \sup_{\nu \in N} \|z\|_{E;\nu}. \ \square$$

Let us define the derivative of a function defined on an open subset of $\mathbb{R}$.

Definition A.79. — *A function f from an open subset of $\mathbb{R}$ into a separated semi-normed space E which is differentiable is also said to be* ***derivable****, and its* ***derivative*** *at a point s, denoted by $f'(s)$, is the element of E defined by*

$$f'(s) \stackrel{\text{def}}{=} \mathrm{d}f(s)(1). \ \square$$

A real-valued function f is said to be **positive** if $f(t) \geq 0$ for all t, in which case we write $f \geq 0$. Similarly, we write $f \geq g$ if $f - g \geq 0$, and $f \leq g$ if $g - f \geq 0$. A real-valued function f on an interval is said to be **increasing** if $s \leq t$ implies $f(s) \leq f(t)$, and **decreasing** if $s \leq t$ implies $f(t) \leq f(s)$.

Let us state some growth properties of a real-valued function on an interval [Vol. 1, Theorem 21.8].

Proposition A.80. — *Let $-\infty \leq a < b \leq \infty$, and let f be a derivable function from (a, b) into $\mathbb{R}$. Then:*

(a) $$f' \geq 0 \quad \Leftrightarrow \quad f \text{ is increasing;}$$

(b) $$f' \leq 0 \quad \Leftrightarrow \quad f \text{ is decreasing. } \square$$

The derivative commutes with linear mappings [Vol. 1, Theorem 19.19].

Proposition A.81. — *Let E, F, and G be three separated semi-normed spaces, U be an open subset of E, T be a mapping from U into F differentiable at a point u, and $L \in \mathcal{L}(F; G)$. Then, $L \circ T$ is differentiable at the point u and*

$$\mathrm{d}(L \circ T)(u) = L \circ \mathrm{d}T(u). \ \square$$

The function $x \mapsto x^s$ is continuous and derivable [Vol. 1, Theorem 22.14].

Proposition A.82. — *Let $s \in \mathbb{R}$. The function $x \mapsto x^s$ is derivable from $(0, \infty)$ into $\mathbb{R}$, and*

$$\frac{\mathrm{d}x^s}{\mathrm{d}x} = s x^{s-1}.$$

If $s > 0$, then it is continuous from $[0, \infty)$ into $\mathbb{R}$. $\square$

Let us define the partial derivatives and gradient of a function.

Definition A.83. — *Let Ω be an open subset of $\mathbb{R}^d$, E be a separated semi-normed space, f be a function from Ω into E, $x \in \Omega$, and $i \in [\![1, d]\!]$ be an integer.*

(a) *We say that f has a* ***partial derivative*** *$\partial_i f(x) \in E$ at the point x if the mapping $x_i \mapsto f(x)$ is derivable at the point x_i with derivative $\partial_i f(x)$.*

(b) *When f has a partial derivative for each i, we say that the* ***gradient*** *of f at the point x is the element of E^d defined by*

$$\nabla f(x) \stackrel{\text{def}}{=} (\partial_1 f(x), \ldots, \partial_d f(x)). \ \square$$

We often write $\partial f / \partial x_i$ instead of $\partial_i f$. We denote by $\mathbf{e}_i$ the i-th basis vector of $\mathbb{R}^d$, i.e. $(\mathbf{e}_i)_i = 1$ and $(\mathbf{e}_i)_j = 0$ otherwise. For $z \in \mathbb{R}^d$ and $u \in E^d$, we write

$$z \cdot u \stackrel{\text{def}}{=} z_1 u_1 + \cdots + z_d u_d.$$

Let us state the expression of the partial derivative after a change of variable [Vol. 2, Theorem 3.12].

Proposition A.84. — *If T is a differentiable function from an open subset Λ of $\mathbb{R}^d$ onto another open subset Ω, and if f is a differentiable function on Ω, then*

$$\partial_i (f \circ T) = \sum_{j=1}^{d} (\partial_j f \circ T) \, \partial_i T_j. \ \square$$

Let us state the **mean value theorem** for a function [Vol. 2, Theorem 2.5].

Proposition A.85. — *Let f be a differentiable function from an open subset Ω of $\mathbb{R}^d$ into a separated semi-normed space E, whose family of semi-norms is denoted $\{\| \, \|_{E;\nu} : \nu \in \mathcal{N}_E\}$.*

Then, for every $x \in \Omega$ and $z \in \mathbb{R}^d$ such that $[x, x+z] \subset \Omega$, and for every $\nu \in \mathcal{N}_E$:

(a) $$\|f(x+z) - f(x)\|_{E;\nu} \leq |z| \sup_{0 \leq t \leq 1} \|\nabla f(x+tz)\|_{E^d;\nu};$$

(b) $$\|f(x+z) - f(x) - z \cdot \nabla f(x)\|_{E;\nu} \leq |z| \sup_{0 \leq t \leq 1} \|\nabla f(x+tz) - \nabla f(x)\|_{E^d;\nu}. \ \square$$

Bibliography

[1] ARCHIMEDES, *Opere omnia*, J.-L. Heiberg, Teubner, 3 vol., 1913–1915.

[2] BANACH, S. and TARSKI, A., Sur la décomposition des ensembles de points en parties respectivement congruentes, *Fundamenta Mat.* 6 (1924), 244–277.

[3] BERNOULLI, J. (Jacob), Analysis problematis antehac propositi. De Inventione Lineæ descendus a corpore gravi percurrendæ uniformiter. *Opera*, 421–426. Cramer and Philibert, Geneva, 1744.

[4] BERNOULLI, J. (Johann), Lectiones matematicæ de methodo integralium allisque (1691–1692). *Opera omnia* III, 385–559. Original edition Bousquet, Lausanne and Geneva, 1742 (= Olms, Hildesheim, 1968).

[5] BOCHNER, S., Integration von Funktionen deren Werte die Elemente eines Vektorraumes sind, *Fundamenta Mat.* 20 (1933), 262–276.

[6] BOLZANO, B., Rein analytischer Beweis der Lehrsatzes, daß zwischen je zwei Werthen, die ein entgegengesetzes Resultat gewähren, wenigstens eine reelle Wurzel der Gleichung liege, *Abhandlungen K. Böhm. Gesell. Wissen.*, (3), 5 (1817), 1–6.

[7] BOREL, É., *Leçons sur la théorie des fonctions*, Gauthiers-Villars, 1898.

[8] BOURBAKI, N., *Intégration*, Hermann, new edition, 1965.

[9] BOURBAKI, N., *Espaces vectoriels topologiques*, Hermann, new edition, 1967.

[10] BOURBAKI, N., *Éléments d'histoire des mathématiques*, Hermann, new edition, 1974.

[11] BOURBAKI, N., *Fonction d'une variable réelle*, Hermann, new edition, 1976.

[12] BRÉZIS, H., *Analyse fonctionnelle*, Masson, 1983.

[13] CAJORI, F., *A History of Mathematical Notations*, two volumes, Open Court (Chicago), 1974.

[14] CAUCHY, A., *Cours d'Analyse de l'École Royale Polytechnique*, Debure (Paris), 1821.

[15] CAUCHY, A., *Résumé des leçons données à l'École Royale Polytechnique sur le calcul infinitésimal*, Debure (Paris), 1823.

[16] CAUCHY, A., *Mémoire sur les intégrales définies*, 1827.

[17] CLARKSON, J.A., Uniformly convex spaces, *Trans. Amer. Math. Soc.*, 40 (1936), 396–414.

[18] DAY, M.M., Some more uniformly convex spaces, *Bull. Amer. Math. Soc.*, 47 (1941), 504–507.

[19] DRONIOU, J., *Intégration et espaces de Sobolev à valeurs vectorielles*, H. A. L., 2023.

[20] DU BOIS-REYMOND, P., Fortsetzung der Erläuterungen zu den Anfangsgründen der Variationsrechnung, *Mathematische Annalen*, 15, no 2 (1879), 564–576.

[21] DUNFORD, N. and SCHWARTZ, J.-T., *Linear Operators*, Interscience, 1958.

[22] EDWARDS, R., *Functional Analysis*, Winston, 1965.

[23] EGOROFF, D., Sur les suites de fonctions mesurables, *C. R. Acad. Sci. Paris*, 152 (1911), 244–246.

[24] EUCLID, *Elementorum Libri* XV, Apud Hieronymum de Marnef & Gulilmum Canellat, Paris, 1573.

[25] EULER, L., *Institutionum calculi integralis, Volumen Primum*, Acad. Imper. Sci. Petropoli, 1768 (*in Opera omnia, Series prima*, XI, Teubner and O. Füssli, 1911 and 1976).

[26] EULER, L., De formulis integrabilibus duplicatis, *Novi comentarii academiæ scientiarum Petropolitanæ*, 14 (1770), 72–103 (*in Opera omnia, Series prima*, XVII, Teubner, 1914).

[27] FISCHER, E., Sur la convergence en moyenne, *C. R. Acad. Sci. Paris*, 144 (1907), 1022–1024.

[28] FOURIER, J., *Théorie analytique de la chaleur*. Didot, Paris, 1822. Reissue, Gabay, Sceaux, 1988.

[29] FRÉCHET, M., Sur les ensembles de fonctions et les opérations linéaires, *C. R. Acad. Sci. Paris*, 144 (1907), 1414–1416.

[30] FRÉCHET, M., Sur les ensembles compacts de fonctions de carrés sommables, *Acta Litt. Sci. Szeged*, 8 (1937), 116–12.

[31] FUBINI, G., Sugli integrali multipli, *Atti Accad. Naz. Lincei*, 16 (1907), 608–614.

[32] GALLOUET, T. and HERBIN, R., *Mesure, intégration, probabilités*, H. A. L., 2013.

[33] HANCHE-OLSEN, H. and HOLDEN, H., The Kolmogorov-Riesz compactness theorem, *Expositiones Mathematicae*, 28 (2010), 385–394.

[34] HANNER, O., On the uniform convexity of L^p and l^p, *Ark. Mat.*, 3 (1956), 239–244.

[35] HARDY, G.H. and LITTLEWOOD, J.E., A maximal theorem with function–theoretic applications, *Acta. Math.*, 54 (1930), 81–116.

[36] HEINE, E., Über trigonometrische Reihen, *J. Reine Angew. Math. (J. de Crelle)*, 71 (1870), 353–365.

[37] HEINE, E., Aus brieflichen Mittelheilungen (namentlich über Variationsrechnung), *Mathematische Annalen*, 2 (1870), 187–191.

[38] HEINE, E., Die Elemente der Functionenlehre, *J. Reine Angew. Math. (J. de Crelle)*, 74 (1872), 172–188.

[39] HÖLDER, O., Über einen Mittelwertsatz, *Nachr. Akad. Wiss. Göttingen. Math.-Phys. Kl.* (1889) 38–47.

[40] KOLMOGOROV, A.N., Über Kompaktheit der Funktionenmengen bei der Konvergenz im Mittel, *Nachr. Ges. Wiss. Göttingen*, 9 (1931), 60–63 (in English: V.M. Tikhomirov (ed.), *Selected Works of A. N. Kolmogorov*, vol. 1, Kluwer, Dordrecht, (1991), 147–150).

[41] KOLMOGOROV, A.N., Zur Normierbarkeit eines allgemeinen topologischen linearen Raumes, *Studia Math.*, 5 (1934), 29–33 (in English: V.M. Tikhomirov (ed.), *Selected Works of A. N. Kolmogorov*, vol. 1, Kluwer, Dordrecht, (1991), 183–186).

[42] LANG, S., *Real Analysis*, Addison–Wesley, 1969.

[43] LEBESGUE, H., Intégrale, longueur, aire, *Ann. di Mat.*, (3), t. VII (1902), 231–359 (= Thesis, Paris).

[44] LEBESGUE, H., *Leçons sur l'intégration et la recherche des fonctions primitives*, Gauthiers-Villars, 1904 (= Reissue, Gabay or Chelsea).

[45] LEBESGUE, H., Sur l'intégration des fonctions discontinues, *Ann. École. Norm. Sup.*, (3), 27 (1910), 361–450.

[46] LERAY, J., Sur le mouvement d'un liquide visqueux emplissant l'espace, *Acta Math.*, 63 (1934), 193–248.

[47] LEVI, B., Sopra l'integrazione delle serie, *Rend. Inst. Lombardo di Sci. e Lett.*, (2), 39 (1906), 775–780.

[48] LIONS, J.-L., *Équations différentielles opérationnelles et problèmes aux limites*, Springer, 1961.

[49] LIONS, J.-L., *Problèmes aux limites dans les équations aux dérivées partielles*, Univ. Montreal Press, 1965.

[50] LUSIN, N., Sur les propriétés des fonctions mesurables, *C. R. Acad. Sci. Paris*, 154 (1912), 1688–1690.

[51] MEIR, A., On the uniform convexity of L^p spaces, $1 < p \leq 2$. *Illinois J. Math.*, 28 (1984), 420–424.

[52] MINKOWSKI, H., *Geometrie der Zahlen, I*, 1896 (2nd edition, Teubner, 1910).

[53] MINKOWSKI, H., *Gesammelte Abhandlungen*, two volumes, Teubner, 1911.

[54] MORAWETZ, C.S., Two L^p inequalities, *Bull. Amer. Math. Soc.*, 75 (1969), 1299–1302.

[55] NEUMANN, J. VON, On complete topological linear spaces, *Trans. Amer. Math. Soc.*, 37 (1935), 1–20.

[56] OKIKIOLU, G.O., *Aspects of the Theory of Bounded Integral Operators in L^p-Spaces*, Academic Press.

[57] PETTIS, B.J., On intégration in vector spaces, *Trans. Amer. Math. Soc.*, 44 (1938), 277–304.

[58] PIER, J.-P., *Histoire de l'intégration*, Masson, 1996.

[59] RADON, J., Theorie und Anwendungen der absolut additiven Mengenfunktionen, *Sitzungsber. Akad. Wissen. Wien.* 122 (1913), 1295–1438.

[60] RENARDY, M. and ROGERS, R., *An Introduction to Partial Differential Equations, Texts in Applied Mathematics*, Springer, 1993.

[61] RIEMANN, B., Über die Darstellbarkeit einer Function durch eine trigonometrische Reihe. *Abh. K. Gesell. Wiss. Göttingen, Math. Classe*, 3 (1866-1867), 87–132 (= *Gesammelte mathematische Werke*, 227–271, Dover, New York, 1953).

[62] RIESZ, F., Sur les systèmes orthogonaux de fonctions, *C. R. Acad. Sci. Paris, sér. I*, 144 (1907), 615–619.

[63] RIESZ, F., Untersuchungen über Systeme integrierbarer Funktionen, *Math. Ann.*, 69 (1910), 449–497.

[64] RIESZ, M., Sur les ensembles compacts de fonctions sommables, *Acta Szeged Sect. Math.*, 6 (1933), 136–142.

[65] SCHWARTZ, L., *Théorie des distributions*, Actual. Sci. et Ind., 1091 and 1122, Hermann, 1950–1951 (Our historical notes refer to the pages numbers of the *Nouvelle édition augmentée*, 1973.)

[66] SCHWARTZ L., Distributions à valeurs vectorielles, I and II, *Ann. Inst. Fourier*, 7 (1957) and 8 (1959).

[67] SCHWARTZ, L., *Analyse, topologie générale et analyse fonctionnelle*, Hermann, 1970.

[68] SIMON, J., Compact sets in the space $L^p(0, T; B)$, *Ann. Mat. Pura Appl.*, 146 (1987), 65–96.

[69] SIMON, J., Representation of distributions and explicit antiderivatives up to the boundary, in *Progress in Partial Differential Equations: The Metz Surveys 2*, M. Chipot ed., Longman, 1993, 201–205.

[70] SIMON, J., Una generalización del teorema de Lions–Tartar, *Bol. Soc. Esp. Mat. Apl.*, 40 (2007), 43–69.

[71] SIMON, J., Distributions et espaces de Lebesgue à valeurs dans un espace vectoriel topologique localement convexe. Submitted in 2009 to *J. Math. Pures Appl.* (rejected).

[72] SIMON, J., *Banach, Fréchet, Hilbert and Neumann Spaces*, Analysis for PDEs set, vol. 1, ISTE and Wiley, London, 2017.

[73] SIMON, J., *Continuous Functions*, Analysis for PDEs set, vol. 2, ISTE and Wiley, London, 2019.

[74] SIMON, J., *Distributions*, Analysis for PDEs set, vol. 3, ISTE and Wiley, London, 2022.

[75] SIMON, J., *Plus sur les distributions*, in progress.

[76] STEGMANN, F.L., *Lehrbuch der Variationsrechnung und ihrer Anwendung bei Untersuchungen über das Maximum und Minimum*, J.G. Luckhardt, Kassel, 1854.

[77] TONELLI, L., Sulla nozione di integrale, *Annali di Mat.*, 1 (1923), 105–145.

[78] VITALI, G., *Sul problema della mesura dei gruppi di punti di una retta*, Bologna, 1905 (= *Opere sull'analisi reale e complessa*, 183–187. Cremonese, Rome 1984).

[79] VITALI, G., Una proprieta delle funzione misurabili, *R. ist. Lombardo Sci. Lett. Rend.*, 2, t. XXXVIII (1905), 599–603.

[80] VITALI, G., Sui gruppi di punti e sulle funzione di variabili reali, *Atti Acad. Sci. Torino*, 43 (1908), 75–92.

[81] YOSIDA, K., *Functional Analysis*, Springer-Verlag, 1980.

[82] YOUNG, W.H., On classes of summable functions and their Fourier series, *Proc. Roy. Soc. London, Ser. A*, 87 (1912), 225–229.

Index

D

E

F

G, H

I

K, L

M

N

Other titles from

ISTE

in

Mathematics and Statistics

2026

LESFARI Ahmed
Complex Manifolds and Geometric Algebraic Analysis

PEYRARD Nathalie, DE SAPORTA Benoîte
A Comprehensive Guide to HSMM: Theory, Software, and Advanced Extensions

2025

DUPUY Jean-François
Generalized Linear Models: Problems with Censored, Missing, and Zero-inflated Data

GAIVORONSKI Alexei A., KNOPOV Pavel S., NORKIN Vladimir I., ZASLAVSKYI Volodymyr A.
Stochastic Modeling and Optimization Methods for Critical Infrastructure Protection 1: Stochastic Modeling
Stochastic Modeling and Optimization Methods for Critical Infrastructure Protection 2: Methods and Tools

PICARD Philippe, LEFÈVRE Claude
Abel-Gontcharoff Pseudopolynomials and Stochastic Applications

2024

BOURLÈS Henri
System Theory – A Modern Approach 1: Linear Ordinary and Functional Differential Equations
(New Mathematical Methods, Systems and Applications Set – Volume 1)

MAKAROV Volodymyr, MAYKO Nataliya
Traditional Functional-Discrete Methods for the Problems of Mathematical Physics: New Aspects

2023

AIT BEN HASSI El Mustapha
The Theory of Distributions: Introduction

KOROLIOUK Dmitri, SAMOILENKO Igor
Asymptotic and Analytic Methods in Stochastic Evolutionary Systems

PARROCHIA Daniel
Graphs, Orders, Infinites and Philosophy

2022

CHAKRAVARTHY Srinivas R.
Introduction to Matrix-Analytic Methods in Queues 1: Analytical and Simulation Approach – Basics
Introduction to Matrix-Analytic Methods in Queues 2: Analytical and Simulation Approach – Queues and Simulation

LESFARI Ahmed
Integrable Systems

RADCHENKO Vadym M.
General Stochastic Measures: Integration, Path Properties and Equations

DE SAPORTA Benoîte, ZILI Mounir
Martingales and Financial Mathematics in Discrete Time

SIMON Jacques
Distributions
(Analysis for PDEs Set – Volume 3)

2021

KOROLIOUK Dmitri, SAMOILENKO Igor
Random Evolutionary Systems: Asymptotic Properties and Large Deviations

MOKLYACHUK Mikhail
Convex Optimization: Introductory Course

POGORUI Anatoliy, SWISHCHUK Anatoliy, RODRÍGUEZ-DAGNINO Ramón M.
Random Motions in Markov and Semi-Markov Random Environments 1: Homogeneous Random Motions and their Applications
Random Motions in Markov and Semi-Markov Random Environments 2: High-dimensional Random Motions and Financial Applications

PROVENZI Edoardo
From Euclidean to Hilbert Spaces: Introduction to Functional Analysis and its Applications

2020

BARBU Vlad Stefan, VERGNE Nicolas
Statistical Topics and Stochastic Models for Dependent Data with Applications

CHABANYUK Yaroslav, NIKITIN Anatolii, KHIMKA Uliana
Asymptotic Analyses for Complex Evolutionary Systems with Markov and Semi-Markov Switching Using Approximation Schemes

KOROLIOUK Dmitri
Dynamics of Statistical Experiments

MANOU-ABI Solym Mawaki, DABO-NIANG Sophie, SALONE Jean-Jacques
Mathematical Modeling of Random and Deterministic Phenomena

SIMON Jacques
Continuous Functions
(Analysis for PDEs Set – Volume 2)

2019

BANNA Oksana, MISHURA Yuliya, RALCHENKO Kostiantyn, SHKLYAR Sergiy
Fractional Brownian Motion: Approximations and Projections

GANA Kamel, BROC Guillaume
Structural Equation Modeling with lavaan

KUKUSH Alexander
Gaussian Measures in Hilbert Space: Construction and Properties

LUZ Maksym, MOKLYACHUK Mikhail
Estimation of Stochastic Processes with Stationary Increments and Cointegrated Sequences

MICHELITSCH Thomas, PÉREZ RIASCOS Alejandro, COLLET Bernard, NOWAKOWSKI Andrzej, NICOLLEAU Franck
Fractional Dynamics on Networks and Lattices

VOTSI Irene, LIMNIOS Nikolaos, PAPADIMITRIOU Eleftheria, TSAKLIDIS George
Earthquake Statistical Analysis through Multi-state Modeling
(Statistical Methods for Earthquakes Set – Volume 2)

2018

AZAÏS Romain, BOUGUET Florian
Statistical Inference for Piecewise-deterministic Markov Processes

IBRAHIMI Mohammed
Mergers & Acquisitions: Theory, Strategy, Finance

PARROCHIA Daniel
Mathematics and Philosophy

2017

CARONI Chysseis
First Hitting Time Regression Models: Lifetime Data Analysis Based on Underlying Stochastic Processes
(Mathematical Models and Methods in Reliability Set – Volume 4)

CELANT Giorgio, BRONIATOWSKI Michel
Interpolation and Extrapolation Optimal Designs 2: Finite Dimensional General Models

CONSOLE Rodolfo, MURRU Maura, FALCONE Giuseppe
Earthquake Occurrence: Short- and Long-term Models and their Validation
(Statistical Methods for Earthquakes Set – Volume 1)

D'AMICO Guglielmo, DI BIASE Giuseppe, JANSSEN Jacques, MANCA Raimondo
Semi-Markov Migration Models for Credit Risk
(Stochastic Models for Insurance Set – Volume 1)

GONZÁLEZ VELASCO Miguel, del PUERTO GARCÍA Inés, YANEV George P.
Controlled Branching Processes
(Branching Processes, Branching Random Walks and Branching Particle Fields Set – Volume 2)

HARLAMOV Boris
Stochastic Analysis of Risk and Management
(Stochastic Models in Survival Analysis and Reliability Set – Volume 2)

KERSTING Götz, VATUTIN Vladimir
Discrete Time Branching Processes in Random Environment
(Branching Processes, Branching Random Walks and Branching Particle Fields Set – Volume 1)

MISHURA Yuliya, SHEVCHENKO Georgiy
Theory and Statistical Applications of Stochastic Processes

NIKULIN Mikhail, CHIMITOVA Ekaterina
Chi-squared Goodness-of-fit Tests for Censored Data
(Stochastic Models in Survival Analysis and Reliability Set – Volume 3)

SIMON Jacques
Banach, Fréchet, Hilbert and Neumann Spaces
(Analysis for PDEs Set – Volume 1)

2016

CELANT Giorgio, BRONIATOWSKI Michel
Interpolation and Extrapolation Optimal Designs 1: Polynomial Regression and Approximation Theory

CHIASSERINI Carla Fabiana, GRIBAUDO Marco, MANINI Daniele
Analytical Modeling of Wireless Communication Systems
(Stochastic Models in Computer Science and Telecommunication Networks Set – Volume 1)

GOUDON Thierry
Mathematics for Modeling and Scientific Computing

KAHLE Waltraud, MERCIER Sophie, PAROISSIN Christian
Degradation Processes in Reliability
(Mathematial Models and Methods in Reliability Set – Volume 3)

KERN Michel
Numerical Methods for Inverse Problems

RYKOV Vladimir
Reliability of Engineering Systems and Technological Risks
(Stochastic Models in Survival Analysis and Reliability Set – Volume 1)

2015

DEVOLDER Pierre, JANSSEN Jacques, MANCA Raimondo
Basic Stochastic Processes

LE GAT Yves
Recurrent Event Modeling Based on the Yule Process
(Mathematical Models and Methods in Reliability Set – Volume 2)

DE SAPORTA Benoîte, DUFOUR François, ZHANG Huilong
Numerical Methods for Simulation and Optimization of Piecewise Deterministic Markov Processes

2014

COOKE Roger M., NIEBOER Daan, MISIEWICZ Jolanta
Fat-tailed Distributions: Data, Diagnostics and Dependence (Mathematical Models and Methods in Reliability Set – Volume 1)

MACKEVIČIUS Vigirdas
Integral and Measure: From Rather Simple to Rather Complex

PASCHOS Vangelis Th
Combinatorial Optimization – 3-volume series – 2nd edition
Concepts of Combinatorial Optimization / Concepts and Fundamentals – Volume 1
Paradigms of Combinatorial Optimization – Volume 2
Applications of Combinatorial Optimization – Volume 3

2013

COUALLIER Vincent, GERVILLE-RÉACHE Léo, HUBER Catherine, LIMNIOS Nikolaos, MESBAH Mounir
Statistical Models and Methods for Reliability and Survival Analysis

JANSSEN Jacques, MANCA Oronzio, MANCA Raimondo
Applied Diffusion Processes from Engineering to Finance

SERICOLA Bruno
Markov Chains: Theory, Algorithms and Applications

2012

BOSQ Denis
Mathematical Statistics and Stochastic Processes

CHRISTENSEN Karl Bang, KREINER Svend, MESBAH Mounir
Rasch Models in Health

DECREUSEFOND Laurent, MOYAL Pascal
Stochastic Modeling and Analysis of Telecom Networks

DEVOLDER Pierre, JANSSEN Jacques, MANCA Raimondo
Stochastic Methods for Pension Funds

2011

MACKEVIČIUS Vigirdas
Introduction to Stochastic Analysis: Integrals and Differential Equations

MAHJOUB Ridha
Recent Progress in Combinatorial Optimization – ISCO2010

RAYNAUD Hervé, ARROW Kenneth
Managerial Logic

2010

BAGDONAVIČIUS Vilijandas, KRUOPIS Julius, NIKULIN Mikhail
Nonparametric Tests for Censored Data
Nonparametric Tests for Complete Data

IOSIFESCU Marius, LIMNIOS Nikoloas, OPRISAN Gheorghe
Introduction to Stochastic Models

VASSILIOU P.C.G.
Discrete-time Asset Pricing Models in Applied Stochastic Finance

2008

ANISIMOV Vladimir
Switching Processes in Queuing Models

FICHE Georges, HÉBUTERNE Gérard
Mathematics for Engineers

HUBER Catherine, LIMNIOS Nikolaos, MESBAH Mounir, NIKULIN Mikhail
Mathematical Methods in Survival Analysis, Reliability and Quality of Life

JANSSEN Jacques, MANCA Raimondo, VOLPE Ernesto
Mathematical Finance

2007

HARLAMOV Boris
Continuous Semi-Markov Processes

2006

CLERC Maurice
Particle Swarm Optimization

Printed and bound by CPI Group (UK) Ltd, Croydon, CR0 4YY

16/09/2026

14963078-0008